Birkhäuser

Static & Dynamic Game Theory: Foundations & Applications

More information about this series at http://www.springer.com/series/10200

Stefan Rass • Stefan Schauer
Editors

Game Theory for Security and Risk Management

From Theory to Practice

Birkhäuser

Editors
Stefan Rass
Institute of Applied Informatics
University of Klagenfurt
Klagenfurt, Kärnten, Austria

Stefan Schauer
Center for Digital Safety & Security
Austrian Institute of Techno GmbH
Klagenfurt, Kärnten, Austria

ISSN 2363-8516 ISSN 2363-8524 (electronic)
Static & Dynamic Game Theory: Foundations & Applications
ISBN 978-3-030-09186-6 ISBN 978-3-319-75268-6 (eBook)
https://doi.org/10.1007/978-3-319-75268-6

Mathematics Subject Classification: 91A35, 90B50, 91A24, 91A80, 91A40

Printed on acid-free paper

This book is published under the imprint Birkhäuser, www.birkhauser-science.com by the registered company Springer International Publishing AG part of Springer Nature.
The registered company address is: Gewerbestrasse 11, 6330 Cham, Switzerland

To our families…

Preface

With the increase in the complexity and prevalence of modern communication technology, security faces bigger challenges than ever. New security concepts and notions have been developed and are continuously seeing deployment in practice. Some of these new security mechanisms root in game theory, whose application to security dates back almost two decades, and which has proven itself as a powerful and fruitful field to tackle the natural competition and complex interaction between those who protect and those who attack assets.

While the idea behind game theory is simple in the sense of optimizing opposing goals and efforts towards them, using the theory for security appears as a natural step. Applying the respective models, however, is a different story, and the challenges arising in security are different from those in economics, where game theory originates. Indeed, exactly this difference is what brings an interesting viewpoint on security, if we no longer consider security as the absence of threats (a state that would not be reachable anyway), but rather as a state in which the expenses for an attack outweigh the gains from it. This *economic view* on security is not new, but somewhat surprisingly, much research on security is still focused on preventing all known attacks (at any cost), rather than optimizing the defender's efforts and limited resources to gain the maximal security achievable. The difficulty of security management is the difficulty of quantifying security. It may not be plausible to claim that security can be measured like a physical quantity, but can it at least be scored (i.e., assigned a number that has no meaning by itself, but lets us compare and rank different situations based on their scores)? The difficulty of finding good security metrics may partly be due to an overly strong requirement implicitly imposed here. Certainly, management would like to talk about numbers that show trends, say to recognize if we are slowly becoming more and more exposed to certain threats, but security is not a physical quantity or measurable in experiments. It is a property of a system that undergoes an evolution and must be continuously checked and kept up. Note that this simple view by no means limits the diversity of what security or vulnerability means. The goals of attacks can be manifold and monetary losses (say, by physical damage or theft of information) are only one possibility. Various games may be played for reputation, where the attacker's goal is destroying the victim's credibility, but without any intention to cause physical damage or steal anything. Risk is therefore an equally diverse term, and risk management is usually a matter of control towards optimizing multiple indicators of interest.

Risk management can be viewed as a certain kind of control problem. The conditions and constraints under which the controller tries to put or keep the system state at a defined level are a matter of choice and decision making, based on numbers that quantify risks (in all aspects of interest). The decision making itself is based on a model that describes the system (typically a whole enterprise, whose internal dynamics is often too complex to be put into equations or other simple terms), and depending on the expressiveness and power of the model, it ships with various parameters. In any case, a good model helps to establish a comprehensive protection strategy that can flexibly adapt itself to changing conditions and new innovations. While the existing theory is huge, practitioners and nonexperts in game theory may face severe difficulties in unleashing the power of game theory in their daily business.

This volume is intended to fill this gap by telling a story in three acts, being parts in the book, and contributed by a total of 38 authors, whom we hereby greatly acknowledge.

Part I is composed of selected models of games and decision making. Here, models can be distinguished according to their components. Let us fix the defender (possibly being a multitude of physical entities, persons, etc.) as a fixed component, playing against an adversary (also physically allowed to appear as multiple and likely collaborating actors). If the adversary acts irrationally, we can consider it as being an act of nature, since there is no incentive to cause harm to the defender, and the defender is simply exposed to some ecosystem and subject to changing environmental conditions. Finding optimal strategies in a changing and uncertain but not hostile environment is subject of *decision theory. Game theory* changes the picture by endowing the environment with its own incentives, which are (in the most typical cases) in conflict with the defender's intentions (though not necessarily opposite). Decision theory can thus, in a simplified perspective, be viewed as optimization of one (or more) goals by one player, while game theory does optimization of usually conflicting goals of at least two or more players.

The temporal aspect adds further distinctions here, since the defender's problem can be finding an optimal decision for now, or an optimal plan to materialize towards a longer term goal in future. The simpler case is obviously making a decision for the moment, and disregarding the consequences that it may have in future. In that case, we arrive at the simplest form of *static games* (against rational opponents) or simple decisions (to deal with nature). If the consequences of an action taken now will matter for the future, then we have either a *control problem* (against an irrational adversary like nature) or a *dynamic game*, if the adversary acts rationally. Extensions and combinations of these types lead to more complex (and hence more powerful) models like Markov decision processes and relatives. All these models are to be treated individually, and the chapters in Part I cover theoretical basics to make a start with some selected candidates.

Specifically, Part I of this volume is divided into the following chapters:

Chapter 1: Utilizing Game Theory for Security Risk Assessment, by L. Rajbhandari and E.A. Snekkenes
The opening of the book is dedicated to making the connection between risk management and game theory explicit. The similarities between a game-theoretic analysis of best behavior and risk management are striking, yet not obvious and this chapter opens the book by making the connection explicit. At the same time, it points at various problems of decision making (covered in Part I of this book) and model building, which the whole Part II of the book is dedicated to. Concrete applications up to tool support for game-theoretic risk management are reported in Chapters 16, 12, 13 and 4, among others.

Chapter 2: Decision Making When Consequences Are Random, by S. Rass
The dynamics in an enterprise or general process that is subject to risk management are rarely open to accurate descriptions in formal terms. Thus, much of risk management is a matter of taking decisions whose consequences are hardly determined or foreseeable to a decent extent. Game theory's hypothesis of rationality being induced by utility maximization is herein generalized towards a decision-making framework that builds upon fuzzy and vague knowledge, and introduces random variables themselves as objects for optimization. The framework established is essentially a *possible* replacement for conventional numbers used in optimization, such as eloquently described in Chapters 5, 10 or 11. Applications of the framework laid out in this chapter are found in Chapters 12, 13, 14, 15 and 16.

Chapter 3: Security Strategies and Multi-Criteria Decision Making, by S. Rass
A considerable deal of attention in risk management is dedicated to an assessment of the attacker's intention or the general incentive as to why an infrastructure may be under attack. If such information is available, then a tailored defense can be defined. However, lacking an idea about who is attacking us or why, the best we can do is using our own incentives as guideline to model the hypothetical adversary. This leads to the concept of security strategies, which, roughly speaking, are the best defense possible against a set of known threats, whose concrete incarnations depend on the unknown incentives of the unknown attacker. Finding such an optimal defense w.r.t. several assets to be protected and several goals of security is the main body of this chapter. The technique established reduces the problem of security strategy computation to a standard equilibrium computation problem, which all other chapters in this book revisit and discuss in different variations.

Chapter 4: A Scalable Decomposition Method for the Dynamic Defense of Cyber Networks, by M. Rasouli, E. Miehling, and D. Teneketzis
Picking up at a similar point as Chapter 3, this also adopts the defender's point of view when a defense against cyber-attacks shall be defined. The uncertainty aforementioned in Chapter 2 is, however, made much more precise here in assuming the defender not having full information about the network status at all times. The additional complexity issue of determining security strategies in large-scale networks is a story on its own, and a core contribution of this chapter is a method to handle the

scalability issue in the computation of worst-case defenses (i.e., security strategies, similar to Chapter 3), obtained from treating the issue as a security *control problem.*

Chapter 5: Factored Markov Game Theory for Secure Interdependent Infrastructure Networks, by L. Huang, J. Chen, and Q. Zhu
The diversity and scalability issues that Chapter 4 talks about are discussed by this chapter from a different angle of view. While networks may grow large in size and thus entail complex models, additional complexity (different in nature) may also arise from the diversity of devices and from extending the view to include cross-layer considerations spanning purely logical but also physical parts of the network. The most typical example of such a system is the internet-of-things (IoT). The cyber-physical perspective generally reveals scalability issues that call for efficient treatment, which this chapter approaches by designated game-theoretic models. Like Chapter 4, decompositions and approximations of problems to handle practically intractable models are in the center of attention here, supported by numeric examples.

Part II begins at the point where the model has been selected, and now we are asking ourselves how to set the parameters, or more generally, how a concrete such model should be defined. Let us consider game-theoretic models as an example to illustrate the issue: suppose that player 1 runs an IT network, whose administrator has recently been informed about a new malware "in the wild." Irrespectively of whether there is an infection already, we are already in a game between the system administrator and the attacker. A game-theoretic model would require three ingredients:

1. The action set of the defender: this is typically a known item, since the system administrator can consult standard catalogues such as those shipping with risk management standards like ISO31000.
2. The action set of the attacker: this is typically more involved to specify, based on domain expertise and experience and supported by catalogues in risk management standards like the "BSI Grundschutzkatalog" of the German Federal Office for Information Security (www.bsi.bund.de)
3. A valuation of the consequences that the actions of the defender and attacker will have. A standard game model asks for this consequence to be described by a number, but how shall we do this? What numeric measure would be accurate to describe the effects of malware in a system. If it causes damage and outages, how would we quantify the loss that the company suffers from this? If the malware spreads in a electricity network and shuts down parts of it, how much would the total damage be in customer's households? If the problem is with a water supplier, who is obliged to inform customers, how would the supplier's reputation be affected (damaged) upon this incident? While the game-theoretic model itself may be useful to describe the management decision challenge, parameterizing the model to accurately describe the effects of actions taken is a different challenge and needs its own treatment and theory.

Part II of this volume is dedicated to work concerned with the instantiation of game-theoretic models and how to define their parameters appropriately. To this end, contributions have been collected from different domains, all dealing with matters of quantifying utilities or shaping game-theoretic models in general.

Specifically, Part II of this volume is divided into the following chapters:

Chapter 6: G-DPS: A Game-Theoretical Decision-Making Framework for Physical Surveillance Games, by A. Alshawish, M.A. Abid, H. de Meer, S. Schauer, S. König, A. Gouglidis, and D. Hutchison
Taking surveillance systems as a showcase example, this chapter demonstrates how to put the abstract theory of risk management (Chapter 1) and the decision making (Chapter 2) to more concrete terms. In essence, the work exemplifies how uncertainty in surveillance can be modeled concretely and how to apply the decision and computational framework laid out in Chapters 2 and 3. Moreover, the decision-making framework approached in this chapter illustrates straightforwardly how the Hybrid Risk Management (HyRiM) process (Chapter 12) can be tailored to specific scenarios and use cases such as surveillance games.

Chapter 7: A Game-Theoretic Framework for Securing Interdependent Assets in Networks, by A.R. Hota, A.A. Clements, S. Bagchi, and S. Sundaram
The initial point of treating security as an economic issue is picked up in this chapter by asking for how much security can be obtained under limited budgets, and what is the minimal expense to achieve a desired level of security. Both problems are analyzed, exposing the solutions to be an instance of moving target defense, whose optimal pattern can be determined from applying game theory. As in Chapters 4 and 5, matters of interdependence are a main aspect to consider, and this chapter (as the entirety of Part II) goes into concrete terms and examples on *how* to model and valuate the interplay of components, particularly in settings with multiple defenders managing different parts of the system. The showcase example is power grids, which complement the so far drawn picture of networks (Chapter 4) and the internet-of-things (Chapter 5).

Chapter 8: Random Damage in Interconnected Networks, by S. König and A. Gouglidis
Large-scale cyber-attacks usually start slowly and barely noticeable, by a single infection occurring somewhere in a system, from which the problem grows slowly and steadily. The question of how much impact such an infection has is natural, yet not easy to answer. This chapter proposes a simulation framework to study and analyze malware infection scenarios, which typically make up the initial phases of advanced persistent threats (APT). The resulting data is a set of possible scenarios rather than guaranteed consequences, and the text shows how to directly use these many possibilities for decision making (based on the methods laid out in Chapter 2). A related APT case study is later revisited in Chapter 13.

Chapter 9: Optimal Dispatch of Electric Transmission Systems Considering Interdependencies with Natural Gas Systems, by T. Hong, F. de Léon, and Q. Zhu
Extending the treatment of interdependencies between layers of an infrastructure of different kinds (where Chapter 5 made a start), it is not always easy for a practitioner to *acquire* the information needed to reach a rational decision (using game theory). Similar to Chapter 8 but with a different focus, co-simulation is the core object of interest here, which this chapter exposes as an indispensable tool to assess the outcome of actions towards optimized defense design. Since any such simulation must be tailored to the application (in general), the technicalities of this chapter relate to modeling the dynamics of power grids (similar methods are later used in Chapter 10). This emphasizes the interdisciplinary flavor of risk management and game theory, which are applicable together, but in any case need to rest on domain expertise.

Chapter 10: Managing Security Risks Interdependencies Between ICT and Electric Infrastructures: A Game Theoretical Analysis, by Z. Ismail, J. Leneutre, D. Bateman, and L. Chen
While Chapter 9 stays at a technical level of power grid dynamics, this chapter covers the more high level management aspects that need consideration in addition and in parallel. Leaving the technical details of simulation shifted to a designated framework and simulation platform, the issue discussed here relates to the question of which actions should be taken on the components of the system for an optimal defense from a more central perspective of the transmission system operator (TSO). Like Chapter 4, this is also the defender's view, but now focused on the "where" and "how" domain expertise can be expressed to go into a game-theoretic model (as parameters).

Part III completes the picture by showing case studies regarding a few selected successful applications of game-theoretic models to real life problems. The core of Part III is thus neither on theoretical basics (Part I) nor on practicalities of the models (Part II), but rather on showcase examples where models have been successfully applied in practice. The boundaries between Part II and Part III are occasionally fuzzy and may overlap in the form of case studies also found in Part II of the book.

Part III of this book is composed of the following contributions:

Chapter 11: Security and Interdependency in a Public Cloud: A Game-Theoretic Approach, by C.A. Kamhoua, L. Kwiat, K.A. Kwiat, J.S. Park, M. Zhao, and M. Rodriguez
Heterogeneity of a system may not exclusively root in diversity within the system, but can also be due to outsourcing and external resources. Cloud computing is a prominent example and a highly complex topic for risk management. Indeed, game theory is applicable to various threat scenarios in this context, and this chapter shows how to define proper games upon virtualization. The insights obtained are driven by an economic perspective, similarly to the introductory thoughts in this preface, and to what Chapter 7 did (only with a different domain of application). The chapter thus

underlines once more the economic view on security that is essential for a practically effective defense. At the same time, the discussion of cloud computing adds to the range of applications so far (electricity networks, internet-of-things, water supply, information and communication networks, and many more).

Chapter 12: A Risk Management Approach for Highly Interconnected Networks, by S. Schauer
With several technical models of interdependence treatment having been described in Part II (Chapters 7 and 9), this chapter continues along the same route as Chapter 10, but staying at the more general level of risk management (thus continuing the topic of Chapter 1). Essentially, it is thus a report about how the theory laid out in Chapters 1, 2, 3, and 8 can be put to practice. This work is later complemented in Chapter 16, culminating in a practical tool support.

Chapter 13: Protecting Water Utility Networks from Advanced Persistent Threats: A Case Study, by A. Gouglidis, S. König, B. Green, K. Rossegger, and D. Hutchison
A major problem of security risk assessments is the intrinsic fuzziness of information. Whenever decisions in risk management refer to controlling interdependent systems (composed of various subsystems of different nature), domain expertise from a single source only would be limited usefulness. The latter is mainly due to the fact that a single person may not be in position to take into consideration all the aspects of an interdependent system. Consequently, the decision maker may have to inquire several experts to get a complete picture of the interdependent system that no single expert could provide. To accomplish this goal, the framework from Chapter 12 is used. Specifically, this chapter presents the applicability of the framework through a water utility case study, where the risk management goal is the avoidance of an APT scenario. The results of the case study are computed using the methods described in Chapters 2 and 3.

Chapter 14: Assessing the Impact of Malware Attacks in Utility Networks, by S. König, A. Gouglidis, B. Green, and A. Solar
The impact of malware is essentially an economic one and depends on what the malware actually does. For a decent understanding of an infection's impact in a system, this chapter studies a concrete network and the ingredients to put the simulation framework of Chapter 8 to work. The parameters for a game-theoretic model of malware infection and insights gained from computing security strategies are the primary items of discussion in this chapter.

Chapter 15: Game-Theoretic Optimization for Physical Surveillance of Critical Infrastructures: A Case Study, by A. Alshawish, M.A. Abid, and H. de Meer
Chapter 6 discussed the surveillance problem in the context of risk management and left the practicalities thereof partly open. This gap is closed in this chapter, where a simulation method for surveillance is described based on the prepared decision making (Chapter 2) and risk management (Chapter 6). In addition, the chapter discusses a form of handling mixed strategies: if there is no purely optimal defense action, then the configuration of defense measures may admit the definition of new

defense strategies from a mix of currently available ones. The surveillance (simulation) framework put forth in this chapter is an example where strategies can be "purified" by proper re-parameterization, an option and method that is appealing for practice and usually left undiscussed elsewhere in the literature.

Chapter 16: Smart SECPLAN: A Process Implementation Tool for Hybrid Risk Management, by A. Zambrano, S. Caceres, and A.I. Martinez
The final chapter closes the loop back to the initial Chapter 1 by presenting a full featured tool support to cover the entire workflow cycle of risk management. It embodies all matters of theory (Part I in this book), concrete model parameter setting (basically being the formulae and methods to put risk in numeric or categorical terms; Part II in this book), and matters of practically doing the risk management (scheduling of actions, etc.). It thus constitutes the closing discussion in this book, by letting all theory converge to a practical solution.

The Overall Picture

Risk management is a complex process that spans various research domains and entails highly diverse tasks. Risk management standards provide a valuable guideline here, and the connection between these and game theory is wrapped up in Chapter 1 of this book. The workflow into which the book's contributions can be integrated is shown on the next page and follows a top-down direction: things start with the decision maker's duty of keeping risk under control, which entails various subtasks and is based on resources that the contributions of this book shall give practical insights to. The picture is in no way meant to quantitatively reflect the amount or importance of research done on any topic, yet it shall illustrate the diversity, complexity, and "size" of the "risk management problem."

After all, all risk management is only about creating a feeling of safety and security, underpinned by the assurance that best practices are adhered to and international standards have been followed. Naturally, one wants to do this in the most economic way possible, and this connects risk management to game theory: since risk management standards prescribe to take certain actions in a certain sequence, but leave wide degrees of freedom in *how* these actions are accomplished, what could be more natural than using optimization theory to find the best way of doing risk management? Concrete steps along this connection all the way from theory to practice make up the content of this book.

Klagenfurt, Austria Stefan Rass
Klagenfurt, Austria Stefan Schauer
December 2017

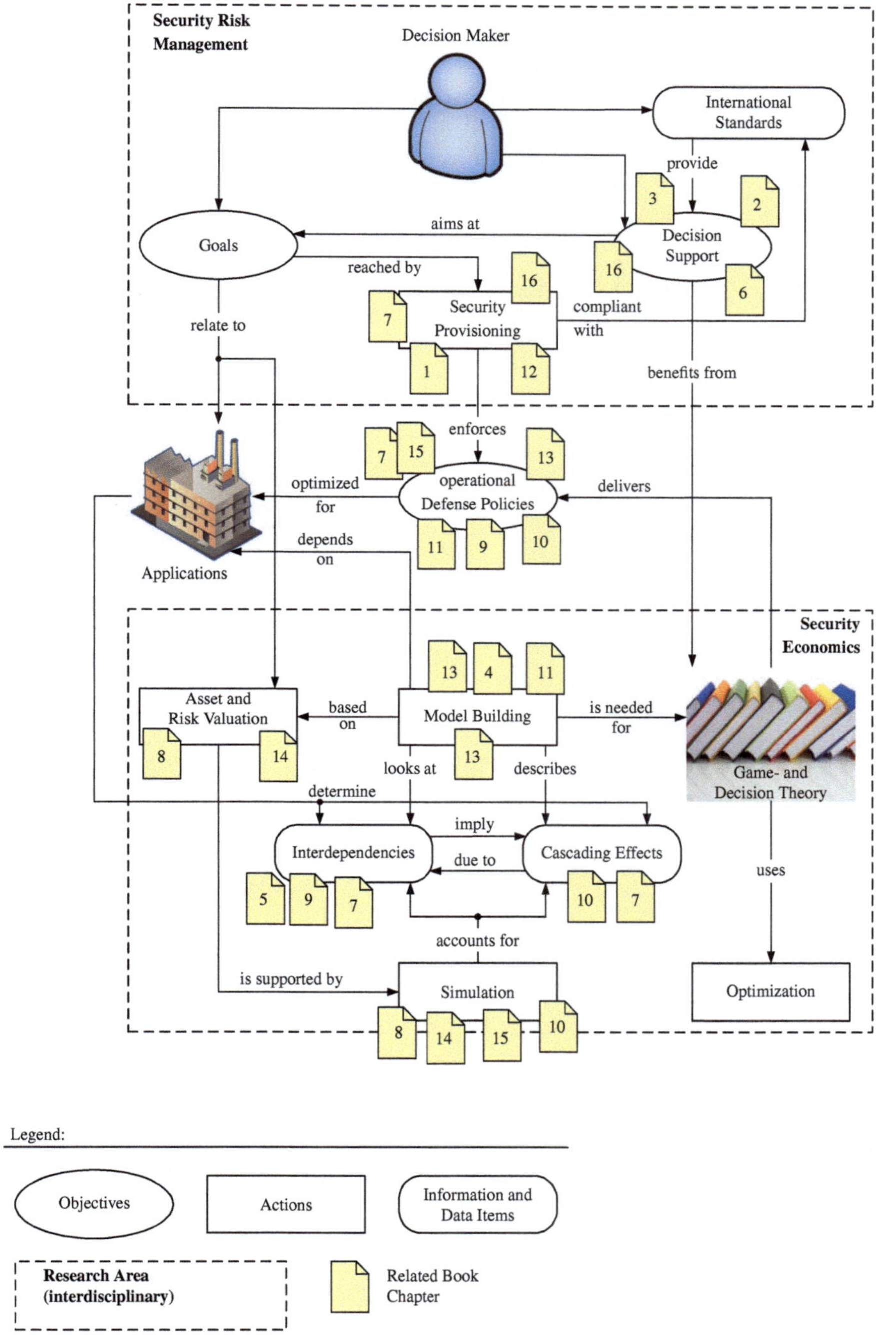

Embedding of the Volume's Contributions in the Landscape of Game Theory-Based Risk Management

Acknowledgements

First of all, the editors would like to give their special appreciation to the authors of the individual chapters for their inputs as well as their efforts and feedback during the review process. This book project would not have been possible in such a short time and with such a high quality without their invaluable contributions.

Secondly, Stefan Rass would like to thank the Alpen-Adria-Universitaẗ Klagenfurt for supporting his work on this book project. In the same way, Stefan Schauer would like to thank the AIT Austrian Institute of Technology for providing the means necessary to carry out the task of compiling and editing this book.

At last, the editors would like to acknowledge that the work on this book project as well as on several chapters therein was financed by European Commission's project no. 608090, HyRiM (Hybrid Risk Management for Utility Networks), under the 7th Framework Programme (FP7-SEC-2013-1) and thank the European Commission for this support.

Contents

Part I Theory

Part III Case Studies

Contributors

Mohamed Amine Abid
Faculty of Computer Science and Mathematics, Chair of Computer Networks and Computer Communications, University of Passau, Innstr. 43, 94032 Passau, Germany
e-mail: amine.abid@uni-passau.de

Ali Alshawish
Faculty of Computer Science and Mathematics, Chair of Computer Networks and Computer Communications, University of Passau, Innstr. 43, 94032 Passau, Germany
e-mail: ali.alshawish@uni-passau.de

Saurabh Bagchi
School of Electrical and Computer Engineering, Purdue University, 465 Northwestern Ave, West Lafayette, IN 47907, USA
e-mail: sbagchi@purdue.edu

David Bateman
EDF, 1 Place Pleyel, 93282 Saint-Denis, France
e-mail: david.bateman@edf.fr

Santiago Caceres
ETRA Investigación y Desarrollo S.A., Calle Tres Forques 147, 46014 Valencia, Spain
e-mail: scaceres.etraid@grupoetra.com

Juntao Chen
Department of Electrical and Computer Engineering, New York University, 2 Metrotech Center, Brooklyn, 11201, USA
e-mail: jc6412@nyu.edu

Lin Chen
University of Paris-Sud 11, 15 Rue Georges Clemenceau, 91400 Orsay, France
e-mail: lin.chen@lri.fr

Abraham A. Clements
School of Electrical and Computer Engineering, Purdue University, 465 Northwestern Ave, West Lafayette, IN 47907, USA
e-mail: clemen19@purdue.edu

Francisco de León
New York University, 5 Metrotech Center, Brooklyn, NY, USA
e-mail: fdeleon@nyu.edu

Hermann de Meer
Faculty of Computer Science and Mathematics, Chair of Computer Networks and Computer Communications, University of Passau, Innstr. 43, 94032 Passau, Germany
e-mail: hermann.demeer@uni-passau.de

Antonios Gouglidis
School of Computing and Communications, InfoLab21, Lancaster University, Lancaster, United Kingdom, LA1 4WA
e-mail: a.gouglidis@lancaster.ac.uk

Benjamin Green
School of Computing and Communications, InfoLab21, Lancaster University, Lancaster, United Kingdom, LA1 4WA
e-mail: b.green2@lancaster.ac.uk

Tianqi Hong
New York University, 5 Metrotech Center, Brooklyn, NY, USA
e-mail: th1275@nyu.edu

Ashish R. Hota
Automatic Control Laboratory, ETH Zürich, Zürich, Switzerland
e-mail: ahota@control.ee.ethz.ch

Linan Huang
Department of Electrical and Computer Engineering, New York University, 2 Metrotech Center, Brooklyn, 11201, USA
e-mail: lh2328@nyu.edu

David Hutchison
School of Computing and Communications, InfoLab21, Lancaster University, Lancaster, United Kingdom, LA1 4WA
e-mail: d.hutchison@lancaster.ac.uk

Ziad Ismail
Télécom ParisTech, Université Paris-Saclay, 46 rue Barrault, 75013 Paris, France
e-mail: ismail.ziad@telecom-paristech.fr

Sandra König
AIT Austrian Institute of Technology GmbH, Centre for Digital Safety & Security, Giefinggasse 4, 1210 Vienna, Austria
e-mail: sandra.koenig@ait.ac.at

Charles A. Kamhoua
Army Research Laboratory, Adelphi, MD, USA
e-mail: charles.a.kamhoua.civ@mail.mil

Kevin A. Kwiat
Air Force Research Laboratory, Cyber Assurance Branch, Rome, NY, USA
e-mail: kevin.kwiat@us.af.mil

Luke Kwiat
Vanderbilt University, Owen Graduate School of Management, Nashville, TN, USA
e-mail: kwiatluke@gmail.com

Jean Leneutre
Télécom ParisTech, Université Paris-Saclay, 46 rue Barrault, 75013 Paris, France
e-mail: jean.leneutre@telecom-paristech.fr

Ana Isabel Martinez
ETRA Investigación y Desarrollo S.A., Calle Tres Forques 147, 46014 Valencia, Spain
e-mail: amartinez.etraid@grupoetra.com

Erik Miehling
University of Michigan, Ann Arbor, MI, USA
e-mail: miehling@umich.edu

Joon S. Park
Syracuse University, School of Information Studies (iSchool), Syracuse, NY, USA
e-mail: jspark@syr.edu

Lisa Rajbhandari
Nets Branch Norway, Oslo, Norway
e-mail: lisa.rajbhandari@outlook.com

Mohammad Rasouli
University of Michigan, Ann Arbor, MI, USA
e-mail: frasouli@umich.edu

Stefan Rass
Universitaet Klagenfurt, Universitaetsstrasse 65–67, 9020 Klagenfurt, Austria
e-mail: stefan.rass@aau.at

Manuel Rodriguez
Air Force Research Laboratory, Cyber Assurance Branch, Rome, NY, USA
e-mail: manuel.rodriguez-moreno.1.ctr@us.af.mil

Karl Rossegger
Linz AG Telekom, Linz, Austria
e-mail: k.rossegger@linzag.at

Stefan Schauer
Center for Digital Safety and Security, Austrian Institute of Technology, B10, Lakeside Science & Technology Park, Lakeside, 9020 Klagenfurt, Austria
e-mail: stefan.schauer@ait.ac.at

Einar Arthur Snekkenes
Norwegian University of Science and Technology, Gjøvik, Norway
e-mail: einar.snekkenes@NTNU.no

Alma Solar
Electrica d'Alginet, Alginet, Spain
e-mail: alma@electricadealginet.com

Shreyas Sundaram
School of Electrical and Computer Engineering, Purdue University, 465 Northwestern Ave, West Lafayette, IN 47907, USA
e-mail: sundara2@purdue.edu

Demosthenis Teneketzis
University of Michigan, Ann Arbor, MI, USA
e-mail: teneketg@umich.edu

Alberto Zambrano
ETRA Investigación y Desarrollo S.A., Calle Tres Forques 147, 46014 Valencia, Spain
e-mail: azambrano.etraid@grupoetra.com

Ming Zhao
Arizone State University, School of Computing and Information Sciences, Tempe, AZ, USA
e-mail: mingzhao@asu.edu

Quanyan Zhu
Department of Electrical and Computer Engineering, New York University, 2 Metrotech Center, Brooklyn, 11201, USA
e-mail: qz494@nyu.edu

Part I
Theory

Chapter 1
Utilizing Game Theory for Security Risk Assessment

Lisa Rajbhandari and Einar Arthur Snekkenes

1.1 Introduction

Organizations are influenced by many factors that may hinder the accomplishment of their business objectives and cause loss of reputation, money, confidential data, etc. In today's sophisticated technological advancement with mobile computing, cloud computing, big data, bring your own device, Internet of things, etc., the importance of information security is even higher. To combat the growing number of threats, organizations need to stay a step ahead. With security risk assessment, organizations may get a credible picture of risks to their information systems and make decisions to allocate their resources to secure their systems against severe risks. Thus, security risk assessment is an approach that is of great value to organizations that want to withstand the current threat environment. If we can identify, estimate, and evaluate risks properly, we can better mitigate or treat them.

In most of the risk assessment approaches (e.g., ISO/IEC 27005:2011 [2]), risk is measured as the combination of consequence and likelihood of an incident. The values of likelihood and consequence are expressed in qualitative or quantitative forms. For instance, a scoring approach is used where qualitative scores such as low, medium, and high or ordinal scale of 1–5 are used in determining the likelihood or consequence of a risk event. However, in the absence of statistical data or the current statistical data being inadequate or irrelevant, the likelihood values are gathered using expert elicitation. In addition, the adaptive nature of the adversary may lead these assessments being based on subjective judgment. Cox also points out that risk

L. Rajbhandari (✉)
Nets Branch Norway, Oslo, Norway
e-mail: lisa.rajbhandari@outlook.com

E. A. Snekkenes
Norwegian University of Science and Technology, Gjøvik, Norway
e-mail: einar.snekkenes@NTNU.no

S. Rass, S. Schauer (eds.), *Game Theory for Security and Risk Management*, Static & Dynamic Game Theory: Foundations & Applications,
https://doi.org/10.1007/978-3-319-75268-6_1

matrices based on "frequency" and "severity" have the following limitations: poor resolution, errors, suboptimal resource allocation, and ambiguous inputs and outputs [9]. Further, in [8] Cox explains the limitations of "Risk= Threat x Vulnerability x Consequence" combination for the analysis of terrorist attacks. The other limitation of classical risk assessment methods is the lack of consideration of opportunity risks. Opportunity risk is "the concern that something desirable might not happen because the other player may not have the incentive to play a certain strategy as he has to bear loss or his gain may be insignificant" [27, 23].

With game theory, we can consider how players with conflicting interest interact in situations of interdependence, the strategies they choose, and how they assess the values of outcomes by choosing those strategies. Using game theory, we can utilize the existing robust mathematical noncooperative game models for risk assessment without the reliance on subjective probabilities. However, this is seldom used in organizations. The reliance on subjective judgment when determining likelihood is one of the main limitations of traditional risk assessment methods. Thus, it is important to investigate how game theory can be integrated with traditional risk assessment methods. Moreover, the reasons behind the occasional use of game theory for risk assessment in organizations might be the difficultly of adapting to a different approach and lack of required skills to implement it, among others as we discuss in Section 1.5.

This chapter shows how three existing risk assessment/management methods, ISO/IEC 27005:2011, NIST Special Publication 800-30 Revision 1 (NIST 800-30r1) [3], and CORAS [5], can be mapped to general risk assessment process and terminology. This shows how most of the traditional risk assessment methods have some common steps. The use of game theory for security has been highlighted in many works as [6, 17]. However, to our knowledge besides [25], no works have been published that show how the game theoretical steps can be mapped with the existing risk assessment approach. In [25], ISO/IEC 27005:2008 was used to show how game theoretical approaches can be used for risk assessment by mapping each game theoretical step with that of the ISO/IEC 27005:2008 and determining where a correspondence was missing. The intent of this chapter is to adapt the mapping to general risk assessment and game theoretical approach to introduce how game theory can be utilized for risk assessment and highlight some key points.

Games mainly between the attackers and defenders are mostly investigated in security as shown in [26, 17]. A cooperative game theoretical model can be used to capture the opportunity risks faced by the organization. However, cooperative game models are seldom considered when addressing security games. For instance, there is an opportunity risk for the organization, as the chief information security officer (CISO) wants to increase security awareness, but a staff has less incentive to go through the security training as he needs to spend time going through the online material and pass the quiz. This setting between the CISO and staff can be modeled as a cooperative game to determine the strategy that the CISO needs to take to address this security problem.

The remainder of this chapter is structured as follows. In Section 1.2, we provide a brief outline of risk assessment process and phases and the mapping between the common risk assessment steps and three selected approaches. Section 1.3 outlines the game theoretical steps for risk assessment with an example and shows the mapping between the game theoretical steps and general risk assessment steps (as derived in Section 1.2). An insight into the cooperative game model to address opportunity risks is explained in Section 1.4. We discuss some of the challenges of using game theory for risk assessment and conclude the chapter in Section 1.5. A literature review is provided in Section 1.6.

1.2 Risk Assessment

Even though risk is often related with the possibility of something bad, it can include both positive and negative aspects. This perspective of risk is defined in the field of economics in general or in project management [12].

There are different definitions for terminologies that are used in the risk management domain. We view risk assessment as a systematic process that consists of three steps – risk identification, risk analysis, and risk evaluation – as considered in the ISO/IEC 27005:2011 standard [2]. Further, risk management program consists of initial preparation (context establishment), risk assessment, risk treatment, risk monitoring and review, and risk communication and consultation.

In the following subsections, we first describe the phases of the general risk assessment. Then, we show how three selected approaches to risk assessment – ISO/IEC 27005:2011, NIST 800-30r1, and CORAS – can be viewed within a common framework. This view is also supported in [28], which provides a framework to show how the different risk assessment methods cover in terms of comprehensiveness of parts of steps. These methods are selected as these relate to information security risk management process. Apart from the selected methods, there are many risk assessment/management standards or methods such as the ISO 31000:2009 standard [1] which provides a common guideline on conducting risk management and Risk IT [14] framework which focuses on IT risk scenarios. The common framework can be extended to ISO 31000:2009 and Risk IT framework as well.

1.2.1 General Risk Assessment Phases

As mentioned above, there are numerous methods to conduct risk assessment, each with their subsequent steps. Even though the assessment can be accomplished in different sequences, they all have some common steps. From the literature review of the existing methods, the broader view of the general risk assessment process can be represented by the following phases: initial preparation, identification of risk scenario (identify threats, vulnerabilities, consequences), estimation of risk (assess likelihood, assess consequence, and determine risk), and evaluation of risk. These four phases are derived to simplify the risk assessment process and create a common framework of the different approaches stated above.

1. Initial Preparation: This phase includes defining scope, risk appetite, and objectives of the organization, identifying assets and their owners, and gathering documentation of the system or application or project in the scope of the risk assessment. One of the key challenges of this step is scope creep which may lead to unnecessary resource utilization and increased workload undermining the successful completion of the risk assessment process. The key stakeholders (apart from the asset owners) and their responsibilities are also identified for establishing communication and cooperation throughout the risk assessment process.

2. Identify Risk Scenarios: This phase consists of three sub-phases: identify threats, identify vulnerabilities, and identify consequences. Threats to the assets determined in the previous step are identified. It is followed by the identification of the vulnerabilities in assets which might be exploited (taking into consideration the existing controls). Then, the consequences if a threat source, e.g., an attacker, successfully exploits a vulnerability are also identified. This provides a clear picture of the risk scenario to be analyzed.

3. Estimate Risk: This phase consists of three sub-phases: assess consequence, assess likelihood, and determine risk. The consequence and likelihood of occurrence of identified risk scenarios are assessed using either a qualitative scale or quantitative values. The likelihood of the risk is assessed considering threats, vulnerabilities, consequences, and currently implemented controls (if any). The consequence to an organization may be determined in the form of financial loss, reputation damage, and loss of customers or in terms of loss of confidentiality, integrity, or availability (CIA). Finally, the risk level is determined usually as a combination of likelihood and consequence as given below.

$$Risk = Likelihood \cdot Consequence$$

4. Evaluate Risk: In this phase, total exposure of all risks of interest is reviewed and prioritized. A risk matrix with the x-axis representing the likelihood and the y-axis representing the consequence and color schemes of red, orange, and green representing high, medium, and low levels of risks may be used for easy visualization and communication.

Risk assessment is followed by the risk treatment phase (which is outside the scope of this chapter). In this phase, the categorization of the identified risks is carried out. The categorization documents whether the risks should be accepted, mitigated, avoided, or transferred [2] based on the cost-benefit criteria and the risk appetite of the organization. The accept option relates with retaining the risk as it is, the mitigate option relates with reducing the risk by applying security control measures, the avoid option relates with excluding the tasks that result in risk, and the transfer option relates with sharing the risk with another party, e.g., buying insurance. Then, the risk control plan is developed. The risk profile should be regularly monitored and reviewed whenever there is a significant change, e.g., in the system or project in scope. Moreover, the entire process should be carried out periodically for effective risk management. In addition, as stated in the ISO/IEC 27005:2011, risk communication and consultation should be carried out throughout the risk management program [2].

1.2.2 Mapping Between the General Risk Assessment and Three Selected Approaches

We map the process and terminology of the following three risk assessment/management standards or methods: ISO/IEC 27005:2011, NIST 800-30r1, and CORAS with the general risk assessment steps mentioned above. The mapping is based on the authors' understanding of the three methods and limited to cover the assessment steps, while the risk treatment, risk communication, and risk monitoring/maintaining phases are excluded.

ISO/IEC 27005:2011 is an international standard for risk management developed by the International Organization for Standardization/International Electrotechnical Commission. The risk management process consists of the following phases: context establishment, risk assessment, risk treatment, risk acceptance, risk communication, and risk monitoring and review.

NIST 800-30r1 developed by National Institute of Standards and Technology provides guidance for conducting risk assessments and supports the NIST 800-39 standard. The risk assessment process consists of the following steps: prepare for the assessment, conduct the assessment, communicate assessment results, and maintain the assessment.

CORAS is a model-based method for conducting security risk analysis. It uses Unified Modeling Language (UML) for modeling threat and risk. It consists of seven steps: introduction, high-level analysis, approval, risk identification, risk estimation, risk evaluation, and risk treatment.

The mapping as depicted in Table 1.1 shows that all the risk assessment steps of the ISO/IEC 27005:2011 standard, NIST 800-30r1, and CORAS can be mapped with the general risk assessment steps. As mentioned above, this point is also supported in [28] which provides a unified framework for different risk assessment methods. The identification of existing controls is implicitly included when identifying vulnerabilities in the general risk assessment steps, so it is not mentioned as a separate step. Apart from the ISO/IEC 27005:2011, other methods also do not mention this step explicitly. Besides, none of these methods consider the identification of opportunities.

Table 1.1: Mapping between the risk assessment processes and terminologies

General risk assessment steps	**ISO/IEC 27005:2011**	**NIST 800-30r1**	**CORAS**
Initial preparation			
Define scope, risk appetite, and objectives of the organization and identify assets and asset owners	Set the basic criteria, define scope and boundaries, organization for information security risk management	Prepare for assessment (identify purpose, scope, assumptions and constraints, information sources, risk model, and analytic approach)	Introductory meeting (identify scope, target, and assets to be protected), high-level analysis, approval
Identify risk scenarios			
Identify threats	Identification of threats	Identify threat sources and events	Identify threats and threat scenarios
Identify vulnerabilities	Identification of vulnerabilities	Identify vulnerabilities and predisposing conditions	Identify vulnerabilities
Identify consequences	Identification of consequences	Included when determining likelihood and impact	Included when modeling the threat diagram
Estimate risk			
Assess consequences	Assessment of consequences	Determine magnitude of impact	Estimate consequences
Assess likelihoods	Assessment of incident likelihoods	Determine likelihood of occurrence	Estimate likelihood values
Determine risk	Level of risk determination	Determine risk	Compute risk values
Evaluate risk			
Prioritize identified risk scenarios	List of risks prioritized	List the number of risks identified for each level of risk	Place risks in the risk evaluation matrix and risk diagram

1.3 Game Theory for Security Risk Assessment

A game-theoretic model includes the players, the strategies they can take, and the payoffs they gain by making the move. In some situations when conducting risk assessment, there is a lack of historical data, or the existing data may be insufficient or irrelevant. However, we may have insight into how the various stakeholders value the outcomes of the different actions they may engage in. Game theory provides a way to convert these payoff values to probabilities, under the assumption that the stakeholders are rational. Thus, game theory may offer a link between information available (strength of preferences) and what is required by typical risk assessment methods (probabilities). The translation from preference strengths to probabilities is obtained by "solving" the game, e.g., by computing the game's Nash equilibrium.

In the following subsections, we provide the game theoretical steps for risk assessment and elaborate the steps with an example of a simple two-player game between an administrator and an attacker. Then, we do the mapping of the game theoretical steps and the general risk assessment steps.

1.3.1 Game Theoretical Steps

The game theoretical steps for risk assessment are given below (adapted from [25]). The steps should be carried out systematically, and the process can be repeated. For each step, we provide a short description and elaborate by explaining how the data is collected (where feasible).

1. Investigate the scenario: The scope and assets that need to be protected along with its owners are identified when investigating the scenario. In addition, the criteria for repeated analysis with alternative strategies or payoffs are determined.

2. Identify the players: The decision-makers whose actions affect each other are identified. These include the player who gets the benefit or must bear loss (risk owner (RO)) and players with conflicting incentives to that of the RO (strategy owners (SO)). We assume the players are rational.

3. For each player, the following data is gathered. To get the full picture of the players, their motivation, capabilities (e.g., resources to implement or defend the attack), and experiences need to be considered:

    a. Determine the information gained: Information the players have when they decide are gathered. In relation to the information they have when they make decisions, the games can be classified into perfect or imperfect and complete or incomplete information games. In perfect information game, each player knows the previous moves of all other players (vice versa for imperfect information game). In complete information game, each player knows both the strategies and payoffs of all the players but might or might not know the previous moves (vice versa for incomplete information game).

b. Determine the strategies: The strategies related to the actions of the players (i.e., RO and SO) are determined. These may include strategies to overcome threats, to cause threats, or to gain opportunities. The actions can be based on an individual or a group. It is important to understand that one player can limit the options of another. Moreover, these options can be negotiated.
c. Identify the preferences: The players may value multiple orthogonal aspects of outcome (e.g., money, reputation, privacy, trust, etc. or confidentiality, integrity, availability (CIA)). These are also referred to as utility factors. These can be obtained by asking how they value the outcomes in workshops, through surveys, or investigating the research in psychology.
d. Represent by payoff/utility: Scale, measurement method, and weight for comparing outcomes are defined. Then, the preferences are ordered according to the obtained rank and represented by utility. Usually, the players have the incentive to maximize their payoff/utility. The utility can be estimated using additive utility function of multi-attribute utility theory (MAUT) [7]. The additive utility function for a player is defined to be the weighted average of its individual utility factors as given below.

$$U(a) = \sum_{k=1}^{m} w_k \cdot a_k \ . \tag{1.1}$$

where
m is the number of utility factors of the player,
w_k is the assigned weight of utility factor a_k, and $\sum_{k=1}^{m} w_k = 1$.

4. Formulate the game: The scenario is formulated, e.g., in normal form as shown in Figure 1.1, assuming that the RO and SO are not aware of the other players' choice when he makes his own choice.

5. Find the optimized strategies/equilibrium: The optimized strategies for each player are identified. The main point with the use of game theory is to bring forward strategies or incentivize the other player so that the best equilibrium is reached. The combination of optimum or best strategies chosen by the players is the pure-strategy Nash equilibrium. The equilibrium specifies the outcome of the game to the players. However, pure-strategy Nash equilibrium may not exist, and the other way of computing the equilibrium solution is to find the mixed-strategy Nash equilibrium which always exists. Utilizing it, we can obtain the probabilities, expected outcome the players get by playing each of the strategy, and the expected outcome of the game [29].

These steps can be iterative until the results are satisfactory (even though this is outside the scope of this chapter, we include it to provide an idea of how the results of game theoretical steps for risk assessment can be assessed). A satisfactory outcome may be the value of the game for the RO (or alternatively the RO payoff in equilibrium) that is within the limit set by the criteria, or "do nothing" may be the best strategy for the RO.

If the outcome is not acceptable, alternative controls may be added and the process repeated. Note, however, that the whole process assumes that both players have common knowledge relating to which strategies are contemplated by the other player. For example, computing the equilibrium involving a strategy not known to the other player breaks the validity of the analysis. Thus, it is recommended to do sensitivity analysis to determine to what extent small changes to player knowledge may influence the outcome of the equilibrium computations.

1.3.2 An Example Elaborating the Game Theoretical Steps

We elaborate the above game theoretical steps with an example of a two-player game.

1. Investigate the scenario: Our example is based on a scenario in which an organization is conducting a risk assessment of one of its system (asset) which has been attacked over the past years. The administrator (asset owner) is responsible for the protection of the system.

2. Identify the players: In our example, the players are an administrator (RO) and an attacker (SO). We assume the players are rational.

3. For each player (here, the administrator and attacker), the following data is gathered:

    a. Determine the information gained: We assume the given scenario between the administrator and attacker as a game of complete but imperfect information.
    b. Determine the strategies: The administrator may limit his actions to use existing controls (i.e., "do nothing") or implement new control measures to mitigate the risks. The implemented controls are categorized as the "do nothing" (here, NotDefend) option. The strategies of the attacker are based on the vulnerabilities he can exploit in the system and threats he can cause to the organization. Thus, the strategy space for the administrator is {Defend, NotDefend} and for the attacker is {Attack, NotAttack}.
    c. Identify the preferences/utility factors: For this scenario, we simply assume the administrator is concerned about the financial loss and reputation of the organization, and the attacker is concerned about his financial gain.
    d. Represent by payoff/utility: For this scenario, we assume w_1 and w_2 as the weights assigned by the administrator for financial loss and reputation of the organization, respectively, where $(w_1 > w_2)$. As the attacker is concerned only about his financial gain, his assigned weight is $w_1 = 1$. The

		Attacker	
		q	$1-q$
		Attack	**NotAttack**
Administrator	p **Defend**	$U_1(x_{1,1})$, $U_2(y_{1,1})$	$U_1(x_{1,2})$, $U_2(y_{1,2})$
	$1-p$ **NotDefend**	$U_1(x_{2,1})$, $U_2(y_{2,1})$	$U_1(x_{2,2})$, $U_2(y_{2,2})$

Fig. 1.1: Normal form representation of the scenario

scale for financial gain/loss is currency unit, and reputation is % which can be obtained from interview as it is assumed that the system has been attacked before. Thus, using the given scales and measurement methods (if any), we can obtain the value of the utility factors for both the administrator and the attacker. The details regarding some of the measurement methods and scales for utility factors like reputation are found in [24]. We apply Equation 1.1 to the attribute vectors produced by each of the strategies, for all the players. This is illustrated in Figure 1.1. Let the number of strategies for the administrator and attacker be r and s. The utility functions for the administrator and attacker are represented in the form $U_1(x_{i,j})$ and $U_2(y_{i,j})$, respectively, where $i = 1..r, j = 1..s$, $x_{i,j}$, and $y_{i,j}$ are utility factor vectors.

4. Formulate the game: We assume that neither RO nor SO can observe the other players' decision before he implements his own decision; thus it is appropriate to frame the setting as a normal-form game as shown in Figure 1.1. As this is the game of imperfect information, the players form belief about the strategies the other players choose. The administrator believes that the attacker plays the strategies Attack and NotAttack with probabilities q and $1-q$, respectively, where $(0 \leq q \leq 1)$. Likewise, the attacker believes that the administrator plays the strategies Defend and NotDefend with probabilities p and $1-p$, respectively, where $(0 \leq p \leq 1)$. Both players have an incentive to win or to maximize their utility. The strategies of the administrator are placed in the rows and the attacker in the columns. The pair of variables in each cell of the matrix represents the utility functions for the administrator and the attacker, respectively, as obtained in the step above.

5. Find the optimized strategies/equilibrium: Let's assume a pure-strategy Nash equilibrium with the strategy profile (Defend, Attack) is the outcome of the game. This suggests that there is a risk of the system being attacked, but by employing the Defend strategy, the administrator can mitigate the risk.

1.3.3 *Mapping Between Risk Assessment and Game-Theoretic Approaches*

In general, both the game theoretical and classical risk assessment methods consist of three phases: data collection, risk assessment/game theoretical model, and decision-making. We do the mapping of the game theoretical steps and the general risk assessment steps (as described in Section 1.2.1). We include the identification of opportunities apart from identification of threats in the risk assessment process.

The mapping as depicted in Table 1.2 shows that all the risk assessment steps are covered in the game theoretical approach. However, the game theoretical steps, such as the information gained by the strategy owners, and their beliefs and incentives are not explicitly considered in the traditional risk assessment. Depending on the used game model, one considers what information the SO might have about the RO in terms of his previous actions, strategies and payoffs, and vice versa. This strategic thinking of the players should be considered in security risk assessment as this will help to get a complete picture of the scenario being analyzed. In general, these corresponds to the lack of explicit consideration of behavior and motives of the opponent when conducting the risk assessment [22]. Even though some methods consider the motives of the opponent at the start of the process to determine threats, these human factors are not explicitly considered during the assessment phase. Further, the optimization of the strategies of the players is not included in the classical risk assessment methods.

In addition, the mapping shows that the probabilities are computed when using game-theoretic process. Thus, it addresses one of the main limitations of traditional risk management approaches. The mapping clearly depicts that game theory can be used for security risk assessment.

1.4 Cooperative Game to Address Opportunity Risks

The incentives of the players may differ according to the situations they face. Likewise, in economics, incentives of the players matter in maintaining security in an organization. Usually, the players have the incentives to win or to maximize their utility. However, the players might negotiate and come up with group incentives besides having their private incentive. Cooperative game theory helps to model the outcome of negotiation as a joint action [29]. It is mainly used to study contractual relations, e.g., a job contract between an employer and a staff in an organization may help to avoid conflicts and align incentives. In a noncooperative game, players' actions are regarded as individual behavior, whereas in a cooperative game, the players can and will collaborate or communicate to form coalitions enforced by an umpire [19] or a contract.

Games mainly between the attackers and defenders are mostly investigated in security [17, 26, 10]. However, cooperative game models are seldom considered

Table 1.2: Mapping between risk assessment and game-theoretic approaches (Adapted from [25])

General risk assessment process and terminology		**Game-theoretic process and terminology**
Initial preparation	Define scope, risk appetite, objectives of the organization	Investigate the scenario, define scope, define criteria for repeated analysis with alternative strategies or payoffs
	Identify assets and asset owners	Identify who owns the asset and who will get the benefit/bear loss (RO) Identify players – RO and SO (i.e., player(s) with opposing incentives to that of the RO)
Identify risk scenarios	Identify threats and opportunities	Determine strategies for the SO; also determine strategies for the RO (control measures already implemented and to be implemented)
	Identify vulnerabilities	Identify options that can be exploited by threats. Included while determining the strategies for the SO
	Identify consequences	Identify how the players value multiple orthogonal aspects of outcomes Identify the preferences or utility factors for each player
Estimate risk	Assess consequences	Define scale, measurement method, and weight for comparing outcomes and ranking preferences Represent by payoff/utility
	Assess likelihoods	Computed probabilities for each strategy of the players
	Determine risk	Computed expected outcome for each of the strategy of the SO is the risk to RO and vice versa
Evaluate risk	Prioritize identified risk scenarios	Prioritize the expected outcome for the players
Not explicitly included for SO		Determine the information gained by the players
Not explicitly included for SO		Determine the beliefs and incentives of the players
Not included		Find the optimized strategies

when addressing security games. As mentioned above, one of the limitations of classical risk assessment methods is the lack of consideration of opportunity risks. Cooperative game theoretical models can be used to capture the opportunity risks faced by the organization.

For instance, there is an opportunity risk for the organization, as the CISO wants to increase security awareness, but a staff has less incentive to go through the security training as he needs to spend time going through the online material and pass the quiz. This gives rise to a situation of interdependence as the behavior of the staff may negatively affect the organization. This setting between the CISO and staff can be modeled as a cooperative game to determine the strategy that the organization needs to take to address this security problem.

The organization can help align the incentive of the staff to go through the training by awarding some points for completing the training based on the obtained quiz scoring instead of taking the other way that may be counterproductive. This strategy of the organization may lead to a win-win situation. Thus, strategic uncertainty of obtaining the opportunity to the organization (increasing security awareness of staff) may be obtained by using the cooperative model that benefits both the players and maximize their utility.

1.5 Discussion and Conclusion

Security risk assessment is a widely used approach to identify and deal with security threats that an organization may face or is facing. The decision to allocate budget to mitigate or treat risk is often based on the severity of the risk. Thus, it is of utmost importance to identify and assess risk scenarios properly. However, the lack of statistical data and the adaptive nature of the adversary may lead the assessment to be based on subjective judgment when using the traditional security risk assessment methods. Besides, most of these methods do not consider opportunity risks.

The increase in security risks to organizations has opened a path for new approaches to guide its security risk assessment and investments. With the limitations of current risk assessment methods, the use of game theoretical models can be beneficial as it is based on mathematical models.

The result from mapping of the different risk assessment methods showed that the different approaches can be generalized to some common steps. Even though three risk assessment methods are mapped in this chapter, the mapping could be extended to cover other approaches such as ISO 31000:2009 [1] and Risk IT framework [14]. Further, the mapping between the common risk assessment process/terminology and that of the game-theoretic steps highlighted some of the game theoretical steps that lacked correspondence with the risk assessment steps. The mapping clearly depicted that game theory can be used for analyzing risk scenarios. However, the use of game theory does not come without challenges.

Many organizations have their own risk assessment methods, so adapting to another method with a completely different approach is a challenge. Moreover, staffs need to acquire the skills to use the method. In game theory, the strategies of a player are based on what he believes the other player might do and vice versa. Thus, modeling real-world scenarios might be complex. It is even harder without the availability of a tool. Even though some tools such as Gambit [18] or GAMUT [30, 20] are available, these tools need to be incorporated with risk assessment to formulate and analyze the risk scenarios. However, the benefits of game theory outweigh the above stated challenges, and considering a practical game-theoretic approach in risk assessment can provide exclusive insight on security risks.

More research needs to be done for organizations to utilize game theory for risk assessment. Further, work on cooperative models to address opportunity risks needs to be investigated which will benefit organizations to focus on both threats and opportunities.

1.6 Chapter Notes and Further Reading

Game theory is used in many disciplines such as economics, biology, political science, and information security. Studies have been conducted for merging or using game theory with traditional risk assessment [11, 4, 13, 10]. Game theory has also been applied in general for information security to capture an attacker's incentive [15] or to quantify security risk [6]. Moreover, it has been specifically used for network security [21, 31, 26, 16, 17].

According to Hausken, game theory can be merged with probabilistic risk analysis considering the individual-collective conflicts that affect risk [11]. Banks et al. highlight that the traditional risk analysis approach is not reliable in most of the cases and have suggested to use statistical risk analysis integrated with game theory [4]. They analyzed strategies for a smallpox attack by modeling it as a zero-sum game with random payoffs and utilized both the minimax and the Bayesian approaches to solve it. The research by Insua et al. involved both game-theoretic concept and statistical risk analysis for solving adversarial risk analysis (ARA) [13]. They also put forward the Bayesian approach to ARA. In [10], Cox states that using game theory models, ARA may be improved in allocating limited resources compared to using the classical risk scoring models.

Liu et al. used the incentive-based method to model attacker intent, objectives, and strategies (AIOS) and developed a game-theoretic approach to interfere AIOS [15]. The QuERIES methodology was developed to quantify cybersecurity risk so that organizations can come up with proper investment strategies [6].

A game-theoretic model has been used for analyzing intrusion detection in mobile ad hoc networks by Patch et al. using a multistage dynamic noncooperative game with incomplete information [21]. Xiaolin et al. proposed a risk assessment model based on Markov game theory for network information system [31], while Maille et al. explain various noncooperative game-theoretic aspects with security games to cover network security issues [16]. The two surveys in [26, 17] show the extensive use of game theory for network security. In [17], Manshaei et al. provide a survey of network security games between attackers and defenders. Roy et al. [26] survey the existing game theoretical solutions applied to network security which falls under noncooperative games.

Some of the tools that exist for game theory applications are Gambit and GAMUT. Gambit is a software comprising of a set of game theory tools with which games can be constructed and analyzed in both the normal and extensive forms [18]. Among other features, there are tools for computing Nash equilibria and quantal response equilibria. GAMUT consists of tools for generating game and testing game-theoretic algorithms [30, 20].

Acknowledgements We would like to thank the anonymous reviewers for their valuable comments and suggestions.

Disclaimer This is an independent research of the first author; thus the view expressed in this book chapter is not associated with any organization she is affiliated with.

References

1. *ISO 31000 Risk management – Principles and guidelines*. 2009.
2. *ISO/IEC 27005 Information technology -Security techniques - Information security risk management*. ISO/IEC, 1st edition, 2011.
3. NIST Special Publication 800-30 Revision 1. Guide for conducting risk assessments. Technical report, 2012.
4. David L. Banks and Steven Anderson. *Combining Game Theory and Risk Analysis in Counterterrorism: A Smallpox Example*. Springer New York, 2006.
5. F. Braber, I. Hogganvik, M. S. Lund, K. Stølen, and F. Vraalsen. Model-based security analysis in seven steps — a guided tour to the coras method. *BT Technology Journal*, 25(1):101–117, January 2007.
6. L. Carin, G. Cybenko, and J. Hughes. Cybersecurity strategies: The queries methodology. *Computer*, 41(8):20–26, Aug 2008.
7. Robert T. Clemen. *Making Hard Decision: An Introduction to Decision Analysis*. Duxbury, second edition, 1996.
8. Jr. Louis Anthony Cox. Some limitations of "Risk = Threat x Vulnerability x Consequence" for risk analysis of terrorist attacks. *Risk Analysis*, 28(6):1749–61, 2008.
9. Jr. Louis Anthony (Tony) Cox. What's wrong with risk matrices? *Risk Analysis*, 28(2):497–512, 2008.
10. Jr. Louis Anthony (Tony) Cox. Game theory and risk analysis. *Risk Analysis*, 29(8):1062–1068, 2009.
11. Kjell Hausken. Probabilistic risk analysis and game theory. *Society for Risk Analysis*, 22, 2002.
12. David Hillson. Extending the risk process to manage opportunities. *International Journal of Project Management*, page 235–240, 2002.

13. David Rios Insua, Jesus Rios, and David Banks. Adversarial risk analysis. *Journal of the American Statistical Association*, 104(486):841–854, Jun 2009.
14. ISACA. *The Risk IT Framework*, 2009.
15. Peng Liu and Wanyu Zang. Incentive-based modeling and inference of attacker intent, objectives, and strategies. In *Proceedings of the 10th ACM Conference on Computer and Communications Security*, CCS '03, pages 179–189, New York, NY, USA, 2003. ACM.
16. Patrick Maillé, Peter Reichl, and Bruno Tuffin. *Of Threats and Costs: A Game-Theoretic Approach to Security Risk Management*, pages 33–53. Springer New York, New York, NY, 2011.
17. Mohammad Hossein Manshaei, Quanyan Zhu, Tansu Alpcan, Tamer Bacşar, and Jean-Pierre Hubaux. Game theory meets network security and privacy. *ACM Computing Surveys (CSUR)*, 45(3):25, 2013.
18. Richard D. McKelvey, Andrew M. McLennan, and Theodore L. Turocy. Gambit: Software tools for game theory, version 16.0.1. http://www.gambit-project.org, 2016. [retrieved: 15-9-2017].
19. John Nash. Non-cooperative games. *Annals of mathematics*, pages 286–295, 1951.
20. Eugene Nudelman, Jennifer Wortman, Yoav Shoham, and Kevin Leyton-Brown. Run the gamut: A comprehensive approach to evaluating game-theoretic algorithms. *Autonomous Agents and Multiagent Systems, International Joint Conference on*, 2:880–887, 2004.
21. Animesh Patcha and Jung-Min Park. A game theoretic formulation for intrusion detection in mobile ad hoc networks. *International Journal of Network Security*, 2:131–137, March 2006.
22. Lisa Rajbhandari. Risk analysis using "conflicting incentives" as an alternative notion of risk, 2013.
23. Lisa Rajbhandari and Einar Snekkenes. Risk acceptance and rejection for threat and opportunity risks in conflicting incentives risk analysis. In *International Conference on Trust, Privacy and Security in Digital Business*, pages 124–136. Springer, 2013.
24. Lisa Rajbhandari and Einar Snekkenes. Using the conflicting incentives risk analysis method. In *IFIP International Information Security Conference*, pages 315–329. Springer, 2013.
25. Lisa Rajbhandari and Einar Arthur Snekkenes. *Mapping between Classical Risk Management and Game Theoretical Approaches*, pages 147–154. Springer Berlin Heidelberg, Berlin, Heidelberg, 2011.
26. Sankardas Roy, Charles Ellis, Sajjan Shiva, Dipankar Dasgupta, Vivek Shandilya, and Wu Qishi. A survey of game theory as applied to network security. In *System Sciences (HICSS), 2010 43rd Hawaii International Conference on*, pages 1–10. IEEE, 2010.
27. Einar Snekkenes. Position paper: Privacy risk analysis is about understanding conflicting incentives. In *IFIP Working Conference on Policies and Research in Identity Management*, pages 100–103. Springer, 2013.

28. Gaute Wangen, Christoffer Hallstensen, and Einar Snekkenes. A framework for estimating information security risk assessment method completeness. *International Journal of Information Security*, pages 1–19, 6 2017.
29. Joel Watson. *Strategy : An Introduction to Game Theory*. W. W. Norton & Company, 2nd edition, 2008.
30. Jenn Wortman, Eugene Nudelman, Mark Chen, and Yoav Shoham. Gamut: Game-theoretic algorithms evaluation suite. http://gamut.stanford.edu/. [retrieved: 15-9-2017].
31. Cui Xiaolin, Tan Xiaobin, Zhang Yong, and Xi Hongsheng. A Markov game theory-based risk assessment model for network information system. In *CSSE '08: Proceedings of the 2008 International Conference on Computer Science and Software Engineering*, pages 1057–1061, Washington, DC, USA, 2008. IEEE Computer Society.

Chapter 2
Decision Making When Consequences Are Random

Stefan Rass

2.1 Introduction

The intricacy of decision making is often due to uncertainty about the data to base a decision upon, and the consequences that the decision implies. Commonly, decision options are rated based on their expected utility. This approach is intuitive and successful in many cases, but has difficulties when the utility to be associated with an action is unknown or at least uncertain. Both problems can be addressed by accepting randomness as an intrinsic part of the utility itself, leading to defining optimal decisions in terms of stochastic orders rather than upon benchmark figures (only). For one such (even total) stochastic order, we will give a complete construction in this chapter, accompanied by examples and procedures how to get a (stochastically optimal) decision.

Mathematical decision making is typically a matter of making an optimal choice w.r.t. some measure of utility or damage. In the simplest case, a decision problem is the question for the best choice among at least two (among finitely many) options $a_1, \ldots, a_n$, the entirety of which is called the *decision space* or *action space* A. Each action has an associated utility $u(a_1), \ldots, u(a_n)$, being a value in some ordered set. The most common choice for that set is $\mathbb{R}$, for its natural ordering that makes it useful in optimization. However, numbers are only one among several possibilities to quantify a decision (i.e., its outcome), and we shall develop probability (distributions) as a full-fledged (and richer) substitute for real numbers here. For the moment, however, let us stick with $\mathbb{R}$ to quantify actions, to keep things simple.

Whenever the utilities associated with the actions can be ordered, the decision problem boils down to taking the action a_i that maximizes $u(a_j)$ among all

---

S. Rass (✉)
Universitaet Klagenfurt, Institute of Applied Informatics, Universitaetsstrasse 65-67, 9020 Klagenfurt, Austria
e-mail: stefan.rass@aau.at

S. Rass, S. Schauer (eds.), *Game Theory for Security and Risk Management*, Static & Dynamic Game Theory: Foundations & Applications, https://doi.org/10.1007/978-3-319-75268-6_2

$j \in \{1,2,\ldots,n\}$. By endowing the action space with a utility function $u : A \to \mathbb{R}$, A becomes an ordered set via $a_i \preceq a_j :\iff u(a_i) \leq u(a_j)$, and looking for optimal decisions becomes an obvious matter of optimizing the function u over the set A.

The existence of u, however, needs to be clarified and the classical von Neumann-Morgenstern approach to this end is axiomatic and based on a set of properties that a choice relation on A should have, which induces the existence of the function u (see [19, Sec.2.2] for a detailed treatment). Here, our focus will not be on assuring u's existence, but rather its concrete specification, which typically is the more challenging issue in practice. While numbers are easy to work with, their specification can create difficulties, since actions are not always obvious to quantify. For example, social risk replies to press releases, or "trust" in general, are difficult to model and measure by crisp numbers. Qualitative security risk management usually speaks about losses in terms of categories rather than precise figures (e.g., a severe damage may happen once at least one person is deadly injured by an incident, no matter how much monetary loss is associated with the incident besides).

The difficulty in fixing a number to measure the consequence of an action a is often due to random influences on the action's outcome. This makes it more natural to model the consequence as a random variable U associated with the action a. In a simple conversion to the numeric setting, we can humbly replace U by a representative number for it (usually the expected value), but this burns lots of information that U contains besides $\mathrm{E}(U)$. A more informed decision making can take into account further moments of U, such as the variance, but the best possibility would be using the whole random variable U itself to measure the utility of the action a. Indeed, random variables may obey so-called stochastic orders [22]. One such order that has particularly nice properties for risk management is developed hereafter.

It is easy to imagine situations where there is no unique optimal decision, say, when the respective utilities are indifferent or the utility is not exclusively determined by our own choice. This is the particular setting of game theory, where a player's choice is also quantified by a utility u, but this depends on the actions of (at least one) opponent too. Games typically assume rational behavior of players so that the additional inputs that determine a player's utility are chosen by others to maximize their own welfare. Decision theory admits a more general view by allowing any kind of influence from outside, and asking only for the best action advisable in a given situation. By letting the outcome of the utility function be random, we can technically replace the utility space $\mathbb{R}$ by the set $\Delta(\mathbb{R})$ of probability distributions supported on $\mathbb{R}$. Moreover, we shall assume that the supports of distributions in $\Delta(\mathbb{R})$ are bounded (and hence compact). This technical restriction is indeed consistent with the subjective perception of utility by individuals; for example, if the gain is monetary, then revenues above, then many individuals may become quite indifferent about their options if they range in utilities above 100,000,000 \$ [19]. While this threshold is clearly different between individuals, and again different for enterprises or whole countries, the assumption of some (problem-specific) upper bound appears nonetheless reasonable for every finite set of actors involved in a specific decision problem.

The idea of letting the utility function be random for fixed actions can equivalently be materialized by letting the actions be random but each giving a deterministic utility. In that setting, we let the decision space be the set $\Delta(A)$ of distributions supported on the action space A, and consider the utility function as a mapping into $\mathbb{R}$ again. This view is justified by the symmetry in a decision problem in which several actors, perhaps unbeknownst to each other, are interacting: each actor takes actions to optimize its own expected utility, but actually sees a randomly different outcome due to other influences to one's own utility. Experience (or learning) may then suggest to take different actions in future situations of the same kind. Formalizing the term "expected utility" here leads to the usual way of ordering utilities under randomized decisions: let P_1, P_2 be two distributions over the same action space, then we *prefer* P_1 over P_2, based on the function $u : A \to \mathbb{R}$ as a loss measure, if and only if

$$\mathrm{E}_{P_1}(u(X)) \leq \mathrm{E}_{P_2}(u(X)), \tag{2.1}$$

when X is the random action chosen from A under either P_1 or P_2, respectively.

Relation (2.1) is convenient as casting the issue of uncertain outcomes back into a scalar (numeric) utilities that can easily be compared. But is this always helpful in decision making? The answer is no, since the expectation can hide information that may be critical for the decision maker.

Example 2.1. Consider the distributions sketched in Figure 2.1, both corresponding to the random quantity $u(Z)$ under two different distributions of Z, hereafter denoted by the symbols $X \sim P_1$ and $Y \sim P_2$. For simplicity, let us assume Gaussian distributions $X \sim \mathcal{N}(\mu = 2, \sigma = 2)$ and $Y \sim \mathcal{N}(3, 1)$. Clearly, the expectation of X is less than that of Y, so (2.1) would point to X as the preferred choice. However, X admits much larger loss than Y in high ranges; specifically, $\Pr(5 \leq Y \leq 9) \approx 0.0228$, while $\Pr(5 \leq X \leq 9) \approx 0.0666$, so if we take prefer a more "stable" outcome at the cost of accepting a slightly larger risk, Y would be preferable, opposed to what (2.1) indicates. Likewise, (2.1) is of no help if the expectations are equal, and we can construct examples like the previous one with equal means but different standard deviations.

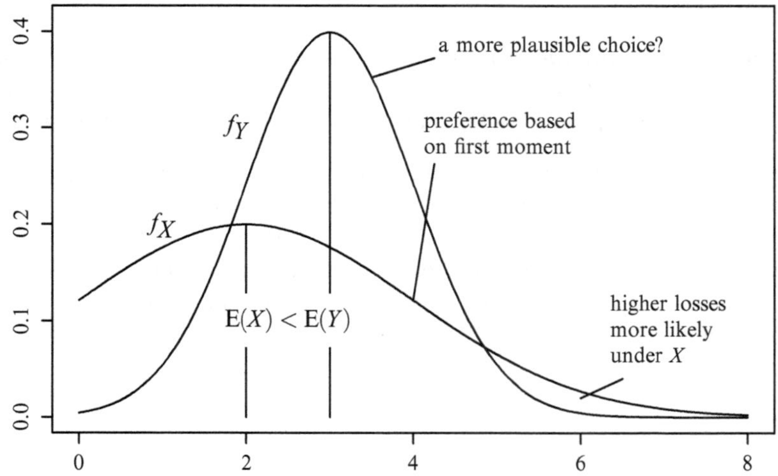

Fig. 2.1: Example Comparison of two Gaussian Distributions

Naturally, one could propose extending the ordering (2.1) to account for standard deviations too. Note that some special forms of uncertainty treatment in decision making such as interval arithmetic appear as special cases here (e.g., compact intervals can be modeled as uniform distributions). However, using variances only reshapes the issue but leaves new problems behind, since skewness could tip the scale when the mean and variances are perceived as secondary (cf. [5]).

A different route is thus offered by ordering the distributions themselves, rather than ordering some derived quantities thereof. The most prominent such *stochastic ordering* is obtained from (2.1) by demanding this relation to hold for *all* nondecreasing functions u instead to only for some fixed utility function. This approach orders two random variables as $X \preceq Y$ if and only if (2.1) holds for all nondecreasing functions u. An equivalent definition of $X \preceq Y$ is

$$\Pr(X \leq x) \leq \Pr(Y \leq x) \quad \text{for all } x \in \mathbb{R}. \tag{2.2}$$

This ordering is not total in general: it is not difficult to construct two (multimodally distributed) random variables X, Y that violate condition (2.2) in either direction.

Our example above went into a similar yet less restrictive direction by arguing that preferences may depend only on utilities with large magnitudes. This "asymptotic" approach will be revisited later.

We refer the reader to [22] for an excellent overview about rich body of literature related to such stochastic orders, and will again take a different route than usual at this point. Let us pick up the idea of using moments for a comparison (like in Example 2.1). Observe that the full sequence of moments uniquely pins down a distribution under quite mild conditions already. To rigorously formulate those, recall that a random variables X may have an associated moment-generating function $m_X(s) := \mathrm{E}(\exp(s \cdot X))$, where the expectation is over the distribution of X (let us assume that this expectation exists; otherwise, we may resort to characteristic functions doing equally well for our purposes). A combination of well-known facts about moment generating functions and Taylor-series expansions thereof delivers the following result:

Proposition 2.1. *Let two random variables X, Y have their moment generating functions $m_X(s), m_Y(s)$ exist within a neighborhood $U_\varepsilon(0) = \{x \in \mathbb{R} : -\varepsilon < x < \varepsilon\}$. Assume that $EX^k = EY^k$ for all $k \in \mathbb{N}$. Then X and Y have the same distribution.*

This suggests that we can use the full moment sequence $(\mathrm{E}(U)^k)_{k \in \mathbb{N}}$ as a valid replacement for a random utility $U = u(a)$, when a is chosen according to some (randomized) decision rule $P \in \Delta(A)$ over the action space A. Alas, the space of sequences is not ordered in general, but this can be fixed with reasonable effort. Before that, however, let us become more specific on the application context that we are aiming at.

2.2 Decision Making for Security: Loss Minimization

Risk management is a broad term that is often associated with financial losses, but also extends to many other fields like security. Although security is a term with a mostly qualitative meaning (with many different technical definitions found in cryptology [8]), it closely relates to economic matters upon taking a proper view on it. While the cryptographic understanding of security is rendering an attack practically impossible (often conditional on certain assumptions), we can alternatively define security as a state where the cost for an attack outweighs its quantitative utility. This is the decision-theoretic understanding of security, where decision-theoretic tools lay the (heuristic) foundations of quantitative security risk management.

Although tempting, there is no obvious way of defining risk in statistical terms, despite the usual understanding of risk as "expected damage" elegantly corresponds to a definition like

$$\text{risk} = \text{impact} \times \text{likelihood}, \tag{2.3}$$

which is commonly used to resemble the term $E_P(u(r))$ in decision making based on (2.1). The difference to the previous discussion is the understanding of "impact," which corresponds to *loss*, defined as negative utility, i.e., $-u$. In fact, it is precisely (2.3), combined with (2.1) as a decision rule that quantitative security risk management [24] is based on. Good reasons to discourage such an approach in risk management have been discussed by [11], giving examples where even solid numeric data can lead to implausible decisions.

Example 2.2 ([11]). From long-term records, it is known that the probability of a lightning strike in the area around Munich (Germany) is approximately 1.26×10^{-6}. Given an estimate of 10,000 $ repair costs, (2.3) leads to an expected damage (risk) of ≈ 0.12 $; a damage that we would certainly not be too concerned about. However, not investing in a lightning protection in a housing for sensitive electronic equipment appears not advisable either.

Due to examples like the above, security risk management is typically recommended to use categorical scales to quantify damage, but also likelihood, in order to avoid numerical imprecisions and the illusion of accuracy where there is none. Unlike in financial risk management, where sophisticated distribution models are available based on underlying stochastic processes, security risks not necessarily correspond to nearly as well understood dynamics and are often subjectively determined by personal experience, perception, reported incidents, and other vague sources of information.

Categorical scales like $\{$ "negligible" $<$ "low" $<$ "medium" $<$ "high" $<$ "very high" $\}$ would (in a rigorous treatment) call for order statistics to define a utility function, or more commonly a loss function in security (we will nonetheless write u to mean utility or loss, where the concrete understanding will become clear from the context). While the term $E_P(u(X))$ remains well-defined if u maps (random) actions into ranks corresponding to loss categories, expression (2.3) no longer makes sense in multiplying ranks (since the likelihood presumably comes on a rank scale too, according to what most risk management standards recommend). As such, (2.3) deteriorates into a (nevertheless plausible) heuristic that can no longer rest on solid decision-theoretic foundations.

To refurbish the framework of quantitative security risk management, stochastic orders can help. The task is constructing a well-defined decision rule that lets us find the best decision when the outcome is uncertain. Whenever there may be no single optimal action, we shall look for randomized decisions that inherently convexify the action space so that the existence of optima is assured again.

2.2.1 A Total Stochastic Ordering Based on Moments

Towards generality, let us consider the action space to be an (in)finite family of distribution functions $A = \{F_1, F_2, F_3, \ldots\}$, where each F_i has a bounded support in $\mathbb{R}$. The decision problem defined hereby is picking the "best" option from A, given that the ranking must (and can) be based only on random outcomes as described by the distribution functions in A. For simplicity, let us consider A as being finite with n elements. Then, each F_i corresponds to a random variable X_i, within some bounded range $-\infty < a_i \le X_i \le b_i < \infty$ (as would, for example, be the case for categorical units of loss, such as are common in security risk management). If X_i measures the loss (whether categorical or continuous), a reasonable choice preference rule should be invariant w.r.t. additive shifts. That is, shifting all scales by the same additive amount should obviously leave the relative preferences unchanged. Under this requirement, we can w.l.o.g., assume all $X_i \ge 1$ and to range within the compact interval $[1, \max\{b_i : i = 1, 2, \ldots, n\}]$. The latter condition assures the existence of moments of all orders for all $X_i \sim F_i$, so that Proposition 2.1 delivers a well-defined characterization of a random variable X_i by its moment sequence $\mathbf{r}_i := (\mathrm{E}X_i^k)_{k\in\mathbb{N}}$. Let us think of the whole sequence $\mathbf{r}_i$ as describing the *risk* associated with the i-th choice, and being described by a full distribution object, rather than an expectation thereof or any other derived quantity (such as variance or similar; note that the third moment seems to play its own distinct role in how preferences are made and risk is perceived [5, 23]).

Given distributions as sequences of numbers, we ought to order them somehow. Most obvious is a standard lexicographic order, putting $\mathbf{r}_i <_{lex} \mathbf{r}_j$ if and only if there is an index $t \in \mathbb{N}$ for which $\mathrm{E}(X_i^k) = \mathrm{E}(X_j^k)$ for $k < t$ and $\mathrm{E}(X_i^t) < \mathrm{E}(X_j^t)$. Obviously, $\mathbf{r}_i \le_{lex} \mathbf{r}_j$ implies (2.1), and also meets the intuition to look at variances or skewness if the lower moments provide no guidance. Still, decisions made under this ordering can be misleading, since the choice made in Example 2.1 remains the same, as does the argument against it.

A different way of ordering sequences while at the same time casting them into arithmetic objects is offered by non-standard calculus [20]. This theory embeds the ordered field $(\mathbb{R}, +, \cdot)$ into a larger and still ordered field $({}^*\mathbb{R}, +, \cdot)$, by associating a number $x \in \mathbb{R}$ by an infinite sequence of the form $(x, x, x, \ldots)$. The set ${}^*\mathbb{R}$ includes all infinite sequences, hereafter denoted as $\mathbb{R}^\infty$ (in analogy to the set of n-dimensional vectors being $\mathbb{R}^n$), and defines arithmetic over them in a canonical way: for two sequences $\mathbf{a} = (a_k)_{k\in\mathbb{N}}, \mathbf{b} = (b_k)_{k\in\mathbb{N}} \in \mathbb{R}^\infty$, we put $\mathbf{a}+\mathbf{b} := (a_k+b_k)_{k\in\mathbb{N}}$ and $\mathbf{a}\cdot\mathbf{b} := (a_k \cdot b_k)_{k\in\mathbb{N}}$. An ordering cannot be defined in the same way, since, for example, the alternating sequences $(0,1,0,1,\ldots)$ and $(1,0,1,0,\ldots)$ obviously do not uniformly satisfy $\leq$ or $\geq$ on their elements. To resolve this, we have to specify which indices matter for the comparison and which are irrelevant. Filters will be the technical vehicle to do this.

A *filter* is a subset $\mathscr{U} \in \mathscr{P}(\mathbb{N})$ so that $\emptyset \notin \mathscr{U}$ and for all $A, B \in \mathscr{U}$ we have $A \cap B \in \mathscr{U}$ (closure under intersection), and $A \subseteq B$ implies $B \in \mathscr{U}$ (closure under set-inclusion). If, whenever $A \notin \mathscr{U}$ we have the complement set $A^{\complement} \in \mathscr{U}$, then $\mathscr{U}$ is an *ultrafilter*. If, in addition, $\mathscr{U}$ contains no finite sets, then $\mathscr{U}$ is called *free*. That is the kind of filter that we will work with in the following: let two sequences $\mathbf{a} = (a_k)_{k\in\mathbb{N}}, \mathbf{b} = (b_k)_{k\in\mathbb{N}}$ and a free ultrafilter $\mathscr{U}$ be given. We define $\mathbf{a} \leq \mathbf{b}$ (modulo $\mathscr{U}$), if and only if $\{i \in \mathbb{N} : a_i \leq b_i\} \in \mathscr{U}$. Equivalence modulo $\mathscr{U}$, denoted as $\equiv_{\mathscr{U}}$, is defined analogously with the equivalence class of a sequence $\mathbf{a}$ being written as $[\mathbf{a}]_{\mathscr{U}}$. The *hyperreal number space* ${}^*\mathbb{R}$ is then defined w.r.t. $\mathscr{U}$ as the quotient space ${}^*\mathbb{R} = \{[\mathbf{a}]_{\mathscr{U}} : \mathbf{a} \in \mathbb{R}^\infty\} = \mathbb{R}^\infty/\mathscr{U}$. For a better understanding how $\equiv_{\mathscr{U}}$ differs from $=$ in ${}^*\mathbb{R}$, reconsider the sequences $\mathbf{a} = (0,1,0,1,0,1,\ldots)$, and $\mathbf{b} = (1,0,1,0,\ldots)$. By definition of the multiplication, $\mathbf{a}\cdot\mathbf{b} = (0,0,0,0,\ldots)$, which is obviously an additively neutral element. So either $\mathbf{a}$ or $\mathbf{b}$ or both must be equivalent to the zero sequence $(0,0,\ldots)$ in ${}^*\mathbb{R}$, though neither is identically zero.

The existence of a free ultrafilter follows from Zorn's lemma (e.g., by going for the $\supseteq$-maximal cover of the Fréchet-Filter $\mathscr{F} = \{A : A^{\complement} \text{ is finite}\}$), but nonconstructively so. Even worse, it is so far unknown if several non-isomorphic models of ${}^*\mathbb{R}$ exist, each of which may therefore have its own (individual) ordering.

Fortunately, however, it can be shown (Theorem 2.1) that the ordering restricted on the set of distributions compactly supported in the range $[1,\infty)$ is ordered in the same way in all models of ${}^*\mathbb{R}$. Let us postpone the proof of this until a little later, to first give the definition of a stochastic order, based on the hyperreal ordering of the representative moment sequences.

Definition 2.1. Take $a > 1$ and let $X \sim F_1, Y \sim F_2$ be two random variables taking values in a (common) compact set $[1,a] \subset \mathbb{R}$. Let the distributions F_1, F_2 be either both continuous or both discrete (or categorical). For such random variables, we define the stochastic order $X \preceq Y$, equivalently denoted as $F_1 \preceq F_2$ (for distributions) or $f_1 \preceq f_2$ (for densities), to hold if and only if $(\mathrm{E}_{F_1}(X^k))_{k\in\mathbb{N}} \leq (\mathrm{E}_{F_2}(Y^k))_{k\in\mathbb{N}}$, where the latter ordering is in the hyperreal space, upon treating the moment sequences as hyperreal numbers.

The totality of this ordering follows from the total ordering of the set ${}^*\mathbb{R}$, but the lack of dependence of the order on the model of ${}^*\mathbb{R}$ is not as obvious. An additional difficulty comes in as we cannot decide the ordering without knowing a free ultrafilter, and today, no such filter is known explicitly. So, in order to work with Definition 2.1, we will convert it into a "more handy" form upon which a variety of useful features of the ordering will become evident.

First, let us study the (in)dependence of Definition 2.1 on models of ${}^*\mathbb{R}$, by looking at how two continuous or two discrete distributions compare by their moment sequences.

Lemma 2.1. *Let X,Y be random variables as in Definition 2.1 with either both continuous or both discrete (categorical) distribution functions. Then, there is an integer K so that either $[\forall k \geq K : E(X^k) \leq E(Y^k)]$ or $[\forall k \geq K : E(X^k) \geq E(Y^k)]$.*

Proof (from [13]). We discuss the case of two continuous distributions first. Let f_1, f_2 denote the densities of the distributions F_1, F_2. Fix the smallest $b^* > 1$ so that $\Omega := [1, b^*]$ covers both the supports of F_1 and F_2. Consider the difference of the k-th moments, given by

$$\Delta(k) := \mathrm{E}X^k - \mathrm{E}Y^k = \int_\Omega x^k f_1(x)dx - \int_\Omega x^k f_2(x)dx = \int_\Omega x^k (f_1 - f_2)(x)dx. \quad (2.4)$$

Towards a lower bound to (2.4), we distinguish two cases:

1. If $f_1(x) > f_2(x)$ for all $x \in \Omega$, then $(f_1 - f_2)(x) > 0$ and because f_1, f_2 are continuous, their difference attains a minimum $\lambda_2 > 0$ on the compact set Ω. So, we can lower-bound (2.4) as $\Delta(k) \geq \lambda_2 \int_\Omega x^k dx \to +\infty$, as $k \to \infty$.
2. Otherwise, we look at the right end of the interval Ω, and define $a^* := \inf\{x \geq 1 : f_1(x) > f_2(x)\}$. Without loss of generality, we may assume $a^* < b^*$. To see this, note that if $f_1(b^*) \neq f_2(b^*)$, then the continuity of $f_1 - f_2$ implies $f_1(x) \neq f_2(x)$ within a range $(b^* - \varepsilon, b^*]$ for some $\varepsilon > 0$, and a^* is the supremum of all these ε. Otherwise, if $f_1(x) = f_2(x)$ on an entire interval $[b^* - \varepsilon, b^*]$ for some $\varepsilon > 0$, then $f_1 \not> f_2$ on Ω (the opposite of the previous case) implies the existence of some $\xi < b^*$ so that $f_1(x) < f_2(x)$, and a^* is the supremum of all these ξ (see Figure 2.2 for an illustration). In case that $\xi = 0$, we would have $f_1 \geq f_2$ on Ω, which is either trivial (as $\Delta(k) = 0$ for all k if $f_1 = f_2$) or otherwise covered by the previous case.
   In either situation, we can fix a compact interval $[a,b] \subset (a^*, b^*) \subset [1, b^*] = \Omega$ and two constants $\lambda_1, \lambda_2 > 0$ (which exist because f_1, f_2 are bounded as being continuous on the compact set Ω), so that the function
   $$\ell(k,x) := \begin{cases} -\lambda_1 x^k, & \text{if } 1 \leq x < a; \\ \lambda_2 x^k, & \text{if } a \leq x \leq b. \end{cases}$$
   lower-bounds the difference of densities in (2.4) (see Figure 2.2), and
   $$\begin{aligned} \Delta(k) = \int_1^{b^*} x^k (f_1 - f_2)(x)dx &\geq \int_1^b \ell(x,k)dx \\ &= -\lambda_1 \int_1^a x^k dx + \lambda_2 \int_a^b x^k dx \\ &= -\frac{a^{k+1}}{k+1}(\lambda_1 + \lambda_2) + \lambda_2 \frac{b^{k+1}}{k+1} \to +\infty, \end{aligned}$$

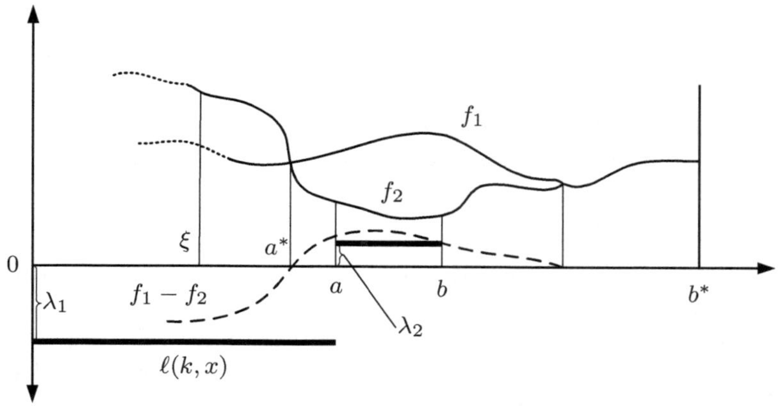

Fig. 2.2: Lower-bounding the difference of densities (the right most region where the densities are identical in this example is irrelevant; the growth of the moment sequence is determined by the difference of the density functions (dashed line), to which we only need a simple (crude) bound being positive only in a sub-region (thick step function ℓ) to verify the divergence of the moment difference for growing indices.

as $k \to \infty$ due to $a < b$ and because λ_1, λ_2 are constants that depend only on f_1, f_2.

In both cases, we conclude that, unless $f_1 = f_2$, $\Delta(k) > 0$ for sufficiently large $k \geq K$ where K is finite. This finishes the proof for continuous distributions.

If both, X and Y are discrete (in fact categorical) variables, then f_1, f_2 give probability masses located at $x = 1, 2, \ldots$. Inspecting the argument made above reveals that the growth of the moment sequence (difference) is determined by the (difference between) the masses put on the highest ranks (since the support is still finite). The argument thus remains the same, and so does the conclusion. □

Lemma 2.1 provides various key-insights, the first of which is useful in proving that the ordering of Definition 2.1 is independent of the model underlying ${}^*\mathbb{R}$:

Theorem 2.1. *Let F_1, F_2 be any two continuous or categorical distributions that are compactly supported on $[1,a] \subseteq \mathbb{R}$ with $a > 1$. Then, the ordering of the respective random variables $X \sim F_1, Y \sim F_2$, based on the canonic embedding $X \mapsto (E(X^k))_{k\in\mathbb{N}} \in {}^*\mathbb{R}$ and $\leq$-ordering in ${}^*\mathbb{R} = \mathbb{R}^\infty/\mathcal{U}$, is invariant of $\mathcal{U}$.*

Proof. Let F_1, F_2 be two probability distributions, and let $R_1 \sim F_1, R_2 \sim F_2$. Lemma 2.1 assures the existence of some $K \in \mathbb{N}$ so that, w.l.o.g, we may define the ordering $F_1 \preceq F_2$ iff $m_{R_1}(k) \leq m_{R_2}(k)$ whenever $k \geq K$. Let L be the set of indices where $m_{R_1}(k) \leq m_{R_2}(k)$, then complement set $\mathbb{N} \setminus L$ is finite (it has at most $K-1$ elements). Let $\mathcal{U}$ be an arbitrary ultrafilter. Since $\mathbb{N} \setminus L$ is finite, it cannot be contained in $\mathcal{U}$ as $\mathcal{U}$ is free. And since $\mathcal{U}$ is an ultrafilter, it must contain the complement a set, unless it contains the set itself. Hence, $L \in \mathcal{U}$, and the claim follows. □

Theorem 2.1 enables working without an ultrafilter, which would practically be unavailable anyway. Using the arguments in the proof of Lemma 2.1 in a different way reveals the physical interpretation of the $\preceq$ ordering:

Theorem 2.2. *Let X, Y have distributions F_1, F_2 with $supp(F_1), supp(F_2) \subseteq [1, \infty)$. Then, if $F_1 \preceq F_2$, then there exists some $x_0 \in supp(F_1) \cup supp(F_2)$ so that*

$$\textit{for all } x \geq x_0, \textit{ we have } \Pr(X_1 > x) \leq \Pr(X_2 > x). \tag{2.5}$$

Proof. Let f_1, f_2 be the density functions of F_1, F_2. Call $\Omega = \text{supp}(F_1) \cup \text{supp}(F_2) = [0, a]$ the common support of both densities, and take $\xi = \inf\{x \in \Omega : f_1(x) = f_2(x) = 0\}$. Suppose there were an $\varepsilon > 0$ so that $f_1 > f_2$ on every interval $[\xi - \delta, \xi]$ whenever $\delta < \varepsilon$, i.e., f_1 would be larger than f_2 until both densities vanish (notice that $f_1 = f_2 = 0$ on the right of ξ). Then the proof of Lemma 2.1 delivers the argument by which we would find a $K \in \mathbb{N}$ so that $\mathrm{E}X_1^k > \mathrm{E}X_2^k$ for every $k \geq K$, which would contradict $F_1 \preceq F_2$. Therefore, there must be a neighborhood $[\xi - \delta, \xi]$ on which $f_1(x) \leq f_2(x)$ for all $x \in [\xi - \delta, \xi]$. The claim follows immediately by setting $x_0 = \xi - \delta$, since taking $x \geq x_0$, we end up with $\int_x^\xi f_1(t)dt \leq \int_x^\xi f_2(t)dt$, and for $i = 1, 2$ we have $\int_x^\xi f_i(t)dt = \int_x^a f_i(t)dt = \Pr(X_i > x)$. □

Theorem 2.2 puts the stochastic order $\preceq$ into the landscape of other stochastic orders, as well as the context of security risk management. Notably, although we did not demand condition (2.5) not on the entire real line, but somewhat recover it asymptotically and coming from a completely different direction (based on an embedding into the space of hyperreals). This is consistent with a common view in risk management to consider potential large damages as more demanding than taking actions against low risks. Since security has mostly an implicit return on investment (in the sense of avoiding costs rather than producing revenue), one would certainly not invest much in guarding against risks that are already low anyway. Risk evaluation is the process of ranking risks in terms of their potential to cause damage. In the vocabulary of distributions, we would thus put more importance on actions against risks with high(er) likelihood for high(er) damages. A risk value computed using formula (2.3) clearly has such a positive correlation with impact (damage) and likelihood, which justifies it as a heuristic decision rule to rank risks. Theorem 2.2 points out $\preceq$ as a decision rule following the same rationale. Reversing the approach, one could think of using condition (2.5) as a starting point to define $\preceq$ as a straightforward generalization of the usual stochastic order (2.2). This method (though left unexplored here) may avoid any appeal to ${}^*\mathbb{R}$, but comes at the cost of having to manually prove all properties of orders that we are interested in (the totality of the order can, however, get lost on this route, as Remark 2.1 will explain later). In letting the ordering root in the order of hyperreal numbers (which, by Theorem 2.1 may be even called canonic), we get all properties of the ordering "for free," since $\preceq$ behaves on the hyperreals like $\leq$ behaves on $\mathbb{R}$. Most importantly, moment sequences present a common representation for both, continuous and discrete random variables. Assuming that there is a meaningful common support for both kinds of distributions (continuous and discrete), the $\preceq$-relation is well-defined even between distributions of different kinds.

2.2.2 Deciding the Stochastic Order

Inspecting the proof of Lemma 2.1 a last time under a different perspective equips us with a variety of simple criteria to decide $\preceq$ between categorical or between two continuous distributions (under some smoothness assumption on the densities).

In the following, let $X \sim F_1, Y \sim F_2$ be two distributions whose supports are $S_1 = [1,a]$ and $S_2 = [1,b]$. Clearly, if $a < b$ then $F_1 \preceq F_2$, since the growth of the moment sequence (difference) is determined by the lot of mass that F_2 puts on the region $[b-a,b]$. Note, however, that an analogous conclusion by an overlap on the left of the interval would be flawed: suppose the supports to be $[a,c]$ and $[b,c]$, where $a < b$. Either, $F_1 \preceq F_2$ and $F_2 \preceq F_1$ is possible under this setting, since what tips the scale is the lot of mass that a distribution puts on the right neighborhood of b, i.e., the interval $(b-\varepsilon, b]$ for $\varepsilon > 0$.

So, the interesting case to decide $\preceq$ occurs when the supports of F_1, F_2 both extend to the same limit $a > 1$ on $\mathbb{R}$. Let $X_1 \sim F_1, X_2 \sim F_2$ throughout the following discussion.

Categorical Distributions

Let the support of F_1, F_2 be an ordered set $\Omega = \{c_n > c_{n-1} > \ldots > c_2 > c_1\}$, i.e., the category of lowest rank (e.g., lowest damage) is c_1. Let $\mathbf{f}_1 = (p_{1,1}, \ldots, p_{1,n}), \mathbf{f}_2 = (p_{2,1}, \ldots, p_{2,n})$ be the respective probability mass functions. Typically, those could be empirical distributions (normalized histograms). The moment sequences are $\mathrm{E}_{F_i}(X_i^k) = \sum_{j=1}^{n} j^k \cdot p_{i,j}$. Suppose (w.l.o.g.) that $p_{1,n} < p_{2,n}$, then the growth of the sum $\mathrm{E}_{F_i}(X_i^k)$ in k is determined by the fastest growing exponential terms $n^k \cdot p_{1,n} < n^k \cdot p_{2,n}$ (note that n is constant here). Thus, $F_1 \preceq F_2$ in that case. Upon equality $p_{1,n} = p_{2,n}$, we can subtract the (now equal) terms $n^k \cdot p_{i,n}$ from both sums, leaving the second-largest probability masses $p_{i,n-1}$ to determine the order in the same way as before. In this special case, the $\preceq$-ordering is thus identical to the lexicographic ordering on the vectors of probability masses: $F_1 \preceq F_2 \iff (p_{1,n}, p_{1,n-1}, \ldots, p_{1,1}) <_{lex} (p_{2,n}, p_{2,n-1}, \ldots, p_{2,1})$

Continuous Distributions

We can invoke Lemma 2.1 to conclude $F_1 \preceq F_2$ if we find some x_0 for which $\Pr_{F_1}(X_1 > x_0) \leq \Pr_{F_2}(X_2 > x_0)$. Finding this value explicitly is quick if the supports overlap as discussed before (e.g., if F_1 lives on $[1,a]$ and F_2 lives on $[1,b]$ with $b > a$, then $x_0 = a$ is the sought threshold). If both random variables range up to the same point $a > 1$, then x_0 can be worked out by intersecting the survival functions of F_1, F_2, which is $S_i(t) := \Pr_{F_i}(X_i > t) = 1 - F_i(t)$ for $i = 1,2$. An admissible choice for x_0 is any value $\geq \inf\{t : S_1(x) < S_2(x) \text{ for all } x \geq t\}$. The shaded areas in Figure 2.3a mark the region $\geq x_0$, which determines the preference.

Since we do not actually need an optimal value for x_0, approximations thereof using numerical evaluations of the survival functions can be used to decide x_0. Alternatively, if the densities of F_1, F_2 are sufficiently smooth, the lexicographic order can be used again:

Lemma 2.2 ([14]). *Let $f, g \in C^\infty([1,a])$ for a real value $a > 1$ be probability density functions. If*

$$((-1)^k \cdot f^{(k)}(a))_{k\in\mathbb{N}} <_{lex} ((-1)^k \cdot g^{(k)}(a))_{k\in\mathbb{N}},$$

then $f \preceq g$.

Proof (from [15]). The argument will look for which distribution is below the other in a right neighborhood of a. To simplify matters, however, let us "mirror" the functions around the vertical line at $x = a$ and look for which of $f(x), g(x)$ grows faster when x becomes larger than a, using an induction argument on the derivative order k. Clearly, whichever function grows slower for $x \geq a$ in the mirrored view is the $\preceq$-preferable one. Furthermore, we may assume $a = 0$ without loss of generality (as this is only a shift along the horizontal line). For $k = 0$, we have $f(0) < g(0)$ clearly implying that $f \preceq g$, since the continuity implies that the relation holds in an entire neighborhood $[0, \varepsilon)$ for some $\varepsilon > 0$. Thus, the induction start is accomplished.

For the induction step, assume that $f^{(i)}(0) = g^{(i)}(0)$ for all $i < k$, $f^{(k)}(0) < g^{(k)}(0)$, and that there is some $\varepsilon > 0$ so that $f^{(k)}(x) < g^{(k)}(x)$ is satisfied for all $0 \leq x < \varepsilon$. Take any such x and observe that

$$\begin{aligned} 0 &> \int_0^x \left(f^{(k)}(t) - g^{(k)}(t)\right) dt = f^{(k-1)}(x) - f^{(k-1)}(0) - \left[g^{(k-1)}(x) - g^{(k-1)}(0)\right] \\ &= f^{(k-1)}(x) - g^{(k-1)}(x), \end{aligned}$$

since $f^{(k-1)}(0) = g^{(k-1)}(0)$ by the induction hypothesis. Thus, $f^{(k-1)}(x) < g^{(k-1)}(x)$, and we can repeat the argument until $k = 0$ to conclude that $f(x) < g(x)$ for all $x \in [0, \varepsilon)$.

For returning to the original problem, we must only revert our so-far mirrored view by considering $f(-x), g(-x)$ in the above argument. The derivatives accordingly change into $\frac{d^k}{dx^k} f(-x) = (-1)^k f^{(k)}(x)$. □

The smoothness assumption made in Lemma 2.2 is practically weaker than it appears at first glance. To see this, consider nonparametric models based on empirical data: Let $\hat{f}$ be a density function that is perhaps discontinuous. Let $K_h \in C^\infty(\mathbb{R})$ be the density function of a Gaussian distribution with zero mean and variance $h > 0$. We define the smoothed version $f := \hat{f} * K_h$, and observe that $f_h \in C^\infty$ for all $h > 0$ by the differentiation theorem of convolution. Furthermore, it is not difficult to verify that by letting $h \to 0$, f_h is L^1-convergent to f, and since the support of $\hat{f}$ is compact, the convergence is uniform.

The Gaussian density offers the particular appeal of letting us work out the derivatives of all orders even analytically, since (by the differentiation theorem), $f^{(k)} = \hat{f} * K_h^{(k)}$, and all we need is the k-th order derivative of the Gaussian density. This is directly given by [17]

$$f^{(k)}(x) = \frac{1}{h \cdot \sqrt{2\pi}} \frac{d^k}{dx^k} \exp\left(-\frac{1}{2h^2}(x_j - x)^2\right), \tag{2.6}$$

in which the k-th derivative of the exponential term can be expressed in closed form using Hermite polynomials. Those are defined recursively by $H_{k+1}(x) := 2xH_k(x) - 2H_{k-1}(x)$ upon $H_0(x) = 1$ and $H_1(x) = 2x$. The k-th derivative in (2.6) can then be computed from the relation

$$(-1)^k \exp\left(\frac{x^2}{2}\right) \frac{d^k}{dx^k} \exp\left(-\frac{x^2}{2}\right) = 2^{-\frac{k}{2}} H_k\left(\frac{x}{\sqrt{2}}\right). \tag{2.7}$$

Most useful is this representation if it is applied to nonparametric kernel density estimates, in which the density in question is a sum of Gaussian densities. The derivatives arising there can be computed via (2.7) (see [17] for details). The ordering should, however, be used with care when the support of the respective densities is a disconnected interval (i.e., has holes in it). In such cases, decisions may be implausible and games can be difficult to construct and solve. We will give examples of such unpleasant effects and develop respective remedies in Chapter 3.

A general issue with parametric losses is their representation of an arbitrary amount of information by a fixed number of parameters. This inevitably incurs a loss of information, and calls for partly sophisticated methods of parameter fitting or similar. On the contrary, nonparametric losses like kernel densities come with the appeal of preserving all information upon which they are constructed, as well as offering the flexibility of allowing for adjustments to model uncertainty in the expert's answers more explicitly. For example, if a set of risk estimates $r_1, r_2, \ldots, r_N$ from N experts is available, and ships with an additional information $\sigma_1, \sigma_2, \ldots, \sigma_N$ about the individual (subjective) certainty of each expert, then a kernel density can be constructed from Gaussian curves, each centered at r_i and with bandwidth σ_i. The resulting model is, strictly speaking, an opinion pool that preserves all information, incorporates all subjective uncertainty, and is perfectly useful with the $\preceq$ ordering and all criteria related to its decision.

2.2.2.1 Comparing Distributions to Numbers (Randomness vs. Determinism)

If some deterministic (fixed) value $a \in \mathbb{R}$ shall be compared to a random outcome $Y \sim F$, the ordering depends on the support $\Omega = [1, b]$ of Y (assuming that Y has a continuous and hence non-degenerate density over Ω). For the constant a, we can easily work out the moment sequence to be $(a^k)_{k\in\mathbb{N}}$. Comparing this to the respective moment sequence of Y then uses only the range of Y's support:

1. If $a < b$: since f is continuous, we can choose a value $0 < \varepsilon < (b-a)/3$ so that f is strictly positive on the interval $[b-\varepsilon, b-2\varepsilon]$. Then, the k-th moment of Y satisfies the lower bound

$$\int_1^b y^k f(y)dy \geq (\inf_{[b-2\varepsilon, b-\varepsilon]} f) \cdot \int_{b-2\varepsilon}^{b-\varepsilon} y^k dy = \frac{1}{k+1}\left[(b-\varepsilon)^{k+1} - (b-2\varepsilon)^{k+1}\right].$$

The bound is positive since f is positive everywhere on the chosen interval. The respective exponential function has a base larger than a, since $b - 2\varepsilon > a$, so $a \preceq Y$ since the moment sequences diverge accordingly.

2. If $a > b$, then Y can never take on values larger than b, which makes its moment sequence necessarily grow slower than that of (the constant) a. Formally, we can upper bound the moment sequence:

$$\int_1^b y^k f(y)dy \leq (\sup_{[1,b]} f) \cdot \int_1^b y^k dy = (\sup_{[1,b]} f) \frac{1}{k+1} b^{k+1}.$$

Since $a > b$, the function a^k grows faster than the upper bound, which gives the ordering $Y \preceq a$.

3. If $a = b$, then we must work out the moment sequence explicitly by virtue of the mean-value theorem: we find some $\xi \in [0, a]$ so that

$$\mathrm{E}Y^k = \int_0^a y^k f(y)dy = \xi^k \underbrace{\int_0^a f(y)dy}_{=1} = \xi^k \leq a^k$$

for all k. Hence, $Y \preceq a$ in that case. Intuitively, this can be explained by outcomes less than a being possible under the random variable Y, which is therefore preferable over the alternative where the maximal loss a is always occurring.

2.2.2.2 Distribution Mixes and Comparing Mixed Types

Distribution mixes of the form $F = \lambda_1 F_1 + \lambda_2 F_2 + \ldots + \lambda_n F_n$, where $\leq \lambda_i \leq 1$ for $i = 1, 2, \ldots, n$ and $\lambda_1 + \ldots + \lambda_n = 1$ require no particular treatment here by virtue of our embedding into the hyperreal space ${}^*\mathbb{R}$. Since $\preceq$ is nothing else than $\leq$ in ${}^*\mathbb{R}$, the properties of this ordering are the same as those of $\leq$ on $\mathbb{R}$ (by the transfer principle [20]), and hence carry over to $\preceq$. This demonstrates the benefit gained by the embedding into ${}^*\mathbb{R}$.

Comparing distributions of mixed type, i.e., comparing categorical to continuous distribution is technically possible, since both are representable by the same objects. The physical meaning of such a comparison, however, is in many cases doubtful, since categorical distributions mostly refer to ranks, while continuous distributions can represent much different quantities. A meaningful comparison appears thus only possible if reals shall be compared to integers; however, given absolute continuity w.r.t. the Lebesgue measure, the masses that the two distributions assign to integers may be too different to lead to any meaningful comparison.

2.2.3 *Distributions with Infinite Support*

For risk management, we require distribution models for extreme events, which calls for heavy, long or fat tails. Common choices are the Gumbel-, Weibull-, or Frechet-distribution, or also the $(a,b,0)$-class of distributions (see the example below). The method of comparing distributions by moments comes with the appeal of being applicable even in cases where there is no analytical expression for the distribution itself (such as happens for stable distributions).

If we compare a distribution with compact support to one with infinite support (such as extreme value distributions or ones with long or fat tails), then the compactly supported distribution is always preferred, by the same argument as used above (and in the proof of the invariance of $\preceq$ w.r.t. the ultrafilter used to construct ${}^*\mathbb{R}$; see [14, Lemma 2.4]).

In some cases, it can be sufficient to look at the tail masses directly, without having to compute any moments explicitly.

Example 2.3. A popular family of loss distributions in risk management is the $(a,b,0)$-class of distributions [10, 9], having their probability masses defined recursively by $\Pr(X=k)=\Pr(X=k-1)\cdot(a+\frac{b}{k})$ for $k=1,2,3,\ldots$. This class includes (only) the Poisson, binomial, and negative binomial distributions. A comparison under $\preceq$ is here particularly easy, depending on the values of a and b. Let $F_1\sim(a_1,b_1,0)$ and $F_2\sim(a_2,b_2,0)$ be two distributions. If $b_1<b_2$ and $a_1=a_2$, then $\Pr(X=k|a,b_1)<\Pr(X=k|a,b_2)$, and $F_1\preceq F_2$. If $a_1<a_2$ and $b_1=b_2$, then $F_1\preceq F_2$ also. More involved conditions allowing for both parameters to be different are not difficult to work out.

The unfortunate occasions are those where:

- both distributions have infinite support, and
- neither Lemma 2.3 nor any direct criterion (like (2.9)) applies, and
- an approximation or truncation (see Section 2.2.3) cannot be done (for whatever reason).

Then we have to work out the moment sequences explicitly. This situation is indeed problematic, as without assuming bounded supports, we can guarantee neither existence nor divergence of the two moment sequences.

Appropriate examples illustrating this problem can easily be constructed by defining distributions with alternating moments from the representation by the Taylor-series expansion of the characteristic function (see [14] for an example). Mixes of such distributions (discussed previously) can perhaps replace an otherwise unhandy model (up to any desired precision; see [19]). Further and important examples relate to catastrophic events, and the class of distributions with heavy, fat, or long tails. Some of these do not even have moment-generating functions (though characteristic functions can be used as well), but all of them have their individual importance and applications in risk management.

Indeed, a compact support is not a necessary circumstance for all moments to exist, as the Gaussian distribution shows. This distribution is characterized entirely by its first two moments, and thus can easily be compared in terms of the $\preceq$-relation.

A compact support is, however, a sufficient condition for the moment sequence to exist, and any distribution with infinite support can be approximated by a truncated distribution. The following arguments are from [14]: Given a random variable X with distribution function F, then the *truncated distribution*

$$\hat{F}(x) = \Pr(X \leq x | a \leq X \leq b).$$

is the distribution of X conditional on X falling into a finite range $[a,b]$. Likewise, the truncated density is scaled by $f(x)/(F(b)-F(a))$ whenever $x \in [a,b]$ and zero elsewhere.

It is easy to see that upon choosing the interval $[a,b]$ large enough, we can approximate every distribution up to a maximal error $\varepsilon > 0$ that we can choose in advance (the size of the interval will of course depend on ε). However, if two distributions F_1, F_2 with infinite supports shall be compared, then both must be truncated on a common interval $[a,b]$, for otherwise, it is straightforward to truncate the distributions differently so that both $\hat{F}_1 \preceq \hat{F}_2$ and $\hat{F}_1 \succeq \hat{F}_2$ are possible. See [14] for a more detailed explanation.

Since the comparison obviously still depends on the chosen interval used for the truncation, we could let ε go to zero and look at a sequence of truncations to approximate the original distribution in the limit. If the sequence of (common) truncations ultimately runs into a fixed $\preceq$-relation, we can define the original distributions to satisfy the same limiting $\preceq$-relationship. This motivates the following definition:

Definition 2.2 (Extended Strict Preference $\prec$). Let F_1, F_2 be distribution functions of nonnegative random variables that have infinite support and continuous density functions f_1, f_2. We *strictly prefer* F_1 *over* F_2, denoted as $F_1 \prec F_2$, if for every sequence $a_n \to \infty$ there is an index N so that the truncations (approximations) $\hat{F}_{i,n}$ on the common interval $[1, a_n]$ for $i = 1,2$ satisfy $\hat{F}_{1,n} \prec \hat{F}_{2,n}$ whenever $n \geq N$.

The $\succ$-relation is defined alike, i.e., the ultimate preference of F_2 over F_1 on any sequence of approximations.

Definition 2.2 is intuitively motivated but not necessarily handy in practice. We can, however, avoid the labor of working out the truncation sequence, as the following Lemma shows:

Lemma 2.3. *Let F_1, F_2 be two distributions supported on $[1,\infty)$ with continuous densities f_1, f_2. Let $(a_n)_{n\in\mathbb{N}}$ be an arbitrary sequence with $a_n \to \infty$ as $n \to \infty$, and let $\hat{f}_{i,n}$ for $i = 1,2$ be the truncated distribution f_i supported on $[1, a_n]$.*

If there is a constant $c < 1$ and a value $x_0 \in \mathbb{R}$ such that $f_1(x) < c \cdot f_2(x)$ for all $x \geq x_0$, then there is a number N such that all approximations $\hat{f}_{1,n}, \hat{f}_{2,n}$ satisfy $\hat{f}_{1,n} \prec \hat{f}_{2,n}$ whenever $n \geq N$.

Proof (adapted from [14]). Let $i \in \{1,2\}$. The truncated distribution density that approximates f_i is $f_i(x)/(F_i(a_n) - F_i(1))$, where $[1, a_n]$ is the common support of

n-th approximation to f_1, f_2. By construction, $a_{n,i} \to \infty$ as $n \to \infty$, and therefore $F_i(a_n) - F_i(1) \to 1$ for $i = 1, 2$. Consequently,

$$Q_n = \frac{F_1(a_n) - F_1(1)}{F_2(a_n) - F_2(1)} \to 1, \quad \text{as } n \to \infty,$$

and there is an index N such that $Q_n > c$ for all $n \geq N$. In turn,

$$f_2(x) \cdot Q_n > f_2(x) \cdot c > f_1(x),$$

and by rearranging terms,

$$\frac{f_1(x)}{F_1(a_n) - F_1(1)} < \frac{f_2(x)}{F_2(a_n) - F_2(1)}, \tag{2.8}$$

for all $x \geq x_0$ and all $n \geq N$. The last inequality (2.8) lets us compare the two approximations easily by the same arguments as have been used in the proof of Lemma 2.1, and the claim follows. □

By virtue of Lemma 2.3, we can decide the strict preference relation to distributions by checking the hypothesis of the lemma but need not work out any truncations.

Remark 2.1. The assumption of bounded supports is crucial for the totality of the resulting order. Even the extension defined above admits pairs of distributions that are not $\prec$- nor $\preceq$-related in either way. A simple counterexample are the densities $f(x) \propto e^{-x}(1 + \sin(x))$ and $g(x) \propto e^{-x}(1 + \cos(x))$. Those alternatingly exceed one another, and it is not difficult to construct a sequence $a_n \to \infty$ for which the truncated densities f_n, g_n satisfy $f_n \prec g_n$ for even n, but $f_n \succ g_n$ for odd n.

The same example also shows that Theorem 2.2 cannot be used as a starting point for a definition to drop the boundedness condition or the assumption that $X \geq 1$ for all random variables of interest. The same sequence a_n as above also provides an unbounded infinitude of x_0 values for which condition (2.5) would fail.

An example showing a case when Lemma 2.3 is inapplicable is the following:

Example 2.4 (from [14]). Take the "Poisson-like" distributions with parameter $\lambda > 0$,

$$f_1(k) \propto \begin{cases} \frac{\lambda^{k/2}}{(k/2)!} e^{-\lambda}, & \text{when } k \text{ is even;} \\ 0, & \text{otherwise.} \end{cases}, \quad f_2(k) \propto \begin{cases} 0, & \text{when } k \text{ is even;} \\ \frac{\lambda^{(k-1)/2}}{((k-1)/2)!} e^{-\lambda}, & \text{otherwise} \end{cases}$$

Obviously, no constant $c < 1$ can ever make $f_1 < c \cdot f_2$ and that all moments exist. However, neither distribution is preferable over the other, since finite truncations to $[1, a_n]$ based on the sequence $a_n := n$ will yield alternatingly preferable results.

A stronger condition that implies the hypothesis of Lemma 2.3 is the following [14]:

$$\lim_{x\to\infty} \frac{f_1(x)}{f_2(x)} = 0. \tag{2.9}$$

To see this, note that if the condition of Lemma 2.3 were violated, then there is an infinite sequence $(x_n)_{n\in\mathbb{N}}$ for which $f_1(x_n) \geq c \cdot f_2(x_n)$ for all $c < 1$. In that case, there is a subsequence $(x_{n_k})_{k\in\mathbb{N}}$ for which $\lim_{k\to\infty} f_1(x_{n_k})/f_2(x_{n_k}) \geq c$. Letting $c \to 1$, we can construct a further subsequence of $(x_{n_k})_{k\in\mathbb{N}}$ to exhibit that $\limsup_{n\to\infty}(f_1(x_n)/f_2(x_n)) = 1$, thus contradicting condition (2.9). Indeed, (2.9) puts $\preceq$ into the proximity of the likelihood ratio order [22] in the sense that this condition implies both, a likelihood ratio and $\prec$-ordering. Note that, however, a likelihood ratio order does not necessarily imply a $\prec$-order, since the former only demands $f(t)/g(t)$ to be increasing, but not a $<$-relation among the densities.

Though the above criteria only relate to strict preferences, there is no conceptual difficulty in defining the $\preceq$-order and equivalence as done in Definition 2.2:

Definition 2.3. Let F_1, F_2 be two distributions supported on the entire nonnegative real half-line $\mathbb{R}^+$ with continuous densities f_1, f_2. Let $(a_n)_{n\in\mathbb{N}}$ be a diverging sequence towards ∞, and let $\hat{F}_{i,n}$ for $i = 1,2$ denote the density F_i truncated to have support $[1, a_n]$. We define $F_1 \preceq F_2$ if and only if for every sequence $(a_n)_{n\in\mathbb{N}}$ there is some index N so that $\hat{F}_{1,n} \preceq \hat{F}_{2,n}$ for every $n \geq N$.

Like before, this definition simply asks for the $\preceq$-relation to hold ultimately on every sequence of common truncations. Given distributions with finite support, a truncation will not do any change once the interval fully overlaps the support. Thus, the sequence of truncations will converge to the original distributions within a finite number of steps, so that the extended $\prec$ and $\preceq$-relations for infinite supports include the same relations for finitely supported distributions as a special case.

2.2.4 Implausible Comparisons

Not all distributions compare equally plausible, and Theorem 2.2 can be even consistent with implausible and unexpected $\preceq$-orders. According to the theory, we would in any case prefer the distribution with smaller support or lighter tails, but this is not necessarily also consistent with our intuition.

Figure 2.3a illustrates the issue using a χ^2-distribution with 3 degrees of freedom for F_1 and a Gaussian distribution with mean 8 and standard deviation 0.4 for F_2. The relation $F_2 \preceq F_1$ is obvious since F_1 assigns more mass to events with larger losses. Precisely, it is the point x_0 where the residual mass left under F_1 equals the residual mass under F_2 (i.e., the survival functions intersect) is the threshold x_0 that Theorem 2.2 speaks about. This limit is $x_0 \approx 8.74$. These apparently rare cases, however, extend beyond the support of distribution F_1, which based on the characterization by a sequence of would clearly let us prefer F_2 over F_1 (Figure 2.3a). Indeed, it is easy to see that such a result is not what we would expect or want in practice.

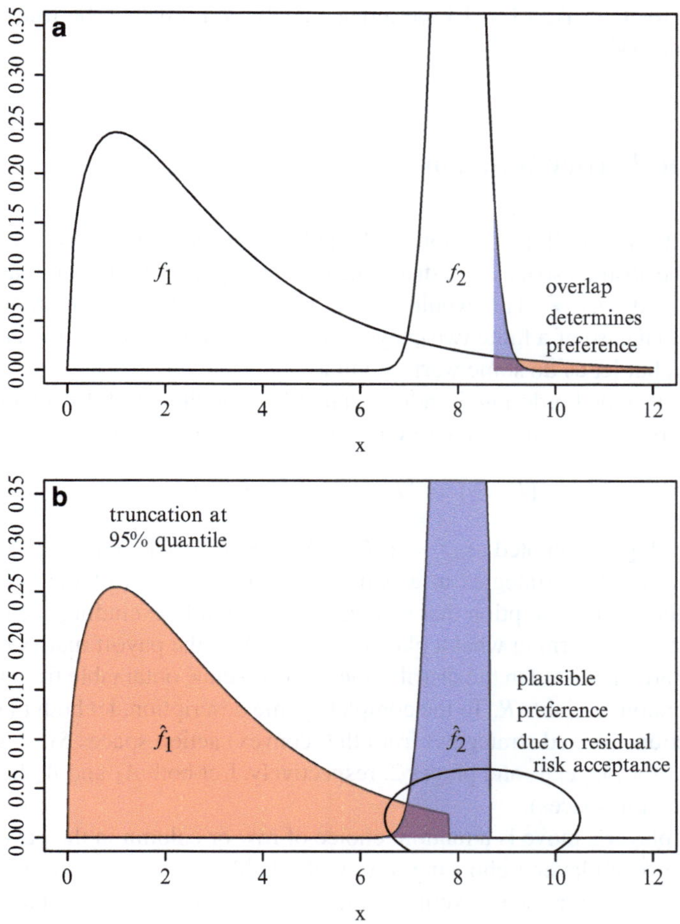

Fig. 2.3: Correcting Implausible Comparisons by Quantile-Based Approximations [15]

Now, we can fix the issue by truncating the loss distributions upon saying that we simply accept events that occur in less than 5% of the cases. This amounts to truncating both distributions at their individual 95% quantiles, giving the picture in Figure 2.3b. Obviously, the implausible preference is hereby corrected, since $\hat{F}_1$ gives less damage in all cases than $\hat{F}_2$.

Note that this *individual and distinct* risk acceptance for F_1, F_2 is seemingly inconsistent with our previous remark that distributions with infinite supports should be compared based on a common truncation. However, we stress that we are, at this point, still involved in the model building. This is where a risk acceptance threshold can be used to truncate the distribution accordingly and to get a model with bounded support. On the contrary, Section 2.2.3 speaks about situations where the appropriate models are (for whatever reason) necessarily ones with infinite support; and those, for the sake of a risk preference analysis, must then be truncated on a common interval.

2.3 Game Theory Based on $\preceq$

Let $\mathscr{F}$ be the set of all distributions of bounded support within $[1,\infty)$. To lift game theory to the abstract space $\mathscr{F}$ of distributions, we require continuity of payoff functionals defined over $\mathscr{F}$. This would be the "expected" payoff distribution, which – for the special case of a finite two-player game, is actually given by the law of total probability, but let us do some warm-up first.

Since $\preceq$ is a total ordering, it induces a topology on the set of distributions given by the family $\mathscr{T}$ of open sets enclosed "between" two distributions F_1, F_2 as

$$(F_1,F_2) := \{F \in \mathscr{F} : F_1 \prec F \prec F_2\},$$

and the topology is denoted as $\mathscr{T} = \{(F_1,F_2) | F_1,F_2 \in \mathscr{F} \text{ where } F_1 \prec F_2\}$.

Suppose that the strategies in a game to form a partition of the action space (which is the mild assumption that strategies are mutually excluding each other but are exhaustive concerning what a player can do). With the payoff matrix being $\mathbf{A} \in \mathscr{F}^{n\times m}$, we are interested in the distribution of the revenue obtainable from the game. Call this random variable R. To the complete game description, let both players take random actions (mixed strategies) from the (convex) action spaces $S_1 = \Delta(A_1)$ and $S_2 = \Delta(A_2)$ for player 1, and player 2, respectively. Let both A_1 and A_2 be compact sets (pure action spaces).

Since a player's move is a random choice of row or column in the payoff matrix $\mathbf{A}$, let us think of player 1 choosing rows with likelihoods as specified by $\mathbf{p}$, and let player 2 draw columns with likelihoods $\mathbf{q}$. Then, the law of total probability gives us the outcome R to satisfy,

$$\Pr(R \le r) = \sum_{i,j} \Pr(R_{ij} \le r | i,j) \Pr(i,j), \tag{2.10}$$

where $\Pr(R_{ij} \le r|i,j)$ is the conditional probability of R_{ij} given a particular choice (i,j), and $\Pr(i,j)$ is the (unconditional) probability for this choice to occur. Assuming independence, we have $\Pr(i,j) = p_i \cdot q_j$ when the mixed strategies of both players over finite action spaces A_1,A_2 are the vectors $\mathbf{p},\mathbf{q}$. Stochastic independence of these choices is a common assumption, but letting them be dependent (and connected by a continuous copula) does not invalidate the upcoming results (we nonetheless leave this direction unexplored here for simplicity).

Denote by $F(\mathbf{p},\mathbf{q})$ the distribution of the game's outcome under strategies $(\mathbf{p},\mathbf{q}) \in S_1 \times S_2$, then $\Pr(R \le r) = F(r)$ depends on $(\mathbf{p},\mathbf{q})$, and (2.10) can be rewritten as

$$\Pr(R \le r) = (F(\mathbf{p},\mathbf{q}))(r) = \sum_{i,j} F_{ij}(r) \Pr_{\mathbf{p},\mathbf{q}}(i,j), \tag{2.11}$$

Optimizing the payoff distribution w.r.t. $\preceq$ means shaping the distribution F of the random damage R via proper playing. According to Theorem 2.2, this means for player one to "push" the mass of R towards lower losses, while player two has the opposite incentive. The game is in that sense "constant sum," though thinking about it as "zero sum" requires care: Obviously, player two is here not rewarded with $-F(\mathbf{p},\mathbf{q})$, since this would not necessarily correspond to a meaningful distribution any more.

An equilibrium can be defined in the natural way, and so can its existence be proven, based on the prior verification that a distribution-valued utility function $F(\mathbf{p},\mathbf{q})$ is continuous in $\mathbf{p}$ and $\mathbf{q}$. To this end, we show that any set in the topology $\mathscr{T}$, i.e., any open set in $\mathscr{F}$, has a preimage under F that is open w.r.t. the product topology. The following lemma establishes the important steps towards this conclusion by exploiting the ordering and arithmetic within $^*\mathbb{R}$.

Lemma 2.4. *Let $r_1,\dots,r_k \in \mathscr{F}$ for $k \geq 1$ be a set of fixed elements, and take $\alpha = (\alpha_1,\dots,\alpha_k) \in \mathbb{R}^k$. If two elements $\ell, u \in \mathscr{F}$ bound the weighted sum $\ell \prec \sum_{i=1}^k \alpha_i r_i = \alpha^T\mathbf{r} \prec u$, then there is some strictly positive $\delta \in \mathbb{R}$ so that $\ell \prec \widetilde{\alpha}^T\mathbf{r} \prec u$ for every $\widetilde{\alpha}$ within a δ-neighborhood of α in $\mathbb{R}^k$.*

Proof (from [14]). Define $\Delta := \min\{\alpha^T\mathbf{r} - \ell, u - \alpha^T\mathbf{r}\} \succ 0$ and $r := \max\{r_1,\dots,r_k\}$. Suppose that we would modify all weights α_i to $\alpha_i + \delta_i = \widetilde{\alpha}_i$. If so, then the so-modified sum differs from the given one by $\left|\widetilde{\alpha}^T\mathbf{r} - \alpha^T\mathbf{r}\right| \leq \sum_{i=1}^k |\delta_i|\, r_i \leq r\cdot\sum_{i=1}^k |\delta_i|$. Now, suppose that all $|\delta_i| \leq \delta$, then the change alters $\alpha^T\mathbf{r}$ by a magnitude of no more than $r\cdot\sum_{i=1}^k \delta_i \leq r\cdot k\cdot\delta$. As k and r are fixed, we can choose δ sufficiently small to satisfy $r\cdot k\cdot\delta \prec \Delta$, in which case we must have $\left|\widetilde{\alpha}^T\mathbf{r} - \alpha^T\mathbf{r}\right| < \Delta$, and therefore $\ell \prec \widetilde{\alpha}^T\mathbf{r} \prec u$ for any choice of $\widetilde{\alpha}$ within an δ-neighborhood of α in the maximum-norm on $\mathbb{R}^k$. □

With Lemma 2.4, the continuity of $F(\mathbf{p},\mathbf{q})$ in $(\mathbf{p},\mathbf{q})$ can be established:

Proposition 2.2. *Let i,j be integers and define the function $D_{ij} : S_1 \times S_2 \to \mathbb{R}$ as $D_{ij}(\mathbf{p},\mathbf{q}) = \Pr_{\mathbf{p},\mathbf{q}}(i,j)$. If D_{ij} is continuous and all F_{ij} are compactly supported in $[1,\infty)$ and have a (continuous or categorical) density function, then the mapping $F : S_1 \times S_2 \to \mathscr{F}$; $(\mathbf{p},\mathbf{q}) \mapsto \sum_{i,j} D_{ij}(\mathbf{p},\mathbf{q})F_{ij}$ is continuous w.r.t. the product topology on $S_1 \times S_2$ and the order topology on $\mathscr{F}$.*

Proof (adapted from [14]). Without a metric on $\mathscr{F}$, we need to show that the preimage of every open set in $\mathscr{F}$ under F is open to prove that F is continuous in $(\mathbf{p},\mathbf{q})$. For that sake, let the open set $(\ell,u) \in \mathscr{T}$ be arbitrary and contain some point $F(\mathbf{p},\mathbf{q})$ (which must exist, for otherwise, the set of preimages would be empty). To ease notation, let us flatten the double-sum $\sum_{i,j}$ into an ordinary sum (say, by introducing a multiindex v) over $k = n\cdot m$ elements, where n,m are the limits in the original expression. Then, the mapping takes the form $F(\mathbf{p},\mathbf{q}) = \sum_{v=1}^k D_v(\mathbf{p},\mathbf{q})F_v$. With the weights α being defined by the individual values of $D_v(\mathbf{p},\mathbf{q})$, we can apply Lemma 2.4 to establish a bound $\delta > 0$ within which we can arbitrarily alter the weights towards $\widetilde{\alpha}$ without leaving the open set (ℓ,u). Since D is continuous on the compact set $S_1 \times S_2$ it is also uniformly continuous, and we can fix a $\delta' > 0$ so that $\|D_V(\mathbf{p}',\mathbf{q}') - D_V(\mathbf{p},\mathbf{q})\| < \delta$ whenever $\|(\mathbf{p},\mathbf{q}) - (\mathbf{p}',\mathbf{q}')\| < \delta'$, independently of the particular point $(\mathbf{p},\mathbf{q})$. The sought pre-image of the open set (ℓ,u) is thus the (infinite) union of open neighborhoods constructed in the way described, and thus itself open. □

Equipped with continuity of the payoff functions, at least for matrix games, all the known results on existence of Nash equilibria, such as the following, apply:

Theorem 2.3 (Glicksberg [6, 7]). *If for a game in normal form, the strategy spaces are nonempty compact subsets of a metric space, and the utility-functions are continuous w.r.t the metric, then at least one Nash-equilibrium in mixed strategies exists.*

Defining games is thus a canonic matter, though this generalized class of games has some unusual properties. We will revisit the matter in more detail in Chapter 3. Note that up to this point, we have only used the compatibility of the ordering with arithmetic (freely shipping with $^*\mathbb{R}$), and different stochastic orders may not come equally handy here.

2.4 Application of $\preceq$ in Risk Management

Risk management in general can be well supported by statistics, but security risk management is different. Unlike in many other branches, there is typically no underlying physical process determining the risk, and much of the assessment is subjective. Predictions by experts usually rely on experience, expertise, and are also influenced by recent information, media, and many other factors. Collecting such information is a challenge on its own that in a traditional approach would be followed by *opinion pooling* [4, 3]. The latter is required for a meaningful comparison of the consensus distilled from the pooling. The use of a stochastic order, however, elegantly avoids this need at all, and thus spares the need to find a consensus. That also allows for expert polls to happen in an essentially different form as usual, since a discussion is not required until a decision recommendation (based on $\preceq$) is available.

The advantages induced by the use of a stochastic order (over a plain numeric order following an opinion pooling) are summarized by the questioning being doable individually, separately, asynchronously, and anonymously. That entails the following features, among an improvement of the data quality as such [12]:

1. Expert interviews can be conducted "offline," i.e., without personal meetings or open discussion. While this saves time for the participant, it also reduces biases and influences occurring in the presence of others (say, induced by cultural, behaviorial habits, superiors/subordinate relations, or similar). Uninformed guesses triggered in meeting situations "just to say something" are thus less likely. Whenever personal meetings are intended, methods of qualitative data collections like the Delphi method remain perfectly applicable.

2. Exploitation of skill diversity: statistics knows a lot of methods to compile models in the presence of missing data. In questioning experts on different aspects, and allowing them to leave out any part of the survey that they have no reliable clue about (also supported since the opinion needs not be uttered against an audience of others), the data collected from multiple people may be individually incomplete but in total draw a still complete picture.
3. Enlargement of the expert pool: since there is no need to involve people in internal (and hence mostly confidential) meetings, it is possible to involve external stakeholders and experts in the risk data collection (say, to measure reputation by collecting customer feedback).
4. In using distributions as the object to manage, we avoid loss of information by compiling a whole data set into a single representative figure. Thus, there is no need for any consensus finding, and all available opinions and data count go into the final decision with the same importance.

An independent advantage of game theory itself lies in the models being independent of any particular details to the strategies. That is, the assessment of risks would normally require a disclosure of the details about threats and countermeasures. Such information is, however, highly confident for good reasons. The models themselves, as well as their analysis, can work in abstract terms so that the labor and information related to risks and countermeasures remains under confidentiality with the customer, while the expertise on the algorithmic details and matters of decision support can rest with the service provider.

Regarding the practical use of the theory laid out here, observe that some of the (abstract) parameters indeed have natural instances in risk management. For example, the *cutoff point a* where distributions are truncated (see Sections 2.2.2 and 2.2.4) corresponds to a *risk acceptance threshold*, i.e., any damage beyond a is considered as so unlikely that the risk is simply taken. We will revisit the issue in part two of this book, when we discuss how to properly parameterize the games for risk management.

Stochastic orders have various applications in risk management, such as for optimized surveillance accounting for uncertainty [16], defense against advanced persistent threats [17] in water and electricity supply networks (Chapters 13 and 14 in this book), protection against malware infections in networks (Chapter 8 in this book), or optimized surveillance (Chapter 15).

2.5 Extensions and Outlook

It is a well-recognized issue of traditional game theory that the perfect rationality induced by a utility maximization assumption is not necessarily a good prediction of human behavior. Experimental verifications of such deviations have motivated the entire field of behavioral game theory [2], which studies behavior induced by other means than utility maximization. The replacement of numeric by stochastic orders may, though this is not verified yet, bring closer together traditional and behavioral game theory by changing the understanding of what utility maximization, or equivalently, loss minimization actually means. Our stochastic order was here developed from a purely technical idea, but naturally carries a meaning that is quite intuitive in risk management, since it pessimistically focuses on extreme outcomes in a strong sense. Consequently, equilibria under the stochastic order on the full distribution objects may be quite different from equilibria computed in numeric orders of averages or other statistics. Studying the extent to which these are accurate in explaining human behavior is a matter of future research and outside the scope of this chapter and book.

In general, the use of stochastic orders can simplify and complicate matters (yet hardly at the same time), and the application or extension of $\preceq$ to other types of games, like in extensive form or continuous games, is theoretically possible (indeed, Glicksberg's extension to Nash's theorem that given above as Theorem 2.3 applies to continuous games). The practical challenges along such aisles are twofold (yet can be overcome): (i) by working in the hyperreal field, we are equipped with well-defined yet partly undoable arithmetic (additions and scalar multiplications are easy to do, yet divisions require an explicit ultrafilter, which we do not have), and (ii) establishing the counterparts to results known from traditional game theory requires care and potentially even new proofs. The transfer principle [20] (or more generally, Łoś theorem; see [1]) directly lifts every first-order logic expression valid in $\mathbb{R}$ to a related claim in $^*\mathbb{R}$. We could hope for this to imply all results on traditional games to hold analogously for distribution-valued games based on the $\preceq$-ordering introduced here. Alas, such hope is wrong due to counterexamples: for instance, iterative algorithms for equilibrium computation, such as fictitious play, are known to converge for games over $(\mathbb{R}, \leq)$ [21]. By the transfer principle, the same convergence claim holds in $^*\mathbb{R}$, but with the subtle yet striking difference: the convergence in $\mathbb{R}$ is along a sequence of integers, whereas convergence in $^*\mathbb{R}$ is, by the transfer principle, along a sequence of *hyper*-integers. The latter extend beyond integer infinity, which means that a practical implementation of fictitious play would be required to count beyond $\mathbb{N}$, which is clearly not achievable. The issue can, however, be circumvented with the help of Lemma 2.2, and Chapter 3 gives the full details on this. In fact, it is possible to convert a distribution-valued game into a traditional game, such that the equilibria in the classical game is an approximation to the $\preceq$-equilibrium, up to arbitrary precision (Theorem 3 in Chapter 3). In using such a conversion, we can also bypass the aforementioned issues of practical arithmetic in $^*\mathbb{R}$, since in converting a $\preceq$-minimizing game into one with payoffs from $\mathbb{R}$, the full armory and results for traditional games are regained as well.

An independent aspect of interest may concern the modeling of uncertainty by information sets. Indeed, if a move in an extensive form game carries uncertainty since a player does not know the exact state of another player, then the outcome may be random (due to this missing information). So, notions like subgame perfectness of equilibria may be reconsidered in stochastic orders, which is a new way of dealing with information sets in extensive form games.

Applications of stochastic orders in game theory are recent and, today, mostly relate to risk management (where the idea originally emerged from). On a purely theoretical level, the framework was recently applied to study the cost of moves when playing a mixed equilibrium. That is, if a player's strategy entails changing a state against some (natural) inertia, then a player may prefer to deviate from a mixed equilibrium, simply because it is perceived as "too expensive" to change the strategies as frequent as the equilibrium would prescribe (thus offering another connection to behaviorial game theory). In a way, this view extends games in a way similar to how algebraic equations are generalized into differential equations by introducing the derivative in the equation. In [18], the same idea of a cost related to changes in a player's behavior (similar to a first-order derivative of a function) is easily modeled and studied within the framework of distribution-valued games.

Stochastic elements are ubiquitous in real life game-theoretic models, and much effort is typically spent on capturing randomness in expressive statistics that game theory can subsequently optimize. In changing the approach into using the randomness "as is," the very same applications could be analyzed within the framework of stochastically ordered payoffs. Whether or not this method would contribute to closing the gap between the predictions of traditional and behaviorial game theory is an unsolved question so far, but surely one with an interesting answer.

References

1. Bell, J.L., Slomson, A.B.: Models and ultraproducts: An introduction, 1. publ., unabr. republ. of 1974 edn. Dover, Mineola NY (2006)
2. Camerer, C.F.: Behavioral game theory: Experiments in strategic interaction. The Roundtable Series in Behavioral Economics. Princeton University Press, s.l. (2011). URL http://gbv.eblib.com/patron/FullRecord.aspx?p=765287
3. Carvalho, A., Larson, K.: A Consensual Linear Opinion Pool. In: Proceedings of the Twenty-Third International Joint Conference on Artificial Intelligence, IJCAI '13, pp. 2518–2524. AAAI Press (2013). URL http://dl.acm.org/citation.cfm?id=2540128.2540491
4. Dietrich, F., List, C.: Probabilistic opinion pooling generalized. Part one: General agendas. Social Choice and Welfare **48**(4), 747–786 (2017). https://doi.org/10.1007/s00355-017-1034-z
5. Eichner, T., Wagener, A.: Increases in skewness and three-moment preferences. Mathematical Social Sciences **61**(2), 109–113 (2011). https://doi.org/10.1016/j.mathsocsci.2010.11.004
6. Fudenberg, D., Tirole, J.: Game Theory. MIT Press, London (1991)
7. Glicksberg, I.L.: A Further Generalization of the Kakutani Fixed Point Theorem, with Application to Nash Equilibrium Points. In: Proceedings of the American Mathematical Society, vol. 3, pp. 170–174 (1952)
8. Goldreich, O.: Foundations of cryptography 1: Basic Tools. Cambridge University Press (2003)
9. Hogg, R.V., Klugman, S.A.: Loss distributions. Wiley series in probability and mathematical statistics Applied probability and statistics. Wiley, New York, NY (1984). https://doi.org/10.1002/9780470316634. URL http://site.ebrary.

com/lib/alltitles/docDetail.action?docID=10344300. Klugman, Stuart A. (VerfasserIn)
10. Klugman, S.A., Panjer, H.H., Willmot, G.E.: Loss models: From data to decisions. A Wiley-Interscience publication. Wiley, New York, NY (1998). URL http://www.loc.gov/catdir/description/wiley031/97028718.html. Panjer, Harry H. (VerfasserIn) Willmot, Gordon E. (VerfasserIn)
11. Münch, I.: Wege zur Risikobewertung. In: P. Schartner, J. Taeger (eds.) DACH Security 2012, pp. 326–337. syssec (2012)
12. Perreault, W.D., Leigh, L.E.: Reliability of Nominal Data Based on Qualitative Judgments. Journal of Marketing Research **26**(2), 135 (1989). https://doi.org/10.2307/3172601
13. Rass, S.: Game-Theoretic Risk Management – Part One: Security Strategies in Non-Deterministic Games (2015). Technical Report internal report of the HyRiM Project
14. Rass, S.: On Game-Theoretic Risk Management (Part One) – Towards a Theory of Games with Payoffs that are Probability-Distributions. ArXiv e-prints (2015). Http://arxiv.org/abs/1506.07368
15. Rass, S.: On Game-Theoretic Risk Management (Part Two) – Algorithms to Compute Nash-Equilibria in Games with Distributions as Payoffs (2015). ArXiv:1511.08591
16. Rass, S., Alshawish, A., Abid, M.A., Schauer, S., Zhu, Q., de Meer, H.: Physical Intrusion Games - Optimizing Surveillance by Simulation and Game Theory. IEEE Access p. 1 (2017). https://doi.org/10.1109/ACCESS.2017.2693425
17. Rass, S., Konig, S., Schauer, S.: Defending Against Advanced Persistent Threats Using Game-Theory. PLoS ONE **12**(1), e0168,675 (2017). https://doi.org/10.1371/journal.pone.0168675. Journal Article
18. Rass, S., König, S., Schauer, S.: On the cost of game playing: How to control the expenses in mixed strategies. In: Proceedings of the 8th International Conference on Decision and Game Theory for Security (GameSec), LNCS 10575, pp. 494–505. Springer, [S.l.] (2017)
19. Robert, C.P.: The Bayesian choice. Springer, New York (2001)
20. Robinson, A.: Non-standard Analysis. Princeton Landmarks in Mathematics and Physics. Princeton University Press, Princeton (1996). URL http://gbv.eblib.com/patron/FullRecord.aspx?p=4626045. Luxemburg, W. A. J. (BeteiligteR)
21. Robinson, J.: An iterative method for solving a game. Annals of Mathematics **54**, 296–301 (1951)
22. Shaked, M., Shanthikumar, J.G.: Stochastic Orders. Springer (2006)
23. Wenner, F.: Determination of Risk Aversion and Moment-Preferences: A Comparison of Econometric models. PhD Thesis, Universität St.Gallen (2002)
24. Wheeler, E.: Security risk management: Building an information security risk management program from the ground up. Syngress, Amsterdam and Waltham, MA (2011). URL http://www.eblib.com/patron/FullRecord.aspx?p=685406

Chapter 3
Security Strategies and Multi-Criteria Decision Making

Stefan Rass

3.1 Introduction

The essence of security is defending assets against an adversary that may behave almost arbitrarily. Game theory can help finding optimal strategies against any possible behavior, provided that the attacker stays within a known action space. This is the typical domain and case of security risk management, where a set of threats is identified, against which a uniformly best defense is sought. In game-theoretic terms, the threat list corresponds to an action space, and the best defense against that list is a *security strategy*. This chapter discusses how such strategies can be computed for single and multiple protection goals, even when the effects of the defense actions are nondeterministic (random). The latter especially admits a treatment of uncertainty in three forms, being about the adversary (form and number), the attacker(s) incentives, and – to a limited extent – also the action space (threat list) itself. Our goal in the following is looking at suitable game-theoretic models and methods to compute best defenses under uncertainty.

In many cases, the information available to a decision maker is uncertain in diverse manners. If at least some information is available, then Bayesian choices [21] appear as the natural way to go, since they aim at minimizing the residual uncertainty given all available information.

What if the information is not uncertain but rather not there at all? In that case, assumptions must be made, but how can we be sure that these are even remotely close to reality? The answer is that this verification problem has no general solution, and cannot be circumvented either. Even the Bayesian approach relies on the specification (assumption) of some a priori distribution (or hyperpriors, in higher

S. Rass (✉)
Universitaet Klagenfurt, Institute of Applied Informatics, Universitaetsstrasse 65-67, 9020 Klagenfurt, Austria
e-mail: stefan.rass@aau.at

S. Rass, S. Schauer (eds.), *Game Theory for Security and Risk Management*,
Static & Dynamic Game Theory: Foundations & Applications,
https://doi.org/10.1007/978-3-319-75268-6_3

order Bayesian methods), which is nothing else but an informed guess about the yet unknown parameters. Alternative to informed guesses that Bayesian decision theory speaks about, minimax decision theory seeks to optimize decisions against any incarnation of the uncertain factors. That is, whatever happens after we made our decision, we have prepared ourselves for the worst and decided for the best in that (worst) case. Such a choice is called a *security strategy*. It comes with the price tag of being presumably pessimistic, and even disregarding much information that would be available perhaps. Indeed, (informed) Bayesian decisions and (uninformed) minimax decisions are closely connected to one another, and both find themselves reflected in the framework of game theory.

In general, any decision being made against a rational opponent or simply against nature (an irrational opponent) can be viewed as a game. The set of actions to choose from makes up the action set AS_1 (synonymously called a strategy set) for the first player. The extent to which that player anticipates what its opponent can do constitutes the action set AS_2 for the second player (in the simplest case). Toward a conventional game, described between two players for simplicity, letting the general case follow soon, let us assume that both players can specify a utility value (payoff) function $u_1, u_2 : AS_1 \times AS_2 \to \mathbb{R}$ that, for each combination of actions, returns a (deterministic) value u_1 for player 1, and u_2 for player 2. This completes the description of a classical game as a triple (N, S, H), in which $N = \{1, 2, \ldots\}$, is the set of players, each of which has an associated action set in $S = \{AS_1, AS_2, \ldots\}$, and another associated utility function in $H = \{u_i : AS_i \times AS_{-i} \to \mathbb{R} : i \in N\}$. The symbol AS_{-i} is the joint action set of i's opponents, being the cartesian product of all A_j for $j \in N \setminus \{i\}$. Typically, theoretical considerations are simplified by assuming a fixed (finite) number of players (in our case, $|N| = 2$), or a fixed (finite) number of strategies. In that case, the game itself is called finite.

Our main interest in this chapter will concern the sources of uncertainty, in light of which security strategies need to be found. So, which would be cases, where any of the three ingredients (or a combination thereof) is uncertain?

- Uncertain number of players: A player surely knows that it is engaging in some sort of competition, but the opponents may not be visible or even classifiable as rational. An example for a game with known players are (classical) auctions, in which all bidders personally face each other (unless the auction happens online). The converse extreme is found in computer security, where attackers are notoriously stealthy, and the exact number of them can fluctuate and is almost always unknown.
- Uncertain strategy spaces: In simple auctions, the game dynamics are clearly defined with precise rules, and a fair auction endows all players with the same strategy spaces, so that there is no uncertainty on this point. Again, computer security is an example of the opposite, where the actions for the defending party are known, while not so for the adversary: from the defender's perspective, there is an unpleasant and even unquantifiable residual chance that the attacker comes up with something completely unexpected. Such an attack is launched "zero days after its discovery," that is, at a time when the defender is still unaware of it. For this reason, this is called a *zero-day exploit*.
- Uncertain utility functions: For the positive extreme, simple auctions allow for an almost exact utility value, which is the same for all players and equal to

the value of the good for which the bidders run. The only uncertainty here is the potentially incorrect pricing of that good, but considering the price itself as the utility, this matter becomes negligible. Computer security games, again, are located at the other end of the spectrum, which becomes evident considering that security has a hardly measurable return on investment. The main purpose of security is to prevent possible losses, rather than to generate revenues. This makes the entire objective of increasing security difficult to argue, and makes things more intricate for decision makers: They cannot choose the action that surely rewards them the most, but must rather find the action that potentially saves them from large cost. In addition, these costs may be difficult to quantify (which is another independent issue).

It is fair to note here that the auctions mentioned above are considered in their simplest form, and the mechanisms and dynamics underneath auctions are sufficiently rich to span all extremes in all three cases above. Our focus in the following, however, will be on computer security, and the related security games. We stress that the coincidence of the naming is what it is here, a mere coincidence. The term "security strategy" has per se nothing to do with (computer) security, and exclusively relates to a minimax choice. That is, a choice against the worst case that could happen. Computer security is only one (among many) natural field of application that we shall use for running illustrations.

A convenient common denominator in the representation of all three of the above cases is obtained by letting the utilities, hereafter also called *payoffs*, be uncertain, or more precisely random variables (r.v.s). With random utility functions, the other two cases become covered:

- Uncertain number of players: If a player faces a varying and unknown number of opponents, it may perceive unexpected payoff fluctuations due to unknown people having taken influence. If the distribution of these fluctuations can be pinned down, then the whole world against which player 1 competes can be viewed as a single player with an unknown action set (physically consisting of many players with individual different capabilities and actions). Here, we assume the adversaries to gain their revenue as a team without negatively impacting each other, so that the payoff for the physical adversary $2,3,\ldots,N$ is only coming from the defending player 1. In reality, this may not be the case, which effectively results in the game being not zero-sum (between player 1 and the "team" acting as player 2). As we will show below (rigorously stated by inequality (3.3)), this violation nonetheless leaves the results and properties of security strategies unchanged and intact.
- Uncertain strategy spaces: Those correspond to unknown (undiscovered) parts in the domain on which the utility functions are defined, thus making them appear random to the extent at which the unknown actions are chosen at random.

The framework of stochastic orders, for example, the one put forth in Chapter 2, can be used for maximal flexibility in replacing the utility functions by random variables. We will switch between talking about real-valued or distribution-valued payoffs, whichever is more convenient.

3.2 Security Games with a Single Objective

Noncooperative players usually look for equilibria, that is, a strategy profile $(x_1^*,\dots,x_n^*)$ for all players $i \in N = \{1,2,\dots,n\}$ so that

$$\forall i \in N : u_i(x_i^*, \mathbf{x}_{-i}^*) \geq u_i(x_i, \mathbf{x}_{-i}^*), \qquad \text{for all } x_i \in AS_i \tag{3.1}$$

that is, no player gains by unilaterally deviating from the optimum x_i^*. In absence of any information about the number of opponents or their utility functions, we will need to view the opponent(s) as one big and vague entity, acting as player 2. Since this makes the payoffs necessarily unknown too, we ought to use the only information that is certain, which is our own payoff. In the best case, the opponent's intentions are completely unrelated to our own ones, in which case we can selfishly maximize our own revenue without anyone interfering (or us disturbing someone else). In the worst case, the intentions of us and the other player are opposite, and we both pull at different ends of the same rope. This is a zero-sum competition, in which we put $u_1 = -u_2 =: u$. For that class of two-player games, let the equilibrium be (x^*,y^*), and condition (3.1) boils down to

$$u(x,y^*) \leq u(x^*,y^*) \leq u(x^*,y), \tag{3.2}$$

for all $x,y \in AS_1 \times AS_2$. The existence of either, the zero-sum or general (Nash) equilibrium above is not assured without additional assumptions on the strategy spaces. Usually, we convexify those by turning to randomized strategies from the set (simplices) $S_i := \Delta(AS_i)$ for all $i \in \mathbb{N}$, and redefine the utilities into expectations, again denoted as $U = \mathrm{E}(u(X,Y))$, where the expectation is w.r.t. the distributions of the random strategies X,Y chosen from AS_1, AS_2.

The intuition of a zero-sum game being a valid worst-case model is an almost immediate consequence of (3.2): let $\Gamma = (\{1,2\},\{S_1,S_2\},\{u_1,u_2\})$ be a general game, and call $\Gamma_0 = (\{1,2\},\{S_1,S_2\},\{u_1,-u_1\})$ its associated zero-sum game from player 1's perspective. That is, player 1 does have no clue whatsoever on the payoff and incentives of player 2, yet the action space of both players is common knowledge. So, the best that player 1 can do is engage in Γ_0, while player 2 is actually playing Γ. Call $v = val(\Gamma_0) = \mathrm{E}(u_1(X^*,Y^*))$ the *saddle-point value* of Γ_0 upon equilibrium strategies X^*,Y^* played there. Since player 2 engages in Γ, it probably has a different equilibrium strategy $Y_\Gamma^* \neq Y^*$ and hence unilaterally deviates from (X^*,Y^*), thus increasing player 1's revenue $v \leq u_1(X^*,Y_\Gamma^*)$. Conversely, from player 2's perspective, player 1 deviates from its optimum $X_\Gamma^* \neq X^*$ and hence can only decrease its own revenue upon this. So, the chain of inequalities continues as $u_1(X^*,Y_\Gamma^*) \leq u_1(X_\Gamma^*,Y_\Gamma^*)$, and in total, leads to

$$v = val(\Gamma_0) \leq u_1(X^*,Y), \tag{3.3}$$

for all $Y \in S_2$ that player 2 could follow. That means that whatever incentives player 2 may have, it can never decrease player 1's revenue below $v = val(\Gamma_0)$, as long as player 1 follows its zero-sum optimal equilibrium strategy X^*. This X^* is the sought *security strategy* for player 1, and it can only fail if the strategy space for player 2, which is necessary to compute X^*, is inaccurate. Likewise, the associated zero-sum game for player 1 is called a *security game*.

Remark 3.1. Assuming the action spaces of both players to be mutual knowledge may appear hard and even unjustified in light of zero-day attacks, whose sole characteristic is the action's *unexpectedness*. To a limited extent, taking payoffs as random variables, the tails of the payoff distribution (see Chapter 2 or [18]) admits losses beyond what the actions would be known to imply. The tail region of the loss distribution is then where zero-day exploits would occur. A concrete valuation for zero-day risks is given by [28].

Remark 3.2. Nash equilibria are typically applied in games with full information, but security is essentially a competition with lack of information on both sides. There are several ways to resolve this seeming issue; one is the use of distributions in the payoff structure, another is the transition to stochastic games themselves (such as partially observable Markov decision processes with full or partial observability [29, 12]). Occasionally, the uncertainty is not about what can happen (the system administrator may have quite a decent idea about the entry points in the system, so that the strategy spaces of both players are not too uncertain), but only about what *will* happen. If the strategy spaces are known, yet only the adversary's incentives are uncertain, then Nash equilibria can be applied in this special case. The point of security strategies is the assumption that the adversary's incentives are opposite to that of the defenders (and hence known to the defender). However, the defender does not even need to be sure that s/he engages in a zero-sum competition, since if the game is not zero-sum, then (3.3) will only become a more pessimistic overestimate of the actually suffered loss.

Definition 3.1 (Single-Goal Security Game). Let $\Gamma = (\{1,2\},\{S_1,S_2\},\{u_1,u_2\})$ be a two-player game. The *security game* (for player 1) associated with Γ is the zero-sum game $\Gamma_0 = (\{1,2\},\{S_1,S_2\},\{u_1,-u_1\})$. If Γ_0 admits a Nash equilibrium $(x^*,y^*) \in S_1 \times S_2$, then x^* is called a *security strategy* (for player 1).

Note we assumed nothing about the strategy spaces (not even finiteness), except for them to admit an equilibrium to exist (one suitable condition is compactness of all S_i and continuity of the payoff functions w.r.t. the same topology [7]).

The bound (3.3) that a security strategy implies is generally sharp, as simple examples show:

Example 3.1 ([14]). Consider the two-person nonzero-sum game with payoff structure as in Figure 3.1.

This game has multiple equilibria with values $v_1 \in E_1 = \left\{2,4,\frac{8}{3},\frac{18}{7},\frac{9}{4},\frac{14}{5}\right\}$ for player 1, and $v_2 \in E_2 = \{2,3\}$ for player 2, with respective strategies and payoffs as listed in Table 3.1. Now, consider the associated security games from player 1, and player 2's perspective (either being the adversary to the other in both cases), having the payoff structures as shown in Figure 3.2.

	Player 2				
Player 1	(2,0)	(2,0)	(1,4)	(3,1)	(2,3)
	(1,1)	(2,3)	(2,1)	(2,3)	(4,2)
	(0,2)	(3,2)	(0,1)	(2,3)	(2,1)
	(0,2)	(4,2)	(1,0)	(0,2)	(1,2)
	(2,3)	(2,1)	(4,3)	(4,1)	(3,0)

Fig. 3.1: Example Nonzero-Sum Game

Security game for player 1:

2	2	1	3	2
1	2	2	2	4
0	3	0	2	2
0	4	1	0	1
2	2	4	4	3

Security game for player 2:

0	0	4	1	3
1	3	1	3	2
2	2	1	3	1
2	2	0	2	2
3	1	3	1	0

Fig. 3.2: Security Games Associated with the bimatrix game in Figure 3.1

Table 3.1: Equilibria (and Security Strategies) for Example 3.1, computed using [2]

#	player 1 equilibrium						player 2 equilibrium					
	x_1^*	x_2^*	x_3^*	x_4^*	x_5^*	u_1^*	y_1^*	y_2^*	y_3^*	y_4^*	y_5^*	u_2^*
1	0	0	0	1	0	2	1/2	1/2	0	0	0	2
2	0	0	0	1	0	4	0	1	0	0	0	2
3	0	0	0	1	0	8/3	0	2/3	0	1/3	0	2
4	0	0	0	1	0	18/7	0	4/7	0	1/7	2/7	2
5	0	0	0	1	0	9/4	1/4	1/2	0	0	1/4	2
6	0	0	0	1	0	14/5	0	3/5	0	0	2/5	2
7	0	0	0	0	1	2	1	0	0	0	0	3
8	0	0	0	0	1	4	0	0	1	0	0	3

(a) Bimatrix Game Equilibrium (Payoffs as in Figure 3.1)

#	player 1 equilibrium						player 2 equilibrium					
	x_1^*	x_2^*	x_3^*	x_4^*	x_5^*	u_1^*	y_1^*	y_2^*	y_3^*	y_4^*	y_5^*	u_2^*
1	2/3	0	0	0	1/3	2	1	0	0	0	0	-2
2	2/3	0	0	0	1/3	2	1/2	1/2	0	0	0	-2
3	0	0	0	0	1	2	1	0	0	0	0	-2
4	0	0	0	0	1	2	1/2	1/2	0	0	0	-2

(b) Security Game Equilibrium for Player 1 (Payoffs as in Figure 3.2)

#	player 1 equilibrium						player 2 equilibrium					
	x_1^*	x_2^*	x_3^*	x_4^*	x_5^*	u_1^*	y_1^*	y_2^*	y_3^*	y_4^*	y_5^*	u_2^*
1	1/6	1/4	0	1/3	1/4	5/3	1/3	1/6	1/6	0	1/3	-5/3

(c) Security Game Equilibrium for Player 2 (Payoffs as in Figure 3.2)

The value of the security game for player 1 is $v_1 = 2 = \min E_1$, so the bound (3.3) is sharp. Switching roles and looking at the security game for player 2, its value is $v_2 = \frac{17}{10} < 2 = \min E_2$, so the bound can be loose as well.

For the sake of simplicity only, let the strategy spaces be finite in the following, so that the optimal randomized actions X^* can be specified as categorical distributions, vectors, $\mathbf{x}^* = (x_1, \ldots, x_{|AS_1|})$ and $\mathbf{y}^* = (y_1, \ldots, y_{|AS_2|})$. The saddle-point value exists under these assumptions and can be rewritten as $v = \max_{\mathbf{x}} \min_{\mathbf{y}} \mathrm{E}(u(X,Y)) = \min_{\mathbf{x}} \max_{\mathbf{y}} \mathrm{E}(u(X,Y))$ with $X \sim \mathbf{x}, Y \sim \mathbf{y}$. This form reveals why we call the point x^*, y^* at which v is attained with equality a *minimax decision*. For finite games with a payoff matrix $\mathbf{A} = (a_{ij})$, we can write $\mathrm{E}(u(\mathbf{x},\mathbf{y})) = \mathbf{x}^T \mathbf{A} \mathbf{y} = \sum_{i,j} x_i y_j a_{ij}$ and $v = (x^*)^T \mathbf{A} \mathbf{y}^*$. Bayesian decisions can be framed into this by letting y^* be a "least favourable distribution," so that the Bayes' optimal decision becomes the minimax decision. While the details of this are intricate, a more intuitive link is discovered by letting the payoffs be random variables. As in Chapter 2, let us replace $u(\mathbf{x},\mathbf{y})$ by a probability distribution $F(\mathbf{x},\mathbf{y})$ of the random revenue R, so that

$$\Pr(R \le r) = F(\mathbf{x},\mathbf{y})(r) = \sum_{i,j}^{n} \Pr(R \le r | i,j) \Pr(i,j) = \sum_{i,j}^{n} x_i y_j F_{ij}(r), \tag{3.4}$$

where $\Pr(i,j)$ is a shorthand for the likelihood of player 1 choosing action i and player 2 taking action j, and F_{ij} is the payoff distribution in the ij-th entry of the payoff matrix in a distribution-valued game. Note the striking similarity of (3.4) with the version for finite (matrix) games mentioned just before.

The beauty of Bayesian decisions lies in their natural capability of improvement upon new information. This corresponds to an a priori distribution π becoming an a posteriori distribution $\pi(\cdot|D)$ upon the data D. The very same concept can be used in games when the payoff is distribution-valued, since there is no conceptual barrier preventing us from calling $\Pr(R \le r) = F(\mathbf{x},\mathbf{y})(r)$ an a priori distribution, and upon new information D coming in, switching to $\Pr(R \le r|D) = F(\mathbf{x},\mathbf{y},D)(r)$ as the a posteriori distribution. A Bayesian decision then goes for a minimization of some loss function applied to the posterior. If that loss function is quadratic, then the Bayes decision is the posterior expectation, which is the same as in regular game theory. Other choices, say, the absolute value loss, yield to the median as a replacement for the expectation. Any such design choice can be avoided at all if we resort to stochastic orders to let the distribution itself be the sole payoff (from which any quantity of interest can be computed afterward).

3.3 Multi-Objective Security Games

Decisions are hardly ever made with only one goal in mind of the defender, but the equilibrium definition cannot straightforwardly be extended to vector-valued payoffs, since those are no longer totally ordered. For any two vectors $\mathbf{u} = (u_1, \ldots, u_n), \mathbf{v} = (v_1, \ldots, v_n) \in \mathbb{R}^n$, we will write $\mathbf{u} \le \mathbf{v}$ iff $u_i \le v_i$ for all

$i = 1, 2, \ldots, n$. The converse relation in which there is at least one index i for which $u_i \geq v_i$, irrespectively of what the other components do, is denoted as $\mathbf{u} \geq_1 \mathbf{v}$. The relations $\geq$ and (its complement) $\leq_1$ are defined in the obvious way. In replacing $\leq$ by $\leq_1$ in (3.1), we obtain a *Pareto-Nash equilibrium*, in which any unilateral deviation from the equilibrium will decrease the payoff for the respective player for at least one of its goals.

Security games and security strategies can be defined by turning the previous observations made for single-goal security games into requirements, toward an axiomatic definition. In this regard, we will demand the bound (3.3) to hold for each goal (we call this *assurance*), and to be optimal as in Example 3.1 (this will be the *efficiency* axiom) [15].

Definition 3.2 (Multi-Goal Security Strategy (MGSS)). A strategy $\mathbf{x}^* \in S_1$ in two-player game with continuous vector-valued payoff $\mathbf{u}_1 : S_1 \times S_2 \to \mathbb{R}^d$ for $d \geq 1$ for player 1. Let us denote the i-th coordinate function in $\mathbf{u}_1$ as $u_1^{(i)} : S_1 \times S_2 \to \mathbb{R}$. The competition in which player 1 engages is called a MGSS with *assurance* $\mathbf{v} = (v_1, \ldots, v_d)$ if two criteria are met:

1. The assurances are the component-wise guaranteed payoff for player 1, that is, for all goals $i = 1, 2, \ldots, d$, we have

$$v_i \leq u_1^{(i)}(\mathbf{x}^*, \mathbf{y}) \qquad \forall \mathbf{y} \in S_2, \tag{3.5}$$

with equality being achieved by at least one choice $\mathbf{y}_i^* \in S_2$.

2. At least one assurance becomes void if player 1 deviates from $\mathbf{x}^*$ by playing $\mathbf{x} \neq \mathbf{x}^*$. In that case, some $\mathbf{y}_0 \in S_2$ exists such that

$$\mathbf{u}_1(\mathbf{x}, \mathbf{y}_0) \leq_1 \mathbf{v}. \tag{3.6}$$

Constraint (3.5) can be stated in a more compact, yet weaker, form by saying that

$$\mathbf{v} \leq \mathbf{u}_1(\mathbf{x}^*, \mathbf{y}), \qquad \forall \mathbf{y} \in S_2. \tag{3.7}$$

The idea to assure existence of a MGSS follows similar lines as before: Let player 1 engage in a hypothetical one-against-all competition where each goal for player 1 relates to its own zero-sum game against a hypothetical opponent. The opponents themselves act independently of each other, and each optimizes only a single goal. This leads to the sibling of the associated zero-sum game Γ_0 from above, which we call the *security game* here to distinguish it from Γ_0 (in previous literature [15], the same concept has been coined an "auxiliary game"; we use the new name here for consistency):

Definition 3.3 (Multi-Objective Security Game (MOSG)). Let Γ be a two-person game, in which only the strategy space and payoff function $\mathbf{u}_1 : S_1 \times S_2 \to \mathbb{R}^d$ for player 1 is known. Let the coordinate functions of $\mathbf{u}_1$ be $u_1^{(1)}, \ldots, u_1^{(d)}$. The MOSG associated with Γ is the game $\mathbf{\Gamma}_0$ composed from the following ingredients:

- A set of $d + 1$ players, in which player 0 is the first player in Γ, having d opponents, each of which corresponds to another of the d goals in Γ.

- An (ordered) multiset of $d+1$ strategy sets being $\{S_1, S_2, S_2, \ldots, S_2\}$
- A set of payoff functions $H = \{\mathbf{f}_0, f_1, \ldots, f_d\}$. Player 0 is the only one with a vector-valued utility $\mathbf{f}_0 = (f_0^{(1)}, f_0^{(2)}, \ldots, f_0^{(d)})$, whose j-th coordinate function is determined by its own action and that of the j-th opponent, that is, $f_0^{(j)} := u_1^{(j)}(\mathbf{x}, \mathbf{y}_j)$. Likewise, the j-th opponent has the scalar payoff $f_j := -u_1^{(j)}$, and the same strategy space S_2 as all other opponents.

Definition 3.3 is made to materialize its foregoing intuition in the way of exhibiting each Pareto-Nash equilibrium (as defined above) in the security game to be an MGSS in the original game. The proof is based on the following result:

Lemma 3.1. *Let Γ be as in Definition 3.3, where the strategy spaces for both players are compact, and let $\mathbf{x}^*$ be a MGSS with assurance $\mathbf{v}$, assuming that one exists. Then, no vector $\widetilde{\mathbf{v}} < \mathbf{v}$ is an assurance for $\mathbf{x}^*$.*

Proof (from [15]). Let $\widetilde{\mathbf{v}} < \mathbf{v}$, put $\varepsilon := \min_{1 \leq i \leq k}\{v_i - \widetilde{v}_i\}$ and observe that $\varepsilon > 0$. The function $\mathbf{u}_1$ is uniformly continuous on $S_1 \times S_2$ (being compact), so a $\delta > 0$ exists with $\|(\mathbf{x},\mathbf{y}) - (\mathbf{x}',\mathbf{y}')\|_\infty < \delta$ implying $\|\mathbf{u}_1(\mathbf{x},\mathbf{y}) - \mathbf{u}_1(\mathbf{x}',\mathbf{y}')\|_\infty < \frac{\varepsilon}{2}$.

Consider the mapping $\mathbf{u}_\mathbf{y} : S_1 \to \mathbb{R}^k, \mathbf{u}_\mathbf{y}(\mathbf{x}) := \mathbf{u}_1(\mathbf{x},\mathbf{y})$, which is as well uniformly continuous on S_1 by the same argument. So, $\|(\mathbf{x}^*,\mathbf{y}) - (\mathbf{x}',\mathbf{y})\|_\infty = \|\mathbf{x}^* - \mathbf{x}'\|_\infty < \delta$ implies $\|\mathbf{u}_\mathbf{y}(\mathbf{x}^*) - \mathbf{u}_\mathbf{y}(\mathbf{x}')\|_\infty = \max_{1\leq i \leq k}\left|u_1^{(i)}(\mathbf{x}^*,\mathbf{y}) - u_1^{(i)}(\mathbf{x}',\mathbf{y})\right| < \frac{\varepsilon}{2} \quad \forall \mathbf{y} \in S_2$. It follows that $\left|u_1^{(i)}(\mathbf{x}^*,\mathbf{y}) - u_1^{(i)}(\mathbf{x}',\mathbf{y})\right| < \frac{\varepsilon}{2}$ for $i = 1, \ldots, k$ and all $\mathbf{y} \in S_2$, and consequently $\max_{\mathbf{y}\in S_2}\left|u_1^{(i)}(\mathbf{x}^*,\mathbf{y}) - u_1^{(i)}(\mathbf{x}',\mathbf{y})\right| < \frac{\varepsilon}{2}$. Now, selecting any $\mathbf{x}' \neq \mathbf{x}^*$ within an δ-neighborhood of $\mathbf{x}^*$, we end up asserting $u_1^{(i)}(\mathbf{x}',\mathbf{y}) \geq u_1^{(i)}(\mathbf{x}^*,\mathbf{y}) - \frac{\varepsilon}{2}$ for every i and $\mathbf{y} \in S_2$.

Using $u_1^{(i)}(\mathbf{x}^*,\mathbf{y}) \geq v_i$, we can continue by saying that $u_1^{(i)}(\mathbf{x}',\mathbf{y}) \geq v_i - \frac{\varepsilon}{2} > v_i - \varepsilon$. By definition of ε, we have $v_i - \widetilde{v}_i \geq \varepsilon$, so that $u_1^{(i)}(\mathbf{x}',\mathbf{y}) > \widetilde{v}_i$ for all i, contradicting (3.6) if $\widetilde{\mathbf{v}}$ were to be a valid assurance vector. □

Theorem 3.1. *Let Γ be as in Lemma 3.1. The vector $\mathbf{x}^*$ constitutes a MGSS with assurance $\mathbf{v}$ for player 1 in the game Γ, if and only if it is a Pareto-Nash equilibrium strategy for player 0 in the MOSG $\mathbf{\Gamma}_0$ according to Definition 3.3.*

Proof (from [15]). Throughout the proof, we will put a bar on top of components, that is, payoff functions, that belong to the security game $\mathbf{\Gamma}_0$, to distinguish these from their counterparts in Γ (showing no horizontal bar accent).

("$\Leftarrow$") Let $\mathbf{s}^* := (\mathbf{s}_0^*, \mathbf{s}_1^*, \ldots, \mathbf{s}_d^*)$ be a Pareto-Nash equilibrium in $\mathbf{\Gamma}_0$, and set the assurances to

$$v_i := u_1^{(i)}(\mathbf{s}_0^*, \mathbf{s}_i^*) \quad \text{for all } i = 1, 2, \ldots, d. \tag{3.8}$$

We prove that $\mathbf{s}_0^*$ is a MGSS with assurance $\mathbf{v}$. Consider the i-th opponent's point of view. By construction (Definition 3.3), we have his utility independent of the other player's deviations. So regardless if any other opponent deviates, as long as player 0 (his sole rival) plays $\mathbf{s}_0^*$, his strategy $\mathbf{s}_i^*$ is (Pareto-)optimal (notice that his payoff is scalar), thus

$$\begin{aligned} -u_1^{(i)}(\mathbf{s}_0^*, \mathbf{s}_i) &= \overline{u}_i(\mathbf{s}_0^*, \mathbf{s}_1^*, \dots, \mathbf{s}_{i-1}^*, \mathbf{s}_i, \mathbf{s}_{i+1}^*, \dots, \mathbf{s}_d^*) \\ &\leq \overline{u}_i(\mathbf{s}_0^*, \mathbf{s}_1^*, \dots, \mathbf{s}_{i-1}^*, \mathbf{s}_i^*, \mathbf{s}_{i+1}^*, \dots, \mathbf{s}_d^*) = -u_1^{(i)}(\mathbf{s}_0^*, \mathbf{s}_i^*) = -v_i \end{aligned}$$

for every $\mathbf{s}_i \in S_2$. As this holds for every $i = 1, \dots, d$, we can conclude $\mathbf{v} \leq \mathbf{u}_1(\mathbf{s}_0^*, \mathbf{s}_2)$ for all $\mathbf{s}_2 \in S_2$. Thus, the first part of Definition 3.2 is verified, since the average outcome of the game cannot undercut its minimum. On the other hand, from player 0's point of view, his strategy $\mathbf{s}_0^*$ is as well Pareto-optimal, that is, by playing $\mathbf{s}_0 \neq \mathbf{s}_0^*$, he ends up with

$$u_1^{(j)}(\mathbf{s}_0, \mathbf{s}_j^*) = \overline{u}_0^{(j)}(\mathbf{s}_0, \mathbf{s}_1^*, \dots, \mathbf{s}_d^*) \leq \overline{u}_0^{(j)}(\mathbf{s}_0^*, \dots, \mathbf{s}_d^*) = u_1^{(j)}(\mathbf{s}_0^*, \mathbf{s}_j^*) = v_j$$

for at least one component j, and condition (3.6) of Definition 3.2 is verified.

("$\Rightarrow$") Put $I := \{1, 2, \dots, d\}$. Let $\mathbf{x}^*$ be a MGSS with assurance $\mathbf{v}$. Let $i \in I$ be arbitrary, and assume that $v_i > \min_{\mathbf{y} \in S_2} u_1^{(i)}(\mathbf{x}^*, \mathbf{y})$. In the light of condition (3.7), this is impossible, for otherwise the i-th opponent could play a strategy $\mathbf{y}_i'$ to enforce an outcome $u_1^{(i)}(\mathbf{x}^*, \mathbf{y}_i') = \min_{\mathbf{y} \in S_2} u_1^{(i)}(\mathbf{x}^*, \mathbf{y}) < v_i$, invalidating $\mathbf{v}$ as the assured outcome. The strategy $\mathbf{y}_i'$ necessarily exists because $u_1^{(i)}$ is continuous. Since Definition 3.2(assurance) implies $\min_{\mathbf{y} \in S_2} u_1^{(i)}(\mathbf{x}^*, \mathbf{y}) \leq v_i$ and strict inequality has been ruled out, we must have equality to the minimum and some $\mathbf{y}_i^*$ exists such that

$$v_i = u_1^{(i)}(\mathbf{x}^*, \mathbf{y}_i^*) = \min_{\mathbf{y} \in S_2} u_1^{(i)}(\mathbf{x}^*, \mathbf{y}) = \max_{\mathbf{y} \in S_2} \underbrace{-u_1^{(i)}(\mathbf{x}^*, \mathbf{y})}_{=\overline{u}_i(\mathbf{x}^*, \mathbf{y})} = \overline{u}_i(\mathbf{x}^*, \mathbf{y}_i^*).$$

Therefore, $\mathbf{y}_i^*$ must be an optimal strategy for the i-th opponent if player 0 acts according to $\mathbf{x}^*$. Put $\mathbf{y}^* := (\mathbf{y}_1^*, \dots, \mathbf{y}_d^*)$. Assume the existence of some MGSS $\mathbf{x}' \neq \mathbf{x}^*$ with uniformly better assurance $\mathbf{v}' > \mathbf{v}$. Then, $\mathbf{v} \leq \mathbf{v}' \leq \mathbf{u}_1(\mathbf{x}', \mathbf{y})$ for all $\mathbf{y} \in S_2$, because (3.7) applies to $\mathbf{x}'$. Take any $\mathbf{x}'' \in S_1$ with $\mathbf{x}'' \neq \mathbf{x}'$. We distinguish two cases: if $\mathbf{x}'' \neq \mathbf{x}^*$, then property (3.6) implies that there is an index j and some $\mathbf{y}$ such that $u_1^{(j)}(\mathbf{x}'', \mathbf{y}) \leq v_j \leq v_j'$. If $\mathbf{x}'' = \mathbf{x}^*$, then by the above argument, we can just use $\mathbf{y}_j^*$ to assert that $u_1^{(j)}(\mathbf{x}'', \mathbf{y}_j^*) = \underbrace{u_1^{(j)}(\mathbf{x}^*, \mathbf{y}_j^*)}_{=v_j} \leq v_j'$ for any index j, thus verifying (3.6). It follows that $\mathbf{v} < \mathbf{v}'$ is as well an assurance for $\mathbf{x}'$, contradicting Lemma 3.1. Hence, there is no $\mathbf{x}'$ for which the assurance $\mathbf{v} = \operatorname{argmin}_{\mathbf{x} \in S_1} \mathbf{u}_1(\mathbf{x}, \mathbf{y})$ (in Pareto's sense) with $\mathbf{y}_i = \operatorname{argmin}_{\mathbf{y} \in S_2} u_1^{(i)}(\mathbf{x}, \mathbf{y})$ is better than for $\mathbf{x}^*$ in Pareto's sense, proving that the profile $(\mathbf{x}^*, \mathbf{y}^*)$ is a Pareto-Nash equilibrium of $\mathbf{\Gamma}_0$. □

Theorem 3.2 ([10]). *Let $\Gamma = (N, S, H)$ be a Multi-Objective Game (MOG), where each $AS_i \in S$ is convex and compact, and each $\mathbf{u}_i \in H$ is continuous. For each player $i \in N$, let every individual payoff $u_i^{(j)}(s_i, \mathbf{s}_{-i})$ for $1 \leq j \leq r_i$ be a concave function of*

s_i on AS_i, whenever the remaining values $\mathbf{s}_{-i}$ are fixed. Then, Γ has a Pareto-Nash equilibrium.

The existence of MGSS is assured under the usual conditions, for example, finiteness of the game (which reproves a known result of [1] by a simple application of Theorems 3.1 and 3.2):

Corollary 1 (Existence of MGSS in matrix games). *Every finite MOSG has a MGSS in mixed strategies.*

Observe that Definition 3.2 is axiomatic and not limited to finite games or games with a finite number of players. In that sense, the characterization Theorem 3.1 can be combined with other existence conditions for (normal) Nash equilibria to extend the existence of MGSS to various other classes of games.

The proof of Theorem 3.2 is "constructive" in the sense of equating the set of Pareto-Nash equilibria to the set of Nash equilibria in a scalarized version of the MOG. Specifically, [10] prescribe the following steps to find a Pareto-Nash equilibrium in a MOG $\mathbf{\Gamma}$, in which there are n players, the i-th of which having a set of r_i goals to optimize:

1. Fix an arbitrary set of real numbers $\alpha_{11}, \alpha_{12}, \ldots, \alpha_{1r_1}, \alpha_{21}, \ldots, \alpha_{2r_2}, \ldots, \alpha_{n1}, \ldots, \alpha_{nr_n}$ that satisfy condition (3.9):

$$\left.\begin{array}{ll} \sum_{k=1}^{r_i} \alpha_{ik} = 1 & \text{for } i = 1,2,\ldots,n, \text{ and} \\ \alpha_{ik} > 0 & \text{for } k = 1,2,\ldots,r_i \text{ and } i = 1,2,\ldots,n. \end{array}\right\} \quad (3.9)$$

2. Form a (scalar) game $\Gamma_s = (N, S, H')$ with $H' = \{f_1, \ldots, f_n\}$ and

$$f_i = \sum_{k=1}^{r_i} \alpha_{ik} u_i^{(k)}. \quad (3.10)$$

3. Find a Nash-equilibrium $\mathbf{x}^* = (x_1^*, \ldots, x_n^*)$ in Γ_s, which is then a Pareto-Nash equilibrium in $\mathbf{\Gamma}$.

Notice that the Nash equilibria found by the above algorithm depend on the particular choice of weights. Indeed, the full set of equilibria is given as the union of all equilibria over all admissible choices of α's in (3.9) [10].

It is not difficult to verify that by letting player 1 be minimizing (up to here, we implicitly assumed a maximizing first player), all arguments work identically after being rephrased in terms of a total stochastic order such as that from Chapter 2. The results are all the same up to obvious (and purely syntactic) changes toward using $\preceq$ in place of $\leq$. Some qualitative similarities, unfortunately, are lost from this point onward, as shown in Section 3.4.2.1, but can be recovered in an approximate form, as we will discuss in Section 3.4.3.

3.4 Computing Equilibria and Security Strategies

The existence of equilibria in single-goal games is assured by Nash's theorem or generalizations thereof, and methods to compute such equilibria, and hence security strategies, are reviewed below. Computing Pareto-Nash equilibria for getting MGSS (via Theorem 3.1) can, with a little more effort, be reduced to the computation of (regular) Nash equilibria thanks to results in [10]. Thus, it suffices to dig into details about how (normal) Nash equilibria are computed, which we do next.

It must be emphasized that the methods to compute equilibria in the following validly apply without any problems for traditional games over $\mathbb{R}$, but when we switch to distribution-valued games (based on a stochastic order), some methods may no longer work. Conversely, the stochastic $\preceq$-order of Chapter 2 includes $\leq$ as a special case, so that the respective algorithms can, w.l.o.g., be stated in terms of $\preceq$, where the respective version for $\mathbb{R}$ can be obtained by the simple syntactic change of $\preceq$ into $\leq$ everywhere. Still, since there are qualitative differences in the use of $\leq$ or $\preceq$ for the optimization, we let the "problematic" procedures use $\preceq$ to point at the issues with that ordering, letting respective solutions follow.

3.4.1 Solution by Linear Programming

Let the zero-sum games of interest be with finite strategy spaces for both players, so that the payoff structure is a matrix $\mathbf{A} \in \mathbb{R}^{n\times m}$, and consider mixed strategies to ensure the existence of equilibria in all cases. Let these random (mixed) equilibrium strategies X^*, Y^* be characterized by their (categorical) distributions $\mathbf{x}^* \in S_1 = \Delta(AS_1) \subset \mathbb{R}^n, \mathbf{y}^* \in S_2 = \Delta(AS_2) \subset \mathbb{R}^m$. It is not difficult to find the saddle-point value of the game to be $u(\mathbf{x}^*, \mathbf{y}^*) = \max_{\mathbf{x}\in S_1} \min_{\mathbf{y}_2\in S_2} \mathbf{x}^T\mathbf{A}\mathbf{y} = \min_{\mathbf{y}_2\in S_2} \max_{\mathbf{x}\in S_1} \mathbf{x}^T\mathbf{A}\mathbf{y}$ (by strong duality).

For security games, we adopt player 1's perspective, and suppose that player 2 has chosen the (pure and minimizing) strategy $\mathbf{y}$. Then $\mathbf{Ay}$ is a vector, and player 1's objective is the maximization $\max_{\mathbf{x}\in S_1} \mathbf{Ay} = \max_{i=1,\dots,n} \mathbf{e}_i^T\mathbf{Ay}$, where $\mathbf{e}_i$ is the i-th unit vector in $\mathbb{R}^n$. Note that we hereby converted the optimization over a continuum into the much simpler task of choosing the best from a set of finite alternatives (as we previously discussed in Chapter 2). The only constraints added were $\mathbf{x} \geq 0$ and $\mathbf{1}^T\mathbf{x} = 1$, where $\mathbf{1}$ is the vector of all 1's. Substitute $v := \max_{i=1,\dots,n} \mathbf{e}_i^T\mathbf{Ay}$, then the saddle point condition directly translates into the linear program that player 1 needs to solve for finding a security strategy:

$$
(P1) \qquad \begin{array}{ll} \max_{\mathbf{X}} v & \\ \text{s.t.} & \\ v \leq \mathbf{e}_i^T\mathbf{Ax} & \forall i = 1,2,\dots,n \\ \mathbf{1}^T\mathbf{x} = 1 & \\ \mathbf{x} \geq 0 & \end{array} \qquad (3.11)
$$

This simple formulation admits an exact computation of an equilibrium even in polynomial time for security games as laid out in Section 3.2. For MGSS, the linear programming approach fails because we are dealing with a $(d+1)$-player game, which includes at least three actors in the simplest multi-goal setting. There, we can resort to iterative methods. Similarly, games defined over stochastic orders may not admit the arithmetics needed to solve Equation 3.11, so iterative (learning) methods are the usual method of choice in that cases too (indeed, the stochastic order $\preceq$ from Chapter 2 comes with the full-fledged arithmetic in the hyperreal space, yet lacking an ultrafilter, we have severe difficulties in doing the calculations practically).

3.4.2 Iterative Solutions by Learning

Iterative methods of computing Nash equilibria by online learning (see [8] for a concrete application) let all players start from a suboptimal strategy, and act according to the best of their so-far recorded knowledge to improve their (randomized) strategies. The usual coupled learning method starts from an initial guess for the optimal strategies and utilities, denoted by $\mathbf{x}_{i,0}, \hat{\mathbf{u}}_{i,0}$ for the i-th player. As the (discrete) time $t \in \mathbb{N}$ goes by, both players choose their respective next moves according to some learning rule (cf. [9, Chp.14])

$$\mathbf{x}_{i,t+1} = \Pi_{i,t}(u_{i,t}, \hat{\mathbf{u}}_{i,t}, \mathbf{x}_{i,t}, \lambda_{i,t}, a_{i,t}), \tag{3.12}$$

and update their corresponding utility estimates as

$$\hat{\mathbf{u}}_{i,t+1} = \Sigma_{i,t}(u_{i,t}, \hat{\mathbf{u}}_{i,t}, \mathbf{x}_{i,t}, \lambda_{i,t}, a_{i,t}), \tag{3.13}$$

where $\Pi_{i,t}, \Sigma_{i,t}$ are learning rules that in the most general form depend on the player i, the current time t, the action $a_{i,t}$, and utility $u_{i,t}$ observed for it, as well as the so-far existing estimates for the utility $\hat{\mathbf{u}}_{i,t}$ and (randomized) actions $\mathbf{x}_{i,t}$ at time t. The remaining parameter $\lambda_{i,t}$ covers additional input, for example, a learning rate (to differently weigh recent against past observations) or similar; it will be of no concrete use for us here but is relevant in several other instances of (3.12) and (3.13). We refer the reader to [9] for an in-depth treatment, and confine ourselves to the simplest learning rule called FP. Other such learning regimes can be studied with help of Lyapunov theory applied to the dynamical system that (3.12) and (3.13) induce [9, Chp.14]. Finally, one should bear in mind that the learning model assumes incentive compatibility of the involved players, so that neither player has an incentive to deviate from the learning rules. Deviations thereof that are observable in practice are studied in behaviorial game theory [3], which is outside of our scope here. The broader area treating techniques like this is algorithmic game theory [11, 24] and learning [4, 6].

Let us instantiate (3.12) and (3.13) for two players, let their action history from time 0 to time step $\ell \in \mathbb{N}$ be $x_0, x_1, \ldots, x_\ell \in AS_1$ for player 1, $y_0, y_1, \ldots y_\ell \in AS_2$. Both players alternatingly (or simultaneously) choose their actions to maximize the so-far long-run average, relative to the recorded behavior of the opponent so far. At time t, player 2 takes its move, followed by player 1 who is assumed to observe what its opponent does. Initially, player 1 takes any choice for a pure strategy as a kickoff:

$$\left.\begin{aligned} y_t &= \text{argmax}_{j\in AS_2} \tfrac{1}{t}\textstyle\sum_{\ell=1}^{t} u_2(y_\ell, j) \\ x_{t+1} &= \text{argmax}_{i\in AS_1} \tfrac{1}{t}\textstyle\sum_{\ell=1}^{t} u_1(x_\ell, i) \end{aligned}\right\} \quad (3.14)$$

where u_1, u_2 are the payoffs for players 1 and 2, respectively. The learning regime (3.14) corresponds to $\Pi_{i,t}$ in (3.12), while the arithmetic means appearing in both expressions correspond to the updating of observed revenues in (3.13). It can be shown that FP via (3.14) converges under alternating moves (as stated here) or synchronous moves (where both players choose their next actions at the same time). Various conditions under which (3.14) converges are known, such as the game having a potential, being zero-sum [23] or being general (nonzero-sum) with $|AS_1| = |AS_2| = 2$. In a practical implementation, a careful distinction must be made regarding convergence of the values vs. convergence of the strategies. While the saddle point approximations (3.13) in FP are assured to converge to each other, this is not necessarily happening for the strategies (3.12) as well. Hence, the convergence threshold used to stop the iteration should be imposed on the so-far averaged payoff(s) u_t, say if u_t differs from u_{t+1} only by a residual amount of some a priori chosen $\varepsilon > 0$ in some norm.

Algorithm 1 shows a version of FP for a minimizing first player, implicitly making player maximizing and assuming a zero-sum competition. For generality, the algorithm is formulated over the stochastic order $\preceq$ from Chapter 2 and distribution-valued games here. The $\preceq$-relation orders two random variables X, Y as $X \preceq Y$ if and only if the moment sequence $(\mathrm{E}Y^k)_{k\in\mathbb{N}}$ "diverges faster" than the moment sequence $(\mathrm{E}Y^k)_{k\in\mathbb{N}}$. Practically, it can be shown that the probability mass assigned to the tails of the distributions of X and Y determines the order, so that $X \preceq Y$ holds if and only if extreme events are less likely to happen for X than to occur under Y (see Theorem 2.2 in Chapter 2).

One reason to look at FP in stochastic orders is that finding equilibria in games over those orders is a widely undiscussed issue in the literature, but *could* offer insights into why real players may not always follow a utility-maximization behavior (either because the utility was not accurately modeled, or the order imposed on the utilities is different from the ordering on $\mathbb{R}$; the latter of which is a yet unverified hypothesis and as such a possible subject of research). Also, it pays to formulate the algorithm in more generality, since this version is capable of solving standard games over $\mathbb{R}$ upon a simple tweak that we will describe and justify after the algorithm. Let us first see how and why it works.

In fact, FP over $\preceq$ works exactly as usual, only having the $\leq$-order on $\mathbb{R}$ being replaced by the stochastic $\preceq$, and imposing a pointwise addition of distribution functions where the standard algorithm would only add payoff values. Note that this pointwise addition is crucial here, and perhaps somewhat counterintuitive, since we do not add random variables as usual by convolution. The pointwise addition is due to the sum occurring in the law of total probability (3.4).

Algorithm 1 Fictitious Play

Require: an $(n \times m)$-matrix $\mathbf{A}$ of payoff distributions $\mathbf{A} = (F_{ij})$
Ensure: an approximation $(\tilde{\mathbf{x}}, \tilde{\mathbf{y}})$ of an equilibrium pair $(\mathbf{x}^*, \mathbf{y}^*)$ and two distributions v_{low}, v_{up} so that $v_{low} \preceq F(\mathbf{x}^*, \mathbf{y}^*) \preceq v_{up}$. Here, $F(\mathbf{x}^*, \mathbf{y}^*)(r) = \Pr(R \leq r) = \sum_{i,j} F_{ij}(r) \cdot x_i^* \cdot y_j^*$.

1: initialize $\mathbf{x} \leftarrow \mathbf{0} \in \mathbb{R}^n$, and $\mathbf{y} \leftarrow \mathbf{0} \in \mathbb{R}^m$
2: $v_{low} \leftarrow$ the $\preceq$-minimum over all column-maxima
3: $r \leftarrow$the row index giving v_{low}
4: $v_{up} \leftarrow$ the $\preceq$-maximum over all row-minima
5: $c \leftarrow$ the column index giving v_{up}
6: $\mathbf{u} \leftarrow (F_{1,c}, \ldots, F_{n,c})$
7: $y_c \leftarrow y_c + 1$ ▷ $\mathbf{y} = (y_1, \ldots, y_m)$
8: $\mathbf{v} \leftarrow \mathbf{0}$ ▷ initialize $\mathbf{v}$ with m functions that are zero everywhere
9: **for** $k = 1, 2, \ldots$ **do**
10: $u^* \leftarrow$ the $\preceq$-minimum of $\mathbf{u}$
11: $r \leftarrow$ the index of u^* in $\mathbf{u}$
12: $v_{up} \leftarrow$ the $\preceq$-maximum of $\{u^*/k, v_{up}\}$ ▷ pointwise scaling of the distribution u^*
13: $\mathbf{v} \leftarrow \mathbf{v} + (F_{r,1}, \ldots, F_{r,m})$ ▷ pointwise addition of functions
14: $x_r \leftarrow x_r + 1$ ▷ $\mathbf{x} = (x_1, \ldots, x_r, \ldots, x_n)$
15: $v_* \leftarrow$ the $\preceq$-maximum of $\mathbf{v}$
16: $c \leftarrow$ the index of v_* in $\mathbf{v}$
17: $v_{low} \leftarrow$ the $\preceq$-minimum of $\{v_*/k, v_{low}\}$ ▷ pointwise scaling of the distribution $\mathbf{v}_*$
18: $\mathbf{u} \leftarrow \mathbf{u} + (F_{1,c}, \ldots, F_{n,c})$ ▷ pointwise addition of functions
19: $y_c \leftarrow y_c + 1$ ▷ $\mathbf{y} = (y_1, \ldots, y_c, \ldots, y_m)$
20: exit the loop upon convergence of the strategy vectors (in some norm)
21: **end for**
22: Normalize $\mathbf{x}, \mathbf{y}$ to unit total sum ▷ turn $\mathbf{x}, \mathbf{y}$ into probability distributions.
23: **return** $\tilde{\mathbf{x}} \leftarrow \mathbf{x}$ and $\tilde{\mathbf{y}} \leftarrow \mathbf{y}$ ▷ $F(\tilde{\mathbf{x}}, \tilde{\mathbf{y}}) \approx F(\mathbf{x}^*, \mathbf{y}^*) = (\mathbf{x}^*)^T \mathbf{A} \mathbf{y}^*$

How can Algorithm 1 be applied to a normal form game over the reals? Simply by conversion into a game with stochastic payoffs and the same equilibria. The trick is the following: let $\mathbf{A} = (a_{ij}) \in \mathbb{R}^{n \times m}$ be the (real-valued) payoff matrix, where we can assume $a_{ij} \geq 1$ w.l.o.g. Put $a^* := \max\{a_{ij}\} \geq 1$, and from a_{ij}, define a corresponding Bernoulli random variable $R_{ij} \sim F_{ij}$ with $\Pr(R_{ij} = 1) = \lambda \cdot a_{ij}$ and $\Pr(R_{ij} = 0) = 1 - \Pr(R_{ij} = 1)$. The factor $\lambda > 0$ is the same for all rows and columns. Why does this work? It has been shown in Chapter 2 that $\preceq$ on categorical distributions (and the Bernoulli distribution is one) is essentially a lexicographic order on the probability mass vector, starting from the highest (rightmost) category in descending order. This renders $\Pr(R_{ij} = 1)$ the relevant quantity to choose best actions and add up into a cumulative sum. Since this probability is proportional to a_{ij} by the same factor $\lambda > 0$ for all elements in the payoff structure, the resulting game, when decided upon $\lambda \cdot a_{ij}$, is strategically equivalent to the original game with payoff matrix A. Thus, it shares the same equilibria. For distributions with more categories, the payoffs are merely vectors, and using $\preceq$ as a lexicographic order is equal to playing FP on a "stack" of games. In the first place, the decision about a best reply is made on the matrix containing the probability masses for the highest loss categories. If the decision can be made (lines 12 and 17 in Algorithm 1), then we are done for this iteration. Upon a tie, the probability mass assigned to the second-highest category counts (in the lexicographic order), and the best reply is sought in this (new) matrix.

If there is a tie again, the next level (third highest matrix of category masses) is taken and so on. The process works just the same for continuous distributions, with the only difference of the stack being made for derivatives of increasing order, starting at the 0th derivative (see Lemma 2.2 in Chapter 2). Figure 3.3 illustrates the stack on which FP is done graphically for the case of continuous (in fact, differentiable) payoff densities f_{ij} in the game.

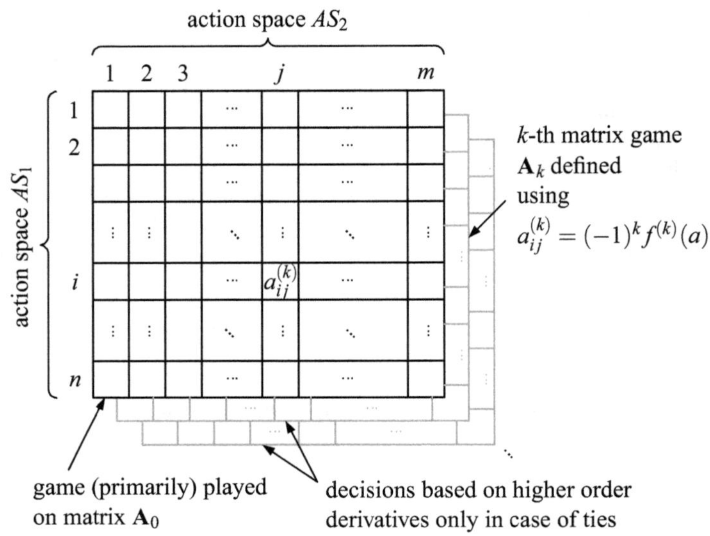

Fig. 3.3: Applying Fictitious Play over a Stochastic Order [18]

The depth of the stack is theoretically unlimited for continuous payoff distributions, thus the algorithm could get stuck within a single iteration during the decision of $\preceq$. In practical applications, we would thus have to fix a finite depth for the stack, and the concrete choice will be discussed later in Section 3.4.2.2.

Example 3.2 ([16]). We construct a 2×2 zero-sum game with payoff matrix $\mathbf{A}$ given in Figure 3.4 and a minimizing player 1.

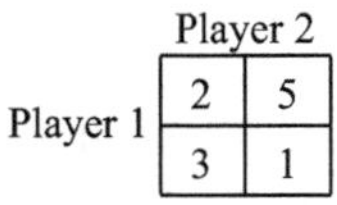

	Player 2	
Player 1	2	5
	3	1

Fig. 3.4: Example Zero-Sum Game; Payoff Structure $\mathbf{A}$

To use Algorithm 1, the respective payoff distributions representing the game would be (all Bernoulli) $F_{11} = \mathscr{B}er(0.8, 0.2), F_{12} = \mathscr{B}er(0.5, 0.5), F_{21} = \mathscr{B}er(0.7, 0.3)$, and $F_{22} = \mathscr{B}er(0.9, 0.1)$, with $\lambda = 1/10$.

Fictitious play starts from the (arbitrary) choice $\mathbf{x}_0 = (1,0)$ for the first (row) player. This choice causes player 2 to choose the second column in the first time step, to reward player 1 with $u_{1,t=1} = 0.5$. Given player two's history of choices being $\mathbf{y}_{t=1} = (0,1)$, player 1 goes for the second row and chooses $\mathbf{x}_{t=1} = (0,1)$. Player 2 updates its records to make its next choice as a best reply to the so-far observed mixed strategy $\mathbf{x} = (0.5,0.5)$. The switch between the two strategies is essentially due to the game having a circular structure.

It is a straightforward matter to compute the sequence of action choices according to (3.14), verifying them to converge to the equilibrium $\mathbf{x}^* = (0.4,0.6)$, $\mathbf{y}^* = (0.8,0.2)$, and $val(\mathbf{A}) = 2.6$.

To verify this as being a security strategy (for player 1), let us assume that player 2 has different incentives causing it to play $\mathbf{y}' = (0.4,0.6)$ or $\mathbf{y}'' = (0.1,0.9)$. For $\mathbf{y}'$, the payoff for player 1 is $\mathbf{x}^*\mathbf{A}\mathbf{y}' = 2.6$, and $\mathbf{y}''$ gives $\mathbf{x}^*\mathbf{A}\mathbf{y}'' = 2$, both of which are damages $\leq val(\mathbf{A}) = 2.6$. So, in these two cases (at least), player 1 receives no more than the assured maximal damage of $val(\mathbf{A}) = 2.6$. Furthermore, the example shows that worst-case strategies for player 1's opponent are not necessarily unique, and that the bound implied by them can be sharp (as is the case for $\mathbf{y}' \neq \mathbf{y}^*$).

3.4.2.1 Failure of FP in Distribution-Valued Zero-Sum Games

Let us consider what happens if we add uncertainty to the payoffs in Example 3.2. According to the initial discussion, this should cover most interesting cases of uncertainty in the game; however, some qualitative properties such as convergence of FP in zero-sum games are lost upon this transition. We show an example to shed light on the issue and its cause.

Example 3.3 ([16]). Concretely, let each payoff value be uncertain within a certain range, where we model a limited amount of uncertainty by an Epanechnikov kernel ($K(x) := \frac{3}{4}(1-x^2)$ for $|x| \leq 1$ and $K(x) := 0$ otherwise) centered around the respective value x_0. The resulting payoff structure in this game with probability-distribution valued is thus a 2×2 matrix of functions displayed in Figure 3.5.

Note that the game has a circular structure, so that the expected behavior of FP *should* roughly be the following: player 1 choosing the first row will make player 2 choose the second column. In turn, player 1 will go for the second row, which player 2 will reply to by choosing the first row, and so on.

The actual FP algorithm, however, runs elsewhere: let the start be made for player 2 by choosing the $\preceq$-maximum in each row, from which player 1 would select the $\preceq$-minimum. This gives F_{21} as an upper bound to the saddle-point value of this game. Likewise, player 1 will choose the $\preceq$-minimum of the $\preceq$-maxima per column, which gives F_{11} as a lower bound to the saddle-point value. Comparing those to the value 2.6 in Example 3.2, both appear plausible, since F_{11} is centered around 2 and F_{21} is centered around 3, with the value 2.6 lying in between. Moreover, since the upper and lower bounds do not coincide, an equilibrium must be in mixed strategies. Unfortunately, FP will not find it, because the iteration gets stuck at choosing $\mathbf{x}_t = (1,0)$ ultimately for all t, since the losses "accumulate" into $\sum_{j=1}^{k} F_{1y_j}$ for player 1, but we

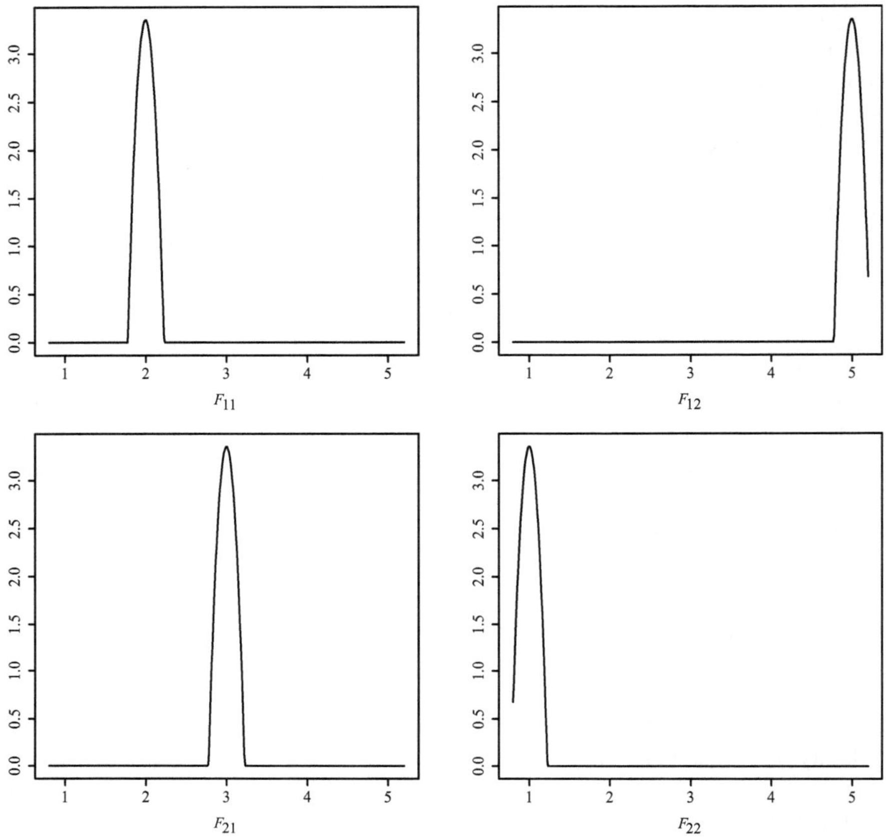

Fig. 3.5: Game from Example 3.2 with uncertainties in the payoffs

have the awkward inequality $F_{11} \preceq \frac{1}{k}u^*$ for all k despite $F_1 1$ and u^* remaining both constant, as Figure 3.6 illustrates! The relation never fails because the tails of the distribution $\frac{1}{k}u^*$ will retain a positive (though decreasing) mass no matter how large k gets; see Figure 3.6 for an illustration. That is, although the losses accumulate, this effect will never justify another choice of strategies, so FP becomes stationary at an incorrect result. Why so? One could think that by the transfer principle [22], the convergence of FP, being a proposition in first-order logic, would equivalently hold in the hyperreals. Indeed, FP *does* converge (as it does classically) by this argument, but for a sequence of *hyperreal* integers, rather than (regular) iterations toward infinity within $\mathbb{N}$. An inspection of the arguments in [23] reveals that the iteration count where convergence occurs is determined by the maximum element in the payoff matrix. Since our distributions are represented by infinite hyperreal numbers, convergence kicks in once the iteration count becomes infinite in the hyperreal sense, which clearly cannot happen in any practical implementation.

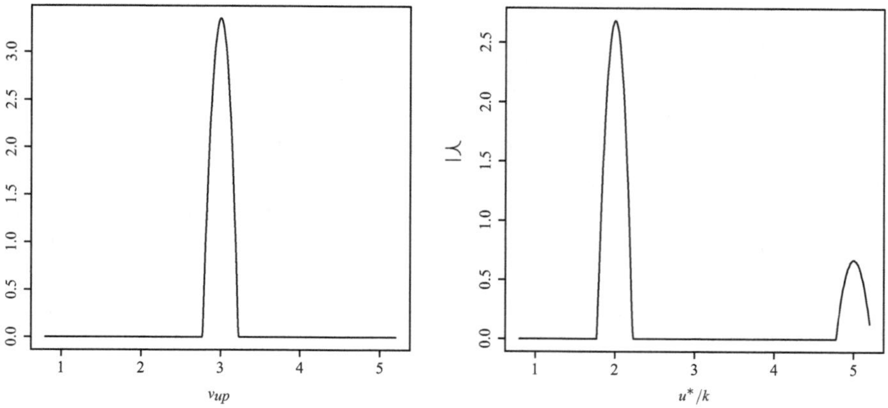

Fig. 3.6: Convergence Failure of FP (situation shown here after $k = 10$ iterations)

The problem outlined in Example 3.3 disappears for distributions with unbounded tails, or if all payoff distributions share the same support $\Omega = [a,b]$ with positive mass assigned in a left neighborhood of b.

3.4.2.2 Restoring Convergence of FP

For the sake of simplicity, let us resort to finite games with continuous payoff distributions, such as the one that caused FP to fail in Section 3.4.2.1. The convergence issue was due to the distribution's tail not reaching out until the point where the stochastic order is decided. Namely, if we consider losses on a bounded scale $[a,b] \subset \mathbb{R}$ (which is a mild and practically handy assumption), the vanishing of the mass located near the end b of the scale along iterations of FP will not be noticed in regions near a (cf. Figures 3.5 and 3.6). To avoid this unpleasant situation, the all relevant distributions must assign strictly positive mass to the entire range $[a,b]$ (so that no "gaps" are near the end or anywhere in the middle of the interval $[a,b]$).

The easiest way of achieving that is convolution by an approximate Dirac mass, say, a Gaussian distribution with small variance, and truncating the resulting density functions. In language of nonparametric statistics, this is nothing else but a standard kernel density estimation (for categorical distributions, a properly discretized Gaussian kernel also works well, but so do more sophisticated methods, e.g., [13, 5], either). In the continuous case, Gaussian kernels come particularly handy for the convolution (see Chapter 2 for the reason), so from here on, we will focus on how and why this also restores convergence of FP. The kernel function is thus $K(x) := \frac{1}{\sqrt{2\pi}} \exp\left(-\frac{1}{2}x^2\right)$, that is, a humble normal density with zero mean and unit variance.

Let us consider the case of two continuous distributions supported on a compact set $[a,b]$ first, and call them $\tilde{f}, \tilde{g}$. We allow both to vanish on entire intervals within the compact set $[a,b]$. Also, let $K_h : \mathbb{R} \to \mathbb{R}$ be a Gaussian density with variance h that we will use as a mollifier to put $f := \tilde{f} * K_h$ and $g := \tilde{g} * K_h$. It is well known that letting $h \to 0$ makes $f_h \to f$ and $g_h \to g$ in the L^1-norm, and since both are supported on a compact set, the convergence is even uniform. Moreover, since K_h is a C^∞-function, f and g have derivatives of all orders, so that we have $f \preceq g$ (being a shorthand for the relation $X \preceq Y$ between the random variables whose distribution densities are f and g), if and only if the derivatives are lexicographically ordered as $\mathbf{f} = ((-1)^k f^{(k)}(b))_{k\in\mathbb{N}} <_{lex} ((-1)^k g^{(k)}(b))_{k\in\mathbb{N}} = \mathbf{g}$. In the following, let us use the shorthand terms $f_k := (-1)^k f^{(k)}(b)$, and $g_k := (-1)^k g^{(k)}(b)$ to ease notation.

We approximate the infinite sequence by a Taylor polynomial $\hat{f}$ of degree N for f,

$$\hat{f}(x) = f(b) + \sum_{k=1}^{N} \frac{f^{(k)}(b)}{k!}(x-b)^k, \tag{3.15}$$

and do the same for the function g. The choice of the degree N will be discussed later. Let the resulting approximations be $\hat{f}$ and $\hat{g}$. Since there are only finitely many coefficients $f_k = \frac{1}{k!}f^{(k)}(b), g_k = \frac{1}{k!}g^{(k)}(b)$ for $k = 0,1,\ldots,N$ taken to represent the continuously differentiable densities f and g, we can find a (common) bound $M > 0$ so that $-M \leq f_i, g_i \leq M$ for all $k = 0,1,2,\ldots,M$. Shifting both by the same amount M puts the numbers $f_i + M, g_i + M$ into the interval $[0,2M]$ and leaves their relative ordering unchanged, so that we can consider them as being in excess representation. Fix a precision and round off all numbers $f_i + M, g_i + M$ up to ℓ bits, giving the approximate numbers $\hat{f}_i, \hat{g}_i$ with a roundoff error of $|f_i - \hat{f}_i|, |g_i - \hat{g}_i| < \varepsilon_M$ for all i. Using $\hat{f}_i, \hat{g}_i$ in the series representation (3.15) for $f^{(i)}(b)$ and $g^{(i)}(b)$, call the resulting approximation polynomials $\hat{f}_\varepsilon$ and $\hat{g}_\varepsilon$. The error from this roundoff is found from (3.15) to be

$$\max_{x\in[a,b]} \left|\hat{f}(x) - \hat{f}_\varepsilon(x)\right| \leq \varepsilon_M + \sum_{k=1}^{\infty} \frac{\varepsilon_M}{k!} b^k = \varepsilon_M \cdot e^b,$$

and the same for the error between $\hat{g}$ and $\hat{g}_\varepsilon$. Observe that ε_M can be made as small as we desire by using a larger bitsize ℓ in the numeric representation, so for any $\varepsilon > 0$ there is an ℓ resulting in a roundoff error ε_M so that $\varepsilon_M \cdot e^b < \varepsilon$. So $\hat{f}$ and $\hat{f}_\varepsilon$ can brought together arbitrarily close. Likewise, in choosing N sufficiently large, we can make the difference between f and $\hat{f}$ as small as we wish, so that the cumulative error by the Taylor polynomial and the roundoff errors can be kept under control.

For a number x, let us write $(x)_2$ to mean its binary representation. Using this notation, define the number $y_f := (\hat{f}_0\hat{f}_1\hat{f}_2\ldots\hat{f}_N)_2 \in \mathbb{R}$ and $y_g := (\hat{f}_0\hat{f}_1\hat{f}_2\ldots\hat{f}_N)_2 \in \mathbb{R}$ by a humble string concatenation of the binary excess representations of the (rounded) coefficients in the Taylor polynomials, assuming that they are all represented with the same number of bits. The resulting bitstring is then again interpreted as a real number. Clearly, the information in y_f and y_g can be chosen to represent f and g at arbitrary accuracy, but the numeric order between y_f and y_g is the same as the lexicographic order between $\mathbf{f}$ and $\mathbf{g}$. This in turn equals the $\preceq$-ordering of the original densities f and g.

Wrapping up, we have found real-valued representatives y_f, y_g for f and g so that $y_f \leq y_g$ "implies" $f \preceq g$, where the quotes are a reminder for the relation to be decided on proxima to the original densities.

The goodness of fit is here determined by the number N of coefficients necessary for an accurate approximation by the Taylor polynomial, and the number of bits ℓ in the excess representation (that controls M and hence ε_M). We can thus think of the so-obtained numbers to act as substitutes in games where payoffs are distribution-valued. In other words, we could convert a game with distribution-valued payoffs into a normal game with real-valued payoffs, at the cost of getting only an approximation of the original game, but at any precision that we desire. This equips us with further methods like linear programming (see Section 3.4.1) to solve these games too. Most importantly, in having transformed a game with distribution-valued payoffs into a regular one over $\mathbb{R}$, convergence of fictitious play now follows from standard arguments again [23].

Practically, the number N of required terms in the Taylor approximation, or the number ℓ of bits may become intractably large to be useful any more. Fortunately, however, there is no need to do either a roundoff, excess representation, or binary concatenation into real values, since we can equally well work with vector representations of the series. Then, we can work at machine precision and can compute the derivatives only on demand and up to the index where the decision can be made (exploiting the lexicographic order to be fixed at the time when the first index with a strict relation between f_i and g_i is obtained. Looking at Figure 3.3, we would thus dig only as deep into the stack as we need to make a choice but no deeper than N). Since we expanded the densities around the point b, in whose neighborhood $\preceq$ is determined, the approximation is expectedly accurate in the region relevant for $\preceq$, even for low orders N, though the Taylor polynomial $\hat{f}$ may badly deviate from the real density f when we get far from b. That is, the decision of $\preceq$ based on smoothing and on-demand computation of derivatives is in many cases quite efficient and accurate.

For discrete distributions, matters are considerably simpler, since the smoothing with a density whose support is the entire line $\mathbb{Z}$ of integers (say, by discretizing a Gaussian density to shift their mass from the continuous interval $[n, n+1)$ to the integer n), the support of the distribution extends until the (category/rank) b, and the lexicographic order kicks in again in replacement for $\preceq$. Like before, it is not difficult to assemble the masses together into a single number whose numeric order equals the $\preceq$ ordering, and all theory related to standard games reapplies.

Summing up our arguments (and framing them in more formal terms) leads the following result that relates distribution-valued games to standard (real-valued) games:

Theorem 3.3 (Approximation Theorem [18]). *Let $\Omega \subset [1, \infty)$ be a compact set (finite or continuous). Let $\Delta(\Omega)$ be the set of all distributions for which a density function exists (and is continuous if Ω is continuous). Then, for every $\varepsilon > 0, \delta > 0$ and every zero-sum matrix game $\Gamma_1 = \mathbf{A} \in (\Delta(\Omega))^{n \times m}$ with distribution-valued payoffs in the set, there is another zero-sum matrix game $\Gamma_2 = \mathbf{B} \in \mathbb{R}^{n \times m}$ so that an equilibrium in Γ_2 is an (ε, δ)-approximate equilibrium in Γ_1 in the following sense:*

- *The equilibrium* $(\tilde{\mathbf{x}}^*, \tilde{\mathbf{y}}^*)$ *in* Γ_1 *differs from the equilibrium* $(\mathbf{x}^*, \mathbf{y}^*)$ *in the matrix game represented by* $\mathbf{A}$ *by* $\|(\mathbf{x}^*, \mathbf{y}^*) - (\tilde{\mathbf{x}}^*, \tilde{\mathbf{y}}^*)\|_1 < \varepsilon$, *where the norm is on* $\mathbb{R}^{|AS_1|+|AS_2|}$,
- *The saddle point* $val(\mathbf{B}) = \tilde{F}^*$ *differs from the saddle point* $val(\mathbf{A}) = F^*$ *by* $\left\|\tilde{F}^* - F^*\right\|_{L^1} < \delta$.

3.4.3 FP for Multi-Goal Security Strategies

For MGSS, it has been shown in [25] that equilibria can be computed by FP for certain one-against-all games, in which a designated player "zero" faces opponents that are acting independently among themselves, but all against player zero. The security game of Definition 3.3 can be modified to fall into this class (cf. [20]).

For a two-player MOG $\boldsymbol{\Gamma}$, let $\boldsymbol{\Gamma}_0$ denote its associated security game. Toward enabling fictitious play in $\boldsymbol{\Gamma}_0$, we need to make it zero-sum. Remember that the defender in $\boldsymbol{\Gamma}$ has $d \geq 1$ goals to optimize, each corresponding to another distinct opponent in the security game $\boldsymbol{\Gamma}_0$. From these, we define the payoffs in a one-against-all *compound game*, while making the scalar payoffs vector-valued to achieve the zero-sum property. The payoff for player 0 is left unchanged, but the payoff for the i-th opponent is "vectorized" into

$$\bar{\mathbf{u}}_i = (0, 0, \ldots, 0, -u_1^{(i)}, 0, \ldots, 0), \tag{3.16}$$

without affecting any equilibria in the game (again, the bar accent on top of $\mathbf{u}$ is to mark this and other items with the same accent to belong to the security game $\boldsymbol{\Gamma}_0$).

To numerically compute an equilibrium in it according to the recipe of [10], we scalarize as follows: to each of player 0's d goals, we assign a weight $\alpha_{01}, \ldots, \alpha_{0d}$ according to (3.9). The scalarization in (3.10) is via

$$\alpha_{ji} := \alpha_{0i} \text{ for } i = 1, 2, \ldots, d \text{ and } j = 1, 2, \ldots, d.$$

With these weights, the payoffs in the scalarized compound game are

- for player 0: $f_0 = \alpha_{01}\bar{\mathbf{u}}_1 + \alpha_{02}\bar{\mathbf{u}}_2 + \cdots + \alpha_{0d}\bar{\mathbf{u}}_d$,
- for the i-th opponent, where $i = 1, 2, \ldots, d$

$$\begin{aligned} f_i &= \alpha_{01} \cdot 0 + \alpha_{02} \cdot 0 + \cdots + \alpha_{0,i-1} \cdot 0 + \alpha_{0i} \cdot (-u_1^{(i)}) + \alpha_{0,i+1} \cdot 0 + \alpha_{0d} \cdot 0 \\ &= -\alpha_{0i} \cdot u_1^{(i)} \end{aligned} \tag{3.17}$$

Concluding the transformation, we obtain a scalar compound game

$$\Gamma_{sc} = (\{0, 1, \ldots, d\}, \{AS_1, AS_2, \ldots, AS_2\}, \{f_0, \ldots, f_d\}) \tag{3.18}$$

from the original two-person MOG Γ with payoffs $u_1^{(1)}, \ldots, u_1^{(d)}$ that can directly be be plugged into expressions (3.16) and (3.17).

Toward a numerical computation of equilibria in Γ_{sc}, we need yet another transformation due to [25]: for the moment, let us consider a general compound game Γ_c as a collection of d two-person games $\Gamma_1, \ldots, \Gamma_d$, each of which is played independently between player 0 and one of its d opponents. With Γ_c, we associate a two-person game Γ_{cr} that we call the *reduced game*. The strategy sets and payoffs of player 0 in Γ_{cr} are the same as in Γ_c. Player 2's payoff in the reduced game is given as the *sum* of payoffs of all opponents of player 0 in the compound game. The following result links the convergence of FP in one-against-all games to convergence in their reduced forms.

Lemma 3.2 ([25]). *A fictitious play process approaches equilibrium in a compound game Γ_c if and only if it approaches equilibrium in its reduced game Γ_{cr}.*

For the reduced version Γ_{scr} of the (by (3.16) vectorized) scalarized security game Γ_{scr}, this sum is always zero. Since FP converges in such games [23], we get the final conclusion:

Theorem 3.4 ([20]). *Fictitious play in the scalarized compound game Γ_{sc} defined by* (3.18) *converges to an equilibrium.*

Any Nash equilibrium obtained in Γ_{sc} upon FP in Γ_{src} is by Theorem 3.1, a Pareto-Nash equilibrium in the security game $\mathbf{\Gamma}_0$ and as such a MGSS in the game that we started from. So, Theorem 3.4 induces the following algorithm to compute multi-criteria security strategies according to Definition 3.2:

Given a two-player MOG $\mathbf{\Gamma}$ with d payoffs $u_1^{(1)}, \ldots, u_1^{(d)}$ for player 1 (and possibly unknown payoffs for player 2), we obtain a MGSS along the following steps:

1. Assign strictly positive weights $\alpha_{01}, \ldots, \alpha_{0d}$ to each goal, and set up the scalarized compound game Γ_{sc} by virtue of expressions (3.18), (3.16), and (3.17). Observe that, as we can choose the weights arbitrarily, these give us a method to *prioritize* different goals.
2. Run the FP Algorithm 1 in Γ_{sc}, stopping when the desirable precision of the equilibrium approximation is reached.
3. The result vector $\mathbf{x}^*$ is directly the sought multi-criteria security strategy, whose assurances are given by the respective expected payoffs of the opponents. In case of matrix games, where the i-th payoff is given by a matrix $\mathbf{A}_i$, the sought assurances are $v_i = (\mathbf{x}^*)^T \mathbf{A}_i \mathbf{y}_i^*$ for $i = 1, 2, \ldots, d$, where $\mathbf{y}_1^*, \ldots, \mathbf{y}_d^*$ are the other player's equilibrium strategy approximations obtained along FP.

Example 3.4. For ease of presentation and an intuitive validation of the results, let us consider a 2×2 MOG with two goals. The payoff structures, shown in Figure 3.7, are composed from categorical (Bernoulli) distributions. For the example purpose, those cover three possible cases of games: 1) classical games with real-valued outcomes (via the aforementioned representation by Bernoulli random variables), 2) games with random payoffs that are converted into classical games by taking expectations, and 3) the general case of probability-distribution-valued games with categorical distributions compared according to $\preceq$.

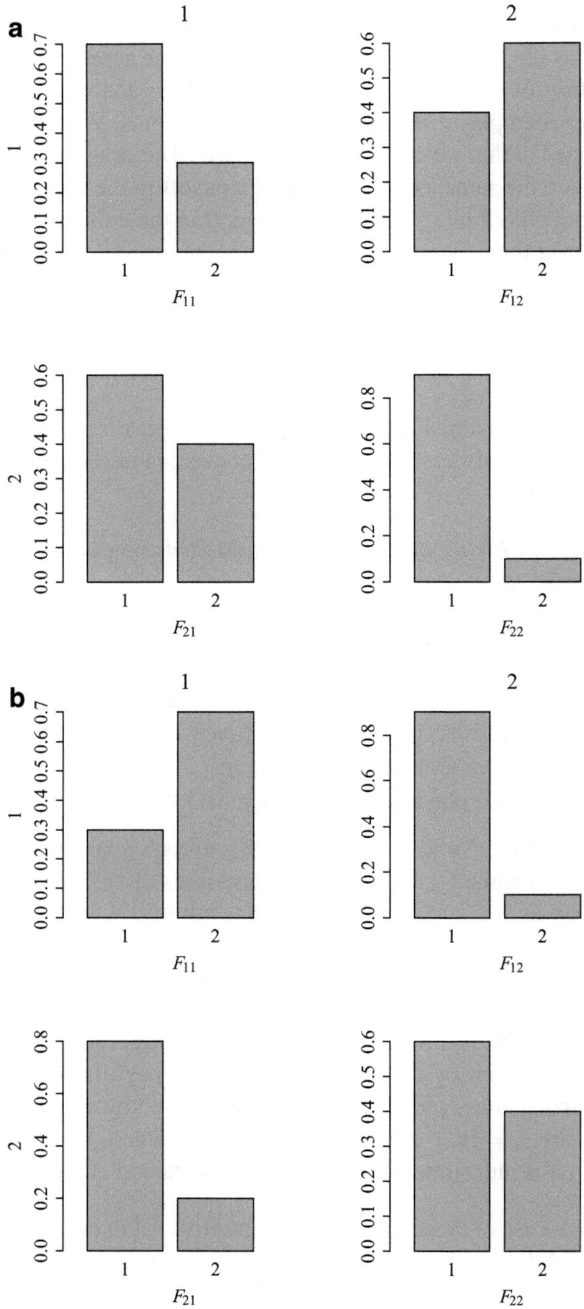

Fig. 3.7: Example Multi-Objective Distribution-Valued 2×2 Game

The following results (and the plots in Figures 3.4 and 3.7) have been obtained with R, version 3.4.4 [27], using the package HyRiM [17], which implements exactly the procedure outlined above with Algorithm 1 at the core. Running Algorithm 1 with equal importance on both goals (i.e., taking the weights $\alpha_{01} = \alpha_{02} = 1/2$) on these games digs up the (approximate) equilibrium $\mathbf{x}^* = (1/4, 3/4)$ and $\mathbf{y}_1^* = (1,0)$ for the first goal, and $\mathbf{y}_2^* = (1/2, 1/2)$ for the second goal. The mixed strategy $\mathbf{x}^*$ is herein the security strategy for player 1, being told the worst-case scenarios for each of his goals to be $\mathbf{y}_1^*$ and $\mathbf{y}_2^*$, respectively. Conditional on the defender playing $\mathbf{x}^*$, the *assurances* are the (Bernoulli) distributions $\mathbf{v}_1 = (0.625, 0.375) = \mathbf{v}_2$ for both goals.

The security strategy is not too sensitive to a change in the goal prioritization. For example, taking $\alpha_{01} = 0.9$ and $\alpha_{02} = 0.1$ to express high importance of the first goal (relative to the second) leaves the security strategy unchanged. Only, the worst-case scenario for the second goal changes into $\mathbf{y}_2^* \approx (0.39, 0.61)$, and its assurance $\mathbf{v}_2$ adapts itself accordingly.

The entire set of equilibria can be discovered by (theoretically) running through all values for the importance weights α_{0i} for $i = 1, 2, \ldots, d$ goals [10]. In a practical setting, one would thus be advised to try different goal priorities in order to find perhaps more plausible equilibria than upon the first try.

3.5 Final Remarks

The assurance offered by a security strategy against whatever behavior of the opponent within its action space is bought at the cost of this being a rather pessimistic approach. As with any minimax decision, this disregards all auxiliary information available to both players, which could improve the decision making. Bayesian decision theory starts from this observation and is developed around the idea of updating loss distributions with incoming data, so as to improve the decisions over time. The same trick, however, can be mounted in game theory, when the game's payoff structures become updated between repetitions. Technically, this makes the games dynamic, but not necessarily stochastic (at least not in the sense of [26]). For distribution-valued games, those can be updated in a Bayesian way, in order to improve the accuracy of the payoff structures. Still, this is not the same as using prior knowledge about the attacker's behavior. However, the same framework allows to integrate that knowledge into the payoff distributions by proper modeling. The details are beyond the scope of this chapter and fall into the domain of general adversary modeling. Hints on how to construct the payoff distributions for several practical cases, however, are subject of Part II of this book. Specifically, the data can be obtained from simulation (Chapters 8, 9, 10, 14, and 15), expert surveys, or other sources. Chapters 8, 14, 15, and 16 report on a practical use of the method, as it is implemented in R [17].

A final remark on security strategies relates to the cost of playing them. Imagine that the equilibrium is mixed and that it prescribes to frequently change configurations or even reset or revert a certain part of the system to some initial state. Frequent such actions may be undesirable and may lead to unreasonably high cost for the defense. Taking into account the cost of playing strategies besides their actual benefits is a matter of multi-objective game theory and can be handled in similar ways as described here. A rigorous treatment of this, however, is beyond the scope of this chapter, but has recently been done in the literature [19].

References

1. Acosta Ortega, F., Rafels, C.: Security Strategies and Equilibria in Multiobjective Matrix Games: Working Papers in Economics
2. Avis, D., Rosenberg, G., Savani, R., von Stengel, B.: Enumeration of nash equilibria for two-player games. Economic Theory (42), 9–37 (2010)
3. Camerer, C.F.: Behavioral game theory: Experiments in strategic interaction. The Roundtable Series in Behavioral Economics. Princeton University Press, s.l. (2011). URL http://gbv.eblib.com/patron/FullRecord.aspx?p=765287
4. Cesa-Bianchi, N., Lugosi, G.: Prediction, learning, and games. Cambridge University Press, Cambridge (2006). URL https://doi.org/10.1017/CBO9780511546921
5. Chesson, P., Lee, C.T.: Families of discrete kernels for modeling dispersal. Theoretical Population Biology pp. 241–256 (2005)
6. Fudenberg, D., Levine, D.K.: The Theory of Learning in Games. MIT Press, London (1998)
7. Fudenberg, D., Tirole, J.: Game Theory. MIT Press, London (1991)
8. Klíma, R., Kiekintveld, C., Lisý, V.: Online Learning Methods for Border Patrol Resource Allocation. In: R. Poovendran, W. Saad (eds.) Decision and Game Theory for Security, *Lecture Notes in Computer Science*, vol. 8840, pp. 340–349. Springer International Publishing, Cham (2014)
9. Lewis, F.L., Liu, D.: Reinforcement Learning and Approximate Dynamic Programming for Feedback Control. John Wiley & Sons, Inc, Hoboken, NJ, USA (2012)
10. Lozovanu, D., Solomon, D., Zelikovsky, A.: Multiobjective Games and Determining Pareto-Nash Equilibria. Buletinul Academiei de Stiinte a Republicii Moldova Matematica **3**(49), 115–122 (2005). ISSN 1024-7696
11. Nisan, N. (ed.): Algorithmic game theory, repr., [nachdr.] edn. Cambridge Univ. Press, Cambridge (2008). URL http://reference-tree.com/book/algorithmic-game-theory?utm_source=gbv&utm_medium=referral&utm_campaign=collaboration
12. Puterman, M.L.: Markov Decision Processes: Discrete Stochastic Dynamic Programming, *Wiley Series in Probability and Statistics*, vol. v.414. John Wiley & Sons Inc, Hoboken (2009). URL http://search.ebscohost.com/login.aspx?direct=true&scope=site&db=nlebk&db=nlabk&AN=294454

13. Rajagopalan, B., Lall, U.: A Kernel Estimator For Discrete Distributions **4**, 409–426 (1995). Gordon and Breach Science Publishers SA
14. Rass, S.: On Information-Theoretic Security: Contemporary Problems and Solutions. Ph.D. thesis, Klagenfurt University, Institute of Applied Informatics (01.01.2009)
15. Rass, S.: On Game-Theoretic Network Security Provisioning. Springer Journal of Network and Systems Management **21**(1), 47–64 (2013). https://doi.org/10.1007/s10922-012-9229-1. URL http://www.springerlink.com/openurl.asp?genre=article&id=doi:10.1007/s10922-012-9229-1
16. Rass, S.: On Game-Theoretic Risk Management (Part Two) – Algorithms to Compute Nash-Equilibria in Games with Distributions as Payoffs (2015). ArXiv:1511.08591
17. Rass, S., König, S.: R package 'HyRiM': Multicriteria risk management using zero-sum games with vector-valued payoffs that are probability distributions (2017). URL https://hyrim.net/software/
18. Rass, S., Konig, S., Schauer, S.: Defending Against Advanced Persistent Threats Using Game-Theory. PLoS ONE **12**(1), e0168,675 (2017). https://doi.org/10.1371/journal.pone.0168675. Journal Article
19. Rass, S., König, S., Schauer, S.: On the cost of game playing: How to control the expenses in mixed strategies. In: Proceedings of the 8th International Conference on Decision and Game Theory for Security (GameSec), LNCS 10575, pp. 494–505. Springer (2017)
20. Rass, S., Rainer, B.: Numerical Computation of Multi-Goal Security Strategies. In: R. Poovendran, W. Saad (eds.) Decision and Game Theory for Security, LNCS 8840, pp. 118–133. Springer (2014)
21. Robert, C.P.: The Bayesian choice. Springer, New York (2001)
22. Robinson, A.: Non-standard Analysis. Princeton Landmarks in Mathematics and Physics. Princeton University Press, Princeton (1996). URL http://gbv.eblib.com/patron/FullRecord.aspx?p=4626045. Luxemburg, W. A. J. (BeteiligteR)
23. Robinson, J.: An iterative method for solving a game. Annals of Mathematics **54**, 296–301 (1951)
24. Roughgarden, T.: Twenty lectures on algorithmic game theory. Cambridge University Press, Cambridge (2016). URL http://dx.doi.org/10.1017/CBO9781316779309
25. Sela, A.: Fictitious play in 'one-against-all' multi-player games. Economic Theory **14**(3), 635–651 (1999). URL http://dx.doi.org/10.1007/s001990050345
26. Shapley, L.S.: Stochastic Games. Proceedings of the National Academy of Sciences **39**(10), 1095–1100 (1953). https://doi.org/10.1073/pnas.39.10.1095. URL http://www.pnas.org/content/39/10/1095.short
27. Team, R.D.C.: R: A Language and Environment for Statistical Computing (2016). URL http://www.R-project.org. ISBN 3-900051-07-0

28. Wang, L., Jajodia, S., Singhal, A., Noel, S.: k-zero day safety: Measuring the security risk of networks against unknown attacks. In: D. Hutchison, T. Kanade, J. Kittler, J.M. Kleinberg, F. Mattern, J.C. Mitchell, M. Naor, O. Nierstrasz, C. Pandu Rangan, B. Steffen, M. Sudan, D. Terzopoulos, D. Tygar, M.Y. Vardi, G. Weikum, D. Gritzalis, B. Preneel, M. Theoharidou (eds.) Computer Security – ESORICS 2010, *Lecture Notes in Computer Science*, vol. 6345, pp. 573–587. Springer Berlin Heidelberg, Berlin, Heidelberg (2010). https://doi.org/10.1007/978-3-642-15497_35
29. White, D.J.: Markov decision processes. Wiley, Chichester (1993). URL http://www.loc.gov/catdir/description/wiley033/92001646.html

Chapter 4
A Scalable Decomposition Method for the Dynamic Defense of Cyber Networks

Mohammad Rasouli, Erik Miehling, and Demosthenis Teneketzis

4.1 Introduction

The defense of computer systems (cyber-security) plays a crucial role in their efficient/normal operation. One class of cyber-security problems concerns the security of networks of computers (cyber networks), which are typically very large. In this work, we investigate the development of defense policies for the security of cyber networks.

There are several approaches to addressing the cyber-security problem. These approaches can be categorized into static vs. dynamic and control theoretic vs. game-theoretic. The static approach considers a one-stage/single-period decision problem where the goal is to determine a defense policy. The dynamic approach considers a multi-period decision problem where the goal is to determine a feedback defense policy that takes into account the evolution of the system and the available information over time. Both the static and dynamic approaches can vary in the assumptions on the attacker's behavior (strategic vs. nonstrategic). Strategic behavior leads to a game, *e.g.*, [1, 15, 16], whereas nonstrategic behavior leads either to an optimization problem in the static case or a control problem in the dynamic case, *e.g.*, [14, 17, 19, 20]. In each of the above categories, there exist various assumptions on the problem's *information structure*, that is, the information each agent possesses at each time instant. The information structure can be symmetric (both agents possess the same information) or asymmetric (agents possess differing information).

In this chapter, we approach the security of cyber networks as a control problem from the defender's point of view. We model the attacker as nature. The security status of the cyber network evolves over time as a function of both the defender's and nature's actions. We assume that the defender does not possess perfect information

M. Rasouli · E. Miehling · D. Teneketzis (✉)
Department of Electrical Engineering and Computer Science, University of Michigan, Ann Arbor, MI, USA
e-mail: rasouli@umich.edu; miehling@umich.edu; teneket@umich.edu

S. Rass, S. Schauer (eds.), *Game Theory for Security and Risk Management*, Static & Dynamic Game Theory: Foundations & Applications,
https://doi.org/10.1007/978-3-319-75268-6_4

of the security status of the network at any given time. Due to the defender's lack of perfect information of the security status, we take a conservative approach to determining a defense policy. Specifically, we seek to minimize the worst-case damage that the attacker/nature can inflict on the cyber network. Therefore, we determine a defense policy as the solution of a minmax control problem with imperfect information. Due to the high dimensionality of the minmax control problem, we cannot solve it precisely. As a result, we develop a scalable approach to its solution, resulting in a suboptimal/approximate solution of the original problem. The approach is based on the concept of an influence graph (which quantifies the functional dependencies among the problem's variables) and uses a clustering algorithm to decompose the original, high-dimensional minmax control problem into a collection of lower-dimensional minmax problems that are computationally feasible.

Our approach captures the dynamic nature of attacks and the fact that the defender does not possess perfect knowledge of the security status of the network. Even though we do not model the attacker as a strategic agent, we compensate (in part) for the lack of this feature by adopting a minmax performance criterion, which leads to a conservative defense approach. This conservative approach differs from the game-theoretic approaches of Chapters 1, 3, 5, 6, 7, 10, 11, and 15.

Our work is distinctly different from the existing literature. From a security perspective, our work falls within the category of *intrusion response systems* (IRSs), where there is a rich literature (see [10, 11, 21] and references therein). The goal of an IRS is to take in security information from the environment and translate it into a defense action, with the goal of interfering with an attacker's objective(s). To the best of the authors' knowledge, our work is the first to investigate the design of an IRS from a minmax control perspective. From a control theory point of view, our model and problem are similar to those of [2, 3, 4, 5, 6, 7, 8, 9, 23, 24], in particular, [3, 5, 7, 8]. The key difference between our model and those of [3, 5, 7, 8] is in the timing of events and the nature of the information gathered by observations (see Figure 4.1) which allows us to capture essential features of cyber-security problems.

The objective of this chapter is to provide a heuristic approach for solving large-scale minmax control problems. This approach is used to determine conservative defense policies in cyber-security domains.

4.1.1 Organization of the Chapter

The chapter is organized as follows. In Section 4.2, we introduce the security model. In Section 4.3, we formulate the defense problem, define the notion of an *information state* for the problem, and describe a sequential decomposition procedure for the problem's solution. In Section 4.4, we describe our approximation approach to the defense problem. This includes defining the influence graph and the process for constructing the local defense problems. We also discuss the scalability of our approach. In Section 4.5, we present an example illustrating some of the concepts used in our approach. In Section 4.6, we discuss our results and provide some concluding remarks.

4.1.2 Notation

The table below (presented for later reference) describes the notation used throughout the chapter.

$\mathscr{X} = \mathscr{X}^1 \times \mathscr{X}^2 \times \cdots \times \mathscr{X}^n$	State space of the problem
$\mathscr{N} = \{1,2,\ldots,n\}$	Set of state elements
$\mathscr{X}^i = \{x^{i,1}, x^{i,2}, \ldots, x^{i,n_x^i}\}$	State space of element $i \in \mathscr{N}$
$\mathscr{T} = \{0,1,\ldots,T\}$	Time horizon of length T
$\mathscr{W} = \{w^1, w^2, \ldots, w^{n_w}\}$	Set of nature's events
$\mathscr{W}(x)$	Set of nature's events admissible from state $x \in \mathscr{X}$
$\mathscr{U} = \mathscr{U}^1 \times \mathscr{U}^2 \times \cdots \times \mathscr{U}^n$	Action-space
$\mathscr{U}^i = \{u^{i,1}, u^{i,2}, \ldots, u^{i,n_u^i}\}$	Action-space of element $i \in \mathscr{N}$
$\mathscr{Y} = \mathscr{Y}^1 \times \mathscr{Y}^2 \times \cdots \times \mathscr{Y}^n$	Set of event observations
$\mathscr{Y}^i = \{y^{i,1}, y^{i,2}, \ldots, y^{i,n_y^i}\}$	Set of event observations for element i
$\mathscr{Z} = \mathscr{Z}^1 \times \mathscr{Z}^2 \times \cdots \times \mathscr{Z}^n$	Set of action observations
$\mathscr{Z}^i = \{z^{i,1}, z^{i,2}, \ldots, z^{i,n_z^i}\}$	Set of action observations for element i
$\mathscr{K} = \{1,2,\ldots,n_k\}$	Set of local defense problems
$\mathscr{N}_k$	State indices of local defense problem k's internal state space
$\mathscr{L}_k$	State indices of local defense problem k's local state space
$\bar{\mathscr{N}}_k$	Exogenous state indices for local defense problem k
$\mathscr{X}^{\mathscr{N}_k}$	Internal state space of local defense problem k
$\mathscr{X}^{\mathscr{L}_k}$	Local state space of local defense problem k
$\mathscr{X}^{\bar{\mathscr{N}}_k}$	Exogenous state space of local defense problem k
$\mathscr{U}^{\mathscr{N}_k}$	Internal action space of local defense problem k
$\mathscr{Y}^{\mathscr{N}_k}$	Internal event observation space of local defense problem k
$\mathscr{Z}^{\mathscr{N}_k}$	Internal action observation space of local defense problem k
m_t^{kl}	Message sent from local defense problem k to l at t
$m_t^{\bar{\mathscr{N}}_k}$	Aggregate message of local defense problem k
$h_t = \{x_0, u_{0:t-1}, z_{0:t-1}, y_{0:t}\}$	Realized history at time t
$h_t^k = \{x_0^{\mathscr{N}_k}, u_{0:t-1}^{\mathscr{N}_k}, z_{0:t-1}^{\mathscr{N}_k}, y_{0:t}^{\mathscr{N}_k}, m_{0:t}^{\bar{\mathscr{N}}_k}\}$	Realized history for local defense problem k at time t
$\mathscr{H}_t$	Space of histories for defender at time t
$\mathscr{H}_t^k$	Space of histories for local defense problem k at time t
$\mathscr{R}$	Space of information states
$\mathscr{B}^{\mathscr{L}_k}$	Space of approximate information states for local defense problem k
ψ	Information state update function
ϕ^k	Approximate information state update of local defense problem k
Γ	Space of admissible defense policies
γ	Element of Γ, admissible defense policy

Γ^k	Space of admissible defense policies for local defense problem k
γ^k	Element of Γ^k, admissible defense policy for local defense problem k
Γ'^k	Space of admissible approximate defense policies for local defense problem k
γ'^k	Element of Γ'^k, admissible approximate defense policy for local defense problem k
$c(x,u)$	State-action cost
$c^i(x^i,u^i)$	State-action cost of element i
$x_{t:t+s}$	The sequence $x_t, x_{t+1}, \dots, x_{t+s}$
$\mathscr{P}(\mathscr{A})$	The powerset of set $\mathscr{A}$
$v^{\mathscr{A}}$	The collection of elements v_i, $i \in \mathscr{A}$, from vector v
$v^{-i}, v^{-(i,j)}$	All elements of v excluding element i, resp. excluding elements i and j

We denote variables by upper-case letters and their realizations by their corresponding lower-case letter, *e.g.*, x is a realization of variable X.

4.2 The Security Model

Consider a system consisting of n elements operating in discrete time. Let $\mathscr{N} := \{1,2,...,n\}$ denote the set of the system's elements. Consider a discrete, finite state space $\mathscr{X} := \mathscr{X}^1 \times \mathscr{X}^2 \cdots \times \mathscr{X}^n$, where $\mathscr{X}^i := \{x^{i,1}, x^{i,2}, \dots, x^{i,n_x^i}\}$ represents the (finite) state space of element $i \in \mathscr{N}$. Let T denote the time horizon over which we consider the system's operation; T may be finite or infinite. Define $\mathscr{T} = \{0,1,\dots,T\}$. The state of the system at any given time t is given by

$$x_t = (x_t^1, x_t^2, \dots, x_t^n) \in \mathscr{X}. \tag{4.1}$$

There are two agents: a controller (the defender) and nature (the adversary). The agents interact according to the following timing diagram, shown in Figure 4.1.

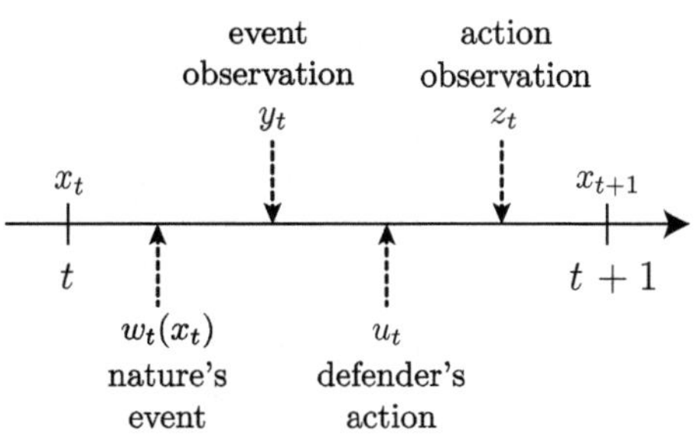

Fig. 4.1: Timeline of events for a given time-step.

The system state, x_t, evolves due to both the defender's decisions (control actions) and nature's events. For a given time-step (following the notation used in Figure 4.1), nature first generates an event, $w_t(x_t)$; nature's set of feasible events at any time t depends on the system's state x_t, hence the notation $w_t(x_t)$. The set of all events due to nature is denoted by $\mathscr{W} := \{w^1, w^2, ..., w^{n_w}\}$. The set of events that are admissible from state x_t is denoted by $\mathscr{W}(x_t) \subseteq \mathscr{W}$. To simplify the notation in the rest of the chapter, we use w_t instead of $w_t(x_t)$. The defender is not able to perfectly observe the event w_t, but instead receives an observation y_t, termed an *event observation*,generated according to the function $\theta : \mathscr{X} \times \mathscr{W} \to \mathscr{Y} = \mathscr{Y}^1 \times \mathscr{Y}^2 \times \cdots \times \mathscr{Y}^n$

$$\begin{aligned} y_t &= \theta(x_t, w_t) \\ &= \big(\theta^1(x_t^1, w_t), \ldots, \theta^n(x_t^n, w_t)\big) \end{aligned} \tag{4.2}$$

where $\theta^i : \mathscr{X}^i \times \mathscr{W} \to \mathscr{Y}^i$, $\mathscr{Y}^i = \{y^{i,1}, y^{i,2}, \ldots, y^{i,n_y^i}\}$, and n_y^i is the total number of possible observations for y_t^i. The defender then takes a defense action

$$u_t \in \mathscr{U} = \mathscr{U}^1 \times \mathscr{U}^2 \times \cdots \times \mathscr{U}^n \tag{4.3}$$

based on its current information, where it is assumed (for simplicity) that the action space decomposes over the elements of the state space. Each space $\mathscr{U}^i$ consists of a finite set of defense alternatives $\mathscr{U}^i = \{u^{i,1}, u^{i,2}, \ldots, u^{i,n_u^i}\}$. We assume that, for a given element i, each action $u^{i,j} \in \mathscr{U}^i$, $j \in \{1, 2, \ldots, n_u^i\}$, only has an effect on the state x^i of element i. The defender incurs a state-dependent cost for each defense action u_t, denoted by

$$c(x_t, u_t) = \sum_{i \in \mathscr{N}} c^i(x_t^i, u_t^i), \tag{4.4}$$

We assume that $|c(x,u)| \leq c^{\max}$ for all $x \in \mathscr{X}$ and $u \in \mathscr{U}$. This cost is incurred immediately after the defense action is selected. The defender then receives an observation z_t, termed an *action observation*, following the defense action, that is generated by the function $\zeta : \mathscr{X} \times \mathscr{W} \times \mathscr{U} \to \mathscr{Z} = \mathscr{Z}^1 \times \mathscr{Z}^2 \times \cdots \times \mathscr{Z}^n$ as

$$\begin{aligned} z_t &= \zeta(x_t, w_t, u_t) \\ &= \big(\zeta^1(x_t^1, w_t, u_t^1), \ldots, \zeta^n(x_t^n, w_t, u_t^n)\big) \end{aligned} \tag{4.5}$$

where $\zeta^i : \mathscr{X}^i \times \mathscr{W} \times \mathscr{U}^i \to \mathscr{Z}^i$, $\mathscr{Z}^i = \{z^{i,1}, z^{i,2}, \ldots, z^{i,n_z^i}\}$, and n_z^i is the total number of possible observations for z_t^i. The action observation provides additional information about the system's state (see [19]). The defense action, $u_t = (u_t^1, \ldots, u_t^n)$, causes the system to transition to the system state x_{t+1} according to the state update function $\pi : \mathscr{X} \times \mathscr{W} \times \mathscr{U} \to \mathscr{X}$, that is,

$$\begin{aligned} x_{t+1} &= \pi(x_t, w_t, u_t) \\ &= \big(\pi^1(x_t, w_t, u_t^1), \ldots, \pi^n(x_t, w_t, u_t^n)\big). \end{aligned} \tag{4.6}$$

where each $\pi^i : \mathscr{X} \times \mathscr{W} \times \mathscr{U}^i \to \mathscr{X}^i$ is the update equation state element i. Note that at any time t, each π^i depends on the global system state x_t, not only on x_t^i, because, as we pointed out above, the set of nature's feasible events at t depends on x_t.

4.3 The Defense Problem

The optimal defense action at any given time-step is dictated by an optimal defense policy. The defense policy at time t, denoted by γ_t, is a function of the defender's information available at time t. This information, termed the *history* and denoted by h_t, consists of the initial state x_0, all previous control actions, $u_0, \ldots, u_{t-1}$ (denoted compactly by $u_{0:t-1}$), and all observations $y_{0:t}$, and $z_{0:t-1}$. Formally, $h_t = \{x_0, u_{0:t-1}, z_{0:t-1}, y_{0:t}\}$, where the initial state x_0 is known by the defender. A defense policy, $\gamma = (\gamma_1, \gamma_2, \ldots, \gamma_T)$, maps the available information at any time t to a defense action u_t, that is, $\gamma_t : \mathscr{H}_t \to \mathscr{U}$, where $\mathscr{H}_t$ is the information space of the controller at time t (the space of histories up to time t). The space of admissible defense policies is given by $\Gamma = \{\gamma = (\gamma_1, \gamma_2, \ldots, \gamma_T) \mid \gamma_t : \mathscr{H}_t \to \mathscr{U} \text{ for all } t \in \mathscr{T}\}$.

An optimal defense policy is a policy γ that solves the following partially observable minmax control problem (P).

$$\min_{\gamma \in \Gamma} \max_{\{X_{\mathscr{T}} \in \mathscr{X}_{\mathscr{T}}^{\gamma}\}} \left\{ \sum_{t \in \mathscr{T}} \beta^t c(X_t, U_t) \,\middle|\, X_0 = x_0 \right\} \tag{P}$$

$$\text{subject to } X_{t+1} = \pi(X_t, W_t, U_t) \tag{P-i}$$

$$Y_t = \theta(X_t, W_t) \tag{P-ii}$$

$$Z_t = \zeta(X_t, W_t, U_t) \tag{P-iii}$$

$$U_t = \gamma_t(H_t) \tag{P-iv}$$

$$H_t = \{x_0, U_{0:t-1}, Z_{0:t-1}, Y_{0:t}\} \tag{P-v}$$

for all $t \in \mathscr{T}$, where β is the discount factor, $0 < \beta < 1$. The set $\mathscr{X}_{\mathscr{T}}^{\gamma}$ denotes the space of all sequences of system states (trajectory) generated by a given defense policy γ. The maximization is taken over all families of state trajectories generated by the defense policy γ, denoted by $\{X_{\mathscr{T}} = \{X_1, X_2, \ldots, X_T\} \in \mathscr{X}_{\mathscr{T}}^{\gamma}\}$. We consider all such families of trajectories (each one associated with a defense policy) and choose the policy that minimizes the highest-cost (worst-case) trajectory among all families.

The remainder of this section is devoted to determining more compact descriptions of the defender's information. Such descriptions may either permit the computation of an optimal defense policy or provide guidelines/insights for computationally tractable approximations of Problem (P).

4.3.1 Information State

In order to prescribe an optimal defense action at time t for Problem (P), we need to determine an *information state* sufficient for performance evaluation. One such information state is the history h_t. Unfortunately, due to the unbounded growth of the domain of h_t (see Figure 4.2), the computation of a defense policy based on h_t is intractable for infinite-time horizon problems (and large finite horizons). This motivates the search for a more compact (albeit still sufficient) summary of the current information (the need for finding a compact information state is even more critical in modern dynamic security environments where the rate of events is high, causing h_t to grow rapidly in size).

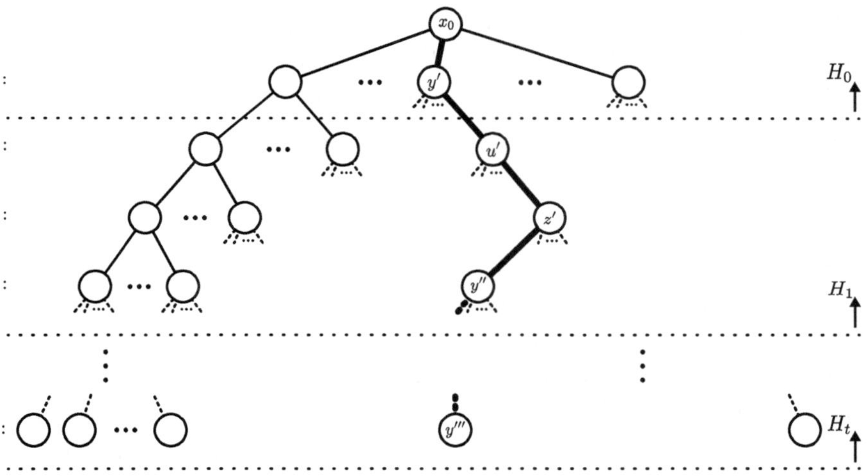

Fig. 4.2: The information, H_t, consisting of the initial state x_0, observations $Y_{0:t}$, $Z_{0:t-1}$, and previous defense actions $U_{0:t-1}$ grows rapidly as a function of t, as can be seen by the tree structure above. The number of information realizations at time t is equal to the number of leaf nodes in the tree at depth t. A *realized information trajectory*, $h_t = (x_0, y', u', z', y'', \ldots, y''')$, is a path from the root of the tree, x_0, to a leaf, in this case y''', as shown by the bolded line.

An alternate information state for Problem (P) can be defined. This alternate information state offers insight into the source of computational difficulty associated with the defender's problem and forms the basis for later approximations that bring the problem into the realm of computational tractability.

We now define this information state, denoted by R_t, in the context of the defense problem and describe its update equations. Consider a realization $h_t := \{x_0, u_{0:t-1}, z_{0:t-1}, y_{0:t}\}$ of the defender's information at time t. Denote by $\{\mathscr{I}_t^1, \ldots, \mathscr{I}_t^{\hat{l}(h_t)}\}$ to be the set of distinct event-action trajectories from 0 to t that are compatible with x_0 and h_t (see Remark 1); $\mathscr{I}_t^i = \{x_0, u_{0:t-1}, {}^i w_{0:t}\}$,

$i = 1,2,\ldots,\hat{l}(h_t)$. Each trajectory $\mathscr{J}_t^i$ leads to a system state ${}^i\hat{x}_t := ({}^i\hat{x}_t^1,\ldots,{}^i\hat{x}_t^{i,n}), i = 1,2,\ldots,\hat{l}(h_t)$. Denote by ${}^i\hat{\kappa}_t$ the cost of reaching ${}^i\hat{x}_t$; ${}^i\hat{\kappa}_t := \sum_{\tau=0}^{t-1}\beta^\tau c({}^i\hat{x}_\tau,u_\tau)$, where ${}^i\hat{x}_\tau$ is the system state at τ due to $\mathscr{J}_t^i$. Define $\hat{R}_t(h_t) := \{({}^1x_t,{}^1w_t,{}^1\hat{\kappa}_t),\ldots,({}^{\hat{l}(h_t)}x_t,{}^{\hat{l}(h_t)}w_t,{}^{\hat{l}(h_t)}\hat{\kappa}_t)\}$. Apply the following *reduction process* to $\hat{R}(h_t)$. If $\hat{R}(h_t)$ contains components $({}^{j_1}x_t,{}^{j_1}w_t,{}^{j_1}\kappa_t),\ldots,({}^{j_q}x_t,{}^{j_q}w_t,{}^{j_q}\kappa_t)$ such that $({}^{j_1}x_t,{}^{j_1}w_t) = \ldots = ({}^{j_q}x_t,{}^{j_q}w_t)$ and ${}^{j_p}\hat{\kappa}_t = \max_{j\in\{j_0,j_1,\ldots,j_q\}}\{{}^j\kappa_t\}$, then omit $({}^{j_i}x_t,{}^{j_i}w_t,{}^{j_i}\kappa_t)$, $j_i \neq j_p$, from $\hat{R}_t(h_t)$. This reduction process results in

$$R_t(h_t) = \{({}^1x_t,{}^1w_t,{}^1\kappa_t),\ldots,({}^{l(h_t)}x_t,{}^{l(h_t)}w_t,{}^{l(h_t)}\kappa_t)\}$$

where $({}^1x_t,{}^1w_t),\ldots,({}^{l(h_t)}x_t,{}^{l(h_t)}w_t)$ are distinct and $R_t(h_t)$ is an alternative information state at t along h_t for Problem (P). From the construction of $R_t(h_t)$, we conclude that for all history realizations h_t and for all t, an information state R_t has the form $\{({}^1x_t,{}^1w_t,{}^1\kappa_t),\ldots,({}^ax_t,{}^aw_t,{}^a\kappa_t)\}$, where $\{({}^1x_t,{}^1w_t),\ldots,({}^ax_t,{}^aw_t)\} \in \mathscr{P}(\mathscr{X}\times\mathscr{W}) = \mathscr{P}(\mathscr{X}^1\times\mathscr{X}^2\times\ldots\times\mathscr{X}^n\times\mathscr{W})$, ${}^i\kappa_t \in [0,\frac{c^{\max}}{1-\beta}]$, for all i, for all t, and $\mathscr{P}(\cdot)$ denotes the power set. We denote by $\mathscr{R}$ the space of information states at any time t, specifically

$$\mathscr{R} = \mathscr{P}(\mathscr{X}\times\mathscr{W})\times\left[0,\frac{c^{\max}}{1-\beta}\right]. \tag{4.7}$$

We now describe how to obtain R_{t+1} from R_t and the new information $H_{t:t+1}$ that becomes available to the defender at time $t+1$. Let h_t be a realization of H_t and $h_{t:t+1} = h_{t+1}\backslash h_t = \{u_t,z_t,y_{t+1}\}$ be a realization of $H_{t:t+1}$. Denote by $\{\mathscr{J}_{t:t+1}^1,\ldots,\mathscr{J}_{t:t+1}^{p(h_{t:t+1})}\}$ the set of distinct action-event trajectories from t to $t+1$ that are compatible with $R(h_t)$ and $h_{t:t+1}$; $\mathscr{J}_{t:t+1}^i := \{u_t,{}^iw_{t+1}\}$, $i = 1,2,\ldots,p(h_{t:t+1})$. Using $R(h_t)$, $\{\mathscr{J}_{t:t+1}^1,\cdots\mathscr{J}_{t:t+1}^{p(h_{t:t+1})}\}$, and the system dynamics, Equation (4.6), we can construct $\hat{R}(h_{t+1}) := \{({}^r\hat{x}_{t+1},{}^r\hat{w}_{t+1},{}^r\hat{\kappa}_{t+1}), r = 1,2,\ldots,a\}$ (see Remark 2) where ${}^r\hat{x}_{t+1} = \pi(({}^rx_t,{}^rw_t,u_t)$, $(u_t,{}^rw_{t+1}) \in \{\mathscr{J}_{t:t+1}^1,\cdots\mathscr{J}_{t:t+1}^{p(h_{t:t+1})}\}$, ${}^rw_{t+1} \in \mathscr{W}({}^r\hat{x}_{t+1})$, ${}^r\hat{\kappa}_{t+1} = {}^r\kappa_t + \beta^t c({}^rx_t,u_t)$, and ${}^r\kappa_t$ is the cost associated with $({}^rx_t,{}^rw_t)$, where $({}^rx_t,{}^rw_t,{}^r\kappa_t) \in R(h_t)$. Apply the above-described reduction process to $\hat{R}_{t+1}(h_{t+1})$ to obtain the alternative information state

$$R_{t+1}(h_{t+1}) = \{({}^1x_{t+1},{}^1w_{t+1},{}^1\kappa_{t+1}),\ldots,({}^{l(h_{t+1})}x_{t+1},{}^{l(h_{t+1})}w_{t+1},{}^{l(h_{t+1})}\kappa_{t+1})\}$$

for Problem (P) at time $t+1$ along h_{t+1}. The recursive update process described above can be summarized by an update equation

$$R_{t+1}(h_{t+1}) = \psi(R_t(h_t),h_{t:t+1}) = \psi(R_t(h_t),u_t,z_t,y_{t+1}). \tag{4.8}$$

The information state described above summarizes the information h_t available to the defender at time t by including all system states at t that are compatible with h_t, the maximum cost that is incurred in order to reach each of these states, and the events in nature that follow each of the possible states at t; these events must be feasible conditioned on the state ${}^rx_t$, that is, ${}^rw_t$ must be in the set $W({}^rx_t)$.

Remark 1. Compatibility implies that the following requirements are satisfied; each event-action trajectory $\{u_{0:t-1}, {}^i w_{0:t}\}$ is consistent with x_0, the observations $y_{0:t}$ and $z_{0:t-1}$ (through θ and ζ, respectively), the system dynamics described by Equation (4.6), and for every $\tau \leq t$, $w_\tau \in \mathscr{W}(x_\tau)$, where x_τ is the system state at τ reached via x_0, $u_{0:\tau-1}$, $w_{0:\tau-1}$.

Remark 2. The number of components a of $\hat{R}(h_{t+1})$ depends on $R(h_t)$, $h_{t:t+1}$, the system dynamics, and the sets $\mathscr{W}(x)$, $x \in \mathscr{X}$.

4.3.2 Sequential Decomposition and Dynamic Programming

We discuss a sequential decomposition for Problem (P) using dynamic programming. To specify the dynamic program for the finite horizon problem, denote by r_t the information state at time $t \leq T$. Define by $V_t(r_t)$, $r_t \in R_t$, the minmax value of Problem (P) from time t on when the information state at t is r_t. Then, when $r_T = (({}^1x, {}^1w, {}^1\kappa), ({}^2x, {}^2w, {}^2\kappa), \ldots, ({}^{l(h_T)}x, {}^{l(h_T)}w, {}^{l(h_T)}\kappa))$,

$$V_T(r_T) = \max_{j \in \{1,2,\ldots,l(h_T)\}} {}^j\kappa \tag{4.9}$$

For $t = 0, 1, \ldots, T-1$,

$$\begin{aligned} V_t(r_t) &= \min_{u_t \in \mathscr{U}} \left[\max_{(x_t, w_t, \kappa_t) \in r_t} V_{t+1}(r_{t+1}) \right] \\ &= \min_{u_t \in \mathscr{U}} \left[\max_{(x_t, w_t, \kappa_t) \in r_t} \left[\max_{w_{t+1} \in \mathscr{W}(\pi(x_t, w_t, u_t))} \right.\right. \\ &\qquad \left.\left. V_{t+1}\Big(\psi_t\big(r_t, u_t, \zeta(x_t, w_t, u_t), \theta(\pi(x_t, w_t, u_t), w_{t+1})\big)\Big)\right]\right]. \end{aligned} \tag{4.10}$$

Equations (4.9) and (4.10) define the dynamic program for the finite (T) horizon Problem (P).

To specify the dynamic program for the infinite horizon problem, we let $r \in \mathscr{R}$ denote the current information state and $V(r)$ denote the minmax value of the infinite horizon Problem (P). Then,

$$V(r) = \min_{u \in \mathscr{U}} \left[\max_{(x, w, \kappa) \in r} \left[\max_{w' \in \mathscr{W}(\pi(x, w, u))} V\Big(\psi\big(r, u, \zeta(x, w, u), \theta(\pi(x, w, u), w')\big)\Big)\right]\right] \tag{4.11}$$

Because of the high dimensionality (see Equation (4.7)), the solution of the finite and infinite-time horizon dynamic programs is computationally intractable. For this reason, in the next section, we provide a scalable approach for the solution of Problem (P).

4.4 Approximation to the Defense Problem

Even though the alternate information state described in Section 4.3.1 above leads to an intractable problem even for small systems (cf. Equation (4.7)), it forms the basis for a scalable approach to the solution of Problem (P). The approach consists of two key steps: (i) using the concept of an *influence graph*, we analyze the functional dependencies among state elements and split elements with weak dependencies, decomposing the original problem (P) into a number of local defense problems (P_k); (ii) we further approximate the solution of each local defense problem in order to permit computation of (suboptimal) local defense policies. We discuss the computational complexity of each of the local defense problems and comment on how to use the features of our approach so as to end up with problems that are compatible with the defender's computational capabilities.

4.4.1 Local Defense Problems

4.4.1.1 Preliminaries

In order to define the local defense problems, denoted by (P_k), we first introduce some necessary notation and describe, at a high level, how the local defense problems interact with one another. Then, in the remaining subsections, we describe in detail how the local defense problems are formed.

Consider a collection of n_k local defense problems, denoted by the set $\mathscr{K} = \{1,2,\dots,n_k\}$. Each local defense problem $k \in \mathscr{K}$ has an associated set of states termed the *internal state space* of problem k, denoted by $\mathscr{X}^{\mathscr{N}_k} \subseteq \mathscr{X}$, where $\mathscr{N}_k \subseteq \mathscr{N}$ is the set of internal state indices for problem k. By construction (described later), $\mathscr{X}^{\mathscr{N}_k}$, $k \in \mathscr{K}$ form a partition of the original state space $\mathscr{X}$, that is, $\mathscr{X}^{\mathscr{N}_i} \cap \mathscr{X}^{\mathscr{N}_j} = \emptyset$ for $i \neq j$ and $\cup_{k\in\mathscr{K}} \mathscr{X}^{\mathscr{N}_k} = \mathscr{X}$. Under this partition, the action and observation spaces for each local defense problem $k \in \mathscr{K}$ are denoted by $\mathscr{U}^{\mathscr{N}_k}$ and $\mathscr{Y}^{\mathscr{N}_k}$, $\mathscr{Z}^{\mathscr{N}_k}$, respectively.

Each local defense problem (P_k) is associated with a local defense policy γ^k. In computing the local defense policies, we assume that the local defense problems can exchange information over time via messages. We denote by m_t^{kl} the message local defense problem k receives from local defense problem l at time t. The message-exchange process occurs immediately before each local defense action is taken, as shown in the timing diagram of Figure 4.3.

The local defense action, $u_t^{\mathscr{N}_k} \in \mathscr{U}^{\mathscr{N}_k}$, causes the internal states of local defense problem k, denoted by $x_t^{\mathscr{N}_k} = \{x_t^j \mid j \in \mathscr{N}_k\}$, to transition to $x_{t+1}^{\mathscr{N}_k}$ according to the state update function $\pi^{\mathscr{N}_k} : \mathscr{X} \times \mathscr{W} \times \mathscr{U}^{\mathscr{N}_k} \to \mathscr{X}^{\mathscr{N}_k}$. The function $\pi^{\mathscr{N}_k}$ is simply defined as the collection of functions π^j, $j \in \mathscr{N}_k$, as described in Equation (4.6). Note that the dynamics of the internal states of each local defense problem (P_k) depend, in general, on the state of the overall system. This means that, without exploiting

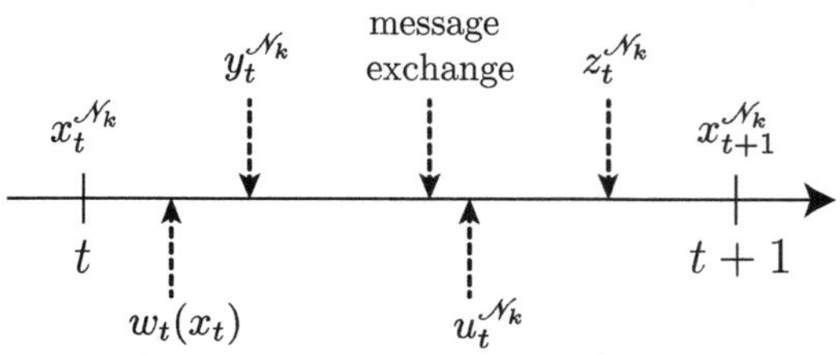

Fig. 4.3: Timeline of events for the local defense problem (P_k) for a given time-step.

any additional structure of the problem, the state space of each local defense problem is just as large as that of the original problem (P). To address this, we define an influence graph (in Section 4.4.1.2) in order to quantify the functional dependencies among state elements. Using the structure of the influence graph, each local defense problem can restrict attention to the state elements that directly influence the evolution of its internal states. The remainder of Section 4.4 is devoted to describing how the local defense problems are formed.

4.4.1.2 Functional Dependencies and the Notion of an Influence Graph

To form the local defense problems, we first analyze the functional dependencies among the state elements $i \in \mathcal{N}$. To this end, we provide the following definition.

Definition 1 (Functional Dependency). State element i is said to have a *functional dependency* on state element j, if there exists an action $u^i \in \mathscr{U}^i$, an event $w \in \mathscr{W}$, and two states $x = (x^1,\dots,x^{j-1},x^j,x^{j+1},\dots,x^n)$, $\hat{x} = (x^1,\dots,x^{j-1},\hat{x}^j,x^{j+1},\dots,x^n)$ differing only in element j, $x^j \neq \hat{x}^j$, such that

$$\pi^i(x,w,u^i) \neq \pi^i(\hat{x},w,u^i),$$

where π^i is the state update function of element i, given by Equation (4.6).

In other words, state element i is said to be functionally dependent on state element j if a change in the state element j influences the update for state element i for some action $u^i \in \mathscr{U}^i$ and some event $w \in \mathscr{W}$. The relationships expressed by Definition 1 can be summarized by a graph, termed the *influence graph*, defined below.

Definition 2 (Influence Graph). The *influence graph*, $\mathscr{G} = (\mathscr{V},\mathscr{E},\xi)$, is a weighted directed graph that consists of nodes $\mathscr{V} = \mathcal{N}$, edge set $\mathscr{E}$, and edge weights ξ. The edge set $\mathscr{E}$ contains edge $e = (i,j)$ if and only if state element j is functionally dependent on state element i, as described by Definition 1. Each edge $e = (i,j) \in \mathscr{E}$ in the influence graph is assigned a weight $\xi_e \in (0,1]$, computed as

$$\xi_e = \frac{1}{d} \sum_{\substack{x^{-(i,j)} \in \mathscr{X}^{-(i,j)} \\ x^i \in \mathscr{X}^i \\ x^j \in \mathscr{X}^j \\ w \in \mathscr{W} \\ u^i \in \mathscr{U}^i \\ \hat{x}^j \in \mathscr{X}^j \setminus \{x^j\}}} \mathbf{1}\big(\{\pi^i(x,w,u^i) \neq \pi^i(\hat{x},w,u^i)\}\big) \tag{4.12}$$

where $\mathbf{1}(A)$ is the indicator function of event A and d is the normalization term; d is the number of all possible terms in the summation of Equation (4.12), and therefore, its value is equal to $d = (\prod_{l \neq i,j} n_x^l) n_x^i n_x^j n_w n_u^i (n_x^j - 1)$.

The influence graph summarizes the dependencies among all state elements for all possible defense actions and events. A directed edge $e = (i, j) \in \mathscr{E}$ exists from element (node) i to element j if there exists an assignment of values to states, events, and actions such that state element j functionally depends on state element i. The weight of edge $(i, j) \in \mathscr{E}$ quantifies the strength of the functional dependency of element j on element i.

4.4.1.3 Formulating the Local Defense Problems

We form local defense problems by partitioning the influence graph into *clusters* of nodes (state elements). The clustering algorithm should place state elements with strong dependencies into a single cluster. Since the weights of the edges in the influence graph quantify the strength of the functional dependencies between state elements, application of the (normalized) min-cut algorithm [12] splits elements with weak functional dependencies, achieving the desired goal. A single application of the (normalized) min-cut algorithm decomposes a graph into two separate graphs using a minimum weight cut. Continued application of the min-cut algorithm results in increasingly finer partitioning of nodes, resulting in clusters with fewer nodes and eventually leading to a partitioning of the state space that permits computation of local defense policies (this procedure is described in detail in the example of Section 4.5). For each cluster of elements, we formulate a local defense problem (P_k) and aim to solve for the corresponding local defense policy.

In order to formulate each local defense problem (P_k), $k \in \mathscr{K}$, we must first describe its associated state space and dynamics. Using the influence graph, one can analyze the functional dependencies between state elements. Specifically, given a set of internal state indices $\mathscr{N}_k$, one can determine the state elements outside the set of internal states, $i \in \mathscr{N} \setminus \mathscr{N}_k$, that influence the evolution of states within $\mathscr{N}_k$ (as described by Definition 1). Formally, this set of elements is defined as

$$\bar{\mathscr{N}_k} = \{i \in \mathscr{N} \setminus \mathscr{N}_k \mid (i, j) \in \mathscr{E}, j \in \mathscr{N}_k\}$$

where $\mathscr{E}$ is the influence graph's set of edges. From this, one can define the set of *local* state element indices as the state elements in $\mathscr{N}_k$ combined with the state elements that directly influence the evolution of any element in $\mathscr{N}_k$, that is,

$$\mathcal{L}_k = \mathcal{N}_k \cup \bar{\mathcal{N}}_k.$$

The state space corresponding to indices $\mathcal{L}_k$ is written as $\mathcal{X}^{\mathcal{L}_k} \subseteq \mathcal{X}$ and is termed the *local state space* of problem k. Since, by construction, there are no edges from any elements in $\mathcal{N} \backslash \mathcal{L}_k$ to elements in $\mathcal{N}_k$, the update of the internal states $x^{\mathcal{N}_k}$ is independent of the states in $\mathcal{X}^{\mathcal{N} \backslash \mathcal{L}_k}$. Therefore, with some abuse of notation, the state update function for the internal states can equivalently be written using the function $\pi^{\mathcal{N}_k} : \mathcal{X}^{\mathcal{L}_k} \times \mathcal{W} \times \mathcal{U}^{\mathcal{N}_k} \to \mathcal{X}^{\mathcal{N}_k}$ as

$$x_{t+1}^{\mathcal{N}_k} = \pi^{\mathcal{N}_k}(x_t^{\mathcal{L}_k}, w_t, u_t^{\mathcal{N}_k}). \tag{4.13}$$

It is important to note that the local state-space $\mathcal{X}^{\mathcal{L}_k}$ of each local defense problem $k \in \mathcal{K}$ is an approximation to the state space that would be necessary for computing an optimal local defense policy. Although the states $x^{\mathcal{N}_k}$ can be updated precisely using the local states $x^{\mathcal{L}_k}$, as evidenced by Equation (4.13), computation of an optimal local defense policy requires knowledge of how $x^{\mathcal{L}_k}$ evolves over time. Due to the computational limitations of the defender, one is unable to take into account the state dynamics associated with the state elements outside its internal state space, $\mathcal{X}^{\mathcal{N}_k}$, for the purpose of computing a local defense policy. As a result, we approximate the state dynamics by introducing a message-exchange procedure in which *neighboring* defense problems communicate relevant elements of their internal states.

The messages that local defense problems exchange consist of the possible states that are consistent with each local defense problem's history (more on this in Section 4.4.2). Before formally defining the messages, define the set

$$\bar{\mathcal{N}}_{kl} = \bar{\mathcal{N}}_k \cap \mathcal{N}_l.$$

The set $\bar{\mathcal{N}}_{kl}$ represents the set of state indices within $\mathcal{N}_l$ that can influence the evolution of a state element in $\mathcal{N}_k$. Note that $\bar{\mathcal{N}}_{kl}$ is only nonempty if there is an edge $(i, j) \in \mathcal{E}$ in the influence graph such that $i \in \mathcal{N}_l$ and $j \in \mathcal{N}_k$. Also, note that $\{\bar{\mathcal{N}}_{kl}, l \in \mathcal{K}\}$ forms a partition of $\bar{\mathcal{N}}_k$. The message that local defense problem k receives from local defense problem l at time t, m_t^{kl}, lives within the set of all possible states of $x^{\bar{\mathcal{N}}_{kl}}$, that is, $m_t^{kl} \in \mathcal{P}(\mathcal{X}^{\bar{\mathcal{N}}_{kl}})$. Local defense problem l constructs this message as the set of possible states that are consistent with its local information (built in part using its imperfect internal observations).

In this sense, local defense problem k receives a summary of local defense problem l's local information that is relevant for taking an internal defense action $u^{\mathcal{N}_k} = \{u^i \mid i \in \mathcal{N}_k\} \in \mathcal{U}^{\mathcal{N}_k} = \prod_{i \in \mathcal{N}_k} \mathcal{U}^i$, permitting local defense problem k to compute a (suboptimal) local defense policy γ^k. The complete set of messages that local defense problem k receives from neighboring local defense problems are combined to form an *aggregate message*, denoted by

$$m_t^{\bar{\mathcal{N}}_k} \in \mathcal{P}\left(\prod_{l \in \mathcal{K}} \mathcal{X}^{\bar{\mathcal{N}}_{kl}}\right) = \mathcal{P}(\mathcal{X}^{\bar{\mathcal{N}}_k}).$$

The aggregate message allows local defense problem k to update its exogenous state elements in a way that is consistent with the exchanged information, that is, $x_{t+1}^{\bar{\mathcal{N}}_k} \in m_{t+1}^{\bar{\mathcal{N}}_k}$. Combined with the state update of its internal states, given by Equation (4.13), the defender is able to (approximately) model the evolution of its local states $x^{\mathcal{L}_k}$.

For each local defense problem $k \in \mathcal{K}$, an optimal local defense action at any given time-step is dictated by an optimal local defense policy. The local defense policy at time t, denoted by γ_t^k, prescribes an action based on its available information at time t. This information, given by $h_t^k = \{x_0^{\mathcal{N}_k}, u_{0:t-1}^{\mathcal{N}_k}, z_{0:t-1}^{\mathcal{N}_k}, y_{0:t}^{\mathcal{N}_k}, m_{0:t}^{\bar{\mathcal{N}}_k}\}$, consists of the initial value of the internal states $x_0^{\mathcal{N}_k}$, all internal defense actions $u_{0:t-1}^{\mathcal{N}_k}$ up to and including $t-1$, all internal action observations $z_{0:t-1}^{\mathcal{N}_k} = \{z_{0:t-1}^i \mid i \in \mathcal{N}_k\}$ up to and including $t-1$, all internal event observations $y_{0:t}^{\mathcal{N}_k} = \{y_{0:t}^i \mid i \in \mathcal{N}_k\}$ up to and including t, and all received messages up to and including t, summarized by the aggregate messages $m_{0:t}^{\bar{\mathcal{N}}_k}$. A local defense policy, $\gamma^k = (\gamma_1^k, \gamma_2^k, \ldots, \gamma_T^k)$, maps the available information at any time t, h_t^k, to a defense action $u_t^{\mathcal{N}_k}$, that is, $\gamma_t^k : \mathcal{H}_t^k \to \mathcal{U}^{\mathcal{N}_k}$, where $\mathcal{H}_t^k$ is the information space of the local defense problem at time t. The space of admissible local defense policies for local defense problem k is given by $\Gamma^k = \{\gamma^k = (\gamma_1^k, \gamma_2^k, \ldots, \gamma_T^k) \mid \gamma_t^k : \mathcal{H}_t^k \to \mathcal{U}^{\mathcal{N}_k} \text{ for all } t \in \mathcal{T}\}$. The optimal local defense policy for local defense problem k, $k \in \mathcal{K}$, is a policy γ^k that solves the following partially observable minimax control problem (P$_k$).

$$\min_{\gamma^k \in \Gamma^k} \max_{\left\{X_{\mathcal{T}}^{\mathcal{L}_k} \in \mathcal{X}_{\mathcal{T}}^{\mathcal{L}_k, \gamma^k}\right\}} \left\{ \sum_{t \in \mathcal{T}} \beta^t c^{\mathcal{N}_k}(X_t^{\mathcal{N}_k}, U_t^{\mathcal{N}_k}) \,\middle|\, X_0^{\mathcal{L}_k} = x_0^{\mathcal{L}_k} \right\} \tag{P$_k$}$$

$$\text{subject to } X_{t+1}^{\mathcal{N}_k} = \pi^{\mathcal{N}_k}(X_t^{\mathcal{L}_k}, W_t, U_t^{\mathcal{N}_k}) \tag{P$_k$-i}$$

$$X_{t+1}^{\bar{\mathcal{N}}_k} \in M_{t+1}^{\bar{\mathcal{N}}_k} \tag{P$_k$-ii}$$

$$Y_t^{\mathcal{N}_k} = \theta^{\mathcal{N}_k}(X_t^{\mathcal{N}_k}, W_t) \tag{P$_k$-iii}$$

$$Z_t^{\mathcal{N}_k} = \zeta^{\mathcal{N}_k}(X_t^{\mathcal{N}_k}, W_t, U_t^{\mathcal{N}_k}) \tag{P$_k$-iv}$$

$$U_t^{\mathcal{N}_k} = \gamma_t^k(H_t^k) \tag{P$_k$-v}$$

for all $t \in \mathcal{T}$, where the local states are defined as collection of internal and exogenous states, $X_t^{\mathcal{L}_k} = (X_t^{\mathcal{N}_k}, X_t^{\bar{\mathcal{N}}_k})$. The functions $\theta^{\mathcal{N}_k}$, $\zeta^{\mathcal{N}_k}$ are defined as the collection of event and action observation functions θ^i, ζ^i, $i \in \mathcal{N}_k$, as defined in Equations (4.2) and (4.5), respectively. The function $c^{\mathcal{N}_k}(x^{\mathcal{N}_k}, u^{\mathcal{N}_k})$ represents the state-action cost of local defense problem k and is defined as the sum of the internal state-action cost functions, that is, $c^{\mathcal{N}_k}(x^{\mathcal{N}_k}, u^{\mathcal{N}_k}) = \sum_{j \in \mathcal{N}_k} c^j(x^j, u^j)$. The set $\mathcal{X}_{\mathcal{T}}^{\mathcal{L}_k, \gamma^k}$ denotes the space of all sequences of local system states under local defense policy γ^k. As described earlier for Problem (P), the optimal policy is the policy that minimizes the worst-case cost over all state trajectories.

Each local defense problem (P$_k$), defined above, is of the same form as Problem (P). Consequently, the information state for the internal state space $\mathcal{X}^{\mathcal{N}_k}$ is of the same form as the one described in Section 4.3.1; that is, the space of information states for Problem (P_k) is $\mathscr{R}_k = \mathscr{P}(\mathscr{X}^{\mathscr{N}_k} \times \mathscr{W}) \times [0, \frac{c^{\max}_{\mathscr{N}_k}}{1-\beta}]$ where $c^{\max}_{\mathscr{N}_k} = \max_{x \in \mathscr{X}^{\mathscr{N}_k}, u \in \mathscr{U}^{\mathscr{N}_k}} c(x,u)$. Such an information space precludes computation of an optimal local defense policy. As a result, we approximate the information state of each local defense problem (P_k) so as to end up with a modified problem that is computationally tractable. This is the topic of the following subsection.

4.4.2 Approximating the Local Defense Problems

We use the information state described in Section 4.4.1 to form an approximate information state for each local defense problem k. For a given realization of the local history $h^k_t \in \mathscr{H}^k_t$, we consider only the *set of states* that are compatible with h^k_t (we omit the set of nature's events and the maximum cost associated with each compatible state). That is, each local defense problem constructs the set of all possible local states, $x^{\mathscr{L}_k} = (x^{\mathscr{N}_k}, x^{\bar{\mathscr{N}}_k})$, consistent with the history of internal defense actions and observations and messages that it has received from neighboring local defense problems. We denote this approximate information state at time t by $b^{\mathscr{L}_k}_t \in \mathscr{B}^{\mathscr{L}_k} = \mathscr{P}(\mathscr{X}^{\mathscr{L}_k})$, where $\mathscr{B}^{\mathscr{L}_k}$ is the space of approximate information states for local defense problem k. Using the new information from time t to $t+1$, given by $h^k_{t:t+1} = \{u^{\mathscr{N}_k}_t, z^{\mathscr{N}_k}_t, y^{\mathscr{N}_k}_{t+1}, m^{\bar{\mathscr{N}}_k}_{t+1}\}$, local defense problem k can update its approximate information state as follows. We note that $b^{\mathscr{L}_k}_t$ has the form $b^{\mathscr{L}_k}_t = \{({}^1x^{\mathscr{N}_k}_t, {}^1x^{\bar{\mathscr{N}}_k}_t), \dots, ({}^lx^{\mathscr{N}_k}_t, {}^lx^{\bar{\mathscr{N}}_k}_t)\}$. We use $({}^1x^{\mathscr{N}_k}_t, \dots, {}^lx^{\mathscr{N}_k}_t)$, $u^{\mathscr{N}_k}_t$, $z^{\mathscr{N}_k}_t$, $y^{\mathscr{N}_k}_{t+1}$ to determine $({}^1x^{\mathscr{N}_k}_{t+1}, \dots, {}^rx^{\mathscr{N}_k}_{t+1})$ according to the update process ψ described in Section 4.3.1. We combine $({}^1x^{\mathscr{N}_k}_{t+1}, \dots, {}^rx^{\mathscr{N}_k}_{t+1})$ with the message $m^{\bar{\mathscr{N}}_k}_{t+1} = ({}^1x^{\bar{\mathscr{N}}_k}_{t+1}, \dots, {}^qx^{\bar{\mathscr{N}}_k}_{t+1})$ to form $b^{\mathscr{L}_k}_{t+1} = \{({}^ix^{\mathscr{N}_k}_{t+1}, {}^jx^{\bar{\mathscr{N}}_k}_{t+1}), i = 1,2,\dots,r;\ j = 1,2,\dots,q\}$. Thus

$$\begin{aligned} b^{\mathscr{L}_k}_{t+1} &= \phi^k(b^{\mathscr{L}_k}_t, h^k_{t:t+1}) \\ &= \phi^k(b^{\mathscr{L}_k}_t, u^{\mathscr{N}_k}_t, z^{\mathscr{N}_k}_t, y^{\mathscr{N}_k}_{t+1}, m^{\bar{\mathscr{N}}_k}_{t+1}). \end{aligned}$$

With this new approximate information state, each Problem (P_k) is approximated by the following minmax control problem (P'_k).

$$\min_{\gamma^k \in \Gamma'} \max_{\{X^{\mathscr{L}_k}_{\mathscr{T}} \in \mathscr{X}^{\mathscr{L}_k,\gamma^k}_{\mathscr{T}}\}} \left\{ \sum_{t \in \mathscr{T}} \beta^t c^{\mathscr{N}_k}(X^{\mathscr{N}_k}_t, U^{\mathscr{N}_k}_t) \,\middle|\, X^{\mathscr{L}_k}_0 = x^{\mathscr{L}_k}_0 \right\} \qquad (\mathrm{P}'_k)$$

$$\text{subject to } X^{\mathscr{N}_k}_{t+1} = \pi^{\mathscr{N}_k}(X^{\mathscr{L}_k}_t, W_t, U^{\mathscr{N}_k}_t) \qquad (\mathrm{P}'_k\text{-i})$$

$$X^{\bar{\mathscr{N}}_k}_{t+1} \in M^{\bar{\mathscr{N}}_k}_{t+1} \qquad (\mathrm{P}'_k\text{-ii})$$

$$Y^{\mathscr{N}_k}_t = \theta^{\mathscr{N}_k}(X^{\mathscr{N}_k}_t, W_t) \qquad (\mathrm{P}'_k\text{-iii})$$

$$Z_t^{\mathscr{N}_k} = \zeta^{\mathscr{N}_k}(X_t^{\mathscr{N}_k}, W_t, U_t^{\mathscr{N}_k}) \tag{P$'_k$-iv}$$

$$U_t^{\mathscr{N}_k} = \gamma_t^k(B_t^{\mathscr{L}_k}) \tag{P$'_k$-v}$$

for all $t \in \mathscr{T}$, where the local states are given by $X_{t+1}^{\mathscr{L}_k} = (X_{t+1}^{\mathscr{N}_k}, X_{t+1}^{\bar{\mathscr{N}}_k})$, and $\Gamma'^k = \{\gamma^k = (\gamma_1^k, \gamma_2^k, \dots, \gamma_T^k) \,|\, \gamma_t^k : \mathscr{B}^{\mathscr{L}_k} \to \mathscr{U}^{\mathscr{N}_k} \text{ for all } t \in \mathscr{T}\}$ represents the set of admissible approximate local defense policies γ^k.

For finite horizon T, we solve Problem (P$'_k$) backward in time via the following set of recursive equations. Let $b_t^{\mathscr{L}_k} \in \mathscr{B}^{\mathscr{L}_k}$ be the approximate information state at t and $V_t^k(b_t^{\mathscr{L}_k})$ denote the minmax value of Problem (P$'_k$) from time t on when the approximate information state at t is $b_t^{\mathscr{L}_k}$, $t = 0, 1, \dots, T+1$. Then, for each $b_{T+1}^{\mathscr{L}_k} \in \mathscr{B}^{\mathscr{L}_k}$,

$$V_{T+1}^k(b_{T+1}^{\mathscr{L}_k}) = 0, \tag{4.14}$$

and for $t = 1, 2, \dots, T$, and each $b_t^{\mathscr{L}_k} \in \mathscr{B}^{\mathscr{L}_k}$,

$$V_t^k(b_t^{\mathscr{L}_k}) = \min_{u_t^{\mathscr{N}_k} \in \mathscr{U}^{\mathscr{N}_k}} \Bigg[\max_{x_t^{\mathscr{L}_k} = (x_t^{\mathscr{N}_k}, x_t^{\bar{\mathscr{N}}_k}) \in b_t^{\mathscr{L}_k}} \Bigg[c^{\mathscr{N}_k}(x_t^{\mathscr{N}_k}, u_t^{\mathscr{N}_k}) +$$
$$\beta \max_{\substack{w_t \in \mathscr{W}(x_t^{\mathscr{L}_k}) \\ m_{t+1}^{\bar{\mathscr{N}}_k} \in \mathscr{P}(\mathscr{X}^{\bar{\mathscr{N}}_k})}} \Bigg[\max_{x_{t+1}^{\bar{\mathscr{N}}_k} \in m_{t+1}^{\bar{\mathscr{N}}_k}} \Bigg[\max_{w_{t+1} \in \mathscr{W}\left(\pi^{\mathscr{N}_k}(x_t^{\mathscr{L}_k}, w_t, u_t^{\mathscr{N}_k}), x_{t+1}^{\bar{\mathscr{N}}_k}\right)}$$
$$V_{t+1}^k\Big(\phi\big(b_t^{\mathscr{L}_k}, u_t^{\mathscr{N}_k}, \zeta^{\mathscr{N}_k}(x_t^{\mathscr{N}_k}, w_t, u_t^{\mathscr{N}_k}), \theta^{\mathscr{N}_k}(\pi^{\mathscr{N}_k}(x_t^{\mathscr{L}_k}, w_t, u_t^{\mathscr{N}_k}, w_{t+1}), m_{t+1}^{\bar{\mathscr{N}}_k})\big)\Big)\Bigg]\Bigg]\Bigg]\Bigg] \tag{4.15}$$

where $\mathscr{W}(x_t^{\mathscr{L}_k})$ is defined as the set of events possible from any state x such that the elements $\mathscr{L}_k$ of the state are equal to $x_t^{\mathscr{L}_k}$.

For the infinite horizon case, we solve Problem (P$'_k$) via the set of equations

$$V^k(b^{\mathscr{L}_k}) = \min_{u^{\mathscr{N}_k} \in \mathscr{U}^{\mathscr{N}_k}} \Bigg[\max_{x^{\mathscr{L}_k} = (x^{\mathscr{N}_k}, x^{\bar{\mathscr{N}}_k}) \in b^{\mathscr{L}_k}} \Bigg[c^{\mathscr{N}_k}(x^{\mathscr{N}_k}, u^{\mathscr{N}_k}) +$$
$$\beta \max_{\substack{w \in \mathscr{W}(x^{\mathscr{L}_k}) \\ m^{\bar{\mathscr{N}}_k} \in \mathscr{P}(\mathscr{X}^{\bar{\mathscr{N}}_k})}} \Bigg[\max_{x^{\bar{\mathscr{N}}_k} \in m^{\bar{\mathscr{N}}_k}} \Bigg[\max_{w' \in \mathscr{W}\left(\pi^{\mathscr{N}_k}(x^{\mathscr{L}_k}, w, u^{\mathscr{N}_k}), x^{\bar{\mathscr{N}}_k}\right)}$$
$$V^k\Big(\phi\big(b^{\mathscr{L}_k}, u^{\mathscr{N}_k}, \zeta^{\mathscr{N}_k}(x^{\mathscr{N}_k}, w, u^{\mathscr{N}_k}), \theta^{\mathscr{N}_k}(\pi^{\mathscr{N}_k}(x^{\mathscr{L}_k}, w, u^{\mathscr{N}_k}, w'), m^{\bar{\mathscr{N}}_k})\big)\Big)\Bigg]\Bigg]\Bigg]\Bigg] \tag{4.16}$$

for all $b^{\mathscr{L}_k} \in \mathscr{B}^{\mathscr{L}_k}$.

Solving the above recursive equations (Equations (4.14) and (4.15) for the finite horizon case, or Equation (4.16) for the infinite horizon case) for each $k \in \mathcal{K}$ yields a set of (suboptimal) local defense policies $\{\gamma^1, \gamma^2, \ldots, \gamma^{n_k}\}$ for Problem (P).

4.4.3 Scalability

Our approach to the solution of Problem (P) consists of two main steps: (i) the partition of the influence graph into clusters and the formulation of approximate local defense problems (P'_k) and (ii) the solution of each problem (P'_k). This approach can provide a (suboptimal) solution to the defense problem (P) associated with networks of arbitrarily large size, as we explain below. Suppose that the designer of the defense policy knows its (limited) computational capability. To implement our approach, the designer must be able to solve the problems associated with the above-described steps.

Forming the influence graph requires the computation of all of the edge weights, ξ_e, as described in Section 4.4.1.2; the complexity of such a computation is of the order of $\mathcal{O}(\sum_{j,i \in \mathcal{N}, j \neq i} |\mathcal{X} \times \mathcal{X}^j \times \mathcal{W} \times \mathcal{U}|) = \mathcal{O}(n^2 n_w n_u)$. Creating clusters requires the use of the min-cut algorithm, the complexity of which is of the order $\mathcal{O}((n_k - 1)n^2)$, where n_k is the number of clusters. The computational complexity of each of the Problems (P'_k) is of the order $\mathcal{O}(|\mathcal{P}(\mathcal{X}^{\mathcal{L}_k}) \times \mathcal{Y}^{\mathcal{N}_k} \times \mathcal{Z}^{\mathcal{N}_k} \times \mathcal{U}^{\mathcal{N}_k} \times \mathcal{W}|) = \mathcal{O}(2^{|\mathcal{L}_k|} n_w \prod_{i \in \mathcal{N}_k} (n_y^i n_z^i n_u^i))$. The problems ($\text{P}'_k$) can be solved in parallel. The above arguments show that the computational complexity associated with the solution of each Problem (P'_k) is the main bottleneck in the application of our approach to the solution of Problem (P). If the computational complexity of each of the problems (P'_k), $k \in \mathcal{K}$, does not exceed the designer's computational capability, then our approach can be used to provide a suboptimal solution to Problem (P).

From the above discussion, it is clear that an increase in number n_k of clusters will on one hand decrease the computational complexity of each (P'_k), as the dimensionality of each $\mathcal{X}^{\mathcal{N}_k}, \mathcal{Y}^{\mathcal{N}_k}, \mathcal{Z}^{\mathcal{N}_k}, \mathcal{W}, \mathcal{U}^{\mathcal{N}_k}$ will decrease, but, on the other hand, will decrease the accuracy of the solution of Problem (P) and increase the complexity of the min-cut algorithm. Therefore, in the application of our approach to Problem (P), one has to explore the above-described tradeoff between computational complexity and solution quality so as to end up with the best approximation that is compatible with the defender's computational capabilities.

Depending on the structure of the influence graph, modifications to the approach can be taken. In some problems, the influence graph may exhibit some sparsity; in this case the computational complexity associated with clustering can be reduced using spectral clustering with non-backtracking matrix [13] (the spectral clustering with non-backtracking matrix algorithm has lower complexity than the min-cut algorithm for super-sparse graphs). In situations where the influence graph is densely connected, one can use approximations, in addition to those described in this chapter, so as to end up with a scalable approximation to Problem (P). We briefly describe such approximations in the conclusion of the chapter (Section 4.6).

4.5 Example

Consider a system of five hosts. Each host can be in one of four security states (a measure of its security level), ranging from the most secure state, s_1, to the least secure state, s_4. At each time-step, the attacker (nature) can choose to attack the hosts through selection of various attack actions. The attacker is assumed to have access to three types of attack actions: a *null* action, corresponding to not attacking the host; *probe* actions, which increment the security state of the attacked host; and *spread* actions, which allow the attacker to use a host in a degraded security state to attack another host. Following the attack action (nature's event), the defender selects its defense action. The defender has access to three types of defense actions: a *null* action, corresponding to not specifying any defense action; a *sense* action, which, if invoked on a host, reveals the true security state of the host to the defender; and a *reimage* action, which resets the security state of the host to s_1.

The five host system described above can be modeled using the security model of Section 4.2. Formally, using the notation of our model, each host corresponds to a state element, that is, $\mathcal{N} = \{1,2,3,4,5\}$. The state space of each element reflects the possible security states that each host can be in, that is, $\mathscr{X}^i = \{s_1, s_2, s_3, s_4\}$. The state space of the problem is $\mathscr{X} = \prod_{i\in\mathcal{N}} \mathscr{X}^i$. The set of attack actions $\mathscr{W}$ is assumed to decompose into attacks on each host, that is, $\mathscr{W} = \prod_{i\in\mathcal{N}} \mathscr{W}^i$. Note that in our model of Section 4.2, the attacks are not necessarily decomposable into attacks on each element; however, for the purposes of our example, we assume (for simplicity) that the system-wide attack can be described as the collection of attacks on each element. Each set $\mathscr{W}^i$ consists of attacks $\mathscr{W}^i = \{w^i_\varnothing, w^i_{p_1}, w^i_{p_2}, w^i_{p_3}, w^{ji}_s, w^{ji}_{s'}\}$ on host i, where $w^i_\varnothing$ represents no attack on host i; $w^i_{p_k}$ represents a probe action, incrementing the security state of host i from $x^i_t = s_k$ to $x^i_{t+1} = s_{k+1}$; and both w^{ji}_s and $w^{ji}_{s'}$ represent spread actions, allowing the attacker to use another host j if it is in state $x^j_t = s_4$ to attack host i. Specifically, w^{ji}_s brings the state of host i from $x^i_t = s_1$ to $x^i_{t+1} = s_3$, and $w^{ji}_{s'}$ brings the state of host i from $x^i_t = s_2$ to $x^i_{t+1} = s_3$. To make the example more interesting (resulting in a more diverse set of weights in the influence graph), we assume that the attacker has limited spreading capabilities, that is, $\mathscr{W}^1$ contains the spreading actions $\{w^{5,1}_s, w^{5,1}_{s'}\}$, $\mathscr{W}^2$ contains $\{w^{1,2}_s, w^{1,2}_{s'}, w^{3,2}_s\}$, $\mathscr{W}^3$ contains $\{w^{1,3}_s, w^{4,3}_s\}$, $\mathscr{W}^4$ contains $\{w^{3,4}_s\}$, and $\mathscr{W}^5$ contains $\{w^{1,5}_s\}$. The set of defense actions $\mathscr{U} = \prod_{i\in\mathcal{N}} \mathscr{U}^i$ is described in terms of the action-space of each element. Specifically, $\mathscr{U}^i = \{u^i_\varnothing, u^i_s, u^i_r\}$, where $u^i_\varnothing$ represents the defender not taking any action on host i, u^i_s represents the sense action on host i, and u^i_r represents the reimage action on host i. Neither the null actions nor the sense action has an effect on the evolution of the state.

In terms of the state update function of Equation (4.6), the evolution of each state element can be written as follows:

$$x^i_{t+1} = \pi^i(x^i_t, w_t, u^i_t) = \begin{cases} s_1 & \text{if } u^i_t = u^i_r \\ s_3 & \text{else if } (w^{ji}_s \in w_t, x^j_t = s_4, x^i_t = s_1) \text{ or} \\ & \qquad (w^{ji}_{s'} \in w_t, x^j_t = s_4, x^i_t = s_2) \\ s_{k+1} & \text{else if } w^i_t = w^i_{p_k}, x^i_t = s_k, k = 1,2,3 \\ x^i_t & \text{otherwise.} \end{cases}$$

We assume that the defender is only able to observe the spreading actions, but cannot observe the attacker's null or probe actions. Formally,

$$y^i_t = \theta^i(x^i_t, w_t) = \begin{cases} w & \text{if for any } w = w^{ji}_s \text{ or } w = w^{ji}_{s'} \text{ in } w_t \\ \varnothing & \text{otherwise.} \end{cases}$$

The defender's action observations are

$$z^i_t = \zeta^i(x^i_t, u^i_t, w_t) = \begin{cases} s_1 & \text{if } u^i_t = u^i_r \\ x^i_t & \text{else if } u^i_t = u^i_s \\ \varnothing & \text{otherwise.} \end{cases}$$

Finally, the (instantaneous) cost that the defender incurs at time t is simply $c(x_t, u_t) = c^1(x^1_t, u^1_t) + \cdots + c^5(x^5_t, u^5_t)$, where $x_t = (x^1_t, x^2_t, \ldots, x^5_t)$ and $u_t = (u^1_t, u^2_t, \ldots, u^5_t)$. The defense problem (P) can now be written. In previous work [19], we were able to obtain a defense policy for a similar problem with $n = 3$ elements using the approximate information state (the set of states compatible with the defender's information at t) described in Section 4.4.2. For larger problems (for instance, the $n = 5$ problem just described), we must employ the decomposition approach proposed in this chapter to permit computation of approximate defense policies.

The influence graph for the example problem can now be constructed. Assume that the computational capability of the defender is such that it can solve Problem (P) for systems consisting of $n^{\max} = 3$ or fewer elements. The quantity $n^{\max}$ is determined by taking into account the defender's computational capability and the computational complexity of Problem (P'_k), as described in Section 4.4.3. In the current example $n = 5 > 3 = n^{\max}$, so we must decompose the problem into local defense problems and determine local defense policies. As described in Section 4.4.1.2, the construction of the influence graph is performed by analyzing the functions and the sets of actions of both the defender and attacker with the weights computed according to Equation (4.12). For illustration purposes, we show how to compute one of the weights, specifically $\xi_{5,1}$. The remaining edge weights are calculated in a similar fashion. To calculate $\xi_{5,1}$, we need to count the cases where the state update of element $i = 1$ functionally depends on the state of element $i = 5$, as described by Definition 1. To do this, we enumerate over all values of x^1, u^1, w^1, x^5, $\hat{x}^5 \neq x^5$, u^5, w^5, $x^{-(1,5)}$, and $w^{-(1,5)}$. The normalization term in Equation (4.12) is $d = (\prod_{l \in \mathcal{N} \setminus \{i,j\}} n^l_x) n^i_x n^j_x (n^j_x - 1) n_w n^i_u = (4^3) \cdot 4 \cdot 4 \cdot (4-1) \cdot (6 \cdot 7 \cdot 6 \cdot 5 \cdot 5) \cdot 3 = 58060800$. Only in the cases that all of the following conditions are satisfied, the event within the indicator function of Equation (4.12) is true: (i) A spread attack is launched from element $j = 5$ to $i = 1$, while the state of element $i = 1$ is such that the spread attack is effective, i.e., $(w^5 = w^{5,1}_s, x^1 = s^1)$ or $(w^5 = w^{5,1}_{s'}, x^1 = s^2)$.

(ii) The state of element $j=5$ allows the attack to be launched in x_5 but does not allow it in $\hat{x}_5$ or vice versa, i.e., $(x^5 = s_4, \hat{x}^5 \in \{s_1,s_2,s_3\})$ or $(x^5 \in \{s_1,s_2,s_3\}, \hat{x}^5 = s_4)$. (iii) There are no similar effective spread attacks on element $i=1$ from other elements $k \in \{2,3,4\}$; since there exists no spread attack from elements $k \in \{2,3,4\}$ to element $i=1$, their state and attack events can take any possible value, i.e., $x^{-(1,5)} \in \mathscr{X}^{-(1,5)}$ and $w^{-(1,5)} \in \mathscr{W}^{-(1,5)}$. (iv) The defender does not deploy a defense action that nullifies the effect of the spread attack from element $j=5$ to element $i=1$, that is, $u^1 \in \{u^1_{\varnothing}, u^1_s\}$. The above conditions ensure the evolution of element $i=1$ conditioned on x^5 is different from its evolution condition on $\hat{x}^5$. The total number of such cases is $2\cdot 2\cdot(1\cdot 3+3\cdot 1)\cdot 4^3\cdot(7\cdot 6\cdot 5\cdot 5) = 1612800$. The normalized weight, computed using Equation (4.12), is thus $\xi_{5,1} = 1612800/58060800 \approx 0.03$. The complete influence graph is depicted in Figure 4.4.

Using the edge weights in the influence graph, the min-cut algorithm [12] can now be applied in order to partition (cluster) the graph. The first application of the min-cut algorithm, shown in Figure 4.4(a), results in the clusters $\{1,2,5\}$ and $\{3,4\}$. Notice that under this clustering, the existence of edges $(3,2)$ and $(4,5)$ would result in one of the local defense problems containing all 5 elements in its local state space, violating the $n^{\max}=3$ limit. As a result, we apply the min-cut algorithm once more, as shown in Figure 4.4(b). With the new cut, we can see that the resulting set of clusters, $\mathscr{N}_1 = \{3,4\}$, $\mathscr{N}_2 = \{2\}$, and $\mathscr{N}_3 = \{1,5\}$, shown in Figure 4.4(c) satisfy the $n^{\max}$ limit. This can be seen by writing the set of state indices for the local defense problems, $\mathscr{L}_1 = \{1,3,4\}$, $\mathscr{L}_2 = \{1,2,3\}$, and $\mathscr{L}_3 = \{1,4,5\}$, and noticing that $|\mathscr{L}_1| = |\mathscr{L}_2| = |\mathscr{L}_3| = 3 = n^{\max}$.

Using the clusters of Figure 4.4(c), the corresponding local defense problems can be defined. The internal state spaces of each problem are defined by the set of nodes within the corresponding cluster, that is, $\mathscr{X}^{\mathscr{N}_1} = \mathscr{X}^3 \times \mathscr{X}^4$, $\mathscr{X}^{\mathscr{N}_2} = \mathscr{X}^2$, and $\mathscr{X}^{\mathscr{N}_3} = \mathscr{X}^1 \times \mathscr{X}^5$. The corresponding state update functions are

$$\begin{aligned}
\pi^{\mathscr{N}_1}_{t+1} &= \pi^{\mathscr{N}_1}(x^{\mathscr{L}_1}_t, w_t, u^{\mathscr{N}_1}_t) \\
&= \left(\pi^3(x^{\mathscr{L}_1}_t, w_t, u^3_t), \pi^4(x^{\mathscr{L}_1}_t, w_t, u^4_t)\right) \\
\pi^{\mathscr{N}_2}_{t+1} &= \pi^{\mathscr{N}_2}(x^{\mathscr{L}_2}_t, w_t, u^{\mathscr{N}_2}_t) \\
&= \pi^2(x^{\mathscr{L}_2}_t, w_t, u^2_t) \\
\pi^{\mathscr{N}_3}_{t+1} &= \pi^{\mathscr{N}_3}(x^{\mathscr{L}_3}_t, w_t, u^{\mathscr{N}_3}_t) \\
&= \left(\pi^1(x^{\mathscr{L}_3}_t, w_t, u^1_t), \pi^5(x^{\mathscr{L}_3}_t, w_t, u^5_t)\right)
\end{aligned}$$

where the local states are $x^{\mathscr{L}_1}_t = \{x^1_t, x^3_t, x^4_t\}$, $x^{\mathscr{L}_2}_t = \{x^1_t, x^2_t, x^3_t\}$, and $x^{\mathscr{L}_3}_t = \{x^1_t, x^4_t, x^5_t\}$ and the internal actions are $u^{\mathscr{N}_1}_t \in \mathscr{U}^{\mathscr{N}_1} = \mathscr{U}^3 \times \mathscr{U}^4$, $u^{\mathscr{N}_2}_t \in \mathscr{U}^{\mathscr{N}_1} = \mathscr{U}^2$, and $\mathscr{U}^{\mathscr{N}_3} = \mathscr{U}^1 \times \mathscr{U}^5$. The cost functions of each local defense problem are $c^{\mathscr{N}_1}(x^{\mathscr{N}_1}_t, u^{\mathscr{N}_1}_t) = c^3(x^3_t, u^3_t) + c^4(x^4_t, u^4_t)$, $c^{\mathscr{N}_2}(x^{\mathscr{N}_2}_t, u^{\mathscr{N}_2}_t) = c^2(x^2_t, u^2_t)$, and $c^{\mathscr{N}_2}(x^{\mathscr{N}_2}_t, u^{\mathscr{N}_2}_t) = c^1(x^1_t, u^1_t) + c^5(x^5_t, u^5_t)$. Similarly, each local defense problem's observations are $y^{\mathscr{N}_1}_t = (y^3_t, y^4_t)$, $z^{\mathscr{N}_1}_t = (z^3_t, z^4_t)$, $y^{\mathscr{N}_2}_t = y^2_t$, $z^{\mathscr{N}_2}_t = z^2_t$, and $y^{\mathscr{N}_3}_t = (y^1_t, y^5_t)$, $z^{\mathscr{N}_3}_t = (z^1_t, z^5_t)$. The message sent from local defense problem l to local defense prob-

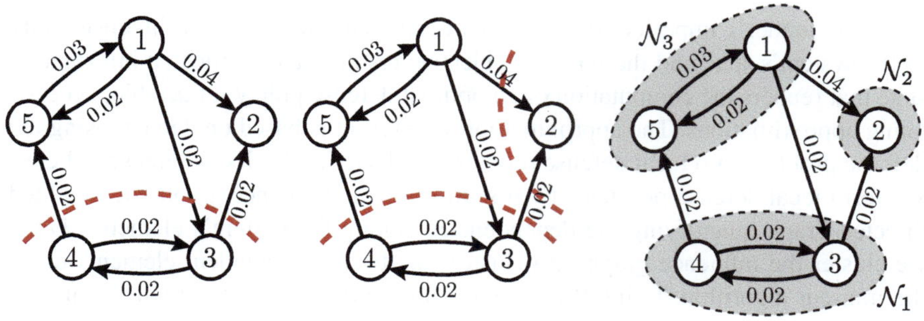

(a) First cut. (b) Second cut. (c) Final set of clusters.

Fig. 4.4: Repeated application of the min-cut algorithm to obtain three clusters, $\mathcal{N}_1 = \{3,4\}$, $\mathcal{N}_2 = \{2\}$, and $\mathcal{N}_3 = \{1,5\}$

lem k at time t, denoted by m_t^{kl}, belongs to the power set of states that can influence the evolution of elements in the internal state space $\mathcal{N}_k$. Specifically, for example, local defense problem $k = 1$ receives a message from problem $k = 3$, described by $m_t^{13} \in \mathscr{P}(\mathscr{X}^{\bar{\mathcal{N}}_{13}}) = \mathscr{P}(\mathscr{X}^1)$ (representing all possible states the element $i = 1$ can be in). Similarly, local defense problem $k = 2$ receives two messages, one from problem $k = 1$, $m_t^{21} \in \mathscr{P}(\mathscr{X}^{\bar{\mathcal{N}}_{21}}) = \mathscr{P}(\mathscr{X}^3)$, and one from problem $k = 3$, $m_t^{23} \in \mathscr{P}(\mathscr{X}^{\bar{\mathcal{N}}_{23}}) = \mathscr{P}(\mathscr{X}^1)$ (these messages can be summarized by the aggregate message $m_t^{\bar{\mathcal{N}}_2} \in \mathscr{P}(\mathscr{X}^3) \times \mathscr{P}(\mathscr{X}^1)$). Lastly, local defense problem $k = 3$ receives a message from problem $k = 1$, $m_t^{31} \in \mathscr{P}(\mathscr{X}^{\bar{\mathcal{N}}_{31}}) = \mathscr{P}(\mathscr{X}^4)$. No other messages are exchanged. The approximate information state spaces for the local defense problems are $\mathscr{B}^{\mathscr{L}_1} = \mathscr{P}(\mathscr{X}^1 \times \mathscr{X}^3 \times \mathscr{X}^4)$, $\mathscr{B}^{\mathscr{L}_2} = \mathscr{P}(\mathscr{X}^1 \times \mathscr{X}^2 \times \mathscr{X}^3)$, and $\mathscr{B}^{\mathscr{L}_3} = \mathscr{P}(\mathscr{X}^1 \times \mathscr{X}^4 \times \mathscr{X}^5)$. The dynamic programs (for the finite horizon defense problem) can be written in a similar fashion to Equations (4.14) and (4.15) and can be addressed by methods in the literature for minmax control problems [7, 24].

4.6 Discussion and Conclusion

We studied a cyber-security problem from the defender's point of view. This is a control problem where the defender's goal is to determine a defense policy to protect the system against an attacker that is modeled by nature. The system's security status evolves dynamically over time; its evolution depends on the defender's actions and the attack events. The defender has imperfect information about the system's security status and takes a conservative approach to the system's defense. Specifically, the defender's goal is to minimize the worst possible damage to the system caused by attack (nature's) events. Therefore, the defender has to solve a minmax control problem with imperfect observations so as to determine an optimal defense policy.

The defender's imperfect observations combined with the high dimensionality of the system's state and the minmax objective result in a complicated information state that renders the computation of an optimal defense policy intractable, necessitating approximations. The approximation we present is based on decomposing the defense problem into local defense problems and solving for local defense policies. We form local defense problems by first forming the influence graph – a weighted directed graph quantifying the dependencies among the system's elements. Next, we cluster the influence graph into clusters of strongly dependent elements using the min-cut algorithm. Using the clusters and the dependencies among them, we form a local defense problem for each cluster. The control of each local defense problem is further approximated by focusing on an approximate information state that allows the computation of a policy sequentially using dynamic programming ideas.

Our approach has two computational requirements: (i) forming the influence graph and clusters and (ii) computing a defense policy for the local defense problem associated with each cluster. Given the defender's computational capabilities, we can address these requirements irrespectively of the system's size (the dimensionality of the system's state) as follows: we can form a large number of clusters so that the size of each local defense problem is compatible with the defender's computational capability. Consequently, the approach to the dynamic defense problem of cyber networks described in this chapter is scalable.

For some instances of our security model, the resulting influence graph may not permit a partitioning into clusters that satisfy the designer's computational capability. For example, consider a completely connected influence graph. In such a graph, each local state space $\mathscr{X}^{\mathscr{L}_k}$ would be equal to the complete state space $\mathscr{X}$. In such a situation, no clustering that satisfies the constraint would exist, and we would not be able to compute local defense policies. However, we believe that, for practical purposes, influence graphs will have at least some sparsity that can be exploited, allowing our decomposition to be applied to obtain a suboptimal defense policy. In the rare event that the influence graph doesn't permit the required clustering, one can employ alternate (more aggressive) approximation techniques that summarize the available information (e.g., collapsing all exogenous elements into a single worst-case element).

In the cyber-security model studied in this chapter, the attacker's behavior is fixed in advance and modeled by the (state-dependent) events that occur in nature. The situation where both the defender and the attacker are strategic and have different objectives is not captured by the model of Section 4.2. Such a situation gives rise to a dynamic game with asymmetric information. Preliminary results on such games can be found in [18, 22].

Acknowledgements This research was partially supported by NSF grant CNS-1238962, ARO MURI grant W911NF-13-1-0421, and ARO grant W911NF-17-1-0232. The authors are grateful to Michael P. Wellman, Hamidreza Tavafoghi, Ouyang Yi, and Ashutosh Nayyar for useful conversations.

References

1. Alpcan T, Başar T (2010) Network security: A decision and game-theoretic approach. Cambridge University Press
2. Baras JS, James MR (1994) Robust and risk-sensitive output feedback control for finite state machines and hidden Markov models. Tech. rep., Institute for Systems Research
3. Bernhard P (1995) Expected values, feared values, and partial information optimal control. In: New trends in dynamic games and applications, Springer, pp 3–24
4. Bernhard P (2000) Max-plus algebra and mathematical fear in dynamic optimization. Set-Valued Analysis 8(1–2):71–84
5. Bernhard P (2003) Minimax – or feared value – $L1/L\infty$ control. Theoretical computer science 293(1):25–44
6. Bertsekas D, Rhodes I (1973) Sufficiently informative functions and the minimax feedback control of uncertain dynamic systems. IEEE Transactions on Automatic Control 18(2):117–124
7. Bertsekas DP (1971) Control of uncertain systems with a set-membership description of the uncertainty. Tech. rep., DTIC Document
8. Bertsekas DP, Rhodes IB (1971) On the minimax feedback control of uncertain dynamic systems. In: 1971 IEEE conference on decision and control. 10:451–455
9. Coraluppi SP, Marcus SI (1999) Risk-sensitive and minimax control of discrete-time, finite-state Markov decision processes. Automatica 35(2):301–309
10. Foo B, Glause MW, Howard GM, Wu YS, Bagchi S, Spafford EH (2008) Intrusion response systems: a survey. In: Information Assurance: Dependability and Security in Networked Systems, Morgan Kaufmann, Burlington, MA, chap 13:377–416
11. Inayat Z, Gani A, Anuar NB, Khan MK, Anwar S (2016) Intrusion response systems: Foundations, design, and challenges. Journal of Network and Computer Applications 62:53–74
12. Johnson EL, Mehrotra A, Nemhauser GL (1993) Min-cut clustering. Mathematical programming 62(1–3):133–151
13. Krzakala F, Moore C, Mossel E, Neeman J, Sly A, Zdeborová L, Zhang P (2013) Spectral redemption in clustering sparse networks. Proceedings of the National Academy of Sciences 110(52):20,935–20,940
14. Ligatti J, Bauer L, Walker D (2005) Edit automata: Enforcement mechanisms for run-time security policies. International Journal of Information Security 4(1):2–16
15. Lye K, Wing JM (2005) Game strategies in network security. International Journal of Information Security 4(1–2):71–86
16. Manshaei MH, Zhu Q, Alpcan T, Başar T, Hubaux JP (2013) Game theory meets network security and privacy. ACM Comput Surv 45(3):1–39

17. Miehling E, Rasouli M, Teneketzis D (2015) Optimal defense policies for partially observable spreading processes on Bayesian attack graphs. In: Proceedings of the Second ACM Workshop on Moving Target Defense, ACM, pp 67–76
18. Ouyang Y, Tavafoghi H, Teneketzis D (2017) Dynamic games with asymmetric information: common information based perfect Bayesian equilibria and sequential decomposition. IEEE Trans Autom Control 62(1):222–237
19. Rasouli M, Miehling E, Teneketzis D (2014) A supervisory control approach to dynamic cyber-security. In: Decision and Game Theory for Security, Springer, pp 99–117
20. Schneider FB (2000) Enforceable security policies. ACM Trans Inf Syst Secur 3(1):30–50
21. Shameli-Sendi A, Ezzati-Jivan N, Jabbarifar M, Dagenais M (2012) Intrusion response systems: survey and taxonomy. International Journal of Computer Science and Network Security 12(1):1–14
22. Tavafoghi H, Ouyang Y, Teneketzis D (2016) On stochastic dynamic games with delayedsharing information structure. In: 2016 IEEE 55th conference on decision and control, IEEE, pp 7002–7009
23. Witsenhausen H (1968) A minimax control problem for sampled linear systems. IEEE Transactions on Automatic Control 13(1):5–21
24. Witsenhausen HS (1966) Minimax control of uncertain systems. PhD thesis, Massachusetts Institute of Technology, Cambridge, Massachusetts 02139

Chapter 5
Factored Markov Game Theory for Secure Interdependent Infrastructure Networks

Linan Huang, Juntao Chen, and Quanyan Zhu

5.1 Introduction

Recent advances in information and communication technologies (ICTs) have witnessed a tight integration of critical infrastructures with sophisticated information technologies (IT) to improve the quality of infrastructure services and the operational efficiency. However, the direct application of off-the-shelf IT systems exposes the critical infrastructures to cyber vulnerabilities, which can compromise the functionalities of the infrastructures and inflict a significant economic loss. For example, the cyberattacks on Ukrainian power systems have successfully disrupted electricity supply and left 230,000 people without power. The WannaCry ransomware attacks have infected thousands of computers worldwide and invalidated critical services such as hospitals and manufacturing plants, causing an estimated loss of $4 billion.

The cyber-physical nature of the interdependent infrastructure systems shown in Figure 5.1 enables the exploitation of the coordinated attacks that leverage the vulnerabilities in both systems to increase the probability of the attacks and the failure rates of the infrastructure. For example, a terrorist can use cyberattacks to compromise the surveillance camera of an airport, government building, or public area and stealthily plant a bomb without being physically detected. The physical damage of infrastructure systems can also assist attackers intrude into cyber systems such as data centers and control rooms. Hence both cyber and physical failures of the infrastructure can create significant consequences. Moreover, the cyber, physical, and logical connectivity among infrastructures create dependencies and interdependencies between nodes and components within an infrastructure and across the infrastructures. As a result, the failure of one component can lead to a cascading failure

---

L. Huang (✉) · J. Chen · Q. Zhu
Department of Electrical and Computer Engineering, New York University, 2 Metrotech Center, Brooklyn, 11201 NY, USA
e-mail: lh2328@nyu.edu; jc6412@nyu.edu; qz494@nyu.edu

S. Rass, S. Schauer (eds.), *Game Theory for Security and Risk Management*, Static & Dynamic Game Theory: Foundations & Applications,
https://doi.org/10.1007/978-3-319-75268-6_5

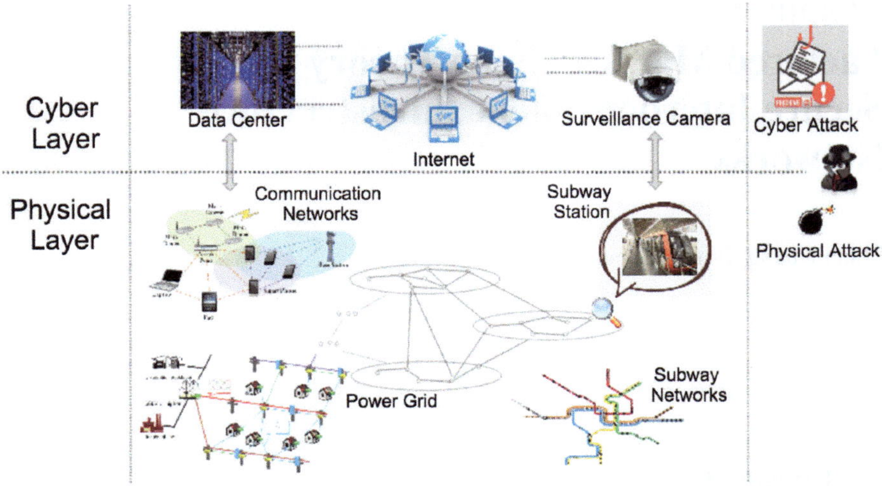

Fig. 5.1: Cyber networks on the top are interdependent with physical systems on the bottom which consists of critical infrastructures such as the power grid, subway, and communication networks. The healthy functioning of components of the physical system, e.g., subway stations, depends on the well-being of other subway stations and cross-layer nodes (e.g., power substations and surveillance cameras). This interdependency allows adversaries to attack different types of nodes to compromise the entire cyber-physical system.

over multiple infrastructures. To mitigate such cyber-physical threats, it is essential to design effective defense mechanisms to harden both the cyber and physical security at the nodes of the infrastructure to protect them from failures.

To this end, we first develop a framework to capture the adversarial interactions between the attack and the defense of the interdependent critical infrastructures (ICIs). Zero-sum games provide a natural framework to model the conflicting objectives of the players. The attacker aims to compromise the cyber and physical components of ICIs that are under his control and inflict maximum loss on the system. The defense of the ICIs seeks to invest resources to minimize the loss by implementing cost-effective defense mechanisms. To capture the dynamics of the ICIs, we use a binary state variable to describe the state of each node. The attacker's strategy can affect the transition probability of a node's state from a normal operation mode to a failure mode. The saddle-point equilibrium analysis of the zero-sum dynamic game provides a systematic way to design defense strategies for the worst-case attack schemes.

In our work, we focus on the class of Markov games whose transition kernel is controlled by the attacker, yet the defender can choose state-dependent actions to mitigate the economic loss or increase attacking costs at each state. The single-controller assumption reduces the computation of the saddle-point equilibrium strategies into a linear program. One challenge in computing security strategies arises from the large-scale nature of the infrastructure systems together with an exponentially growing number of global states. To address it, we use linear function approximation techniques for the value function and exploit the sparse network structure to formulate the factored Markov game to reduce the computational complexity. Another challenge is the implementability of the security strategies. The global stationary policies of both players are difficult to implement since the knowledge of the global state of each infrastructure is not often accessible. Hence we restrict the security strategies to a decentralized and local information structure and use the factored Markov game framework to compute approximately distributed policies for each node in the multilayer infrastructure networks.

Our analytical results show that the optimal attacker's policy obtained in the dual of the exact LP is pure, and the suboptimal attacker's policy is distributed assuming a restricted information structure of the defender. Besides, the computation complexity is provided for each linear program. Numerical results illustrate the implementable distributed policies, significant computation reductions, reasonable accuracy losses, and impacts of different information structures and the interdependent networks.

Firstly, we observe that fewer attacks happen when defenders are present in the system because attacks tend to avoid attacking nodes equipped with safeguard procedures. The security strategy for the infrastructure defender developed using the game framework yields a proactive protection as the nodes mitigate their losses even at the working state when their neighbors are observed to be attacked. Secondly, the numerical experiments have shown that the approximation scheme yields a significant reduction of the complexity while maintaining a reasonable level of accuracy. Thirdly, we observe that a node can improve its security performance with more information about the global state of the multilayer infrastructures. Besides, when strengthening every node is too costly, we choose to consolidate every other node in a ring network to mitigate cascading failures as shown in Section 5.4.4.

5.2 Mathematical Model

This section introduces in Subsection 5.2.1 a zero-sum Markov game model over interdependent infrastructure networks to understand the interactions between an attacker and a defender at the nodes of infrastructures. The solution concept of the saddle-point equilibrium strategies is presented in Subsection 5.2.2, and the computational issues of the equilibrium are discussed in Subsection 5.2.3.

5.2.1 Network Game Model

The dynamic and complex infrastructure networks can be represented by nodes and links. For example, in an electric power system, a node can be a load bus or a generator, and the links represent the transmission lines. Similarly, in a water distribution system, a node represents a source of water supply, storage or users, and

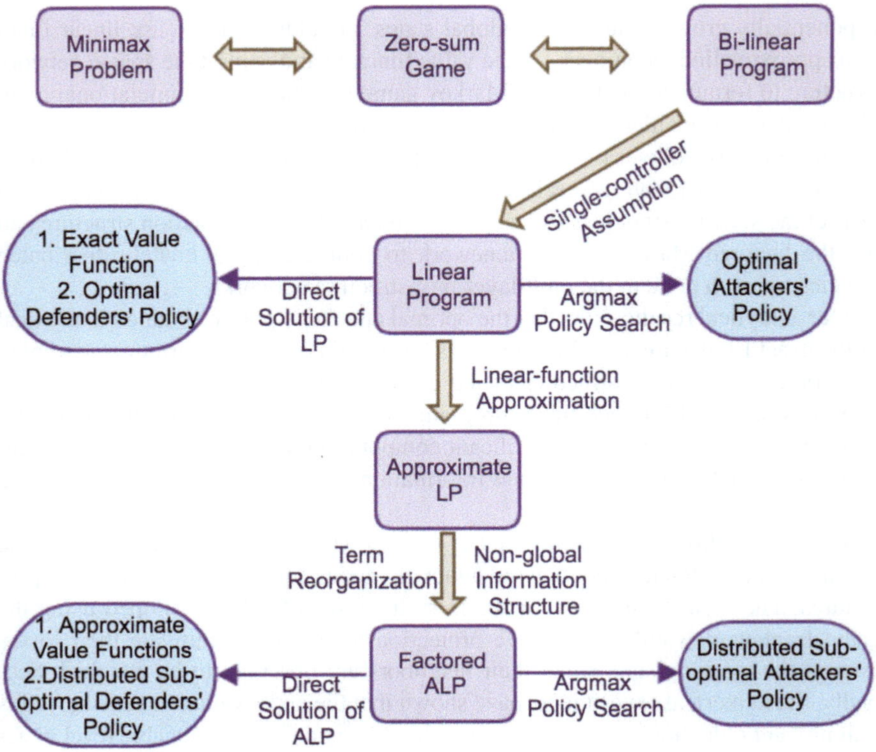

Fig. 5.2: In this overall structural diagram, blue squares show a sequence of techniques used in the problem formulation. The linear programming technique yields the exact value functions and the optimal defender's policy. The factored approximate linear program yields an approximate value function and distributed suboptimal defender's policy. The greedy search method solves for the attacker's policy.

the links can represent pipes for water delivery. Consider a system of I interdependent infrastructures. Let $\mathscr{G}^i = (\mathscr{N}^i, \mathscr{E}^i)$ be the graph representation of infrastructure $i \in \mathscr{I} := \{1, 2, \cdots, I\}$, where $\mathscr{N}^i = \{n_1^i, n_2^i, \cdots, n_{m_i}^i\}$ is the set of m_i nodes in the infrastructure and $\mathscr{E}^i = \{e_{j,k}^i\}$ is the set of directed links connecting nodes n_j^i and n_k^i. The directed link between two nodes indicates either physical, cyber, or logical influences from one node to the other. For example, the state of node n_j^i in the electric power system can influence the state of node n_k^i through the physical connection or the market pricing. The dependencies across the infrastructures can be captured by adding interlinks. Let $\mathscr{E}^{i,j}$ be the set of directed interlinks between nodes in infrastructure i and infrastructure j. In particular, let $\varepsilon_{n_k^i, n_l^j} \in \mathscr{E}^{i,j}$ denote the interlink between n_k^i and n_l^j. Hence, the composed network can be represented by the graph $\mathscr{G} = (\mathscr{N}, \mathscr{E})$, where $\mathscr{N} = \cup_{i=1}^I \mathscr{N}^i$ and $\mathscr{E} = \left(\cup_{i=1}^I \mathscr{E}^i\right) \bigcup \left(\cup_{i \neq j} \mathscr{E}^{i,j}\right)$.

Denote by $X_j^i \in \mathscr{X}_j^i$ the state of node n_j^i that can take values in the state space $\mathscr{X}_j^i$. We let $\mathscr{X}_j^i = \{0,1\}$ be binary random variables for all $i = 1,2,\cdots,I$ and $j \in \mathscr{N}^i$. Here, $X_j^i = 1$ means that node n_j^i is functional in a normal mode; $X_j^i = 0$ indicates that node n_j^i is in a failure mode. The state of infrastructure i can be thus denoted by $X^i = (X_1^i, X_2^i, \cdots, X_{m_i}^i) \in \mathscr{X}^i := \prod_{j=1}^{m_i} \mathscr{X}_j^i$, and the state of the whole system is denoted by $X = (X^1, X^2, \cdots, X^I) \in \prod_{i=1}^{I} \mathscr{X}^i$. The state transition of a node n_j^i from state $x_j^{i\prime} \in \mathscr{X}_j^i$ to state $x_j^i \in \mathscr{X}_j^i$ is governed by a stochastic kernel $\Pr_{i,j}(x_j^{i\prime}|x, d_j^i, a_j^i) := \Pr(X_j^i = x_j^{i\prime}|X = x, d_j^i, a_j^i)$, which depends on the protection policy $d_j^i \in \mathscr{D}_j^i$ adopted at node n_j^i as well as the adversarial behavior $a_j^i \in \mathscr{A}_j^i$, where $\mathscr{D}_j^i, \mathscr{A}_j^i$ are feasible sets for the infrastructure protection and the adversary, respectively. The state transition of a node depends on the entire system state of the interdependent infrastructure. It, in fact, captures the interdependencies between nodes in one infrastructure and across infrastructures. The infrastructure protection team or defender determines the protection policy with the goal of hardening the security and improving the resilience of the interdependent infrastructure. On the other hand, an adversary aims to create damage on the nodes that he can compromise and inflict maximum damage on the infrastructure in a stealthy manner, e.g., creating cascading and wide-area failures. Let $\mathscr{M}_a^i \subseteq \mathscr{N}^i$ and $\mathscr{M}_d^i \subseteq \mathscr{N}^i$ be the set of nodes that an adversary can control, and the system action vector of the adversary is $\mathbf{a} = (a_j^i)_{j \in \mathscr{M}_a^i, i \in \mathscr{I}} \in \mathscr{A} := \prod_{i \in \mathscr{I}} \prod_{j \in \mathscr{N}^i} A_j^i$ with $|\mathscr{M}_a^i| = \bar{m}_{a,i}$. The system action vector for infrastructure protection is $\mathbf{d} = (d_j^i)_{j \in \mathscr{M}_d^i, i \in \mathscr{I}} \in \mathscr{D} := \prod_{i \in \mathscr{I}} \prod_{j \in \mathscr{N}^i} D_j^i$ with $|\mathscr{M}_d^i| = \bar{m}_{d,i}$. At every time $t = 1,2,\cdots$, the pair of action profiles $(\mathbf{d}_t, \mathbf{a}_t)$ taken at t and the kernel Pr defined later determine the evolution of the system state trajectory. Here, we use add subscript t to denote the action taken time t. The conflicting objective of both players can be captured by a long-term cost J over an infinite horizon:

$$J := \sum_{i \in \mathscr{I}, j \in \mathscr{N}^i} \sum_{t=1}^{\infty} \gamma^t c_j^i(X_t, d_{j,t}^i, a_{j,t}^i), \tag{5.1}$$

where $\gamma \in (0,1)$ is a discount factor, $X_t \in \mathscr{X}$ is the system state at time t, and $c_j^i : \mathscr{X} \times \mathscr{D}_j^i \times \mathscr{A}_j^i \to \mathbb{R}_+$ is the stage cost function of the node n_j^i. Let $\mathscr{U}_j^i, \mathscr{V}_j^i$ be the sets of admissible strategies for the infrastructure and the adversary, respectively. Here, we consider a feedback protection policy $\mu_j^i \in \mathscr{U}_j^i$ as a function of the information structure $F_{j,t}^i$, i.e., $d_{j,t}^i = \mu_j^i(F_{j,t}^i)$. Likewise, we consider the same class of policies for the adversary, i.e., $a_{j,t}^i = \nu_j^i(F_{j,t}^i), \nu_j^i \in \mathscr{V}_j^i$.

The policy can take different forms depending on the information structure. For example, if $F_{j,t}^i = X_t$, i.e., each node can observe the whole state across infrastructures, then the policy is a global stationary policy, denoted by $\mu_j^{i,\mathrm{GS}} \in \mathscr{U}_j^{i,\mathrm{GS}}$, where $\mathscr{U}_j^{i,\mathrm{GS}}$ is the set of all admissible global stationary policies. If $F_{j,t}^i = X_{j,t}^i$, i.e., each node can only observe its local state, then the policy is a local stationary policy, denoted by $\mu_j^{i,\mathrm{LS}} \in \mathscr{U}_j^{i,\mathrm{LS}}$, where $\mathscr{U}_j^{i,\mathrm{LS}}$ is the set of all admissible local stationary policies. If $F_{j,t}^i = X_t^i$, i.e., each node can observe the infrastructure-wide state, then the policy is an infrastructure-dependent stationary policy, denoted by $\mu_j^{i,\mathrm{ID}} \in \mathscr{U}_j^{i,\mathrm{ID}}$, where $\mathscr{U}_j^{i,\mathrm{ID}}$ is the set of all admissible infrastructure-dependent stationary policies. Similarly, an adversary chooses a policy $v^i_{j,t}$, i.e., $a^i_{j,t} = v^i_j(F^i_{j,t})$. Denote by $\mu^i = (\mu^i_1, \mu^i_2, \cdots, \mu^i_{m_i})$, $v^i = (v^i_1, v^i_2, \cdots, v^i_{m_i})$ the protection and attack policies for infrastructure i, respectively, and let $\boldsymbol{\mu} = (\mu^1, \mu^2, \cdots, \mu^I)$ and $\mathbf{v} = (v^1, v^2, \cdots, v^I)$. Note that although both policies are determined only by the information structure and are independent of each other, the total cost function J depends on them both because of the coupling of the system stage cost $c(X_t, \mathbf{d}, \mathbf{a}) := \sum_{i,j} c^i_j(X_t, d^i_{j,t}, a^i_{j,t})$ and the system state transition probability $\Pr(X' = x'|X = x, \mathbf{d}, \mathbf{a}) := \prod_{i \in \mathscr{I}, j \in \mathscr{N}^i} \Pr_{i,j}(x^{i'}_j | x, d^i_j, a^i_j)$. Therefore, with $\mathscr{U} = \prod_{i \in \mathscr{I}, j \in \mathscr{N}_i} \mathscr{U}^i_j$ and $\mathscr{V} = \prod_{i \in \mathscr{I}, j \in \mathscr{N}_i} \mathscr{V}^i_j$, the total cost function $J : \mathscr{X} \times \mathscr{U} \times \mathscr{V} \to \mathbb{R}_+$ starting at initial state x^0 can be written as the expectation of the system stage cost regarding the system state transition probability, i.e.,

$$J(x^0, \boldsymbol{\mu}, \mathbf{v}) := \sum_{t=0}^{\infty} \gamma^t E_{\boldsymbol{\mu}, \mathbf{v}, x^0}[c(X_t, \mathbf{d}, \mathbf{a})]. \tag{5.2}$$

Remark: Notice that there is a difference between policy $\boldsymbol{\mu}, \mathbf{v}$ and action $\mathbf{d}, \mathbf{a}$. A policy or strategy is a mapping and an action is the outcome of the mapping. Besides, since the information structure is the state information available to attackers or defenders, we can abstract it from the entire state information X_t at time t. Given a policy and an information structure, we can uniquely determine the action. Therefore, we write $\mathbf{d}, \mathbf{a}$ instead of $\boldsymbol{\mu}, \mathbf{v}$ in the RHS of (5.2). We use the same terminology in the following equations such as (5.6) where the solution provides us the optimal action pair $\mathbf{d}^*, \mathbf{a}^*$ at every state x. With the knowledge of the mapping outcome and corresponding information structure as the input of the mapping, the policy functions $\boldsymbol{\mu}^*$, $\mathbf{v}^*$ are uniquely defined.

Hence a security strategy for the infrastructure protection achieves the optimal solution $J^*(x^0)$ to the following minimax problem, which endeavors to minimize the system cost under the worst attacking situation $\max_{\mathbf{v} \in \mathscr{V}} J(x^0, \boldsymbol{\mu}, \mathbf{v})$, i.e.,

$$J^*(x^0) = \min_{\boldsymbol{\mu} \in \mathscr{U}} \max_{\mathbf{v} \in \mathscr{V}} J(x^0, \boldsymbol{\mu}, \mathbf{v}). \tag{5.3}$$

5.2.2 Zero-Sum Markov Games

The noncooperative objective function (5.3) leads to the solution concept of *saddle-point equilibrium* in game theory.

Definition 1. A saddle-point equilibrium (SPE) $(\boldsymbol{\mu}^*, \mathbf{v}^*) \in \mathscr{U} \times \mathscr{V}$ of the discounted zero-sum Markov games with two players satisfies the following inequalities:

$$J(x^0, \boldsymbol{\mu}, \mathbf{v}^*) \geq J(x^0, \boldsymbol{\mu}^*, \mathbf{v}^*) \geq J(x^0, \boldsymbol{\mu}^*, \mathbf{v}), \forall \mathbf{v} \in \mathscr{V}, \boldsymbol{\mu} \in \mathscr{U}, \forall x^0 \in \prod_{i=1}^{I} \mathscr{X}^i. \tag{5.4}$$

The value $J^*(x^0)$ achieved under the saddle-point equilibrium of the game (5.3) for a given initial condition x^0 is called the value function of a two-player zero-sum game, i.e.,

$$J^*(x^0) := J(x^0,\boldsymbol{\mu}^*,\boldsymbol{\nu}^*) = \min_{\boldsymbol{\mu}\in\mathscr{U}} \max_{\boldsymbol{\nu}\in\mathscr{V}} J(x^0,\boldsymbol{\mu},\boldsymbol{\nu}) = \max_{\boldsymbol{\nu}\in\mathscr{V}} \min_{\boldsymbol{\mu}\in\mathscr{U}} J(x^0,\boldsymbol{\mu},\boldsymbol{\nu}). \tag{5.5}$$

By focusing on the class of global stationary policies, i.e., $\mu_j^{i,\mathrm{GS}} \in \mathscr{U}_j^{i,\mathrm{GS}}$ and $\nu_j^{i,\mathrm{GS}} \in \mathscr{V}_j^{i,\mathrm{GS}}$, the value function $J^*(x^0)$ can be characterized using dynamic programming principles. The action pairs $\mathbf{d}^*,\mathbf{a}^*$ with $d_j^{i*} = \mu_j^{i*,\mathrm{GS}}(x)$ and $a_j^{i*} = \nu_j^{i*,\mathrm{GS}}(x)$ satisfy the following Bellman equation:

$$J^*(x) = c(x,\mathbf{d}^*,\mathbf{a}^*) + \gamma \sum_{x'\in\prod_{i=1}^{I}\mathscr{X}^i} \Pr(x'|x,\mathbf{a}^*,\mathbf{d}^*)J^*(x'), \forall x. \tag{5.6}$$

The first term is the reward of current stage x, and the second term is the expectation of the value function over all the possible next stage x'. The optimal action pairs $(\mathbf{d}^*,\mathbf{a}^*)$ guarantee that the value function starting from x equals the current stage cost plus the expectation starting at the next stage x'. By solving the Bellman equation (5.6) for every state x, we can obtain the saddle-point equilibrium strategy pairs $(\boldsymbol{\mu}^*,\boldsymbol{\nu}^*)$ in global stationary policies.

The global stationary saddle-point policies in pure strategies may not always exist. The Bellman equation (5.7) can be solved under mixed-strategy action spaces. Let the mixed-strategy actions for the attacker and the defender be $\phi^a(x,\mathbf{a})$ and $\phi^d(x,\mathbf{d})$, where $\phi^d(x,\mathbf{d})$ (resp., $\phi^a(x,\mathbf{a})$) denotes the probability of taking action $\mathbf{d}$ (resp., $\mathbf{a}$) at the global state x for a defender (or an attacker). The saddle-point mixed-strategy action pair $(\phi^{a*}(x,\mathbf{a}),\phi^{d*}(x,\mathbf{d}))$ satisfies the following generalized Bellman equation:

$$J^*(x) = \sum_{\mathbf{a}\in\mathscr{A}} \phi^{a*}(x,\mathbf{a}) \sum_{\mathbf{d}\in\mathscr{D}} \left[c(x,\mathbf{d},\mathbf{a}) + \gamma \sum_{x'\in\prod_{i=1}^{I}\mathscr{X}^i} \Pr(x'|x,\mathbf{a},\mathbf{d})J^*(x') \right] \phi^{d*}(x,\mathbf{d}), \forall x. \tag{5.7}$$

The existence of the mixed-strategy action pair is guaranteed when the action spaces $\mathscr{A}$ and $\mathscr{D}$ are finite. Hence solving (5.7) for every state x, we can obtain the mixed-strategy saddle-point equilibrium strategy pairs $(\hat{\boldsymbol{\mu}}^*,\hat{\boldsymbol{\nu}}^*)$ in global stationary policies, where $\hat{\boldsymbol{\mu}},\hat{\boldsymbol{\nu}}$ are the mixed strategy extension of $\boldsymbol{\mu},\boldsymbol{\nu}$, respectively.

5.2.3 Mathematical Programming Perspective

One way to compute the mixed-strategy equilibrium solutions for zero-sum games is to use a mathematical programming approach. Given a defender's policy $\phi^d(x,\mathbf{d})$, the attacker solves the following maximization problem for every state x:

$$J^*(x) = \max_{\phi^a(x,\mathbf{a})} \sum_{\mathbf{a}\in\mathscr{A}} \phi^a(x,\mathbf{a}) \sum_{\mathbf{d}\in\mathscr{D}} \left[c(x,\mathbf{d},\mathbf{a}) + \gamma \sum_{x'} \Pr(x'|x,\mathbf{a},\mathbf{d}) J^*(x') \right] \phi^d(x,\mathbf{d}), \forall x. \tag{5.8}$$

Define $f(x,\mathbf{a}) := \sum_{\mathbf{d}\in\mathscr{D}} \left[c(x,\mathbf{d},\mathbf{a}) + \gamma \sum_{x'\in\prod_{i=1}^{I}\mathscr{X}^i} \Pr(x'|x,\mathbf{a},\mathbf{d}) J^*(x') \right] \phi^d(x,\mathbf{d})$ and $f^*(x,\mathbf{a})$ when the defender's policy is optimal. We have the following lemma:

Lemma 1. *The optimal attacker's policy $\phi^{a*}(x,\mathbf{a})$ of* (5.8) *is a pure policy $\phi^a(x,\mathbf{a})\mathbb{1}_{\{\mathbf{a}=\mathbf{a}^*\}}$ when the defender's policy is given, where $\mathbf{a}^* \in \arg\max_{\mathbf{a}} f(x,\mathbf{a})$.*

Proof. There exists an optimal action $\mathbf{a}^* \in arg\max_{\mathbf{a}} f(x,\mathbf{a})$ and $f(x,\mathbf{a}^*) \geq f(x,\mathbf{a}), \forall \mathbf{a}$. As a probability measure, all elements of $\phi^a(x,\mathbf{a})$ are positive and $\sum_{\mathbf{a}\in\mathscr{A}} \phi^a(x,\mathbf{a}) = 1, \forall x$. Multiply both sides of the equation $f(x,\mathbf{a}^*) \geq f(x,\mathbf{a})$ by $\phi^a(x,\mathbf{a})$ and sum over all possible $\mathbf{a}$, we arrive at

$$\begin{aligned} \sum_{\mathbf{a}\in\mathscr{A}} \phi^a(x,\mathbf{a}) f(x,\mathbf{a}) &\leq \sum_{\mathbf{a}\in\mathscr{A}} \phi^a(x,\mathbf{a}) f(x,\mathbf{a}^*) = 1 \cdot f(x,\mathbf{a}^*) \\ &= \sum_{\mathbf{a}\in\mathscr{A}} \phi^a(x,\mathbf{a}) \mathbb{1}_{\{\mathbf{a}=\mathbf{a}^*\}} f(x,\mathbf{a}), \quad \forall \mathbf{a}. \end{aligned}$$

Therefore, the optimal attacker's policy is deterministic, i.e., $\phi^a(x,\mathbf{a})\mathbb{1}_{(\mathbf{a}=\mathbf{a}^*)}$.

Lemma 1 is true for arbitrary defender's policy, thus true for the optimal one. Therefore, $J^*(x) = f^*(x,\mathbf{a}^*) \geq f^*(x,\mathbf{a}), \forall \mathbf{a}$. Now, we can form a bilinear program:

$$\min_{J^*(x),\phi^d(x,\mathbf{d})} \sum_{x\in\prod_{i=1}^{I}\mathscr{X}^i} \alpha(x) J^*(x)$$

subject to

$$(a)\ J^*(x) \geq \sum_{\mathbf{d}\in\mathscr{D}} \left[c(x,\mathbf{d},\mathbf{a}) + \gamma \sum_{x'\in\prod_{i=1}^{I}\mathscr{X}^i} \Pr(x'|x,\mathbf{a},\mathbf{d}) J^*(x') \right] \phi^d(x,\mathbf{d}),\ \forall x,\mathbf{a} \tag{5.9}$$

$$(b)\ \sum_{\mathbf{d}\in\mathscr{D}} \phi^d(x,\mathbf{d}) = 1, \qquad \forall x$$

$$(c)\ \phi^d(x,\mathbf{d}) \geq 0, \qquad \forall x,\mathbf{d}$$

Constraints $(b)(c)$ reflect $\phi^d(x,\mathbf{d})$ as a probability measure. Constraint (a) guarantees that (5.8) is achieved under the optimal defender's policy. State-dependent weights $\alpha(x)$ are positive and satisfy $\sum_x \alpha(x) = 1$. Solutions of this program provide us the value function $J^*(x)$ and the optimal defender's policy $\phi^{d*}(x,\mathbf{d})$.

5.2.4 Single-Controller Markov Game

In the single-controller game, one player's action entirely determines transition probabilities. This structure captures the fact that the failure probability of a node in the infrastructure depends on the action taken by the attacker once the node is attacked.

The single-controller assumption fits the infrastructure protection application because of the deficiency in real-time attack countermeasure after infrastructure networks are designed. Thus, defenders may not be capable of decreasing the probability of node failures under attacks once the network is established. However, the protection term can positively enhance the system security by mitigating the attack loss or increase the cost of an attacker. For example, defenders can apply for the cyber-insurance for high-risk nodes or set up "honeypot" to increase the cost of the adversaries once trapped.

We focus on an attacker-controlled game Γ^a where the stochastic kernel for each node possesses $\Pr_{i,j}(x_j^{i'}|x, d_j^i, a_j^i) = \Pr_{i,j}(x_j^{i'}|x, a_j^i), \forall x_j^{i'}, x, d_j^i, a_j^i$ and the system transition probability $\Pr(X' = x'|X = x, \mathbf{d}, \mathbf{a}) = \Pr(X' = x'|X = x, \mathbf{a})$. Because the system transition probability is independent of $\mathbf{d}$ and $\sum_{\mathbf{d}} \phi^{d*}(x, \mathbf{d}) \equiv 1$, the bilinear program (5.9) can be reduced into a linear program (LP) where the primal LP is described as follows:

$$
\begin{aligned}
&\min_{J^*(x),\phi^d(x,\mathbf{d})} \quad \sum_{x' \in \prod_{i=1}^I \mathscr{X}^i} \alpha(x')J^*(x') \\
&\text{subject to} \\
&(a)\ J^*(x) \geq \sum_{\mathbf{d}\in\mathscr{D}} c(x,\mathbf{d},\mathbf{a})\phi^d(x,\mathbf{d}) + \gamma \sum_{x' \in \prod_{i=1}^I \mathscr{X}^i} \Pr(x'|x,\mathbf{a})J^*(x') \quad \forall x, \mathbf{a} \\
&(b)\ \sum_{\mathbf{d}\in\mathscr{D}} \phi^d(x,\mathbf{d}) = 1, \quad \forall x \\
&(c)\ \phi^d(x,\mathbf{d}) \geq 0, \quad \forall x, \mathbf{d}
\end{aligned}
\tag{5.10}
$$

After solving (5.10), we obtain the value functions $J^*(x')$ and the optimal defender's policy $\phi^{d*}(x, \mathbf{d})$, and we resort to the dual LP for the attacker's policy:

$$
\begin{aligned}
&\max_{z(x),\phi^a(x,\mathbf{a})} \quad \sum_{x \in \prod_{i=1}^I \mathscr{X}^i} z(x) \\
&\text{subject to} \\
&(d)\ \sum_{\mathbf{a}\in\mathscr{A}} \phi^a(x',\mathbf{a}) - \sum_{x \in \prod_{i=1}^I \mathscr{X}^i} \sum_{\mathbf{a}\in\mathscr{A}} \gamma \Pr(x'|x,\mathbf{a})\phi^a(x,\mathbf{a}) = \alpha(x'),\ \forall x' \\
&(e)\ z(x) \leq \sum_{\mathbf{a}\in\mathscr{A}} \phi^a(x,\mathbf{a})c(x,\mathbf{d},\mathbf{a}) \quad \forall x, \mathbf{d} \\
&(f)\ \phi^a(x,\mathbf{a}) \geq 0, \quad \forall x, \mathbf{a}
\end{aligned}
\tag{5.11}
$$

We normalize $\phi^{a*}(x, \mathbf{a}) = \frac{\phi^a(x,\mathbf{a})}{\sum_{\mathbf{a}} \phi^a(x,\mathbf{a})}$ to obtain the optimal policy for attacker. Analogous to the optimality principle of the value function (5.6), constraint (d) in the dual LP can be interpreted as the occupancy equality. The total occupancy frequency of state x', $\sum_{\mathbf{a}\in\mathscr{A}} \phi^a(x', \mathbf{a})$, is equal to the initial probability distribution of state

x', $\alpha(x')$, plus the discounted expected visit from any other state x to state x', i.e., $\sum_{x\in\prod_{i=1}^{I}\mathscr{X}^i}\sum_{\mathbf{a}\in\mathscr{A}}\gamma\Pr(x'|x,\mathbf{a})\phi^a(x,\mathbf{a})$.

Theorem 1. *The optimal policy of attacker* $\phi^a(x,\mathbf{a})$ *solved by* (5.11) *is a pure policy, i.e., for each system state x,* $\phi^a(x,\mathbf{a}^*) > 0$ *and* $\phi^a(x,\mathbf{a}) = 0, \forall \mathbf{a} \neq \mathbf{a}^*$. *The explicit form is*

$$\mathbf{a}^* = arg\max_{\mathbf{a}\in\mathscr{A}}\left[\sum_{\mathbf{d}\in\mathscr{D}} c(x,\mathbf{d},\mathbf{a})\phi^{d*}(x,\mathbf{d}) + \gamma\sum_{x'}\Pr(x'|x,\mathbf{a})J^*(x')\right].$$

Proof. Lemma 1 has shown that the optimal policy is deterministic, and thus here we only need to show that $\phi^{a*}(x,\mathbf{a}) = \frac{\phi^a(x,\mathbf{a})}{\sum_{\mathbf{a}}\phi^a(x,\mathbf{a})}$ is the optimal policy for the attacker. Following the proof of [6], we show that $\phi^{a*}(x,\mathbf{a})$ is the saddle point of the zero-sum game (5.4).

First, $\phi^{a*}(x,\mathbf{a})$ is well defined since the constraint (d) shows that $\sum_{\mathbf{a}}\phi^a(x,\mathbf{a}) \geq \alpha(x'), \forall x'$. By the complementary slackness of the dual linear program, we require $J^*(x)$ strictly equal to $\sum_{\mathbf{d}\in\mathscr{D}} c(x,\mathbf{d},\mathbf{a})\phi^{d*}(x,\mathbf{d}) + \gamma\sum_{x'}\Pr(x'|x,\mathbf{a})J^*(x')$ for all state x and the corresponding action $\mathbf{a}$ such that $\phi^a(x,\mathbf{a})$ is strictly positive, which is equivalent to $\phi^{a*}(x,\mathbf{a}) > 0$. Then, by multiplying both side by $\phi^{a*}(x,\mathbf{a})$ and summing over $\mathbf{a} \in \mathscr{A}$, we obtain the vector equation $\mathbf{J}^* = J(x^0,\boldsymbol{\mu}^*,\boldsymbol{\nu}^*)$. Next, we multiply an arbitrary $\phi^a(x,\mathbf{a})$ to both sides of constraints (a), sum over $\mathbf{a}$, and obtain a vector inequality $\mathbf{J}^* \geq J(x^0,\boldsymbol{\mu}^*,\boldsymbol{\nu})$. Therefore, we arrive at the right hand side (RHS) of saddle-point condition $J(x^0,\boldsymbol{\mu}^*,\boldsymbol{\nu}) \leq J(x^0,\boldsymbol{\mu}^*,\boldsymbol{\nu}^*)$. Similarly, the complementary slackness of the primal LP together with constraints (e) leads to $c(x,\mathbf{a}^*,\mathbf{d}^*) \leq c(x,\mathbf{a}^*,\mathbf{d})$. Because the transition probability is independent of defender's policy, we can obtain the left hand side (LHS) of the saddle-point condition by computing (5.1).

The major challenge to solve the LP is the large-scale nature of the infrastructure networks, which is known as the curse of dimension. Take (5.10) for an instance, we have $|\prod_{i=1}^{I}\mathscr{X}^i|$ variables in the LP objective and a constraints number of $|\prod_{i=1}^{I}\mathscr{X}^i|\times|\mathscr{A}| + |\prod_{i=1}^{I}\mathscr{X}^i| + |\prod_{i=1}^{I}\mathscr{X}^i|\times|\mathscr{D}|$. If we have n nodes in the network of CIs and all nodes can be attacked and defended, then we will have $N := 2^n$ variables and N^2+N+N^2 constraints, which both grow exponentially with the number of nodes. The high computation cost prohibits the direct computation using the LP with a large number of nodes.

5.3 Factored Markov Game

To address the issue of the combinatorial explosion of the state size or the curse of dimensionality, we develop a factored Markov game framework in this section by leveraging the sparsity of the transition kernel. We first use factor graphs to represent the sparse structure of the probability transition matrix. Next, we introduce an approximation method for the value function and then reorganize terms and eliminate variables by exploiting the factored structure. We focus on the linear programming formulation of the attacker-controlled game. However, the technique can be extended to a bilinear form for the general zero-sum game to reduce computational complexity. Finally, we refer our reader to an overall structure diagram of this work in Figure 5.2.

5.3.1 Factored Structure

Define Ω_l as the set that contains all the parent nodes of node l. Parent nodes refer to the nodes that affect node l's state at the following time step through physical, cyber, or logic interactions. The network example in Figure 5.3 is a bidirected graph that represents a 3-layer interdependent critical infrastructures. Then, Ω_l contains node l itself and all its neighbors, e.g., $\Omega_{1,1} = \{n_1^1, n_2^1, n_1^2, n_7^3\}$. Node l can affect itself because if, for instance, node l fails at time t, then it remains faulty in probability one without proper actions at next time step $t+1$. Note that we do not distinguish the dependence within (links in black) and across (links in blue) layers when considering the stochastic kernel. Recall m_i as the total number of nodes in layer i. We use a global index l to unify the 2D index of $\{i, j\}$, e.g., $l := \sum_{i'=1}^{i} i' m_{i'} + j$, which transforms the multilayer network into a larger single network with $n = \sum_{i \in \mathscr{I}} m_i$ nodes. In this way, we can write $\Omega_{1,1} = \{n_1^1, n_2^1, n_1^2, n_7^3\}$ as $\Omega_1 = \{n_1, n_2, n_6, n_{19}\}$ and $\Pr_{i,j}(x_j^{i\,\prime} | x, d_j^i, a_j^i), \forall i \in \mathscr{I}, j \in \mathscr{N}^i$ equivalently as $\Pr_l(x_l{}' | x, d_l, a_l), \forall l = 1, 2, \cdots, n$. Define $x_{\Omega_l} := (x_l)_{l \in \Omega_l}$ as the state vector of the nodes inside set Ω_l, e.g., $x_{\Omega_1} = (x_1, x_2, x_6, x_{17})$. Then, each node's kernel will be $\Pr_{i,j}(x_j^{i\,\prime} | x, d_j^i, a_j^i) = \Pr_{i,j}(x_j^{i\,\prime} | x_j^i, x_{\Omega_{i,j}}, d_j^i, a_j^i)$ due to the sparsity, or in the global index $\Pr_l(x_l{}' | x, d_l, a_l) = \Pr_l(x_l' | x_l, x_{\Omega_l}, d_l, a_l)$.

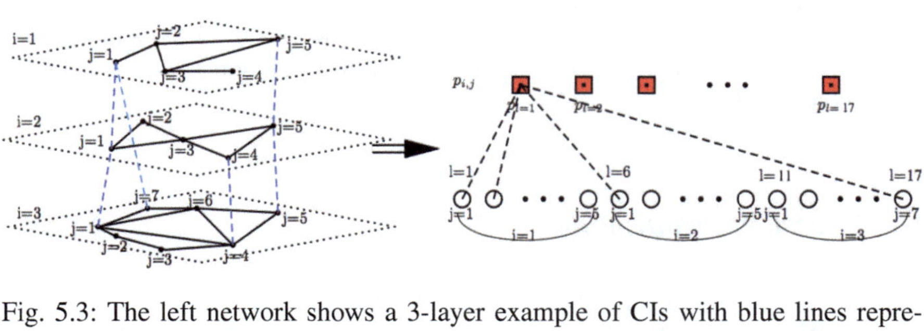

Fig. 5.3: The left network shows a 3-layer example of CIs with blue lines representing the interdependencies across layers. The right bipartite graph shows a factor graph representation of the sparse transition probability. The total node number $n = \sum_{i=1,2,3} m_i = 5 + 5 + 7 = 17$.

5.3.2 Linear Function Approximation

We first approximate the high-dimensional space spanned by the cost function vector $\mathbf{J} = (J^*(x'))_{x' \in \prod_{i=1}^{I} \mathscr{X}^i}$ through a weighted sum of basis functions $h_l(x'), l = 0, 1, \cdots, k$, where k is the number of "features" and $h_0(x') \equiv 1, \forall x'$. Take infrastructure networks as an example. We choose a set of basis which serves as an indicator function of each node n_j^i's working state, e.g., $h_{i,j}(x') = x_j^{i\,'}, \forall i \in \mathscr{I}, j \in \mathscr{N}_j^i$. We unify the index with $l := \sum_{i'=1}^{i} i' m_{i'} + j$ and k equal to n, the total node number in the network. To this end, we can substitute $J^*(x') = \sum_{l=0}^{k} w_l h_l(x')$ into (5.10) to obtain an approximate linear programming (ALP) with k variables $w_l, l = 0, 1, \cdots, k$.

$$
\begin{aligned}
&\min_{\mathbf{w}, \phi^d(x,\mathbf{d})} \sum_{x' \in \prod_{i=1}^{I} \mathscr{X}^i} \alpha(x') \sum_{l=0}^{k} w_l h_l(x') \\
&\text{subject to} \\
&(a) \sum_{l=0}^{k} w_l h_l(x) \geq \sum_{\mathbf{d} \in \mathscr{D}} c(x, \mathbf{d}, \mathbf{a}) \phi^d(x, \mathbf{d}) + \gamma \sum_{x' \in \prod_{i=1}^{I} \mathscr{X}^i} \Pr(x'|x, \mathbf{a}) \sum_{l=0}^{k} w_l h_l(x'), \ \forall x, \mathbf{a} \\
&(b) \sum_{\mathbf{d} \in \mathscr{D}} \phi^d(x, \mathbf{d}) = 1, \qquad \forall x \\
&(c)\ \phi^d(x, \mathbf{d}) \geq 0, \qquad \forall x, \mathbf{d}
\end{aligned}
\tag{5.12}
$$

The feature number k is often much smaller than the system state number 2^n. Hence the ALP reduces the involving variables in the LP objective. However, the exponentially growing number of constraints still makes the computation prohibitive. To address this issue, we further reduce the computational complexity in the following sections with similar techniques in [8].

Remark: The ALP approximates $\min_{\boldsymbol{\mu} \in \mathscr{U}} \max_{\boldsymbol{\nu} \in \mathscr{V}} J(x^0, \boldsymbol{\mu}, \boldsymbol{\nu})$. The minimax strategy yields the optimal defensive strategy for the worst-case attacks, which is achieved by searching the entire feasible attackers' actions of all possible system states in constraint (a) of (5.12). Thus, the approximate solution $\sum_{l=0}^{k} w_l h_l(x')$ is an upper bound to $J^*(x')$.

5.3.3 Term Reorganization

The system transition matrix $\Pr(x'|x, \mathbf{a})$ has the dimension of $N \times N \times |\mathscr{A}|$ in constraint (a) of (5.10). Here, we choose indicator functions of each node $h_l(x') = x_l, \forall x', l = \{1, 2, \cdots, n\}$ as the set of basis functions, which yields a good trade-off between the accuracy and computation complexity as shown in Section 5.4. We observe that the rightmost term of constraint (a) of (5.10) can be rewritten as follows:

$$
\begin{aligned}
&\sum_{x' \in \prod_{i=1}^{I} \mathscr{X}^i} \Pr(x'|x,\mathbf{a}) \sum_{l=0}^{n} w_l h_l(x') \\
&\stackrel{(1)}{=} w_0 + \sum_{l=1}^{n} w_l \left[\sum_{x_1',\cdots,x_n'} \prod_{k=1}^{n} \Pr_k(x_k'|x_k,a_k) x_l \right] \\
&\stackrel{(2)}{=} w_0 + \sum_{l=1}^{n} w_l \left[\sum_{x_l'} \Pr_l(x_l'|x_l,x_{\Omega_l},a_l) x_l \sum_{\{x_1',\cdots,x_n'\}\backslash\{x_l'\}} \prod_{k=1,k\neq l}^{n} \Pr_k(x_k'|x_k,a_k) \right] \\
&\stackrel{(3)}{=} w_0 + \sum_{l=1}^{n} w_l \left[\sum_{x_l'} \Pr_l(x_l'|x_l,x_{\Omega_l},a_l) x_l \prod_{k=1,k\neq l}^{n} \sum_{x_k'} \Pr_k(x_k'|x_k,a_k) \right] \\
&\stackrel{(4)}{=} w_0 + \sum_{l=1}^{n} w_l \left[\sum_{x_l'} \Pr_l(x_l'|x_l,x_{\Omega_l},a_l) x_l \right] \\
&= w_0 + \sum_{l=1}^{n} w_l \left[\Pr_l(x_l' = 1|x_l,x_{\Omega_l},a_l)\cdot 1 + \Pr_l(x_l' = 0|x_l,x_{\Omega_l},a_l)\cdot 0 \right] \\
&= w_0 + \sum_{l=1}^{n} w_l \left[\Pr_l(x_l' = 1|x_l,x_{\Omega_l},a_l) \right] := w_0 + \sum_{l=1}^{n} w_l g_l(x_l,x_{\Omega_l},a_l),
\end{aligned}
$$

where $g_l(x_l,x_{\Omega_l},a_l) := \Pr_l(x_l' = 1|x_l,x_{\Omega_l},a_l)$.

Equation (1) represents the vector x' with the set of its elements $\{x_i'\}$, writes the system transition probability in its factored form, and separates the first constant item w_0. The symbol $\sum_{\{x_1,\cdots,x_n\}\backslash\{x_l\}}$ in Equation (2) means the summation over all variables except x_l. Equation (3) exchanges the summation and multiplication, and Equation (4) is true because $\sum_{x_k'} \Pr_k(x_k'|x_k,a_k) \equiv 1$. To this end, we reduce $N = 2^n$ summations over the huge dimension system transition matrix into $n+1$ summations over the local stochastic kernel.

5.3.4 Restricted Information Structure

The second step is to deal with $\sum_{\mathbf{d}} c(x,\mathbf{d},\mathbf{a})\phi^d(x,\mathbf{d})$ in constraint (a) of (5.10). The saddle-point strategies studied in Section 5.2.2 belong to a class of global stationary policies in which the actions taken by the players are dependent on the global state information. The implementation of the policies is often restricted to the local information that is specific to the type of the infrastructure. For example, the Metropolitan Transportation Authority (MTA) may not be able to know the state of nodes in the power grid operated by Con Edison. Thus, MTA cannot make its policy based on the states of power nodes. Therefore, one way to approximate the optimal solution is to restrict the class of policies to stationary policies with local observations. We consider a time-invariant information structure of the defender $F_{j,t}^i \equiv F_j^i$.

By unifying with the global index in Section 5.3.1, we let $l := \sum_{i'=1}^{i} i' m_{i'} + j$ and $F_l := F_j^i$. Define $\phi_l^d(x, d_l)$ as the probability of node l choosing d_l at state x. Therefore, $\phi^d(x, \mathbf{d}) = \prod_{l=1}^{n} \phi_l^d(x, d_l) = \prod_{l=1}^{n} \phi_l^d(F_l, d_l)$ and $F_l = (x_{\bar{\Omega}_l})$, where $\bar{\Omega}_l$ is the set of nodes which node l can observe. Note that not all nodes can be protected, i.e., $|\mathscr{D}| \leq N$. We let $d_l \equiv 0$ if node l cannot be defended.

$$
\begin{aligned}
\sum_{\mathbf{d} \in \mathscr{D}} c(x, \mathbf{d}, \mathbf{a}) \phi^d(x, \mathbf{d}) &= \sum_{\mathbf{d} \in \mathscr{D}} \sum_{k=1}^{n} c_k(x_k, d_k, a_k) \prod_{l=1}^{n} \phi_l^d(F_l, d_l) \\
&= \sum_{k=1}^{n} \left[\sum_{d_w, w=1,\cdots,|D|} c_k(x_k, d_k, a_k) \phi_k^d(F_k, d_k) \prod_{l=1, l \neq k}^{n} \phi_l^d(F_l, d_l) \right] \\
&= \sum_{k=1}^{n} \left[\sum_{d_k} c_k(x_k, d_k, a_k) \phi_k^d(F_k, d_k) \prod_{l=1, l \neq k}^{n} \sum_{d_l} \phi_l^d(F_l, d_l) \right] \\
&= \sum_{k=1}^{n} \left[\sum_{d_k \in \{0,1\}} c_k(x_k, d_k, a_k) \phi_k^d(F_k, d_k) \right].
\end{aligned}
\tag{5.13}
$$

Therefore, the ALP with the restricted information structure can be further rewritten as follows to form the factored ALP:

$$
\begin{aligned}
&\min_{\mathbf{w}, \phi_l^d(F_l, d_l)} \sum_{l=0}^{n} \alpha(w_l) w_l h_l(x) \\
&\text{subject to} \\
&(a)\ 0 \geq \sum_{k=1}^{n} \sum_{d_k \in \{0,1\}} c_k(x_k, d_k, a_k) \phi_k^d(F_k, d_k) + \sum_{l=0}^{n} w_l [\gamma g_l(x_l, x_{\Omega_l}, a_l) - h_l(x)], \ \forall x, a_l \\
&(b)\ \sum_{d_l \in \{0,1\}} \phi_l^d(F_l, d_l) = 1, \qquad \forall l, F_l \\
&(c)\ 0 \leq \phi_l^d(F_l, d_l) \leq 1, \qquad \forall l, F_l, d_l
\end{aligned}
\tag{5.14}
$$

To this end, the number of constraints (b) $n \times |F_l|$ and (c) $n \times |F_l| \times 2$ relates only to the node number n and the domain of each node's information structure.

Remark: For a general zero-sum game with bilinear programming formulation (5.9), we can extend constraint (a) as follows with the same factored technique:

$$
\begin{aligned}
0 \geq &\sum_{k=1}^{n} \sum_{d_k \in \{0,1\}} c_k(x_k, d_k, a_k) \phi_k^d(F_k, d_k) \\
&+ \sum_{l=0}^{n} w_l [\gamma \sum_{d_l \in \{0,1\}} g_l(x_l, x_{\Omega_l}, a_l) \phi_l^d(F_l, d_l) - h_l(x)], \quad \forall x, a_l,
\end{aligned}
$$

where the second term is bilinear in the variables of w_l and $\phi_l^d(F_l, d_l)$.

5.3.5 Variable Elimination

Constraint (a) of (5.14) can be further rewritten as one nonlinear constraint using the variable elimination method (see Section 4.2.2 of [7]) as follows:

$$0 \geq \max_{a_1,\cdots,a_n} \max_{x_1,\cdots,x_n} \sum_{k=1}^{n} \sum_{d_k \in \{0,1\}} c_k(x_k,d_k,a_k)\phi_k^d(F_k,d_k) + \sum_{l=0}^{n} w_l[\gamma g_l(x_l,x_{\Omega_l},a_l) - h_l(x)]. \tag{5.15}$$

For simplicity, we have provided above an inequality for the case of a local information structure $\phi_l^d(F_l,d_l) = \phi_l^d(x_l,x_{\Omega_l},d_l)$ and $|F_l| = 2^{|\Omega_l|+1}$.

First, we eliminate the variables of the attackers' action. Define $f_l(x_l,x_{\Omega_l},a_l) := w_l[\gamma g_l(x_l,x_{\Omega_l},a_l) - h_l(x_l)] + \sum_{d_l} c_l(x_l,d_l,a_l)\phi_l^d(x_l,d_l), l = 1,2,\cdots,n$. We separate w_0, the weight of the constant basis, to the left-hand side and (5.15) becomes

$$\begin{aligned}(1-\gamma)w_0 &\geq \max_{x_1,\cdots,x_n} \max_{a_1,\cdots,a_n} \sum_{l=1}^{n} f_l(x_l,x_{\Omega_l},a_l) \\ &= \max_{x_1,\cdots,x_n} \sum_{l=1}^{n} \max_{a_l} f_l(x_l,x_{\Omega_l},a_l) \\ &:= \max_{x_1,\cdots,x_n} \sum_{l=1}^{n} e_l(x_l,x_{\Omega_l}).\end{aligned} \tag{5.16}$$

To achieve the global optimal solution of (5.14), we impose the following constraints for each l:

$$e_l(x_l,x_{\Omega_l}) \geq f_l(x_l,x_{\Omega_l},a_l), \quad \forall x_l,x_{\Omega_l},a_l. \tag{5.17}$$

Note that if node n_l cannot be attacked, we take $a_l \equiv 0$ and arrive at a simplified form:

$$e_l(x_l,x_{\Omega_l}) = f_l(x_l,x_{\Omega_l},1), \quad \forall x_l,x_{\Omega_l}. \tag{5.18}$$

The second step is to eliminate the variable of each node's state following a given order of $\mathcal{O} = \{p_1,p_2,\cdots,p_n\}$, where $\mathcal{O}$ is a permutation of $\{1,2,\cdots,n\}$. The RHS of (5.16) is rewritten as

$$\begin{aligned}&\max_{x_1,\cdots,x_n} \sum_{l=1}^{n} e_l(x_l,x_{\Omega_l}) \\ &= \max_{x_{p_2},\cdots,x_{p_n}} \sum_{l=\{1,\cdots,n\}\setminus\mathcal{K}} e_k(x_k,x_{\Omega_k}) + \max_{x_{p_1}} \sum_{k\in\mathcal{K}} e_k(x_k,x_{\Omega_k}) \\ &= \max_{p_2,\cdots,p_n} \sum_{l=\{1,\cdots,n\}\setminus\mathcal{K}} e_k(x_k,x_{\Omega_k}) + E_1(\mathcal{E}),\end{aligned} \tag{5.19}$$

where the set $\mathcal{K} := \{k : p_1 \in \{\Omega_k \cup \{k\}\}\}$ and E_1's domain $\mathcal{E} := \{x_j : j \in \{\{\cup_{k\in\mathcal{K}}\Omega_k\} \cup \{k\} \setminus \{p_1\}\}\}$. The variable x_{p_1} is eliminated, and similar new constrains are generated to form the new LP, i.e., $E_1(\mathcal{E}) \geq \sum_{k\in\mathcal{K}} e_k(x_k,x_{\Omega_k})$, for all variables included in $\mathcal{E}$.

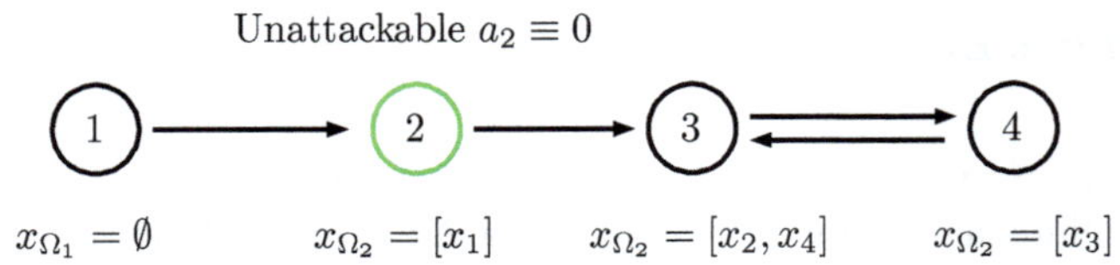

Fig. 5.4: A four node example with node 2 unattackable. Assume a local information structure for each node $F_l = x_l, l = 1,2,3,4$.

We repeat the above procedure of variable eliminations and constraints generation for n times following the order $\mathscr{O}$ and finally reach the equation $(1-\gamma)w_0 \geq E_n$, where E_n is a parameter independent of state and action variables. This method is suitable for a sparse network where each e_l has a domain involving a small set of node variables.

Example 1. Consider a four-node example in Figure 5.4 for the illustration of the variable elimination. With node 2 being immune to attacks, (5.18) can be reduced to $e_2(x_1,x_2) = f_1(x_1,x_2,0), \forall x_1,x_2$. For node 1, (5.17) leads to four new inequality constraints $e_1(x_1) \geq f_1(x_1,a_1), \forall x_1,a_1$. Similarly, we have $2^4 = 16$ inequalities for node 3, i.e., $e_3(x_2,x_3,x_4) \geq f_3(x_2,x_3,x_4,a_3), \forall x_2,x_3,x_4,a_3$ and $2^3 = 8$ for node 4, i.e., $e_4(x_3,x_4) \geq f_3(x_3,x_4,a_4), \forall x_3,x_4,a_4$. After that, we eliminate all action variables and (5.16) becomes

$$(1-\gamma)w_0 \geq \max_{x_1,x_2,x_3,x_4} e_1(x_1) + e_2(x_1,x_2) + e_3(x_2,x_3,x_4) + e_4(x_3,x_4). \tag{5.20}$$

With an elimination order $\mathscr{O} = \{3,2,4,1\}$, the RHS of (5.20) can be rewritten as

$$\begin{aligned} &\max_{x_1,x_2,x_4} e_1(x_1) + e_2(x_1,x_2) + \max_{x_3} e_3(x_2,x_3,x_4) + e_4(x_3,x_4) \\ &= \max_{x_1,x_2,x_4} e_1(x_1) + e_2(x_1,x_2) + E_1(x_2,x_4). \end{aligned}$$

The new constraints are generated, i.e., $E_1(x_2,x_4) \geq e_3(x_2,x_3,x_4) + e_4(x_3,x_4)$, $\forall x_2,x_3,x_4$. Then, we can repeat the above process and eliminate x_2,x_4,x_1 in sequence, i.e.,

$$\begin{aligned} &\max_{x_1,x_2,x_4} e_1(x_1) + e_2(x_1,x_2) + E_1(x_2,x_4) \\ &= \max_{x_1,x_4} e_1(x_1) + \max_{x_2} E_1(x_2,x_4) + e_2(x_1,x_2) \\ &= \max_{x_1,x_4} e_1(x_1) + E_2(x_1,x_4) \\ &= \max_{x_1} \max_{x_4} e_1(x_1) + E_2(x_1,x_4) \\ &= \max_{x_1} E_3(x_1) = E_4. \end{aligned}$$

Along with the above process, new constraints appear: $E_2(x_1,x_4) \geq E_1(x_2,x_4) + e_2(x_1,x_2), \forall x_1,x_2,x_4$; $E_3(x_1) \geq e_1(x_1) + E_2(x_1,x_4), \forall x_1,x_4$; and $E_4 \geq E_3(x_1), \forall x_1$. Finally, (5.20) becomes $(1-\gamma)w_0 \geq E_4$.

The new LP in this example contains 51 constraints, while the original constraint (a) includes $2^{(4+3)} = 128$ inequalities. With the increase of the node number and a sparse topology, our factored framework greatly reduces the exponential computation complexity. Note that the order of $\{1,2,3,4\}$ introduces the least number of constraints in this case, yet choosing the optimal order is shown to be NP-hard.

5.3.6 Distributed Policy of Attacker

Similar to Lemma 1, we search for the approximate saddle-point policy of the attacker as follows:

$$\mathbf{a}^* \in \arg\max_{a_1,\cdots,a_n} \sum_{k=1}^{n} \sum_{d_k \in \{0,1\}} c_k(x_k,d_k,a_k)\phi_k^{d*}(F_k,d_k) + \sum_{l=0}^{n} w_l \gamma g_l(x_l,x_{\Omega_l},a_l), \forall x_1,\cdots,x_n$$

Separate w_0 in the second term, and we obtain

$$\mathbf{a}^* \in \gamma w_0 + \arg\max_{a_1,\cdots,a_n} \sum_{k=1}^{n} \sum_{d_k} c_k(x_k,d_k,a_k)\phi_k^{d*}(F_k,d_k) + w_k \gamma g_k(x_k,x_{\Omega_k},a_k), \forall x_1,\cdots,x_n.$$

Exchanging the argmax and the summation, we arrive at

$$\mathbf{a}^* \in \gamma w_0 + \sum_{k=1}^{n} \arg\max_{a_k} \sum_{d_k} c_k(x_k,d_k,a_k)\phi_k^{d*}(F_k,d_k) + w_k \gamma g_k(x_k,x_{\Omega_k},a_k), \forall x_1,\cdots,x_n.$$

Therefore, we can obtain a distributed attack policy of node k which is fully determined by the state of itself and its parent nodes x_k, x_{Ω_k} and the state of nodes observable for the defender F_k, i.e.,

$$a_k = arg\max_{a_k} \sum_{d_k \in \{0,1\}} c_k(x_k,d_k,a_k)\phi_k^{d*}(F_k,d_k) + w_k \gamma g_k(x_k,x_{\Omega_k},a_k), \forall x_k,x_{\Omega_k},F_k.$$

Note that the approximate policy can be different from the optimal policy in Theorem 1. However, as long as the computation reduction surpasses the approximation error of the value function, it is worthwhile to equip with this suboptimal policy.

Remark: Under a local information structure with $F_l = x_l$, the defender decides its action at node l based on x_l, and yet the attacker requires the state information of x_l and x_{Ω_l}. The difference in the structures of the policies is caused by the distinct factored structures of the cost function and the attacker-controlled transition probability matrix. The former $c_k(x_k,d_k,a_k)$ contains only x_k, and the latter $g_l(x_l,x_{\Omega_l},a_l)$ contains both x_l and x_{Ω_l}.

5.3.7 Approximate Dual LP

We compute the dual of the ALP (5.12) by the Lagrange function. Our objective is to find a function $l(\mathbf{w},\phi^a(x,\mathbf{a}),z(x))$ such that $l(\mathbf{w},\phi^a(x,\mathbf{a}),z(x)) = 0$ when the constraints of (5.12) is satisfied and unbounded otherwise. Then, the following equation is equivalent to (5.12) :

$$\mathscr{L}(\mathbf{w},\phi^a(x,\mathbf{a}),z(x)) = \min_{\mathbf{w}} \left[\sum_{x'} \alpha(x') \sum_{l=1}^{k} w_l h_l(x') + \max_{\phi^a(x,\mathbf{a}),z(x)} l(\mathbf{w},\phi^a(x,\mathbf{a}),z(x)) \right].$$

Let

$$l(\mathbf{w},\phi^a(x,\mathbf{a}),z(x)) = \sum_x z(x)(1 - \sum_{\mathbf{d}\in\mathscr{D}} \phi^d(x,\mathbf{d})) +$$

$$\sum_x \sum_{\mathbf{a}} \phi^a(x,\mathbf{a}) \left[\sum_{\mathbf{d}\in\mathscr{D}} c(x,\mathbf{d},\mathbf{a})\phi^d(x,\mathbf{d}) + \sum_{x'} \gamma \Pr(x'|x,\mathbf{a})] \sum_{l=1}^{k} w_l h_l(x') - \sum_{l=1}^{n} w_l h_l(x) \right], \tag{5.21}$$

where $\phi^a(x,\mathbf{a}) \geq 0, \forall x,\mathbf{a}$ are multipliers for the inequality constraint (a). Next, we reorganize the term and follow the minimax theorem to obtain

$$\begin{aligned} \mathscr{L}(\mathbf{w},\phi^a(x,\mathbf{a}),z(x)) &= \max_{z(x)} \sum_x z(x) \\ &+ \max_{\phi^a(x,\mathbf{a}),z(x)} \{ \sum_x \sum_{\mathbf{d}} \phi^d(x,\mathbf{d}) [\sum_{\mathbf{a}} \phi^a(x,\mathbf{a}) c(x,\mathbf{d},\mathbf{a}) - z(x)] \\ &+ \min_{\mathbf{w}} \sum_l w_l [\sum_{x'} \alpha(x') h_l(x') \\ &+ \gamma \sum_x \sum_{\mathbf{a}} \phi^a(x,\mathbf{a}) \sum_{x'} \Pr(x'|x,\mathbf{a}) h_l(x') - \sum_x \sum_{\mathbf{a}} \phi^a(x,\mathbf{a}) h_l(x)] \}. \end{aligned} \tag{5.22}$$

Finally, we can obtain the dual of (5.12) as follows:

$$\begin{aligned} &\max_{z(x),\phi^a(x,\mathbf{a})} \sum_{x\in\prod_{i=1}^{I}\mathscr{X}^i} z(x) \\ &\text{subject to} \\ &(a) \sum_x \alpha(x) h_l(x) + \gamma \sum_x \sum_{\mathbf{a}} \phi^a(x,\mathbf{a}) \sum_{x'} \Pr(x'|x,\mathbf{a}) h_l(x') = \sum_x \sum_{\mathbf{a}} \phi^a(x,\mathbf{a}) h_l(x), \ \forall l \\ &(b)\ z(x) \leq \sum_{\mathbf{a}} \phi^a(x,\mathbf{a}) c(x,\mathbf{d},\mathbf{a}), \qquad \forall x,\mathbf{d} \\ &(c)\ \phi^a(x,\mathbf{a}) \geq 0, \qquad \forall x,\mathbf{a} \end{aligned} \tag{5.23}$$

The dual of the ALP reveals the fact that constraint (a) approximates constraint (d) of (5.10) while the objective and other constraints remain the same. The term $\gamma \sum_x \sum_{\mathbf{a}} \phi^a(x,\mathbf{a}) \sum_{x'} \Pr(x'|x,\mathbf{a}) h_l(x')$ sums over both x and x' in the same domain of

$\prod_{i=1}^{I} \mathscr{X}^i$, and thus we can exchange x and x' in this term. Let $x^{(i)}, i = 1, \cdots, N$ be $N = 2^n$ possible values of the system state and $\mathbf{h}_l = (h_l(x^{(i)}))_{i=1,\cdots,N}$. Define $q^i(x^{(i)}) := \alpha(x^{(i)}) + \gamma \sum_{\mathbf{a}} \phi^a(x^{(i)}, \mathbf{a}) \sum_{x'} \Pr(x^{(i)}|x', \mathbf{a}) - \sum_{\mathbf{a}} \phi^a(x^{(i)}, \mathbf{a})$ and $\mathbf{q} := (q^1(x^{(1)}), \cdots, q^N(x^{(N)}))^T$.

Then, constraint (a) can be rewritten in matrix form as $\mathbf{Hq} = \mathbf{0}$, where $\mathbf{H} = (\mathbf{h}_1, \mathbf{h}_2, \cdots, \mathbf{h}_k)^T \in \mathscr{R}^{k \times N}$, and we can regard (5.23) as a relaxation of (5.11). If we select $k = N$ basis functions $h_l(x), l = \{1, 2, \cdots, |\prod_{i=1}^{I} \mathscr{X}^i|\}$ to be an indicator function of each possible value of the system state $x^{(l)}$, i.e., $h_l(x) = \mathbb{1}_{\{x = x^{(l)}\}}, l = 1, \cdots, N$, matrix $\mathbf{H}$ turns out to be an $N \times N$ identity matrix. Then, we arrive at $\mathbf{q} = \mathbf{0}$, i.e., N constraints $q^i(x^{(i)}) = 0, \forall i = 1, \cdots, N$, which is the same as constraint (a) in (5.11). Actually, as long as $k = N$ and $\mathbf{H}$ is of full rank, we have $\mathbf{q} = \mathbf{0}$. However, we obtain k constraints if we choose k less than N in the approximation form (5.23) with a reduced number of constraints. For each equation, according to the basis function selection, the corresponding elements in $\mathbf{q}$ sum up to 0.

Remark: Analogous to the explanation of (5.11), constraint (a) in (5.23) achieves the occupancy equality for each feature rather than at each system state. For example, with the choice of the basis functions as $h_l(x') = x_l, \forall x', l = \{1, 2, \cdots, n\}$, the lth equation of constraint (a) in (5.11) becomes $\sum_{x^{(i)} \in \mathscr{X}} q^i(x^{(i)}) = 0$.

5.4 Numerical Experiments

We implement our framework of the factored single-controller game and investigate the LP objective function as well as the policy of the attack and defender. Besides, we compare the approximation accuracy and the computation time. The LP objective shows in average the accuracy of the value functions starting at different initial states, which reflects the security level of the system. This risk analysis can have applications in areas such as cyber-insurance where risky systems have high premium rates. We use the pseudocode to summarize the algorithm for computing the saddle-point equilibrium policies for the factored single-controller game framework as follows.

5.4.1 Transition Probability and Cost

To illustrate the algorithm, we take one node's failure probability proportional to the failure number of its neighboring nodes. After one node is attacked, it can infect the connecting nodes and increase their failing risks. Besides, a node has a larger failure probability if it is targeted directly by attackers. In an attacker-controlled game, the defender cannot change the failure probability yet can positively affect the cost function.

---

procedure INITIALIZE
Initialize topology $\mathcal{G}$, elimination order $\mathcal{O}$, vector *aflag* (and *dflag*) to indicate whether a node is controllable by attackers (and defenders).

Define ALP variables $w = \{w_0, w_1, \cdots, w_n\}$ and $\phi^d = \{\phi_i^d(x_i, d_i)\}_{i=1,\cdots,n}$.
Note that $\phi_i^d(x_i, d_i)$ *is a LP variable whose value depends on the value of* x_i, d_i. *Thus, we set up a* $n \times n$ *matrix to list all possible values for each* $\phi_i^d(x_i, d_i)$.

Determine the domain of $g = \{g_i(x_i, x_{\Omega_i}, a_i)\}, i = 1, \cdots, n$, based on the topology $\mathcal{G}$.

Set up an n-dimensional cell for functions $f_i(x_i, x_{\Omega_i}, a_i), i = 1, \cdots, n$.

loop *over each cell i:*

Create a table of f_i's value based on the value of variables involved in f_i's domain, i.e., x_i, x_{Ω_i}, a_i.

Compute the value of functions g_i, h_i, c_i according to $f_i(x_i, x_{\Omega_i}, a_i) = w_l[\gamma g_l(x_l, x_{\Omega_l}, a_l) - h_l(x_l)] + \sum_{d_l} c_l(x_l, d_l, a_l)\phi_l^d(x_l, d_l)$ in Section 5.3.5.

  if $aflag(i) = 0$ (or $dflag(i) = 0$) **then**
    $a_i \leftarrow 0$ (or $d_i \leftarrow 0$)
  end if
  goto *loop*.

Eliminate action variables a_i.
Generate n new LP variables $e_i, i = 1, \cdots, n$ and set up a table based on the value of variables in its domain. Add constraints (5.17) or (5.18) according to *aflag*.

Eliminate state variables x_i according to the elimination order $\mathcal{O}$.
Generate another n new LP variables $E_i, i = 1, \cdots, n$ and setup a table based on the value of variables in its domain. Add constraints (5.19).

Solve the new ALP (5.14) to get the value function and the optimal defender's policy.

Use greedy search for the distributed attacker's policy (5.3.6).
end procedure

The system stage cost is the sum of the local stage cost of each node $c(x, \mathbf{a}, \mathbf{d}) = \sum_{l=1}^{n} c_l(x_l, a_l, d_l)$, where $c_l(x_l, a_l, d_l) = \xi_1(1 - x_l) - \xi_2 a_l + \xi_3 d_l - \xi_4 a_l d_l$. The explicit form consists of four terms: the loss for faulty nodes, a cost of applying attacks, protection costs, and a reward of protecting a node which is being attacked. Since c_l is the cost function of node l in the defender's perspective and weights $\xi_i > 0, i = 1, 2, 3, 4$, the second and fourth terms are negative. The ordering of $\xi_1 > \xi_4 > \xi_3 > \xi_2$ is assumed because the functionality of nodes serves as our primary goal. Protections are more costly than attacks; however, once an adversary attacks the node that possesses defensive strategies, e.g., a honeypot, it will create a significant loss for the attacker.

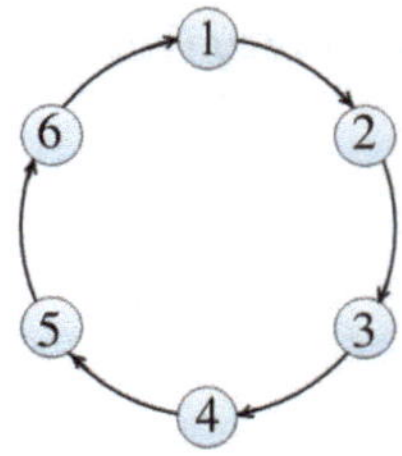

Fig. 5.5: Directed ring topology of six nodes with index 1 to 6.

Table 5.1: Time cost (units: seconds) for the directed ring with a increasing node number.

Network Size	2	3	4	5	6	7
Exact LP	0.214229	0.629684	3.329771	34.51808	178.6801	1549.016
ALP	2.664791	2.755704	2.749961	2.769759	3.238778	3.534943

5.4.2 Approximation Accuracy

We use a directed ring topology as shown in Figure 5.5 to show the accuracy of the linear function approximation under the local information structure assumption. The comparison is limited to a network with 7 or fewer number of nodes due to the state explosion of the exact LP as shown in Table 5.1. The computational time is recorded by tic and toc function in MATLAB and indicates the efficiency of the approximate algorithm as node number increases.

Figure 5.6 illustrates the fact that the growth of the network size causes an increase of the absolute error $obj(ALP) - obj(exact) \geq 0$. This increasing absolute error is inevitable due to the growth of difference $2^n - n$ as n grows. In particular, the linear growth of the ALP variables $w_i, i \in \{0, 1, ..., n\}$ may not catch up with the exponential growth of the exact LP variables $v(\mathbf{x}), x \in \mathscr{X}$.

However, the linear function approximation remains suitable when we take a look at the relative error $(obj(ALP) - obj(exact))/obj(exact)$. We observe a decrease in the value of the objective function when the number of nodes in the network is larger than 3. Therefore, the error becomes negligible with a massive node number, which serves well for our large-scale infrastructure networks.

Besides accuracy, we see that for the ring topology, increasing the network size brings a higher cost to the attacker. Exponential function $f(x) = 18.25e^{0.6178x}$ provides a good fitting to the green line with the root mean squared error (RMSE) of 10.64.

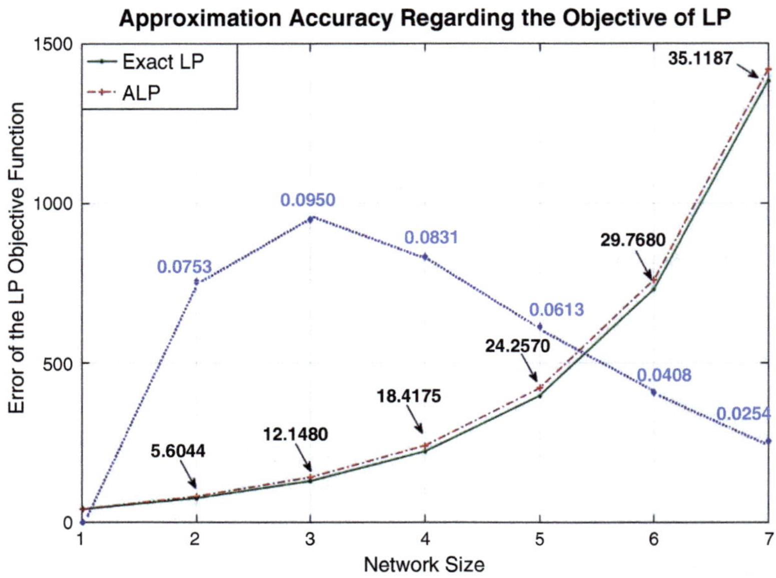

Fig. 5.6: Approximation accuracy for a directed ring topology. The red and green lines are the value of the objective function of the exact and approximate LP, $obj(exact)$ and $obj(ALP)$, respectively. The black arrow shows the value of the absolute error, while the blue number is the percentage of the relative error. The ALP achieves the upper bound for the exact LP as the size of network grows, i.e., $obj(ALP) \geq obj(exact)$ for the same network size.

5.4.3 Various Information Structure

In Figure 5.7, we compare the influence of global and local information structure of the defender to the exact LP. Recall that the y-axis shows the optimal cost of the system and a smaller value introduces a more secure system. Then, a local information structure in red brings a higher system cost than a global information structure in green for all initial states.

It shows that more knowledge can help defender better respond to the threat from the attacker. We can understand this with an example of the information structure of its neighboring nodes. Since the failure of its neighboring nodes increases its risk of being attacked, it tends to defend itself even when it is still working, yet all his neighbors fail. Apparently, a defender with local information structure cannot achieve that. Besides, with the increasing of node number, the difference grows between global and local information structures.

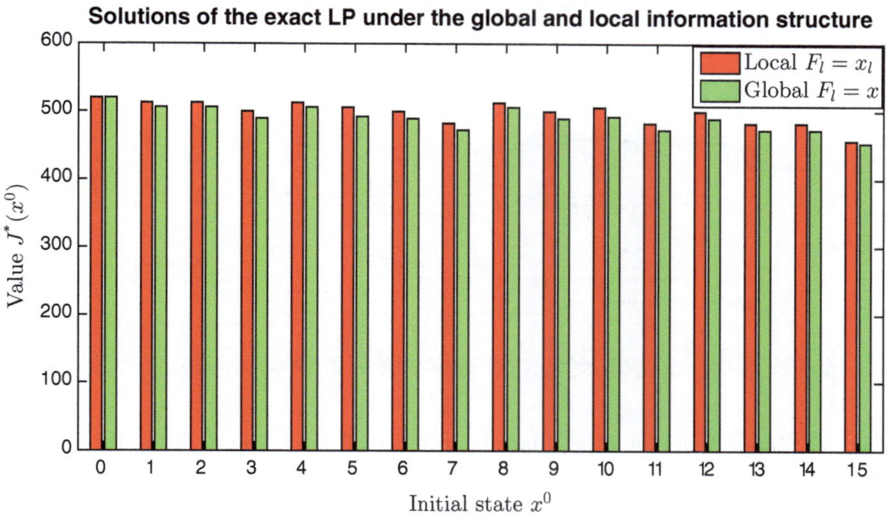

Fig. 5.7: Value functions of different initial states in a four-node-directed ring topology. State $0, 1, \cdots, 15$ is a decimalization of 2^4 different states from $(0,0,0,0)$ to $(1,1,1,1)$. Because the topology is symmetric, the number of working nodes determines the value. For example, state $1,2,4,9$ share the same value in either global or local information structure because they all just have one working node. Besides, a better initial state $(1,1,1,1)$ with all nodes working causes less loss of the system.

5.4.4 Network Effect

We reorganize the value-initial state pair $(J^*(x^0), x^0)$ of a 6-node ring topology in the top of Figure 5.8 in an increasing order. Then, we see that the number of faulty nodes dominates the order of value. However, when the number of failures is the same, the location of the failure has an impact on $J^*(x^0)$, and a high degree of the failure aggregation results in a less secure system. For example, $J^*(x^0 = (1,1,1,0,0,0)) > J^*(x^0 = (1,1,0,0,1,0)) > J^*(x^0 = (1,0,1,0,1,0))$ because the dense pattern of the state vector $(1,1,1,0,0,0)$ is more likely to cause a cascading failure in the network than a sparse one $(1,0,1,0,1,0)$. These results suggest an alternating node protection if we cannot consolidate every node due to a limited budget. Specifically as shown in Figure 5.5, we choose to consolidate every other connecting node in the 6-node ring network, i.e., node 1,3,5.

5.4.5 Optimal Policy

The global stationary policies of defenders and attackers for a 6-node ring topology is shown in Figure 5.8 in red and green, respectively. We observe that the size of

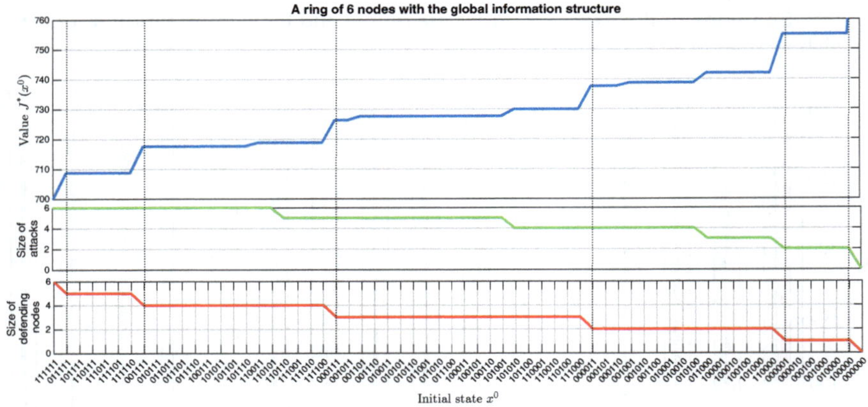

Fig. 5.8: Value function $J^*(x^0)$ and the number of defending nodes at the optimal policy for different initial states x^0 in a 6-node ring example. From the value function (the blue line), the size of failures (the number of failure nodes) and the location of the failures affect the security level of the system. At the equilibrium policy, the size of defenses (the red line) is proportional to the number of the working nodes in the network. The attacker (the green line) decreases the number of nodes to attack as more nodes have been taken down, but the green line is not aligned with the top two figures. The initial states between dotted lines share the same number of working nodes.

defense is proportional to the number of working nodes in the network, while the attacker compromises fewer nodes when the failure size increases.

Remark: Since the defender can only affect the system through the reward function, the defense's policy follows an opposite pattern of the value function. The attacker, on the other hand, has a more irregular pattern because it can also influence the transition of the system.

Other results of the approximated policy are summarized below. The local stationary defender's policy is to defend a normal node with a higher probability. The defender does not defend the faulty nodes since the recovery of a failed node cannot mitigate the loss. Furthermore, if we reduce the cost of state failure ξ_1 or increase the defense cost ξ_3, we observe that the defender is less likely to defend.

The suboptimal distributed attacker's policy avoids attacking node l when nodes in Ω_l are working. With an increase in ξ_4, the total number of attacks decreases to avoid attacking protected nodes. Thus, the presence of the defender results in fewer attacks. Besides, if node k cannot be attacked, then, naturally node k will not be defended, and attacker tends to decrease attack levels at the parent nodes of k.

5.5 Conclusion

In this work, we have formulated a zero-sum dynamic game model to design protection mechanisms for large-scale interdependent critical infrastructures against cyber and physical attacks. To compute the security policies for the infrastructure designers, we have developed a factored Markov game approach to reduce the computational complexity of the large-scale linear programming (LP) problem by leveraging the sparsity of the transition kernel and the network structure. With techniques such as linear function approximations, variable elimination, and the restriction of the local information structures, we have significantly reduced the computational time of the defender's saddle-point policy. The saddle-point strategy of the attacker can be computed likewise using the dual LP.

Numerical experiments have shown that the defender's policy can successfully thwart attacks. The lack of defenders gives rises to the attack number because the attack cost is negligible comparing to the system loss. As more nodes equip with protections, the attack number decreases. Besides, attackers avoid attacking nodes with healthy neighboring nodes because they have a larger probability of survival and are also more likely to be protected. The global stationary policy of defender of each state depends on the security level at that state because of the single-controller assumption.

Moreover, with more information or observations of the system states available to the defender, the infrastructure is shown to be more secure under the saddle-point equilibrium security policy. Finally, a ring topology example has illustrated an increase in approximation accuracy when the number of nodes grows as well as an acceptable approximation error introduced by the localized information structure.

Future work would incorporate the design of resiliency mechanism to enable the infrastructures to recover after the attack. It would be of interest to investigate the inherent trade-off between the security and resiliency design objectives by jointly studying the defender-controlled recovery process into the attacker-controlled failure process. Another important research direction is on the development of the theoretical foundations on the approximation schemes and the extension of the framework to nonzero-sum games.

5.6 Chapter Notes and Further Reading

A lot of works have been devoted to understand the interdependent networks by concept identification [16]; dependency classification including physical, cyber, geographic, and logical types [20]; and input-output or agent-based model construction [19]. The authors in [3, 1] have also proposed a game-theoretic framework to capture the decentralized decision-making nature of interdependent CIs. To analyze and manage the risks of CIs due to the interdependencies, various models have been proposed, e.g., based on network flows [12], numerical simulations [9], dynamic coupling [21], and the ones summarized in [17].

Game-theoretic methods have been extensively used to model the cyber security with applications to infrastructures [15, 23, 25, 26]. Zhu et al. have proposed a proactive feedback-driven moving target defense mechanism to secure the computer networks [22]. In [18], a *FlipIt* game framework has been used to model the security in cloud-enabled cyber-physical systems. The authors in [2, 4] have addressed the multilayer cyber risks management induced by attacks in Internet of things through a contract-theoretic approach. In [24, 25], Markov games model have been used to deal with network security. Our work differs from the previous works by proposing a factored Markov game framework and developing computational methods for the dynamic protection policies of large-scale interdependent CIs.

The computation limitation caused by the curse of dimension urges researchers to find scalable methods. A number of works have focused on the linear programming formulation of Markov decision processes (MDP) and complexity reduction of the objective and constraints of the linear programming [7, 14, 11]. In [5], the authors have reduced the number of constraints by proper sampling and derived its error bound. [13] has formulated a linear program of the asymmetric zero-sum game and reduced its computational complexity to polynomial time. In [10], the authors have used a factored approach to approximate the value function.

References

1. Chen J, Zhu Q (2016a) Interdependent network formation games with an application to critical infrastructures. In: American Control Conference (ACC), pp 2870–2875
2. Chen J, Zhu Q (2016b) Optimal contract design under asymmetric information for cloud-enabled internet of controlled things. In: International Conference on Decision and Game Theory for Security, Springer, pp 329–348
3. Chen J, Zhu Q (2016c) Resilient and decentralized control of multi-level cooperative mobile networks to maintain connectivity under adversarial environment. In: IEEE Conference on Decision and Control (CDC), pp 5183–5188
4. Chen J, Zhu Q (2017) Security as a service for cloud-enabled internet of controlled things under advanced persistent threats: A contract design approach. IEEE Transactions on Information Forensics and Security, 12(11): 2736–2750
5. De Farias DP, Van Roy B (2004) On constraint sampling in the linear programming approach to approximate dynamic programming. Mathematics of operations research 29(3):462–478
6. Filar J, Vrieze K (2012) Competitive Markov decision processes. Springer Science & Business Media
7. Guestrin C, Koller D, Parr R, Venkataraman S (2003) Efficient solution algorithms for factored mdps. Journal of Artificial Intelligence Research 19:399–468

8. Huang L, Chen J, Zhu Q (2017, April) A factored MDP approach to optimal mechanism design for resihent large-scale interdependent critical infrastructures. In Modeling and Simulation of Cyber-Physical Energy Systems (MSCPES), 2017 Workshop on, IEEE, pp 1–6
9. Korkali M, Veneman JG, Tivnan BF, Hines PD (2014) Reducing cascading failure risk by increasing infrastructure network interdependency. arXiv preprint arXiv:14106836
10. Lagoudakis MG, Parr R (2003) Learning in zero-sum team markov games using factored value functions. In: Advances in Neural Information Processing Systems, pp 1659–1666
11. Lakshminarayanan C, Bhatnagar S (2015) A generalized reduced linear program for markov decision processes. In: AAAI, pp 2722–2728
12. Lee II EE, Mitchell JE, Wallace WA (2007) Restoration of services in interdependent infrastructure systems: A network flows approach. IEEE Transactions on Systems, Man, and Cybernetics, Part C (Applications and Reviews) 37(6):1303–1317
13. Li L, Shamma J (2014) Lp formulation of asymmetric zero-sum stochastic games. In: Decision and Control (CDC), 2014 IEEE 53rd Annual Conference on, IEEE, pp 1930–1935
14. Malek A, Abbasi-Yadkori Y, Bartlett P (2014) Linear programming for large-scale markov decision problems. In: International Conference on Machine Learning, pp 496–504
15. Manshaei MH, Zhu Q, Alpcan T, Bacşar T, Hubaux JP (2013) Game theory meets network security and privacy. ACM Computing Surveys (CSUR) 45(3):25
16. Moteff J, Parfomak P (2004) Critical infrastructure and key assets: definition and identification. LIBRARY OF CONGRESS WASHINGTON DC CONGRESSIONAL RESEARCH SERVICE
17. Ouyang M (2014) Review on modeling and simulation of interdependent critical infrastructure systems. Reliability engineering & System safety 121:43–60
18. Pawlick J, Farhang S, Zhu Q (2015) Flip the cloud: Cyber-physical signaling games in the presence of advanced persistent threats. In: 6th International Conference on Decision and Game Theory for Security, GameSec 2015, Springer Verlag
19. Pederson P, Dudenhoeffer D, Hartley S, Permann M (2006) Critical infrastructure interdependency modeling: a survey of us and international research. Idaho National Laboratory 25:27
20. Rinaldi SM, Peerenboom JP, Kelly TK (2001) Identifying, understanding, and analyzing critical infrastructure interdependencies. IEEE Control Systems 21(6):11–25
21. Rosato V, Issacharoff L, Tiriticco F, Meloni S, Porcellinis S, Setola R (2008) Modelling interdependent infrastructures using interacting dynamical models. International Journal of Critical Infrastructures 4(1–2):63–79

22. Zhu Q, Başar T (2013) Game-theoretic approach to feedback-driven multi-stage moving target defense. In: International Conference on Decision and Game Theory for Security, Springer, pp 246–263
23. Zhu Q, Basar T (2015) Game-theoretic methods for robustness, security, and resilience of cyberphysical control systems: Games-in-games principle for optimal cross-layer resilient control systems. IEEE control systems 35(1):46–65
24. Zhu Q, Li H, Han Z, Basar T (2010a) A stochastic game model for jamming in multi-channel cognitive radio systems. In: Communications (ICC), 2010 IEEE International Conference on, IEEE, pp 1–6
25. Zhu Q, Tembine H, Başar T (2010b) Network security configurations: A nonzero-sum stochastic game approach. In: American Control Conference (ACC), 2010, IEEE, pp 1059–1064
26. Zhu Q, Fung C, Boutaba R, Basar T (2012) Guidex: A game-theoretic incentive-based mechanism for intrusion detection networks. IEEE Journal on Selected Areas in Communications 30(11):2220–2230

Part II
Practice

Chapter 6
G-DPS: A Game-Theoretical Decision-Making Framework for Physical Surveillance Games

Ali Alshawish, Mohamed Amine Abid, Hermann de Meer, Stefan Schauer, Sandra König, Antonios Gouglidis, and David Hutchison

6.1 Introduction

Critical infrastructure protection becomes increasingly a major concern in governments and industries. Besides the increasing rates of cybercrime, recent terrorist attacks bring organizations operating in such environments in a critical state. Therefore, risk management plays an important role in operations of such infrastructures (see Chapter 1). The risk management process mostly requires a deep understanding of the system's functions, processes, and assets as well as their mutual dependencies. Therefore, situational awareness turns out to be a core component of risk management approaches since it provides the means to develop a constantly updated understanding of the current state of the system of interest. Situational awareness enables involved security operators or risk managers to keep track of what is currently

A. Alshawish · M. A. Abid (✉) · H. de Meer
Faculty of Computer Science and Mathematics, Chair of Computer Networks and Computer Communications, University of Passau, Innstr. 43, 94032 Passau, Germany
e-mail: ali.alshawish@uni-passau.de; amine.abid@uni-passau.de; hermann.demeer@uni-passau.de

S. Schauer
Center for Digital Safety and Security, Austrian Institute of Technology, B10, Lakeside Science & Technology Park, Lakeside, 9020 Klagenfurt, Austria
e-mail: stefan.schauer@ait.ac.at

S. König
AIT Austrian Institute of Technology GmbH, Centre for Digital Safety & Security, Giefinggasse 4, 1210 Vienna, Austria
e-mail: sandra.koenig@ait.ac.at

A. Gouglidis · D. Hutchison
School of Computing and Communications, InfoLab21, Lancaster University, Lancaster LA1 4WA, UK
e-mail: a.gouglidis@lancaster.ac.uk; d.hutchison@lancaster.ac.uk

S. Rass, S. Schauer (eds.), *Game Theory for Security and Risk Management*, Static & Dynamic Game Theory: Foundations & Applications,
https://doi.org/10.1007/978-3-319-75268-6_6

happening and to understand it or interpret it depending on what happened in the past time in order to foresee what could happen in the future and thus to be able to make decision and take a proper action [16]. Thus, understanding the current situation has a considerable impact on the proper prediction of future events and subsequently on the operator's decision-making. In the context of security, a proper anticipation of the potential adversary's actions can give the security personnel a good advantage to get ahead of the adversary in a security game and to eventually stop the whole attack chain at a very early stage. Nevertheless, the irrationality level of the adversary has a significant impact on the perfectness of our knowledge on his attack preferences. Generally, the substance that glues past, present, and future phases of the situational awareness process together is data, which varies across multiple scales in space and time (i.e., historical and real-time data). Further, data is an important element for accomplishing a precise risk assessment process. Broadly speaking, one of the key approaches to collect and obtain data and information is monitoring and observation, henceforth referred to as surveillance. Therefore, most critical infrastructures, in particular those operating large industry complexes, incorporate increasingly different surveillance technologies to operate as an early incident warning or even prevention systems [42]. Surveillance systems, such as video surveillance and malicious behavior detection, have been long used for perimeter protection as a first line of defense. Traditional perimeter security solutions typically monitor the outer boundary structures and lines, thus ignoring threats from the inside [6]. Moreover, the deterrent effect of surveillance systems like closed-circuit television (CCTV) becomes considerably less important due to the inflexibility induced by their fixed installations. Hence, an infrastructure's surveillance policy is more predictable, and a potential adversary has a better opportunity to observe and bypass it subsequently. Therefore, it is important to maintain situational awareness within such environments so that potential intruders can still be detected. In this chapter, the main focus is laid on physical surveillance scenarios, in which mobile agents perform repetitive spot-checks within the infrastructure boundaries to improve flexibility and intrusion detection probabilities. These mobile agents, conducting random inspection activities, play an important role in ensuring a persistent monitoring and on-site observation. However, this requires an effective planning of inspection schedules in order to reach the envisaged goals taking into account challenges such as uncertainty (due to unforeseen events or dynamic system nature) and potential existence of multiple competing goals or criteria that need to be optimized simultaneously. This problem already has a natural reflection in game theory known as "cops-and-robbers" game, but these models always assume a deterministic outcome of the gameplay. Regardless of whether personnel (e.g., security guards, etc.) or technical solutions (e.g., cameras, etc.) are applied, surveillance systems have an imperfect detection rate, leaving an intruder with the potential to cause some damage to the infrastructure. Hence, the core problem is to find surveillance configurations that could provide an optimal solution in minimizing the damage caused to the organization. Therefore, we present a decision-making framework, which assesses possible choices and alternatives toward finding an optimal surveillance configuration and hence minimizing addressed risks. The decision is made by means of a game-theoretic model for optimizing physical surveillance systems and minimizing the potential damage caused by an intruder with respect to the imperfect detection rates of surveillance technology. With our approach, we have the advantage of using categorical (or continuous) distributions instead of a single numerical value to capture the uncertainty in describing the potential damage of an intruder as in Chapter 2. This gives us the opportunity to model the imperfection of surveillance systems and to optimize over large collections of empirical or simulated data without losing valuable information during the process.

The structure of the chapter is as follows. In Section 6.2, a general overview of surveillance technologies is introduced. In Section 6.3, the term of physical surveillance games is presented. Section 6.4 describes the game-theoretic approach for risk minimization. The six-step decision-making framework for physical surveillance games (G-DPS framework) is comprehensively explained in Section 6.5. Section 6.6 illustrates the application of G-DPS framework to minimize the risk of fare evasion in public transportation systems. Finally, chapter notes and further reading are discussed in Section 6.7, before Section 6.8 concludes the chapter.

6.2 Overview of Surveillance

Surveillance is commonly described as the careful watching of objects, persons, and areas, due to a crime that has happened or is expected to happen. Surveillance has been explained as "*the systematic investigation or monitoring of the actions or communications of one or more persons. Its primary purpose is generally to collect information about them, their activities, or their associates. There may be a secondary intention to deter a whole population from undertaking some kinds of activity*" [13]. The process of surveillance consists basically of five activities, as depicted in Figure 6.1. These activities include sensor selection and placement, which helps identify the set of proper surveillance devices matching the envisioned purposes and objectives of the surveillance system of interest. There is a wide spectrum of existing surveillance technologies including, but not limited to, visual surveillance, auditory surveillance, biometric surveillance, location-based surveillance, and ubiquitous surveillance, among others. In addition to sensor selection, the deployment layout of sensors and surveillance entities represents a major step toward achieving the predefined goals of any surveillance system. The subsequent activity is data collection, which is more relevant to the events of interest. Since collected data can be provided by sources of different types, the process of data integration or data fusion is therefore of vital importance, in particular, for hybrid surveillance systems, in which traditional and most-advanced monitoring techniques are leveraged simultaneously. Data fusion is required in order to bring the collected data into the same semantic and, hence, to extract more meaningful information. The next phase refers to the activity of employing the available data into further processing procedures, referred to as decision-making process, which identifies and assesses possible choices and alternatives based on the available and organized information from the preceding phase. As a result, "first response" activities can be managed and coordinated more effectively toward minimizing potential risks.

6.2.1 Categorization of Surveillance Systems

Currently, a wide range of surveillance technologies is used in order to provide end users with different levels of functionality. Hence, for us to identify the various surveillance technologies, we conducted a systematic literature review in [19]. This review method is capable of providing a valid and comprehensive categorization of existing technologies and also helps in overcoming the difficulty of assigning various technologies with homogeneous groups. Specifically, the analysis resulted in the identification of six main categories of surveillance technologies, namely, biometrics, dataveillance, visual surveillance, communication surveillance, location tracking, and ubiquitous surveillance. Brief information about each of these categories is provided in the following:

- Biometrics is concerned with automated methods in order to identify or recognize the identity of a living person, based on his/her physiological and/or behavioral characteristics [50, 49, 30]. Their main objective is to identify, verify, or authorize an individual, which is accomplished through the application of pattern-matching algorithms on a set of gathered data.
- Visual surveillance technologies are characterized by their wide variety of technologies, e.g., video, imaging scanners, photography, satellites, or unmanned aerial vehicles (UAVs), and are closely coupled with the concept of territorial privacy [10].
- Dataveillance technologies are mostly utilized in the context of data systems that collect personal information. This information could be used subsequently in the investigation or monitoring of the actions or communications of one or more persons. Hence, dataveillance technologies are highly applicable by security agencies and bodies to perform pattern recognitions and predictions [7].
- Communication surveillance is used to monitor, intercept, collect, preserve, and retain information that has been communicated, relayed, or generated over communication networks to a group of recipients by a third party [41]. These types of technologies have introduced a high level of ambiguity, mostly when considering them as a mean to protect nations against terrorism.
- Location tracking surveillance technologies are used to monitor position and movements, e.g., proximity sensing, scene analysis, and triangulation [22].

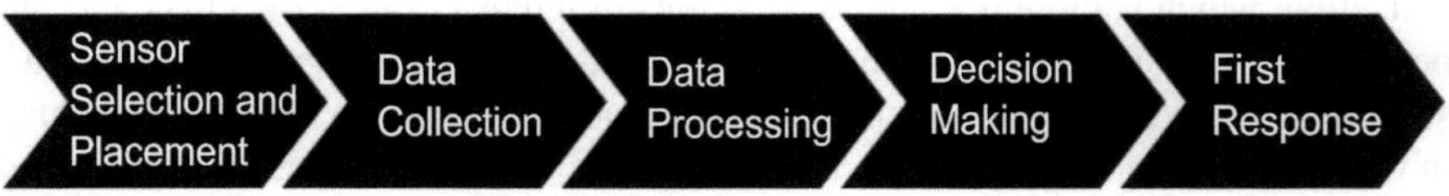

Fig. 6.1: A general overview of the basic activities of a surveillance system

Such technologies appear to be useful in military operations, espionage, or policing.

- Ubiquitous surveillance is related to the unilateral gathering of data on people in their living environment through the use of various embedded sensors [31]. The application areas of ubiquitous surveillance and computing devices are numerous since they are applicable to any type of objects (e.g., on people).

6.2.2 Limitations of Traditional Surveillance Systems in Critical Infrastructures

Generally, there are several challenges facing surveillance systems including performance, accuracy, or working in harsh environments. For example, many infrastructures (e.g., supply networks and large-scale enterprises) spread over a large geographic area and across long distances connecting regions which are geographically very far apart. Hence, full coverage of large-scale areas is a very challenging aspect, as well. Surveillance coverage is strictly limited by the number of available resources such as sensors, processing devices, or human resources. Therefore, available surveillance resources have to be strategically allocated to achieve envisaged goals.

Such organizations constantly tend to extend beyond their physical existence to include other entities like vendors, business partners, service providers, or even costumers. Thus, it is very common to have external entities inside their sites such as temporary workers, interns, independent contractors and subcontractors, or even visitors. In general, people are more interested in getting their job done, and if the security measures and the security perimeter are slowing them down or limiting them, they will certainly find ways to circumvent it. Moreover, it is very likely that employees behave and perform inappropriately, resulting in direct breaches of an organization's security policy. For example, adversaries can exploit the fact that the issued badges of terminated employees, temporary visitors, or workers are not always timely recovered before leaving the site and the access of stolen or lost badges are similarly not revoked in a timely manner. As a consequence, the perimeter-centric physical security measures such as traditional surveillance technologies (e.g., closed-circuit television (CCTV) systems or access control solutions) that use static surveillance devices mounted at specific locations are not adequate to detect and prevent such potential intruders [32]. Due to the inflexibility and fixed installation of these systems, their deterrent effect will be considerably less, and hence an intruder's chance of successfully circumventing security controls located at the perimeter is significantly higher. Therefore, it is important to maintain situational awareness within the industrial complexes of such infrastructures so that the potential intruders can still be detected.

To cope with the dynamic nature of critical infrastructure and to achieve adequate level of situational awareness in such large-scale areas taking into account the limited available resources (e.g., security guards and checking devices), surveillance strategies have to exhibit two key features, namely, risk-based and on-demand strategies. Risk-based strategies are necessary to allocate and focus resources and capabilities in the high sensitive areas and against severe threats and therefore to effectively and efficiently mitigate risks [27, 28]. On-demand strategies, on the other hand, are required to randomize security and inspection checks to improve flexibility and detection probability and hence to enhance the organization's security posture. These features can be achieved by mobilizing the required surveillance resources (e.g., security personnel).

6.3 Physical Surveillance Games

Critical infrastructure systems can be characterized by closed structural environments (e.g., power plant, refinery, and airports) or open structural environments (e.g., public transit systems and nation's borders). In either case, the disruptions of these systems could have widespread and devastating consequences on the national economy and public health. Therefore, safety, security, and service continuity of these systems are of utmost importance. Although there are several advanced biometric and access control techniques, which can be used to secure facilities of interest, visual monitoring and on-site observation are still indispensable practices to ensure persistent surveillance in such environments. However, covering a moderate-sized environment requires a substantial number of static cameras, which induces a heavy monitoring activity for security personnel behind monitoring screens, leading to poor efficiency due to potential fatigue [8]. In this chapter, we target mainly scenarios in which mobile agents can be deployed in the environment for surveillance applications. In such scenarios, while potential adversaries are seeking for causing a maximum damage to the target infrastructure, the defenders or first responders are to the contrary seeking optimal resource allocation in an attempt to thwart any potential adversarial plans. Mostly, the security resources (mobile agents) are not adequate to track all targets at once. Thus, these resources have to be strategically assigned to maximize the benefits for the system's defenders. This problem has already been reflected in several game-theoretical models, but current models always assume a deterministic outcome of the gameplay. However, the decision-making process in such application has to consider uncertainty since even if the defender and the adversary share the same site, there is a probability that the defender misses the adversary inducing randomness in the player's outcomes. Modeling this randomness based on domain knowledge usually culminates in an expected payoff (e.g., a success rate for the patroller, average damage for the attacker) for the players, but this is basically a reduction of information from the full-fledged probabilistic model (a distribution) back to a real value.

Therefore, throughout this chapter, we understand physical surveillance games as distribution-valued games that model the interaction between at least two players (i.e., defenders/first responders/security personnel and potential adversaries/attackers/criminals) each equipped with a finite action set (i.e., strategies). Additionally, the chance is deemed as a "hypothetical" third player that induces randomness in the real player's outcome. Thus, a distribution-valued game takes the random outcome distribution as the payoff itself to avoid any information reduction [3]. That is, instead of computing the behavior that maximizes a numeric revenue, we can compute the behavior that shapes the payoff distribution at best (cf. Chapter 2). Further, the equilibrium strategy of a game will deliver the defenders with optimal surveillance policies and strategic allocation of the available resources within the environment of interest. Regarding the general setting of physical surveillance games, we consider a large environment, e.g., an industry complex of a utility provider, consisting of several areas/sectors/working lines of different importance and having a number of security guards, who are patrolling these areas to detect potential violations. Broadly speaking, physical surveillance games have several important real-life manifestations such as physical border patrolling, scheduling random security checkpoints, mobile robot path planning, public transit security, and fare enforcement planning, among others.

6.3.1 Overview of Risk Management Challenges

A core difficulty in managing risk is the inherent uncertainty and fuzziness of the information that is available to risk managers; see Chapters 1 and 2. Knowing the assets that require protection and assuming to have limited resources to perform this task, a core issue is resource allocation in terms of scheduling the route and frequency of patrol inspections. Obviously, the frequency of inspections should correlate with the value of the asset. In other words, if highly sensitive business assets are stored at location X, while relatively less sensitive data resides at location Y, then it makes sense to check X more often than Y, at frequencies proportional to the value of the respective goods. Extending the problem to a whole infrastructure calling for an all-encompassing protection quickly induces the need for a surveillance strategy which performance can be optimized.

When thinking of physical surveillance, this can be done at different locations and at different levels of granularity (e.g., ranging from quickly inspecting to thoroughly examining an area, with the latter being more time consuming) and variable rates (e.g., hourly, every two hours, or every six hours). Intruders will in turn react on the surveillance patterns by allocating their efforts to places that are (currently) not under surveillance. The game-theoretic model of our surveillance games is essentially a simplified version of a pursuit-evasion ("cops-and-robbers") game [33, 9], in which the security guard is the "cop" and the intruder is the "robber." However, the issue in real-world scenarios is that an intruder may not always be detected by the surveillance system (Section 6.2.2 discusses the limits in detail). Hence, let us here confine ourselves and state that there is an intrinsic likelihood of missing the intruder in every round of the game and thus for the intruder to be able to cause a certain amount of damage in the specific area. In essence, if some zones are known to be under stronger surveillance than others, the natural reaction would be to focus intrusion efforts on spots with weakest supervision and detection mechanisms. Therefore, the overall goal is to avoid damage suffered from intrusions by managing the surveillance activities accordingly. Consequently, the performance of surveillance has to be quantified in terms of damage prevention to make surveillance activities comparable.

Quantifying the damage expected from an intrusion is usually the most difficult part in a practical application of game theory in the context of physical surveillance. Obviously, we cannot simply define the effect of a successful intrusion as a payoff being equal to the negative value of the stolen good, simply because this value may be unknown or difficult to quantify. Likewise, we cannot straightforwardly assign a non-negative payoff upon thwarting an intrusion, as this event may not even be noticed in practice. Often, we end up with a purely nominal quantification of both, value and probability, according to fuzzy terms like "damage is high if the intruder enters a high-security area; however, this is expected only with very low probability." For setting up a game-theoretic model to optimize the surveillance system's configuration, we require something crisper and more reliable. The latter is achieved by querying a maximum of available sources of information and aggregating the results.

Combining the multiplicity of potential sources usually leads to a detailed and thus difficult picture to manage risk minimization. For example, cameras may raise alarms upon detection of unusual behavior or even classify the current image sequence in terms of criticality (e.g., if a person is showing up at some place at a time when this place is supposed to be empty or if a car remains parked when all others left the place). This information and its classification are by themselves subject to some errors and presented to human operators to decide upon taking action or not. Additionally, a purely static surveillance system cannot avoid having dead angles or shadowed spots so that the static surveillance data is usually combined with "dynamic" information obtained from the security staff patrolling the premises. The immediate question here is concerned with how to do the surveillance optimally, i.e., where to place the surveillance equipment, what data to collect and how often, etc. Assuming that every such choice is among finitely many alternatives only, we can rephrase the issue as a game-theoretic (i.e., an optimization) problem.

6.3.2 Our Contribution

As pointed out in Section 6.2.2, real-life surveillance systems have limitations and fuzziness in their detection mechanisms. An accurate description often relies on a collection of categorical values describing different performance possibilities. Hence, to set up a standard game-theoretic model using the performance of a surveillance system as a measure of payoff, an ambiguous (fuzzy or probabilistic) description has to be converted into a representative number. For example, taking the expected detection rate amounts to averaging ranks of the respective categories. The major drawback of such a conversion is its implied loss of information (the results of the surveillance can become somewhat blurred) and potential deviations between the semantics of the original model and the representative numerical figure.

To avoid these issues, we present a comprehensive decision-making framework based on a game-theoretic model that ensures the calculation of optimal security measures being physical surveillance systems against adversaries. This will lead to minimizing the potential damage caused by an adversary (i.e., an intruder) and thus provide a strategy for risk minimization. We will explicitly address the uncertainty in assessing the potential damage caused by the adversary by making use of empirical data (i.e., diverging expert opinions, inaccuracies of detection mechanisms, random misclassification of incidents, etc.). As such, our approach falls into the category of empirical game theory but with the particularity that the game is played over a function space instead of the (usual) real numbers. In more detail, we apply a specially tailored framework for game theory over abstract spaces of probability distributions (cf. Chapters 2 and 3). This framework allows us to integrate uncertainty and allows the use of distribution-valued payoffs in game theory, by playing games toward utility maximization (or risk minimization) over stochastic orders, rather than over real numbers, as well as allows optimizing over different goals (e.g., damage caused by the adversary, costs for security measures, acceptance of the security measures by the end users). This is especially interesting and relevant in the use-case of surveillance systems, some of which are "privacy-friendly," whereas others may have impacts that are perceived as uncomfortable by people (in terms of their privacy). Likewise, costs are different depending on which type of surveillance is done. Therefore, the actual problem of optimal surveillance is more diversified than asking for a pure maximal detection rate. In fact, additional factors have to be taken into account, e.g., the maximum must be achieved at minimum cost or maximal privacy for the honest users. Since not all of these goals are measurable in numeric terms (such as privacy), a model capable of handling categorical data in the game seems to be necessary. To detail the picture on the different aspects that influence the gameplay, Section 6.2 will discuss the notion of surveillance and its various types to illustrate their impact and relevant aspects.

Furthermore, this model explicitly avoids assumptions on the attacker's behavior apart from potential ways to attack. In particular, we do not model his preferences with regard to which attacks are more likely than others (as it is common practice) since such predictions are often wrong but may significantly influence the result of the analysis. Rather, we assume a worst case scenario saying that the adversary tries to cause as much damage as possible (i.e., we play a zero-sum game). Besides the optimal defense strategy and the worst case damage, the computed equilibrium returns the optimal strategy for an attacker. This gives some hints how the attacker might behave (i.e., which of his strategies he is likely to choose). However, there is no guarantee that he will indeed act rational. Also if we optimize several goals simultaneously, he may have to follow different strategies at the same time (i.e., he has one optimal strategy per goal) which is not possible. Luckily, any deviation from his optimal behavior yields a lower damage to the defender as long as he plays according to his optimal strategy.

6.4 Game-Theoretic Approach for Risk Minimization

There has been several research works done in the field of game theory and surveillance. Hence, before going into detail on our risk minimization approach respecting the uncertainty in surveillance systems, we start with sketching a more basic game-theoretic model for this topic (cf. Section 6.3 for further details on the general setting of physical surveillance games).

6.4.1 *Basic Game-Theoretic Model of Physical Surveillance Games*

It is convenient to think of the infrastructure environment as a finite undirected graph $G = (V, E)$ with V being the set of nodes corresponding to physical areas (buildings or vehicle trips in the context of public transit systems) and E the set of edges representing connection paths among them. Without loss of generality, we may assume edges to be without surveillance, since we can always model any path (e.g., an aisle) under surveillance as another node in the middle of the edge. More formally, if areas A and B are connected by an aisle and that aisle is under surveillance (e.g., by a camera), then it is treated as a third place C with the graph model having the edge sequence $A - C - B$, instead of the single edge $A - B$ in which the aisle would be assumed without any protection or detection mechanism. In this view, the intruder may (randomly) walk on the graph in an attempt to reach his goal (the area with the valuable business assets) while avoiding meeting the security personnel at any node. In case the intruder is captured, it gets kicked out of the area (removed from the graph), and the gameplay starts afresh again.

Putting this in a more formal way, let a single pure strategy in the standard model be a circle in the infrastructure graph G so that the strategy space of the surveillance person is a (not necessarily minimal) set of circles $C_1, \ldots, C_n$ that spans G. Likewise, let the attacker's action set be a set of paths $P_1, \ldots, P_m$ which, without loss of generality, all end at a specific valuable target node $v_0 \in V$. In the classical version of the pursuit-evasion game, the payoff in the game would correspond to the outcome of the detection of the intruder. In this case, the game itself becomes a simple matrix game, whose payoffs are stochastic in the sense that the payoff matrix $A = (A_{ij})_{(i,j=1)}^{(n,m)}$ is one of the Bernoulli random variables $A_{ij} \sim Ber(p_{ij})$ with the semantic that:

$$A_{ij} := \begin{cases} 0 & \text{if the intruder is missed;} \\ 1 & \text{if the intruder is caught.} \end{cases} \tag{6.1}$$

in which the parameter p_{ij} tells how likely a detection of the path P_j along the tour C_i is. Packing all temporal matters and detection errors into the simulation or other assessment methodologies (as discussed in Section 6.5), it is an easy yet laborious matter of working out the specific distributions. Solutions in the sense of Nash equilibria of the resulting "nondeterministic" game can be obtained in various ways. The most obvious one is to convert the matrix of random variables into a real-valued matrix by taking the expectation per element. This results in a real-valued matrix $B = (p_{ij})_{i,j=1}^{n,m} = (\mathrm{E}[A_{ij}])_{i,j=1}^{n,m}$ that can be treated with the entire well-known machinery of game theory (von Neumann's minimax theorem and linear optimization).

6.4.2 Game-Theoretic Model Using Uncertainty

The basic model sketched in the previous section deviates from reality for exactly the reasons already mentioned in Section 6.2.2 above. In real-world surveillance systems, there are several practicalities and imperfections that can significantly result in a fluctuating detection performance of the system. There are pieces of uncertainty that must be reflected in a good model.

To describe the uncertainty stemming from these various limitations of surveillance systems, we assume the payoff of our game not to be quantified by a single number. Rather, it is described by a set of possible outcomes that either stem from simulations, surveys, or expert interviews. In any case, a real-valued payoff matrix, similar to matrix B and based on the Bernoulli random variables from matrix A in Equation (6.1), is no longer appropriate. And we need to resort to a more expressive categorical distribution to avoid information loss.

Putting this in a more formal way, we assume that $T_1, T_2, \cdots, T_{Max}$ are different types of areas tagged with their respective security demands. Accordingly, let a single pure strategy in the model be a set of frequencies $f = (f_{T_1}, f_{T_2}, \cdots, f_{T_{Max}})$ representing the amount of times a security guard is performing a security check in the different security demand areas, respectively. Hence, the strategy space is the collection $f_1 \ldots f_n$ of all admissible (i.e., practically reasonable and doable) frequency tuples. Accordingly, the adversary's strategy space comprises paths to the set of target security zones $Z_1 \ldots Z_m$, where the adversary wants to cause some damage. Suppose that either by simulation or by other means of assessments (expert domain knowledge, crowd sourcing, penetration testing, etc.), we have obtained a collection of data dat_{ij} that refers to the effectiveness of defense strategy i against attack strategy j. This information may include the aforementioned indicators like detection events, correct incident recognition, correct classification, or similar. From this data, we can construct the payoff matrix $A = (A_{ij})_{(i,j=1)}^{(n,m)}$ by specifying probability distributions as payoffs instead of single numbers. An easy (nonparametric) choice is kernel density estimates F_{ij}, based on dat_{ij}, which make the random payoff A_{ij} to be

$$A_{ij} \sim F_{ij}(dat_{ij}). \tag{6.2}$$

Note that this approach can also be described in the terms introduced in Section 6.4.1, where circles $C_1, \ldots, C_n$ represent the tour of the security guard and $P_1, \ldots, P_m$ represent the intruder's paths. The set of frequencies $f = (f_{T_1}, f_{T_2}, \cdots, f_{T_{Max}})$ can be translated to a sequence of areas the security guard has to check, thus corresponding to a circle C_i in the infrastructure graph. On the other hand, an intruder often has to pass several security areas before he reaches his target Z. This set of areas he has to pass can be translated to an attack path, P_j, which is a strategy in the game model (determining the random outcome distributed according to F_{ij} if the defender plays its i-th move to protect).

To preserve all the information provided in the probability distribution F_{ij}, we invoke the more flexible framework put forth in Chapters 2 and 3 alternatively to the standard minimax and optimization approach described in Section 6.4.1 above. This allows us to play the game directly with the distribution-valued payoffs rather than having to convert them into "representative" real numbers. Moreover, we can add several more dimensions to the gameplay optimization, such as cost to traverse round trip C_i, i.e., to go to a specific zone and perform the security checks therein, or of the inconvenience caused by unwanted and too frequent identity checks (since they might interrupt the current work of a person or might not be possible immediately). However, the most important benefit from directly working with the distribution is gained when the Bernoulli distribution is replaced by a more general, categorical, or even continuous distribution model over the categorical damage scale that applies to the different indicators (e.g., detection rates, privacy infringement, comfort, etc.).

For convenience of the reader, we will relate the basic notions put forth in Chapters 2 and 3 to the context of physical surveillance games. Let the random variable X, which can be continuous, discrete, or categorical, represent a(ny) payoff in the matrix structure, and assume that X is supported on a compact set (and has a continuous probability density function in case that X has an infinite yet compact support within $\mathbb{R}$). We represent X by the sequence of its moments, treating this sequence as a hyperreal number $\mathbf{x} = (E(X^n))_{(n \in N)}$. It is an easy matter to verify that X and, respectively, its distribution function F_X are uniquely represented by the sequence of moments and that any two variables are $\leq$-ordered in the hyperreal space ${}^*\mathbb{R}$. Transferring this ordering to random variables X_1, X_2 with distributions F_1, F_2, we write $X_1 \preceq X_2$, respectively, $F_1 \preceq F_2$, if the corresponding hyperreal representatives are $\mathbf{x}_1 \leq \mathbf{x}_2$ (cf. Chapter 2). Under this embedding of distributions into ${}^*\mathbb{R}$, we can play the game "as usual," only bearing in mind that the gameplay itself is now over a new algebraic structure. Things are, however, greatly simplified in the sense that we do not have to deal with hyperreal arithmetic, based on the following facts (see Chapter 2 for proofs):

- Two distributions can be compared by looking at their tails. Specifically,
  - if the distributions are categorical, written as $F_1 = (p_1 \dots p_n)$ and $F_2 = (q_1 \dots q_n)$, where both distributions are ordered along descending ranks, then $F_1 \preceq F_2$ if and only if the vector F_1 is less or equal to the vector F_2 in terms of the usual lexicographic ordering;
  - if the distributions are discrete or continuous (with compact support), then we can truncate them to become supported on a compact set. Truncated discrete distributions then compare as categorical distributions, and continuous distributions compare lexicographically under a slightly more complicated representation that we do not look at here (as being not required for our current application);

  - if one of the two distributions is degenerate, say $X_1 = a$ is a constant (say, a deterministic outcome in the game) and X_2 is random and ranges within the set $[x_1, x_2] \subset [1, \infty)$, then $X_1 \preceq X_2$ if and only if $a < x_2$ (conversely, $X_2 \preceq X_1$ if and only if $x_2 \leq a$).

  In any case, the decision $F_1 \preceq F_2$ can be made without using any hyperreal arithmetic.
- There are modified versions of the fictitious play (FP) algorithm to solve zero-sum matrix games with probability distribution-valued payoffs (note that [12] gives an example demonstrating that regular FP like in [43] can fail to converge although the game is zero-sum; cf. Chapter 3).

It should be noted that the special case of Bernoulli distributions is canonically covered by the framework using uncertainty [39], since the lexicographic ordering on this distribution (with only two categories) equals the natural ordering of the real-valued expectations. Thus, the simple approach of converting 0–1-valued random values into their averages for a game-theoretic treatment is an easy special case of the framework that we use.

To facilitate analysis of such game with uncertain payoffs in risk management applications, all of the functionalities of the framework have been implemented in `R` [46]. The `HyRiM` package allows applying the game-theoretic framework for risk manager, keeping away the burden of data aggregation or consensus finding. Using an implementation of the generalized fictitious play algorithm in the `R` package HyRiM (see Chapter 3 or [38]), a risk manager can conveniently rely on theory and algorithms to support his decisions purely based on all the available data.

6.5 Decision-Making Framework for Physical Surveillance Games

Risk management based on surveillance involves a decision-making process, which identifies and assesses possible choices and alternatives toward finding an optimal usage pattern of surveillance and hence minimizing risks of a scenario of interest. In this section, we describe a six-step decision-making framework (in short, G-DPS framework) that applies the game-theoretic approach described in Section 6.4.2 to find an optimal solution for risk minimization through playing surveillance games with stochastic outcomes. The G-DPS framework illustrates how the generic HyRiM Risk Management Process described in Chapter 12 can be tailored to specific scenarios and application cases. A schematic representation of the G-DPS framework is depicted in Figure 6.2. For the sake of clarity, an illustrative application of the framework, including the usage of games over distribution spaces, is briefly described in Section 6.6. An extensive and detailed view of the application of the G-DPS framework is presented in Chapter 15.

6.5.1 Context Establishment

The first step toward risk management is to understand the environment of interest as well as the different objectives that should be achieved. This involves (*i*) identifying the boundaries of the environment and hence the overall scope of the risk management framework, (*ii*) identifying the different parties involved in the game (i.e., potential game players), and (*iii*) identifying the different functions, units, processes, and resources relevant to the system under investigation. Broadly speaking, this step represents the basis for identifying possible exposures to risks of any kind and mitigation actions and strategies preventing those risks in later steps. Techniques, such as business process analysis, ethnographic studies, vulnerability assessment, or organizational architecture analysis, can be used to acquire the aforementioned relevant information.

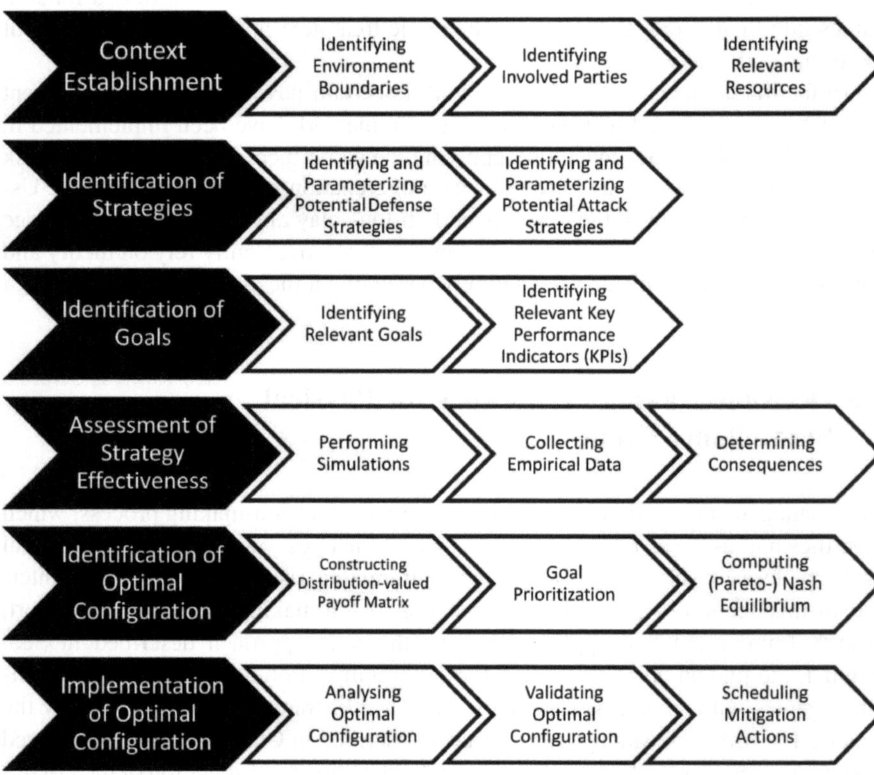

Fig. 6.2: Schematic representation of the game-theoretic decision-making framework for physical surveillance games (G-DPS)

6.5.2 Identification of Strategies

This phase involves identification and parameterization of possible configurations, layouts, and operational patterns for the surveillance infrastructure. This is a purely technical issue and is based on domain knowledge about the infrastructure, enterprise, premises, environment, etc. at hand. The outcome of this step should describe the action sets of the various players involved in the game (i.e., defensive and offensive strategies).

6.5.3 Identification of Goals

Identification of relevant indicators related to the surveillance technology may include but are not limited to the following indicators (the nature of the indicator (quantitative or qualitative) is given in brackets):

- Probabilities to detect/prevent an intrusion (quantitative indicator);
- Costs induced by implementing surveillance strategies (quantitative indicator): it indicates the expenses of conducting surveillance activities relevant to the identified strategies (e.g., base salaries of inspectors). Generally, cost indicator aims at measuring the efforts of enforcing different configurations and strategies;
- Potential privacy breaches for end users (qualitative indicator): the more data is collected, the more private information may accumulate (and leak); since the division of data into "sensitive" and "insensitive" uses already qualitative terms, measuring privacy infringements usually happens on a nominal scale (in absence of a quantitative measure of privacy);
- Willingness to support (or maybe also leverage) upon the surveillance system and health-related issues (qualitative indicator). If people feel uncomfortable upon the surveillance, then they will either avoid the system, find work-arounds, or ask for deactivation (at least temporarily). Similarly with respect to privacy, the gained comfort by security balances against the feeling of being watched. The only viable measure here may as well be a qualitative score (e.g., ranging from 1 = "no problem" up to 10 = "unacceptably uncomfortable");
- Legal regulations that concern employees (qualitative scale). For example, if surveillance is either forbidden by law or the collection of too much data is against some internal regulations of the company.

6.5.4 Assessment of Strategy Effectiveness

For each known configuration (i.e., strategies identified in Section 6.5.2), the effectiveness with regard to all aspects identified in Section 6.5.3 is determined. Given the wide variety of possible goals, there are different qualitative, quantitative, and semi-qualitative assessment methodologies of the various scenarios. For example, since the response dynamics of the game, e.g., people's reactions, etc., may be unknown, the damage/outcome of a certain situation in the game can be assessed with aid of a "soft" indicator. In other words, if simulations of surveillance systems are possible and feasible, a more or less reliable risk estimate (e.g., given in terms of probabilities) may be achievable, but not necessarily so for all goals of interest. A "soft" indicator like the degree to which end users appreciate the surveillance or feel uncomfortable upon such monitoring is one example of a goal that may not be measurable by simulation. In such cases, empirical data, e.g., coming from classical surveys and expert and stakeholder opinions, or historical and statistical data may be the way to go and measure the effects of certain configurations. For example, the end users may be asked how they feel upon having installed cameras somewhere or whether or not they would be willing to have their own devices become part of the surveillance infrastructure. Even if a user consents, an involved surveillance device (e.g., a mobile device) may not always be connected, may be out of power, etc., which adds an intrinsic element of randomness to the outcome in every scenario.

In either case, the damage quantified in this step is a categorical (or continuous) probability distribution including all available information but must be constructed under the following constraints:

1. All assessments are made in the same scale. This is required for the multicriteria optimization to work (using scaling of the vector-valued payoffs). Numeric indicators are thus discretized onto a common categorical scale (that all categorical indicators use as well). For technical reasons the scale used should not contain values smaller than 1;
2. The data source is reliable with regard to the intended risk assessment.

6.5.5 Identification of Optimal Configuration

This step involves basically the process of computing equilibria (more specifically, Pareto-Nash equilibria) and can be implemented based on standard theory (referenced in Section 6.4.2). More precisely, the stochastic outcome of the assessment process of the strategy effectiveness (cf. Section 6.5.4) should be leveraged to construct the distribution-valued payoff matrix of the applied game-theoretical model. Finding Nash equilibria in games with distribution-valued payoffs works similarly as in classical game theory but with a few qualitatively differences in the behavior of these games. For example, under a proper modification of payoff distributions, fictitious play (FP) can be used to solve distribution-valued games (Section 6.4.2 and the literature cited therein deliver more details on solving such games). In the presence of multiple goals, the Nash equilibrium is replaced by a Pareto-Nash equilibrium, meaning that any unilateral deviation from the equilibrium will result in a degeneration of at least one of the payoff measures for the deviating player. Technically, Pareto-Nash equilibria are computed by scalarizing the game into one with a single goal and computing (normal) Nash equilibria there [29]. That scalarization is nothing else than a weighted sum of all goal payoffs, where the weights can be set to reflect priorities of each goal, under the sole constraint of the weights to be all strictly positive (for a zero weight, we can simply exclude the respective goal from the analysis completely). Observe the neat side effect here: the scalarization induces a set of variables for theoretical reasons, yet these variables have a perfectly meaningful practical use in the specification of the importance of each goal. This is an independent benefit of the particular method applied here to compute multi-goal optimal surveillance configuration.

6.5.6 Implementation of Optimal Configuration

Having found an optimal solution to the configuration of the surveillance system, such as optimal surveillance routes and frequencies, etc., the daily business requires to implement the static precautions, e.g., building the surveillance system according to its optimal layout and configuration and adhering to random reconfigurations and daily operation. However, in some cases, the assessment process (cf. Section 6.5.4), applied to assess effectiveness of involved players' strategies (cf. Section 6.5.2), could be leveraged at this step to analyze and validate the efficiency and feasibility of the obtained optimal solution. For example, if the assessment is conducted using simulation, the game equilibrium strategy can be similarly implemented in the developed simulation environment and then contrasted with results obtained in early steps. Toward a practical implementation of equilibrium strategies, remember that all we require is a certain frequency of actions to happen over repetitions of the game. To this end, let us fix a time unit, say T hours, and then if the equilibrium prescribes action a_1 to happen with probability p_1, this means an average of $p_1 \cdot T$ actions during a day. Taking the pauses between repetitions of action a_1 as exponentially distributed with rate parameter $1/p_1$, it is a simple matter of drawing exponentially distributed pause times to get the time when action a_1 is to be launched next. In turn, the number of actions is a Poisson distributed variable with the same rate parameter, as desired to play the equilibrium. For the other strategies, the procedure works analogously and ultimately gives a (randomized) schedule of actions that assures the optimal frequencies as prescribed by the equilibrium [3].

6.6 Illustrative Scenario: Risk of Fare Evasion in Public Transportation Systems

For the sake of better understanding of our presented decision-making framework, we will devote the present section to apply it on an illustrative use-case. We choose to consider the case of a "public transportation system" (PTS) already described in [3], where the ultimate goal is to make the best decision on how to schedule fare checking/inspections in a way that we minimize the risk of fare evasion while mastering induced costs. For this purpose, we will apply our presented framework, step by step, until we come up with the best way to realize this goal.

As aforementioned, our framework presents six main steps that we need to go through. Thus, we structure the present section into six subsections, corresponding to these respective steps (cf. Section 6.5).

6.6.1 PTS' Context Establishment

Our considered PTS is a basic public transportation network illustrated in Figure 6.3. It consists of 24 stops served by 4 lines named *A*, *B*, *C*, and *D*. The lines exhibit statistically different utilization rates, depicted by the thickness of the edges representing them (the thicker the edge, the higher the average passenger volume over this line). This means Line A is on average the most crowded line, while Line D has the least passenger volume.

After presenting the architecture of our PTS, we will move to analyze its business process. The administrative entity running this transportation system has basically two main concerns that are tightly related: (i) how to master overall costs and (ii) how to stop fare evasion as it is causing high losses (e.g., in 2016, the revenue loss of the Massachusetts Bay Transportation Authority has been estimated around $42 million annually simply because of fare evasion on various public transport means [15]).

In most public transportation systems, customers are required to purchase a ticket and carry it during the trip as a proof of paying a proper fare. To detect potential evaders, transportation companies traditionally employ inspectors to perform physical checking on "randomly" selected trips. The number of employed inspectors depends basically on two factors, namely, the size of the network and the passenger volume. Based on available knowledge in this domain, it is valid to assume that the fraction of potential offenders and ticket-less passengers is generally proportional to the passenger volume on each line since crowded vehicles give a potential fare evader a pretty decent chance of avoiding being caught. Therefore, trips with higher passenger volumes are checked more frequently to discourage potential violations and to compensate the revenue losses due to fare evasion. As a result, the more inspection activities performed on a certain line, the more likely passengers purchase valid tickets [47]. However, increasing the rate of checks and/or number of employed inspectors incurs higher expenses. Therefore, an effective scheduling mechanism should take into account the aforementioned trade-off between revenue and cost. For the sake of simplicity, we will suppose that the maximum number of inspectors to be deployed by our transportation system is 4. Note that this scenario can also be described in the terms introduced in Section 6.4.1, where V represents the set of single trips of the transportation systems' vehicles (e.g., bus, train, or metro). Additionally, the circles $C_1, \dots, C_n$ represent the tour of the fare inspectors and $P_1, \dots, P_m$ represent the fare dodgers paths.

6.6.2 PTS' Identification of Strategies

In this step, we seek to list the different possible strategies to be adopted by our two players (i.e., potential evaders vs. inspectors). Strategy parameters can include, but not limited to:

- Inspection frequency: the frequency at which the inspector will inspect a specific line or station.
- Inspection duration: the time spent by the inspector to check specific line or trip in terms of number of stops.
- Line selection: How an inspector chooses the next line to be checked? Here, we can distinguish between two basic methods, `UNIFORM` and `MOST-CROWDED`. The latter indicates that the lines with higher passengers' volume will be checked more frequently than others (recall that the risk of fare evasion is presumably proportional to the expected passenger volume on each line or trip). For this reason, statistical data is used to help define a probability distribution over the whole transportation grid. The former, i.e., `UNIFORM` method, stipulates that all lines share the same importance. Hence, the inspectors will have no preference of a certain line over others.
- Total number of inspectors: it indicates the number of inspectors employed to check the whole transportation network.
- Collaboration: it indicates whether the inspectors are conducting their inspection activities individually or collaboratively as a team.

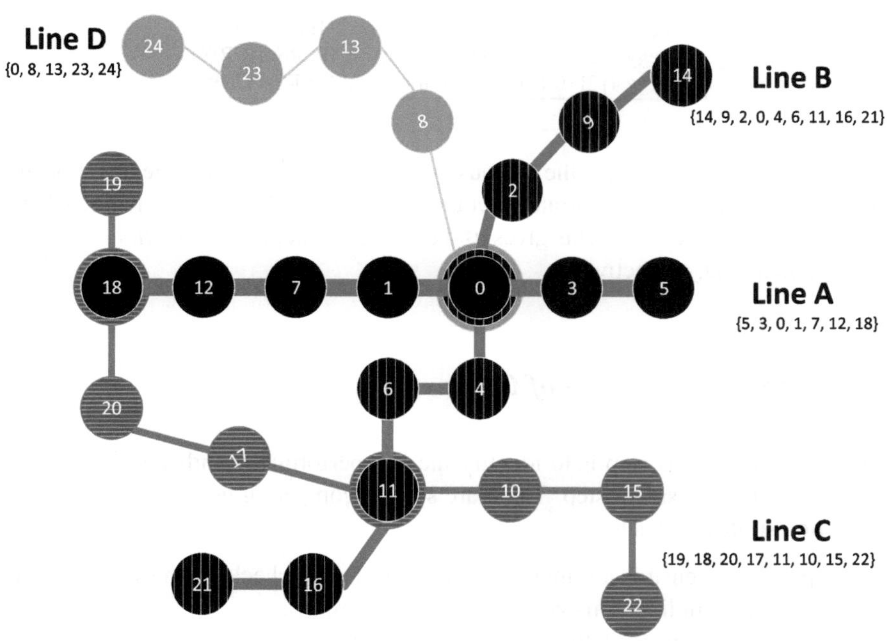

Fig. 6.3: A basic public transportation network with 4 lines

Further parameters can be defined, such as type of clothing (uniform or plain clothes), movement of the team during checks, and number of possible switches between different lines in each inspection schedule. For this illustrative example, we will limit ourselves to the following strategy parameters:

- Total number of inspectors (maximum 4): $\{2, 4\}$ inspectors having a daily duty of 8 inspection hours. This parameter is denoted in the strategy label by *xPERS*, where x is the number of inspectors.
- Collaboration: the inspectors have the possibility to carry out the inspection actions either individually (denoted by I) or jointly as a team (denoted by T).
- Line selection: we define $P(A)$, $P(B)$, $P(C)$, and $P(D)$ the respective probabilities of selecting Line A, B, C, or D. *UR* denotes the application of `UNIFORM` method (i.e., $P(A) = P(B) = P(C) = P(D) = 25\%$). *CRW*, in turn, refers to `MOST-CROWDED`, in which the following probabilities' profile is applied $P(A) = 45\%, P(B) = 30\%, P(C) = 17\%, P(D) = 8\%$. `MOST-CROWDED` strategies can be seen as risk-based strategies with the goal of inspecting lines with higher passengers' volume at a higher rate.

Combining only these parameters leads us to the action set of the inspectors (i.e., inspection strategies), summarized in Table 6.1.

Table 6.1: List of possible ticket inspection strategies

#No.	Strategy label	#No.	Strategy label
Strategy 1	2PERS-T-UR	Strategy 5	4PERS-T-UR
Strategy 2	2PERS-I-UR	Strategy 6	4PERS-I-UR
Strategy 3	2PERS-T-CRW	Strategy 7	4PERS-T-CRW
Strategy 4	2PERS-I-CRW	Strategy 8	4PERS-I-CRW

Next, we need to identify the various possible ticket evasion strategies. For the sake of simplicity, an evader strategy would be simply to choose a single line where he will never pay the fares. This gives us 4 different strategies labeled *line A*, *line B*, *line C*, and *line D*, respectively.

6.6.3 PTS' Identification of Goals

The main goal of this step is to identify the key performance indicators (KPIs) to be used in the assessment step. There are several competing goals relevant to fare inspection strategies, such as:

- Inspection Intensity: the number (volume) of spot-checking missions carried out on a certain line. This goal should be maximized.
- Costs: the expenses of the spot-checking activities (inspectors' base salaries, additional bonuses, etc.). This goal should be minimized.

- Detection intensity: the number of penalty fares claimed from ticket-less passengers. It is noteworthy that detection intensity is not linearly related to inspection intensity indicator due to varying passenger volumes on the different lines. This goal should be maximized.

6.6.4 PTS' Assessment of Strategies

We aim now at assessing the identified strategies with respect to our fixed goals. Even if the assessment could be done in variant ways (i.e., simulation, experts and stakeholders opinions, historical and statistical data, or social surveys), we choose to rely on evaluations done by a set of experts using the nominal rating {VL= very low, L=low, LM=low to medium, M=medium, MH= medium to high, H= high, and VH= very high}. The assessment results (from 15 experts) are included in the Appendix, in Tables 6.2, 6.3, and 6.4, with respect to the goals inspection intensity, detection intensity, and costs, respectively.

6.6.5 PTS' Optimal Configuration

This step is about playing the game itself. First, the collected data is used to define the distribution-valued payoffs required by our framework (recall here that data will not be aggregated to avoid any loss of information). Then, priorities among goals need to be set. For our case, we assume that all goals are equally important. Finally, we apply HyRiM risk management tool implemented as R package [38] (cf. Section 6.4.2 for further details on the model) which results on the outcome (i.e., Pareto-Nash equilibrium) presented by Figure 6.4. It corresponds to a nontrivial mixed strategy inducing a probability distribution over the different identified inspection strategies. Figure 6.4 shows that the most effective strategies are those with non-zero probabilities (i.e., "4PERS-I-UR," "4PERS-T-CRW," and "4PERS-I-CRW"). This implies that we should avoid the set of strategies with a zero probability since they are useless with regard to the addressed goals. This mixed strategy should correspond to the best way to schedule trip inspections as it should avoid making the inspectors predictable by potential evaders.

6.6.6 PTS' Implementation of Optimal Configuration

For the implementation part, the transportation system authority should schedule inspections with respect to this resulting mixed strategy. One possible way of implementation would be randomly affecting one of these three effective strategies to the inspectors such that the probabilities of selecting each of them corresponds to the probabilities given by the equilibrium. In this way, and after a long run, the applied schedule should be aligned with our optimal mixed strategy.

6.7 Chapter Notes and Further Reading

In the past years, there have been several approaches introducing game-theoretic concepts and algorithms in the general field of security and risk management (cf. [2, 34, 35, 40], as well as the literature cited therein). Similarly, the application of game theory to optimize surveillance has been subject to a quite considerable amount of prior work. This includes observing evading targets [8], optimal surveillance resource allocation under imperfect information for the attacker [5], sensor and mobile ad hoc network surveillance [45, 21], purely camera-based pursuit-evasion models [44], or the more general area of counter-terrorism [51], to mention only a few. Furthermore, several Stackelberg games have been employed to schedule randomized patrol schedules to ensure security and fare collection in public transportation systems [14, 11, 24, 48]. All these approaches have two main aspects in common: first, the modeling is always crisp, and outcomes are measurable in numbers. Second, the focus is purely on the game theory side, leaving out the specifics

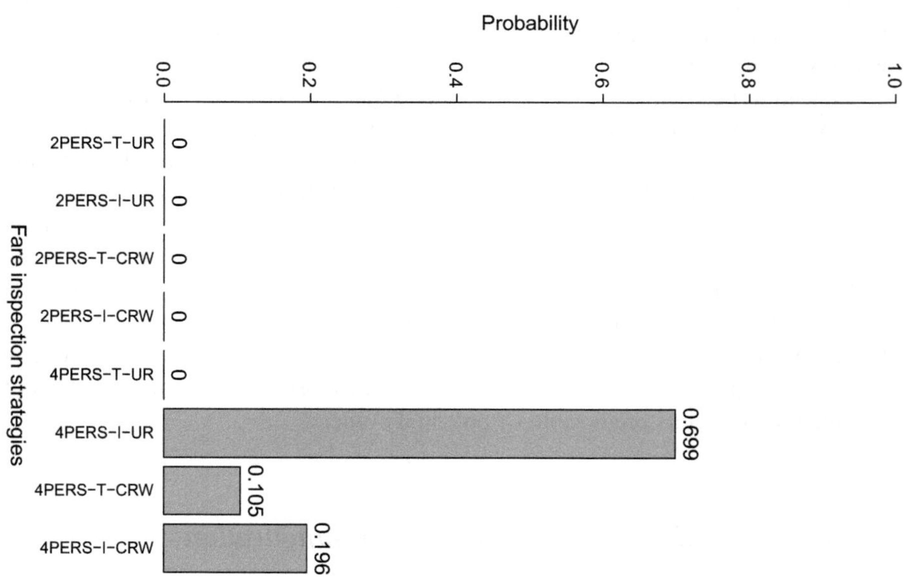

Fig. 6.4: The optimal inspection (mixed) strategy (equilibrium strategy for a multi-objective inspection game)

and limitations of surveillance systems that can dramatically change the gameplay due to their imperfections (cf. Section 6.2.2 for details). Our approach presented in this chapter differs from these two aspects. We assume that the impact of surveillance systems cannot be expressed completely in a numeric utility to the defender or the attacker. Further, we take the specifics of current surveillance technologies into account and tailor the game-theoretic model to the specifically fuzzy terms in which the quality of the surveillance is usually expressed. In this context, we apply the game-theoretic framework based on distribution-valued payoffs, which has been developed along precursor work [13, 12, 39] (cf. Section 6.4.2, Chapters 2, and 3 for further details).

In the context of game theory and surveillance, the general cops-and-robbers game has been studied in a variety of different forms. These include asking for the minimal number of cops to catch one (or more) robber(s) [20], relating structural properties of the graph to winning strategies for either party [1], or discussing the benefit of (in)visibility for either player [26]. Given the vast amount of available research, we refer to surveys such as [18, 4] as well as the references in the cited literature. Further interesting applications outside the scope of this work have also been reported, e.g., in the domain of robotics [23, 12]. We leave this application area aside here but mention reachability games [25] since they are closely related to the described surveillance games (cf. Section 6.3). In reachability games, a sliding token in a graph shall reach a particular goal position before getting caught by the other player. This is conceptually close to our settings but usually assumes perfectness of payoffs (even though not necessarily assuming perfect information), whereas we are dealing with imperfect and uncertain payoffs. Different to most previous work, we are more constrained in not having the freedom to choose the graph (the topology). Moreover, in light that the decision about winning strategies depending on the number of cops is known to be NP-complete [17], we ought to work with whatever number of surveillance people is available and on how to act optimally in the given infrastructure. Thus, our focus is on laying out the surveillance technology optimally, using cops-and-robbers as the optimization framework to account for intrinsically detecting errors.

6.8 Conclusion

People become increasingly dependable on public services supplied by critical infrastructure systems. Hence, public safety and security of such systems are of utmost importance. This implies that risks like terrorism, criminal offenses, and business revenue loss should be managed and kept at a minimum. Risk management, in its turn, involves maintaining a high level of situational awareness. Therefore, the work presented in this chapter addresses the possibility of enhancing situational awareness by means of surveillance and on-site observation. In addition, game theory concepts are leveraged toward finding an optimal configuration and usage pattern of the surveillance system of interest. Modeling surveillance as a pursuit-evasion game (cops-and-robbers) is quite common; however, assuming perfectness of detection or a crisp assessment of the occurring damage appears to be an overly strong assumption to really match reality. Real-life surveillance systems have fuzziness in their detection mechanisms. For example, every surveillance camera system has blind spots, and not every person in an inspected zone may be caught or available for a quick automated identity check. Emergency and unforeseen events, such as human errors or undisciplined inspection staff, and irregular (random) behavior of potential intruders are all factors that can significantly affect the ability of inspectors to adhere to planned schedules as well as the ability to deterministically assess the effectiveness and performance of specific surveillance (i.e., inspection) strategy, resulting in noticeable performance fluctuations and stochastic strategy effects. To turn game theory into a practically effective tool that accurately describes real surveillance scenarios, the modeling needs to account for the characteristics and technical details of the surveillance system in charge. Doing so generically appears out of reach, since the diversity of surveillance systems is far too large to be captured by a single model. To address this issue, we are proposing to include the intrinsic uncertainty of surveillance systems and of the respective risk assessment into the game-theoretic model itself. Therefore, we presented a game-theoretic framework capable of dealing with random payoffs and showed how it can be applied in a standard surveillance scenario. This additional degree of freedom provided by the game-theoretic framework allows us to work with several and not equiprobable outcomes in the same scenario of attack and defense. Hence, we have the advantage to model the practical imperfections of surveillance systems and to account for the subjectivity of expert opinions. Furthermore, the game-theoretic framework optimizes not only over the whole distributions characterizing this uncertainty but also over different security goals like damage, costs, detection rate, privacy aspects, or end-user acceptance. Thus, the optimal security strategies resulting from the presented game-theoretic approach provide risk managers with the information they need to make better decisions and take several aspects into account at the same time. Finally, issues such as implementing and validating optimal security strategies are further discussed in Chapter 15.

Acknowledgements This work was supported by the European Commission's Project No. 608090, HyRiM (Hybrid Risk Management for Utility Networks) under the 7th Framework Programme (FP7-SEC-2013-1). The authors would like to thank Dr. Stefan Rass from University of Klagenfurt for his helpful suggestions and his valuable comments.

Appendix

Table 6.2: Expert judgments of the identified strategies with regard to expected inspection intensity

	Line A					Line B					Line C				Line D			
	expert ID																	
Strategies	*1*	*2*	*6*	*7*	*10*	*3*	*8*	*9*	*12*	*14*	*2*	*5*	*10*	*11*	*12*	*13*	*14*	*15*
2PERS-T-UR	VL	L	L	VL	L	M	L	VL	L	L	L	LM	VL	VL	L	M	L	M
2PERS-I-UR	M	LM	LM	L	L	M	LM	L	L	L	L	M	L	L	L	M	LM	M
2PERS-T-CRW	M	LM	M	LM	LM	LM	LM	L	L	L	VL	L	VL	VL	VL	VL	L	VL
2PERS-I-CRW	M	LM	M	M	LM	LM	LM	L	LM	M	L	LM	VL	VL	VL	L	L	VL
4PERS-T-UR	VL	L	LM	VL	L	LM	LM	VL	L	L	LM	LM	L	VL	VL	LM	L	M
4PERS-I-UR	H	H	LM	M	MH	M	LM	LM	LM	LM	M	LM	LM	M	MH	H	MH	M
4PERS-T-CRW	VH	H	H	M	M	M	LM	H	M	LM	LM	LM	L	VL	L	L	L	LM
4PERS-I-CRW	VH	H	VH	MH	MH	MH	VH	VH	MH	M	LM	L	L	L	LM	L	L	M

Table 6.3: Expert judgments of the identified strategies with regard to expected detection intensity

	Line A					Line B					Line C				Line D			
	expert ID																	
Strategies	*1*	*2*	*6*	*7*	*10*	*3*	*8*	*9*	*12*	*14*	*2*	*5*	*10*	*11*	*12*	*13*	*14*	*15*
2PERS-T-UR	VL	L	L	VL	L	M	L	VL	L	L	L	LM	VL	VL	L	M	L	M
2PERS-I-UR	L	LM	LM	L	LM	M	LM	LM	LM	LM	L	LM	L	VL	L	M	L	M
2PERS-T-CRW	L	L	VL	L	L	M	VL	M	L	L	L	M	L	VL	M	M	M	LM
2PERS-I-CRW	M	M	H	M	LM	MH	LM	M	L	L	VL	L	VL	LM	L	L	L	VL
4PERS-T-UR	M	M	L	M	LM	M	LM	L	H	M	LM	L	M	M	M	L	L	L
4PERS-I-UR	LM	M	LM	M	L	M	VL	M	M	M	LM	M	M	LM	M	MH	M	M
4PERS-T-CRW	H	MH	MH	VH	L	LM	M	LM	M	LM	VH	H	MH	LM	L	M	M	MH
4PERS-I-CRW	M	H	H	MH	LM	MH	H	H	H	H	H	H	MH	M	L	M	M	M

Table 6.4: Expert judgments of the identified strategies with regard to expected incurred costs

	Line A					Line B					Line C				Line D			
	expert ID																	
Strategies	*1*	*2*	*6*	*7*	*10*	*3*	*8*	*9*	*12*	*14*	*2*	*5*	*10*	*11*	*12*	*13*	*14*	*15*
2PERS-T-UR	L	LM	L	VL	LM	M	L	LM	LM	L	VL	LM	L	VL	VL	LM	L	M
2PERS-I-UR	L	LM	LM	L	LM	M	LM	LM	LM	LM	L	LM	L	VL	L	M	L	M
2PERS-T-CRW	L	LM	L	VL	LM	M	LM	LM	LM	L	VL	LM	VL	VL	VL	M	L	M
2PERS-I-CRW	L	LM	LM	L	LM	M	LM	LM	LM	LM	LM	LM	L	VL	L	M	LM	M
4PERS-T-UR	H	MH	M	MH	M	M	LM	MH	M	M	L	M	MH	H	LM	M	LM	H
4PERS-I-UR	H	MH	M	MH	MH	M	LM	MH	M	MH	L	M	H	H	LM	M	L	M
4PERS-T-CRW	H	H	MH	MH	H	H	M	MH	M	MH	LM	L	VL	VL	VL	VL	L	M
4PERS-I-CRW	VH	H	H	VH	MH	MH	M	MH	H	H	LM	LM	L	VL	L	VL	VL	L

References

1. Aigner, M., Fromme, M.: A game of cops and robbers. Discrete Applied Mathematics **8**(1), 1–12 (1984)
2. Alpcan, T., Başar, T.: Network security: A decision and game-theoretic approach. Cambridge University Press (2010)
3. Alshawish, A., Abid, M.A., Rass, S., de Meer, H.: Playing a Multi-objective Spot-checking Game in Public Transportation Systems. In: Proceedings of The SHCIS'17, p. 6. ACM, Neuchâtel, Switzerland, June 21–22, 2017 (2017). https://doi.org/10.1145/3099012.3099019
4. Alspach, B.: Searching and sweeping graphs: a brief survey. Le matematiche **59**(1, 2), 5–37 (2006)
5. An, B., Kempe, D., Kiekintveld, C., Shieh, E., Singh, S., Tambe, M., Vorobeychik, Y.: Security games with limited surveillance. Ann Arbor **1001**, 48,109 (2012)
6. Arata, M.J.: Perimeter security. McGraw-Hill (2006)
7. Bellanova, R., Friedewald, M.: Deliverable 1.1: Smart Surveillance—State of the Art. SAPIENT. FP7 Sapient Project, Brussels. http://www.sapientproject.eu/docs D **1** (2011)
8. Bhattacharya, S., Başar, T., Falcone, M.: Surveillance for Security as a Pursuit-Evasion Game. In: Proceedings of The International Conference on Decision and Game Theory for Security, pp. 370–379. Springer (2014)
9. Borie, R.B., Tovey, C.A., Koenig, S.: Algorithms and Complexity Results for Pursuit-Evasion Problems. In: IJCAI, vol. 9, pp. 59–66 (2009)
10. Brassil, J.: Technical challenges in location-aware video surveillance privacy. In: Protecting Privacy in Video Surveillance, pp. 91–113. Springer (2009)
11. Brown, M., Saisubramanian, S., Varakantham, P.R., Tambe, M.: STREETS: game-theoretic traffic patrolling with exploration and exploitation (2014)
12. Chung, T.H., Hollinger, G.A., Isler, V.: Search and pursuit-evasion in mobile robotics. Autonomous robots **31**(4), 299–316 (2011)
13. Clarke, R.: Information technology and dataveillance. Communications of the ACM **31**(5), 498–512 (1988)
14. Delle Fave, F.M., Jiang, A.X., Yin, Z., Zhang, C., Tambe, M., Kraus, S., Sullivan, J.P.: Game-theoretic patrolling with dynamic execution uncertainty and a case study on a real transit system. Journal of Artificial Intelligence Research **50**, 321–367 (2014)
15. Dungca, N.: Commuter rail cited as source of most t fare evasion (2016)
16. Endsley, M.R.: Toward a theory of situation awareness in dynamic systems. Human factors **37**(1), 32–64 (1995)
17. Fomin, F., Golovach, P., Kratochvl, J.: On tractability of cops and robbers game. In: Proceedings of The Fifth Ifip International Conference On Theoretical Computer Science—Tcs 2008, pp. 171–185. Springer (2008)
18. Fomin, F.V., Thilikos, D.M.: An annotated bibliography on guaranteed graph searching. Theoretical computer science **399**(3), 236–245 (2008)

19. Gouglidis, A., Hutchison, D., Alshawish, A., de Meer, H.: D4.1 - Physical and cyber risk prediction modeling using surveillance systems. Tech. rep., Public deliverable, The HyRiM project (FP7 grant agreement no. 608090) (2015)
20. Hahn, G., MacGillivray, G.: A note on k-cop, l-robber games on graphs. Discrete mathematics **306**(19), 2492–2497 (2006)
21. Hao, D., Ren, Y., Sakurai, K.: A game theory-based surveillance mechanism against suspicious insiders in MANETs. Trusted Systems **6802**, 237–252 (2010)
22. Hightower, J., Borriello, G.: Location systems for ubiquitous computing. Computer **34**(8), 57–66 (2001)
23. Isler, V., Kannan, S., Khanna, S.: Randomized pursuit-evasion in a polygonal environment. IEEE Transactions on Robotics **21**(5), 875–884 (2005)
24. Jiang, A.X., Yin, Z., Johnson, M.P., Tambe, M., Kiekintveld, C., Leyton-Brown, K., Sandholm, T.: Towards Optimal Patrol Strategies for Fare Inspection in Transit Systems. In: AAAI Spring Symposium: Game Theory for Security, Sustainability, and Health (2012)
25. Kehagias, A., Konstantinidis, G.: Cops and Robbers, Game Theory and Zermelo's Early Results. arXiv preprint arXiv:1407.1647 (2014)
26. Kehagias, A., Mitsche, D., Prałat, P.: The role of visibility in pursuit/evasion games. Robotics **3**(4), 371–399 (2014)
27. Kiekintveld, C., Jain, M., Tsai, J., Pita, J., Ordez, F., Tambe, M.: Computing optimal randomized resource allocations for massive security games. In: Proceedings of The 8th International Conference on Autonomous Agents and Multiagent Systems-Volume 1, pp. 689–696. International Foundation for Autonomous Agents and Multiagent Systems (2009)
28. Klíma, R., Kiekintveld, C., Lisý, V.: Online learning methods for border patrol resource allocation. In: Proceedings of The International Conference on Decision and Game Theory for Security, pp. 340–349. Springer (2014)
29. Lozovanu, D., Solomon, D., Zelikovsky, A.: Multiobjective games and determining pareto-nash equilibria. Buletinul Academiei de Ştiinţe a Republicii Moldova. Matematica (3), 115–122 (2005)
30. Miller, B.: Everything you need to know about biometric identification. Personal Identification News 1988 Biometric Industry Directory. Washington DC: Warfel & Miller. Inc., Washington DC (1988)
31. Oulasvirta, A., Pihlajamaa, A., Perkiö, J., Ray, D., Vähäkangas, T., Hasu, T., Vainio, N., Myllymäki, P.: Long-term effects of ubiquitous surveillance in the home. In: Proceedings of the 2012 ACM Conference on Ubiquitous Computing, pp. 41–50. ACM (2012)
32. Pang, H., Jiang, L., Yang, L., Yue, K.: Research of android smart phone surveillance system. In: Computer Design and Applications (ICCDA), 2010 International Conference on, vol. 2, pp. V2–373. IEEE (2010)
33. Parsons, T.D.: Pursuit-evasion in a graph. Theor. Appl. Graphs. In: Proc (1976)
34. Rajbhandari, L., Snekkenes, E.A.: Mapping between classical risk management and game theoretical approaches. In: Proceedings of The IFIP International Conference on Communications and Multimedia Security, pp. 147–154. Springer (2011)

35. Rass, S.: On game-theoretic network security provisioning. Journal of network and systems management pp. 1–18 (2013)
36. Rass, S.: On Game-Theoretic Risk Management (Part One)-Towards a Theory of Games with Payoffs that are Probability-Distributions. arXiv preprint arXiv:1506.07368 (2015)
37. Rass, S.: On Game-Theoretic Risk Management (Part Two)-Algorithms to Compute Nash-Equilibria in Games with Distributions as Payoffs. arXiv preprint arXiv:1511.08591 (2015)
38. Rass, S., König, S.: R package 'hyrim': Multicriteria risk management using zero-sum games with vector-valued payoffs that are probability distributions (2017). https://hyrim.net/software/
39. Rass, S., König, S., Schauer, S.: Uncertainty in Games: Using Probability-Distributions as Payoffs. In: Proceedings of The International Conference on Decision and Game Theory for Security, pp. 346–357. Springer (2015)
40. Rass, S., Rainer, B., Vavti, M., Göllner, J., Peer, A., Schauer, S.: Secure communication over software-defined networks. Mobile Networks and Applications **20**(1), 105 (2015)
41. Regan, L.: Electronic Communications Surveillance **66**(3), 32–O2–12 (2014). 10.14452/MR-066-03-2014-07_2. URL www.privacyinternational.org/?q=nod
42. Schauer, S., König, S., Rass, S., Gouglidis, A., Alshawish, A., de Meer, H.: Risk Minimization in Physical Surveillance: Playing an Uncertain Cops-and-Robbers Game. In: Decision and Game Theory for Security: 7th International Conference, GameSec 2016, New York, NY, USA, November 2–4, 2016, Proceedings, vol. 9996, p. 471. Springer (2016)
43. Sela, A.: Fictitious play in one-against-all multi-player games. Economic Theory **14**(3), 635–651 (1999)
44. Singh, V.K., Kankanhalli, M.S.: Adversary aware surveillance systems. IEEE Transactions on Information Forensics and Security **4**(3), 552–563 (2009)
45. Szajowski, K.: Multi-variate Quickest Detection of Significant Change Process. In: GameSec, pp. 56–66. Springer (2011)
46. Team, R.D.C.: R: a language and environment for statistical computing R Foundation for Statistical Computing, 2.13. Vienna, Austria, 2011. Tech. rep., ISBN 3-900051-07-0, url http://www.R-project.org
47. Thorlacius, P., Clausen, J., Brygge, K.: Scheduling of inspectors for ticket spot checking in urban rail transportation. Trafikdage på Aalborg Universitet (2008)
48. Varakantham, P., Lau, H.C., Yuan, Z.: Scalable randomized patrolling for securing rapid transit networks. In: IAAI (2013)
49. Wayman, J., Jain, A., Maltoni, D., Maio, D.: An introduction to biometric authentication systems. Biometric Systems pp. 1–20 (2005)
50. Wayman, J.L.: National Biometric Test Center: Collected Works 1997–2000. Biometric Consortium of the US Government interest group on biometric authentication) San Jose State University, CA (2000)
51. Wilson, A.G., Wilson, G.D., Olwell, D.H.: Statistical methods in counterterrorism. Springer Science+ Business Media **250**, 281 (2006)

Chapter 7
A Game-Theoretic Framework for Securing Interdependent Assets in Networks

Ashish R. Hota, Abraham A. Clements, Saurabh Bagchi, and Shreyas Sundaram

7.1 Introduction

The prevalence of networked engineered systems in the twenty-first century has made it increasingly challenging to ensure their security and resiliency. While the growing interdependency between cyber and physical entities has led to improved system performance, it has also led to new avenues for attackers to target a large number of entities by exploiting those interdependencies. The magnitude, sophistication, and scope of such cyber-attacks have seen rapid growth; examples include a cyber-attack on the power grid in Ukraine [14] and a distributed denial of service (DDOS) attack launched via Internet of Things devices [36].

In a preliminary version of this work [20], we only investigated the security risk minimization game, and considered different sets of case studies.

Abraham Clements is supported by Sandia National Laboratories. Sandia National Laboratories is a multimission laboratory managed and operated by National Technology and Engineering Solutions of Sandia LLC, a wholly owned subsidiary of Honeywell International Inc. for the U.S. Department of Energy's National Nuclear Security Administration under contract DE-NA0003525. SAND2017-10889 B.
This material is based in part upon work supported by the National Science Foundation under Grant Number CNS-1548114. Any opinions, findings, and conclusions or recommendations expressed in this material are those of the authors and do not necessarily reflect the views of the National Science Foundation.

A. R. Hota
Automatic Control Laboratory, ETH Zürich, Zürich, Switzerland
e-mail: ahota@control.ee.ethz.ch

A. A. Clements · S. Bagchi · S. Sundaram (✉)
School of Electrical and Computer Engineering, Purdue University, West Lafayette, IN, USA
e-mail: clemen19@purdue.edu; sbagchi@purdue.edu; sundara2@purdue.edu

S. Rass, S. Schauer (eds.), *Game Theory for Security and Risk Management*, Static & Dynamic Game Theory: Foundations & Applications,
https://doi.org/10.1007/978-3-319-75268-6_7

These large-scale attacks share several important characteristics: these attacks (i) proceed in multiple stages, (ii) exploit interdependencies between diverse entities (e.g., vulnerabilities in devices made by third-party vendors have been exploited in an attack [36]), (iii) are stealthy, and (iv) often exploit unknown or *zero-day* vulnerabilities. The existing literature has investigated approaches to protect against individual attack characteristics mentioned above. Vulnerabilities and their interdependencies are often modeled as *attack graphs* [42]. Bayesian networks are often used to determine how to defend against attacks within the attack graph representation [33]. The authors in [43] quantify security risks due to zero-day vulnerabilities in multistage attacks. Mathematical models of stealthy attacks include the framework of FlipIt games [40]. Dynamic defense strategies, such as *moving target defense (MTD)* [23], are increasingly being deployed to prevent stealthy [3] and multistage attacks [18, 44]. Note that under MTD, the system being protected is reconfigured either periodically or based on some events so that it is harder to penetrate by an external adversary.

Most of the existing literature has focused on network security aspects from the perspective of a single or centralized defender. However, large-scale cyber-physical systems are seldom managed by a single entity, and instead, they are operated by multiple self-interested stakeholders. For instance, different independent system operators (ISOs) are responsible for managing different portions of the power grid. Nonetheless, assets that belong to these different stakeholders remain interdependent, a fact which is exploited by attackers to increase the magnitude and spread of their attacks. There is a growing body of work that applies tools and ideas from game theory in network security in order to model decentralized decision-making by multiple stakeholders. We provide a short summary of the existing literature that is relevant for our setting. A more comprehensive discussion of the literature is beyond the scope of this chapter, and for this, we refer interested readers to [1, 26, 39].

Most of the existing work can be classified into two distinct paradigms. In the first class of games, referred to as interdependent security games [25, 26], each node in the network is treated as an independent decision-maker responsible for protecting itself. In the context of interdependent security games, the existing literature has investigated inefficiency of equilibria [24], effectiveness of cyber insurance [37], impacts of behavioral decision-making [19], and applications in industrial control systems [2], among others. In the second class of games, there are typically two players, an attacker and a defender, who compete over attacking and defending multiple targets. Game-theoretic models in this second framework include Stackelberg security games [39], Colonel Blotto games [34], and network interdiction games [21]. Applications of these models include protecting physical assets [39], analyzing military conflict [34], and securing cyber [9] and cyber-physical systems [15].

However, these models do not adequately capture interactions between multiple defenders each protecting multiple nodes in the network while simultaneously facing strategic adversaries. In order to bridge this gap, we present a game-theoretic framework that (i) incorporates essential features of both the above paradigms, (ii) systematically captures the characteristics of sophisticated attacks discussed above, and (iii) allows us to quantify the security risk of interdependent assets under both centralized and decentralized (game-theoretic) allocation of defense resources. While there are a few recent papers on multi-defender security games [27, 28], these papers assume that the strategy space of a defender is discrete, leading to very different analysis compared to our work.

We model the interdependencies between the assets that belong to possibly different defenders as a directed graph referred to as an *interdependency graph*. We present two complementary game-theoretic formulations. In both settings, the defenders assign defense resources to reduce attack success probabilities on the edges of the interdependency graph but with different objectives. In the first class of games, referred to as the *security risk minimization game*, the defenders minimize their expected loss (formally defined in Section 7.2) due to cyber-attacks on a subset of assets that they own (or are valuable to them). In the second class of games, referred to as the defense cost minimization game, each defender minimizes the cost of defense allocation subject to a maximum security risk (referred to as its *risk tolerance*) it is willing to tolerate on each asset it values. In this second class of games, the set of feasible strategies for a defender is a function of the strategies of other defenders, which makes it an instance of a generalized Nash equilibrium problem (see the chapter appendix for a discussion on this class of problems).

We establish the existence of a pure Nash equilibrium (PNE) in the security risk minimization game and a generalized Nash equilibrium (GNE) (Definition 1 in the appendix to the chapter) in the defense cost minimization game. For both settings, we show that a defender can compute its best response (i.e., its optimal defense allocation for a given allocation by other defenders) by solving appropriately defined convex optimization problems. We demonstrate how our framework can be used to identify certain important aspects of MTD deployment, specifically, how frequently the configurations should be updated to meet security requirements.

We illustrate the application of our framework in two case studies arising in diverse applications. First we consider the IEEE 300 bus power grid network topology with three ISOs who manage different subsets of the buses. We compare the Nash equilibrium outcomes in both games with the outcomes where a central authority minimizes the sum of expected losses (and defense cost) of all defenders. For the security risk minimization game, we show that as the total budget decreases, the total expected losses under centralized and Nash equilibrium defense allocations increase exponentially. For the defense cost minimization game, we show that as the risk tolerance decreases, the total defense cost increases much faster under a Nash equilibrium allocation compared to the centralized allocation. We also study the increase in total defense cost at an equilibrium when multiple assets are supplied by a common vendor that can be compromised directly by an attacker. The second case study is on an e-commerce system adapted from [29]. We compute optimal MTD deployment by applying our framework and investigate how security risk varies as a function of the defense budget.

7.2 Model

In this section, we introduce the mathematical framework that captures the different network security scenarios discussed above. We introduce the notion of an **interdependency graph** to model network interactions at different levels of abstractions.

For example, interdependency graphs capture essential features of *attack graphs* [17] where a node represents a single attack step or vulnerability that can be exploited. Similarly, the nodes can also correspond to cyber or physical entities, such as firewalls or Human Machine Interfaces (HMIs), present in enterprise networks and industrial control systems. Edges capture whether two nodes communicate with each other. At a higher level of abstraction, the interdependency graph can model large-scale networks such as the electric power grid where nodes represent buses, and edges represent physical interconnections between them.

Formally, an **interdependency graph** is a directed graph $G = \{V, E\}$. We refer to each node $v \in V$ as an *asset*. The presence of a directed edge $(v_j, v_i) \in E$ (with index $j \neq i$) indicates that when the asset v_j is compromised, it can be used to launch an attack on asset v_i. In the absence of any defense action, this attack succeeds with a probability $p^0_{j,i} \in (0,1]$. The success of attack on an edge is independent of the success of attacks on other edges.

We consider strategic attackers who launch sophisticated cyber-attacks, such as advanced persistent threats (APTs), into the network. These tools exploit the interdependencies between the assets to move within the network and compromise valuable assets. Without loss of generality, let s be the source node from which an attacker launches the attack from outside the network. If there are multiple entry points to the network, we can effectively replace them by a single entry point s by adding edges from s to all neighbors of all entry points in the original graph (with attack probabilities on these edges same as the original graph).

For an asset $v_i \in V$, let $\mathscr{P}_i$ be the set of directed paths from the source s to v_i on the graph; a path $P \in \mathscr{P}_i$ is a collection of edges $\{(s, u_1), (u_1, u_2), \ldots, (u_l, v_i)\}$ where $u_1, \ldots, u_l \in V$. The probability that v_i is compromised due to an attacker exploiting a given path $P \in \mathscr{P}_i$ is $\prod_{(u_m,u_n) \in P} p^0_{m,n}$ which is the product of probabilities (due to our independence assumption) on the edges that belong to the path P.

Remark 1. There are systematic ways to assign initial attack probabilities depending on the application. For instance, in the attack graph representation, initial attack probabilities are typically defined based on Common Vulnerability Scoring System (CVSS) scores [33]. The CVSS score is a widely adopted metric for assessing the severity of computer system security vulnerabilities. It incorporates the factors of how a vulnerability may be exploited, how difficult it is to exploit a vulnerability, what level of authentication is needed by an adversary, and which of the security dimensions of confidentiality, integrity, and availability are affected by the exploit.

Strategic Defender(s): Let $\mathscr{D}$ be the set of defenders; we use the index k to represent a defender. Defender $D_k \in \mathscr{D}$ is responsible for the security of a set $V_k \subseteq V \setminus \{s\}$ of assets. For each asset $v_m \in V_k$, there is a financial loss $J_m \in \mathbb{R}_{\geq 0}$ that defender D_k incurs if v_m gets compromised. If an asset v_m is not considered valuable, we can set $J_m = 0$. A defender allocates its resources to reduce the attack probabilities on the edges interconnecting different assets on the interdependency graph.

Let the feasible (defense) strategy set of defender D_k be $\mathbb{R}^{n_k}_{\geq 0}$, where n_k represents the (finite) different dimensions of responses that the defender can deploy. The defense resources reduce the attack probabilities on the edges of the interdependency graph. Accordingly, we introduce a transformation matrix $\mathbf{T}_k \in \mathbb{R}_{\geq 0}^{|E| \times n_k}$ which maps a feasible defense strategy $\mathbf{x}_k \in \mathbb{R}_{\geq 0}^{n_k}$ to a defense allocation on edges. By appropriately defining the matrix $\mathbf{T}_k$, we can capture very general classes of defense strategies. We discuss two such examples.

Edge-Based Defense Strategy: In this case, a defender D_k allocates defense resources on a subset of edges $E_k \subseteq E$ of the graph G, and accordingly $n_k = |E_k|$. For example, E_k can represent the set of all the edges that are incoming to nodes in V_k, i.e., defender D_k can reduce the attack probabilities on all the edges that are incoming to the nodes under its ownership. In general, it is not required to assume E_k's to be mutually disjoint; certain edges can potentially be managed by multiple defenders.

Node-Based Defense Strategy: In this case, a defender D_k allocates defense resources to the set of nodes in V_k, and accordingly, $n_k = |V_k|$. Specifically, the defense resource x_i^k being allocated to node v_i implies that all the incoming edges to v_i in the graph G have a defense allocation x_i^k. Here $\mathbf{T}_k$ maps the allocation on a node into the edges that are incoming to it. An example of node-based defense strategy is IP-address randomization, an MTD technique where x_i^k potentially captures how frequently the IP-address on v_i is updated.

We now illustrate the concepts defined above. Consider the interdependency graph shown in Figure 7.1 with a source node and three nodes or assets. Let there be two defenders; defender 1 is responsible for assets 1 and 3, while defender 2 is responsible for asset 2. Under edge-based defense, let E_1 be all edges incoming to nodes 1 and 3, and accordingly, $n_1 = 4$, while under node-based defense, $n_1 = 2$. The respective transformation matrices are

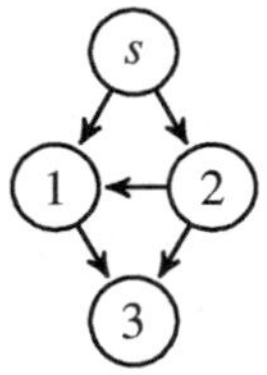

Fig. 7.1: Interdependency graph

$$\mathbf{T}_{e,1} = \begin{bmatrix} 1\,0\,0\,0 \\ 0\,1\,0\,0 \\ 0\,0\,0\,0 \\ 0\,0\,1\,0 \\ 0\,0\,0\,1 \end{bmatrix}, \qquad \mathbf{T}_{n,1} = \begin{bmatrix} 1\,0 \\ 1\,0 \\ 0\,0 \\ 0\,1 \\ 0\,1 \end{bmatrix}.$$

Both matrices have five rows corresponding to the edges in the graph in the order $(s,1),(2,1),(s,2),(1,3)$, and $(2,3)$. Note that under edge-based defense, defender 1's defense resources are not applied on the $(s,2)$ edge. Under node-based defense, both incoming edges to node 1 (as well as 3) receive identical defense resources.

We now introduce our assumptions behind defense effectiveness and cost.

Defense Effectiveness: Let $\mathbf{x} = (\mathbf{x}_1, \mathbf{x}_2, \ldots, \mathbf{x}_{|\mathscr{D}|})$ be a joint defense strategy of the defenders. The attack success probability of an edge (v_j, v_i) under this joint defense strategy is denoted by $p_{j,i}(\mathbf{x})$. Note that, in our framework, it is possible to have multiple defenders simultaneously reducing the attack probability on a single edge. We make the following assumption on $p_{j,i}(\mathbf{x})$:

$$p_{j,i}(\mathbf{x}) := p_{j,i}^0 \exp(-\sum_{k=1}^{|\mathscr{D}|} \mathbf{t}_{j,i}^k \mathbf{x}_k), \tag{7.1}$$

where exp is the exponential function and $\mathbf{t}^k_{j,i} \in \mathbb{R}^{1\times n_k}$ is the row vector in the transformation matrix $\mathbf{T}_k$ that maps the defense allocation $\mathbf{x}_k$ to the edge (v_j, v_i). In the example pertaining to Figure 7.1, $\mathbf{t}^1_{1,3} = \begin{bmatrix}0 & 0 & 1 & 0\end{bmatrix}$ under edge-based defense.

Under our assumption, the marginal reduction in attack probability decreases with increasing security investment.

Defense Cost: For a defender D_k and feasible defense strategy $\mathbf{x}_k$, we define the cost of defense allocation

$$c_k(\mathbf{x}_k) := \sum_{i=1}^{n_k} g^k_i(x^k_i). \tag{7.2}$$

We assume that g^k_i is strictly increasing and convex for every defender D_k and every $i \in \{1, 2, \ldots, n_k\}$ and $g^k_i(0) = 0$. The convexity assumption captures increasing marginal cost of deploying more effective mitigation strategies.

Remark 2. Our assumptions on the defense cost and effectiveness are motivated by a recurring assumption in the security economics literature that probability of successful attack is decreasing and convex (similar to (7.1)) as a function of security investment [12, 24]. Under this interpretation, x represents investments in monetary or dollar amount, and it suffices to assume that $g(x) = x$. Our motivation behind choosing $g(x)$ to be increasing and convex is twofold.

1. It enables us to indirectly capture a broader class of defense effectiveness functions than (7.1). For example, suppose every edge is defended by at most one defender, and security investment reduces attack probability as $p(x) = p^0 \exp(-\sqrt{x})$. We can capture such a scenario indirectly by defining $w = \sqrt{x}$ as the defense resource and cost function $g(w) = w^2$.
2. On the other hand, x could represent the unit of defense resource deployed, and $g(x)$ is the (possibly nonmonetary) cost to the system. For example, in the context of IP-address randomization, x might represent the rate at which the IP-addresses are updated, while $g(x)$ could capture certain types of implementation overhead that are often nonlinear in x; examples of convex overhead costs include probability of genuine connection loss [6] and decrease in bandwidth [41].

Security Risk of an Asset: For an asset v_m, we define its security risk as

$$r_m(\mathbf{x}) := \max_{P \in \mathscr{P}_m} \prod_{(v_j, v_i) \in P} p_{j,i}(\mathbf{x}). \tag{7.3}$$

In other words, the security risk of an asset is given by the highest probability of attack on any path from the source to that asset on the interdependency graph. This is motivated by practical cyber-physical systems that face sophisticated adversaries and APTs and the security maxim that any interdependent system is only as secure as its weakest link. Our choice of defining security in terms of the worst case attack probabilities on an asset in (7.3) implicitly captures strategic attackers who aim to compromise valuable assets and choose a plan of attack that has the highest probability of success for each asset.

We consider two complementary problems that the defenders face.

Security Risk Minimization: In this problem, a defender minimizes its expected loss, where security risk on every asset is defined in Equation (7.3), subject to a budget constraint on defense allocation. Let $\mathbf{x}_{-k}$ denote the defense allocation profile of all defenders except D_k. Then, the objective of D_k is to

$$\underset{\mathbf{x}_k \in \mathbb{R}^{n_k}_{\geq 0}}{\text{minimize}} \quad \phi_k(\mathbf{x}_k, \mathbf{x}_{-k}) := \sum_{v_m \in V_k} J_m \cdot \left(\max_{P \in \mathscr{P}_m} \prod_{(v_j, v_i) \in P} p_{j,i}(\mathbf{x}_k, \mathbf{x}_{-k}) \right) \tag{7.4}$$

$$\text{subject to} \quad \sum_{i=1}^{n_k} g_i^k(x_i^k) \leq b_k, \tag{7.5}$$

where $b_k > 0$ is the security budget of D_k. Note that the feasible strategy set $X_k := \{\mathbf{x}_k \in \mathbb{R}^{n_k}_{\geq 0} | \sum_{i=1}^{n_k} g_i^k(x_i^k) \leq b_k\}$ is nonempty, compact, and convex. Furthermore, the cost function $\phi_k(\mathbf{x}_k, \mathbf{x}_{-k})$ depends on the strategy profile of all defenders. In Section 7.3.2, we analyze the existence of pure Nash equilibria (PNE) of the game between multiple defenders and show how to compute the best response of a player.

Defense Cost Minimization: In this problem, a defender minimizes its cost of defense allocation subject to constraints on the security risk on each asset it values. Let $\theta_m \in (0, 1]$ be the *risk tolerance* of asset $v_m \in V_k$; it captures the maximum security risk (3) defender D_k is willing to tolerate on v_m. A smaller value of θ_m indicates that the defender prefers v_m to have a smaller security risk and must choose its defense allocation accordingly. When $\theta_m = 1$ for an asset, the defender is essentially indifferent to whether the asset is attacked or remains secure. Thus, the defender can choose to not defend a subset of assets, e.g., by defining $\theta_m = 1$ for an asset v_m with $J_m = 0$. Note that $\theta_m \neq 0$ since the probability of successful attack is always nonzero under our assumptions.

Let $\mathbf{x}_{-k}$ denote the vector of defense allocation by all defenders other than D_k. The objective of D_k is to

$$\underset{\mathbf{x}_k \in \mathbb{R}^{n_k}_{\geq 0}}{\text{minimize}} \quad f_k(\mathbf{x}_k) := \sum_{i=1}^{n_k} g_i^k(x_i^k) \tag{7.6}$$

$$\text{subject to} \quad r_m(\mathbf{x}_k, \mathbf{x}_{-k}) \leq \theta_m, \qquad v_m \in V_k. \tag{7.7}$$

In other words, D_k has a risk tolerance for every asset it owns (denoted by the vector θ), and it wants to allocate the defense resources with minimum cost to achieve the desired risk tolerance.

Note that the cost function for defender D_k is independent of the strategies of other players, but the set of constraints in (7.7) is a function of the strategies of all other players. This class of problems is referred to as *generalized Nash equilibrium problems* (GNEPs). A brief overview is presented in the appendix to the chapter. In Section 7.4, we establish the existence of a generalized Nash equilibrium (GNE) in the game between defenders and discuss how to compute the best response of a defender.

7.3 Security Risk Minimization Game

The analysis in this section relies on establishing the convexity of the optimization problem defined in Equations (7.4) and (7.5). We start by introducing certain auxiliary variables. We define the *length* or distance of an edge (v_j, v_i) in terms of the attack probability under the given joint defense allocation $\mathbf{x}$ as,

$$l_{j,i}(\mathbf{x}) := -\log(p_{j,i}(\mathbf{x})) \geq 0, \tag{7.8}$$

where $p_{j,i}(\mathbf{x})$ is given by (7.1). A higher probability of an attack on an edge leads to smaller length for the edge. It follows from (7.1) that the modified length of the edge under a joint strategy profile $\mathbf{x}$ is given by

$$l_{j,i}(\mathbf{x}) := l^0_{j,i} + \sum_{D_k \in \mathscr{D}} \mathbf{t}^k_{j,i}\mathbf{x}_k := l^0_{j,i} + x_{j,i}, \tag{7.9}$$

where $l^0_{j,i} := -\log(p^0_{j,i})$ and $x_{j,i} = \sum_{D_k \in \mathscr{D}} \mathbf{t}^k_{j,i}\mathbf{x}_k$ captures the total defense allocation on the edge (v_j, v_i). Recall that $\mathbf{t}^k_{j,i}$ is the row vector corresponding to edge (v_j, v_i) in the transformation matrix $\mathbf{T}_k$. We denote the vector of modified lengths of the graph under joint defense strategy $\mathbf{x}$ as $\mathbf{L}(\mathbf{x}) = \mathbf{L}^0 + \sum_{D_k \in \mathscr{D}} \mathbf{T}_k\mathbf{x}_k$, where $\mathbf{L}^0$ is the vector of lengths in the absence of any defense allocation.

With this additional notation, we can express the probability that a node v_m is compromised via a given $P \in \mathscr{P}_m$ by

$$\prod_{(v_j,v_i) \in P} p_{j,i}(\mathbf{x}) = \exp\left(-\sum_{(v_j,v_i) \in P} l_{j,i}(\mathbf{x})\right). \tag{7.10}$$

Accordingly, the security risk on asset v_m is given by

$$r_m(\mathbf{x}) = \max_{P \in \mathscr{P}_m} \prod_{(v_j,v_i) \in P} p_{j,i}(\mathbf{x}) = \exp\left(-\min_{P \in \mathscr{P}_m} \sum_{(v_j,v_i) \in P} l_{j,i}(\mathbf{x})\right). \tag{7.11}$$

In other words, the path with the largest probability of successful attack is the path that has the smallest length under the transformation stated in Equation (7.8). This observation enables us to utilize concepts from shortest path problems on graphs, discussed subsequently.

7.3.1 Existence of a Pure Nash Equilibrium

We are now ready to show the existence of a PNE in the game between multiple defenders.

Proposition 1. *The strategic game with multiple defenders where a defender minimizes its cost $\phi_k(\mathbf{x}_k, \mathbf{x}_{-k})$ defined in (7.4), subject to $\mathbf{x}_k \in X_k$ defined in (7.5), possesses a pure Nash equilibrium.*

Proof. From our transformation of attack probabilities into lengths on edges given in (7.8) and (7.9), the probability of successful attack on a node $v_m \in V_k$ due to a path $P \in \mathscr{P}_m$ and joint defense strategy $\mathbf{x}$ is equal to

$$\prod_{(u_j,u_i)\in P} p_{j,i}(\mathbf{x}) = \exp\left(-\sum_{(v_j,v_i)\in P}\left[l^0_{j,i} + \sum_{D_k\in\mathscr{D}} \mathbf{t}^k_{j,i}\mathbf{x}_k\right]\right).$$

Following (7.11), we can express the cost function of a defender D_k, defined in (7.4), as a function of its strategy $\mathbf{x}_k$ and the joint strategy of other defenders $\mathbf{x}_{-k}$ as

$$\phi_k(\mathbf{x}_k, \mathbf{x}_{-k}) = \sum_{v_m\in V_k} J_m \exp\left(-\min_{P\in\mathscr{P}_m} \sum_{(v_j,v_i)\in P} (l_{j,i}(\mathbf{x}_{-k}) + \mathbf{t}^k_{j,i}\mathbf{x}_k)\right), \tag{7.12}$$

where $l_{j,i}(\mathbf{x}_{-k}) = l^0_{j,i} + \sum_{D_l\in\mathscr{D}, l\neq k} \mathbf{t}^l_{j,i}\mathbf{x}_l$ for an edge (v_j, v_i).

Note that $\sum_{(v_j,v_i)\in P}\left[l_{j,i}(\mathbf{x}_{-k}) + \mathbf{t}^k_{j,i}\mathbf{x}_k\right]$ is an affine and, therefore, concave function of $\mathbf{x}_k$. The minimum of a finite number of concave functions is concave [5]. Finally, $\exp(-z)$ is a convex and decreasing function of z. Since the composition of a convex decreasing function and a concave function is convex, $\phi_k(\mathbf{x}_k, \mathbf{x}_{-k})$ is convex in $\mathbf{x}_k$ for any given $\mathbf{x}_{-k}$. Furthermore, the feasible strategy set $X_k = \{\mathbf{x}_k \in \mathbb{R}^{n_k}_{\geq 0} | \sum_{i=1}^{n_k} g^k_i(x^k_i) \leq b_k\}$ is nonempty, compact, and convex for every defender D_k. As a result, the game is an instance of a *concave game* and has a PNE following Theorem 1 of [35]. □

7.3.2 Computing the Best Response of a Defender

The best response of D_k at a given strategy profile $\mathbf{x}_{-k}$ of others is defined as $\mathbf{x}^*_k := \operatorname{argmin}_{\mathbf{x}_k\in X_k} \phi_k(\mathbf{x}_k, \mathbf{x}_{-k})$. While the previous proposition shows that $\phi_k(\mathbf{x}_k, \mathbf{x}_{-k})$ is convex in $\mathbf{x}_k$, the cost function in (7.4) is non-differentiable. We now present an equivalent formulation of the problem below with a smooth cost function. Let $\mathbf{L}(\mathbf{x}_{-k}) = \mathbf{L}^0 + \sum_{D_r\in\mathscr{D}, r\neq k} \mathbf{T}_r\mathbf{x}_r$ be the vector of edge lengths under defense allocation $\mathbf{x}_{-k}$. For a given $\mathbf{x}_{-k}$, consider the following convex optimization problem:

$$\underset{\mathbf{y}\in\mathbb{R}^{|V|}_{\geq 0},\mathbf{x}_k\in\mathbb{R}^{n_k}_{\geq 0}}{\text{minimize}} \quad \sum_{v_m\in V_k} J_m e^{-y_m} \tag{7.13}$$

$$\text{subject to} \quad \mathbf{B}\mathbf{y} - \mathbf{T}_k\mathbf{x}_k \leq \mathbf{L}(\mathbf{x}_{-k}), \tag{7.14}$$

$$y_s = 0, \tag{7.15}$$

$$\sum_{i=1}^{n_k} g_i^k(x_i^k) \leq b_k, \tag{7.16}$$

where **B** is the node-edge incidence matrix of the graph G. Note that the constraint in (7.14) is affine. This formulation is motivated by similar ideas explored in the shortest path interdiction games literature [21, 38].

Remark 3. For a directed graph G, its incidence matrix is $\mathbf{B} \in \mathbb{R}^{|E|\times|V|}$, where the row corresponding to the edge (v_j, v_i) has entry -1 in the jth column and 1 in the ith column.

Remark 4. We refer to the vector $\{y_u\}_{u\in V}$ as a *feasible potential* if it satisfies (7.14) for every edge in the graph. We make the following observations.

1. The inequality in (7.14) for an edge is precisely the inequality that the Bellman-Ford algorithm tries to satisfy in every iteration. As shown in (7.8), the length of every edge is nonnegative in our setting. Therefore, the Bellman-Ford algorithm terminates with a feasible potential [8]. Note that we don't actually use the Bellman-Ford (or Dijkstra's) algorithm in solving the above problem.
2. Consider a path P from s to a node $v \in V$. Then, $y_v - y_s \leq \sum_{(v_j,v_i)\in P} l_{j,i}(\mathbf{x}_k, \mathbf{x}_{-k})$. In other words, when $y_s = 0$, y_v is a lower bound on the length of every path (and consequently the shortest path) from s to v.
3. In the absence of negative cycles, there always exists a feasible potential where y_v is *equal* to the length of the shortest path from s to v [8, Theorem 2.14] for every $v \in V$ (the solution of the Bellman-Ford algorithm).

We now prove the following result.

Proposition 2. *A defense strategy* $\mathbf{x}_k^* \in \mathbb{R}^{n_k}_{\geq 0}$ *is the optimal solution of the problem defined in Equations* (7.13) *to* (7.16) *if and only if it is the minimizer of* $\phi_k(\mathbf{x}_k, \mathbf{x}_{-k})$ *defined in* (7.4) *subject to constraint* (7.5).

Proof. Consider a strategy profile $\mathbf{x}_{-k}$ of all defenders other than D_k. Consider a feasible defense allocation vector $\mathbf{x}_k$ satisfying the constraint in (7.16). The joint strategy profile $\mathbf{x} = (\mathbf{x}_k, \mathbf{x}_{-k})$ defines a modified length vector $\mathbf{L}(\mathbf{x}_k, \mathbf{x}_{-k}) = \mathbf{L}(\mathbf{x}_{-k}) + \mathbf{T}_k\mathbf{x}_k$ on the edges of G. Let $\{y_u^{\mathbf{x}}\}_{u\in V}$ be the feasible potential where $y_u^{\mathbf{x}}$ is equal to the length of the shortest path from s to u under the joint defense allocation $\mathbf{x}$ for every $u \in V$. Now consider a path P from s to $v_m \in V_k$, and let P^* be a path of shortest length s to v_m. From Remark 4, we have

$$y^{\mathbf{x}}_{v_m} \leq \sum_{(u_j,u_i)\in P} l_{j,i}(\mathbf{x}_k,\mathbf{x}_{-k}) = -\sum_{(u_j,u_i)\in P} \log(p_{j,i}(\mathbf{x}))$$
$$\implies e^{-y^{\mathbf{x}}_{v_m}} \geq \prod_{(u_j,u_i)\in P} p_{j,i}(\mathbf{x}), \tag{7.17}$$

with equality for the path P^*. Accordingly, if $\mathbf{x}^*_k$ is optimal for the problem in Equations (7.4) and (7.5), $\{\mathbf{x}^*_k, \{y_u^{(\mathbf{x}^*_k,\mathbf{x}_{-k})}\}_{u\in V}\}$ is feasible for the problem in Equations (7.13) to (7.16), and both have identical cost. Therefore, the optimal cost for the problem in Equations (7.13) to (7.16) is at most the optimal cost of Equations (7.4) and (7.5).

Now let $\{\mathbf{x}^*_k, \{y^*_u\}_{u\in V}\}$ be the optimal solution of the problem defined in Equations (7.13) to (7.16) for a given $\mathbf{x}_{-k}$. We claim that $y^*_{v_m}$ is equal to the length of the shortest path from s to v_m for every v_m with $J_m > 0$.

Assume on the contrary that $y^*_{v_m}$ is strictly less than the length of the shortest path from s to v_m, under the defense allocation $\mathbf{x}^*_k$. From Remark 4 we know that there exists a feasible potential $\{\hat{y}_u\}_{u\in V}$ such that $\hat{y}_{v_m}$ is equal to the length of the shortest path from s to v_m for every node $v_m \in V_k$ with length of every edge (u_j,u_i) given by $l_{j,i}(\mathbf{x}^*_k,\mathbf{x}_{-k})$. As a result, we have $y^*_{v_m} < \hat{y}_{v_m}$, and the objective is strictly smaller at $\hat{y}_{v_m}$, contradicting the optimality of $\{\mathbf{x}^*, \{y^*_u\}_{u\in V}\}$.

Therefore, at the optimal $\{\mathbf{x}^*_k, \{y^*_u\}_{u\in V}\}$, the cost in (7.13) is equal to the cost in (7.4) with defense allocation $\mathbf{x}^*_k$ (following similar arguments as the above paragraph). Furthermore, $\mathbf{x}^*_k$ is feasible for the problem in Equations (7.4) and (7.5). Accordingly, the optimal cost for the problem in Equations (7.4) and (7.5) is at most the optimal cost of Equations (7.13) to (7.16). Combining both observations, we have the required result. □

We now discuss the security risk minimization problem from the perspective of a central authority.

Centralized Defense Allocation to Minimize Security Risk: The security risk minimization problem for a central authority is to find a defense allocation $\mathbf{x}^{\mathtt{OPT}} \in \{\mathbb{R}_{\geq 0}^{\sum_{D_k\in\mathscr{D}} n_k} \mid \sum_{D_k\in\mathscr{D}} \sum_{i=1}^{n_k} g^k_i(x^k_i) \leq \sum_{D_k\in\mathscr{D}} b_k\}$ which minimizes $\sum_{D_k\in\mathscr{D}} \phi_k(\mathbf{x})$. This problem can also be solved via an analogous reformulation as Equations (7.13) to (7.16), and the equivalence result from Proposition 2 applies for this case as well. In our case study in Section 7.6, we compare the security risks under both centralized and game-theoretic defense allocations.

Nash Equilibrium Computation: We compute the PNE strategy profile by iteratively computing the best responses for the defenders. This family of algorithms is referred to as *best response dynamics* [11]. Specifically, we apply the *sequential best response dynamics* in our case studies, and this scheme converges in all considered instances. However, proving theoretical guarantees on the convergence of best response-based update schemes is challenging for the following reasons. First, the expected loss of a defender represented in (7.12) is non-differentiable. Second, in the equivalent formulation Equations (7.13) to (7.16), the players' cost minimization problems are coupled through their constraints which makes it an instance of a GNEP. Analysis of best response schemes for GNEPs is challenging with few algorithms that provide convergence guarantees. Therefore, a theoretical investigation of convergence of best response dynamics is beyond the scope of this chapter.

7.4 Defense Cost Minimization Game

In this section, we analyze the defense cost minimization game between multiple defenders. We start by showing that the risk tolerance constraints (7.7) are equivalent to a set of affine constraints in the defense allocation vector $\mathbf{x}$, and this fact will be useful in our proofs. Consider a node $v_m \in V_k$. Let $P_m \in \mathscr{P}_m$ be a path from the source node s to v_m. Let $r^0_{P_m} := (\prod_{(v_j,v_i)\in P_m} p^0_{j,i})^{-1}$.

Now consider the transformation matrix $\mathbf{T}_k$ for a defender D_k. Let $\mathbf{t}^k_{j,i}$ be the row vector that corresponds to the edge (v_j, v_i) as before. Furthermore, let $\mathbf{t}^k_{P_m} := \sum_{(v_j,v_i)\in P_m} \mathbf{t}^k_{j,i}$. We assume that for every node $v_m \in V_k$, and every path $P_m \in \mathscr{P}_m$, $\mathbf{t}^k_{P_m}$ has at least one nonzero entry, i.e., for every path from s to v_m, there exists at least one edge that D_k can defend. We compute

$$
\begin{aligned}
& r_m(\mathbf{x}) = \max_{P_m\in\mathscr{P}_m} \prod_{(v_j,v_i)\in P_m} p_{j,i}(\mathbf{x}) \leq \theta_m \\
\iff & \prod_{(v_j,v_i)\in P_m} p_{j,i}(\mathbf{x}) \leq \theta_m, \qquad \forall P_m \in \mathscr{P}_m \\
\iff & \left(\prod_{(v_j,v_i)\in P_m} p^0_{j,i}\right) \exp\left(-\sum_{(v_j,v_i)\in P_m}\sum_{D_l\in\mathscr{D}} \mathbf{t}^l_{j,i}\mathbf{x}_l\right) \leq \theta_m, \qquad \forall P_m \in \mathscr{P}_m \\
\iff & \exp\left(-\sum_{D_l\in\mathscr{D}} \mathbf{t}^l_{P_m}\mathbf{x}_l\right) \leq \theta_m r^0_{P_m}, \qquad \forall P_m \in \mathscr{P}_m \\
\iff & \sum_{D_l\in\mathscr{D}} \mathbf{t}^l_{P_m}\mathbf{x}_l \geq -\log(\theta_m r^0_{P_m}), \qquad \forall P_m \in \mathscr{P}_m.
\end{aligned}
\tag{7.18}
$$

Therefore, each constraint in (7.7) can be expressed as a set of affine constraints.

7.4.1 Existence of a Generalized Nash Equilibrium

We now prove the existence of a GNE. Note that in the chapter appendix, we have formally defined the notion of a GNE and provided a general result on the existence of a GNE (Theorem 1). First observe that Theorem 1 requires each X_k (for defender D_k) to be compact, while $\mathbb{R}^{n_k}_{\geq 0}$ is unbounded. In the proof, we define an appropriate compact subset of $\mathbb{R}^{n_k}_{\geq 0}$ for every player that contains the optimal defense allocation irrespective of the strategies of others.

Proposition 3. *The defense cost minimization problems contains a GNE.*

Proof. Let $\mathbf{x}_k^0$ be the optimal defense allocation of defender D_k when the allocation by every other player is 0. Let $\beta_k \in \mathbb{R}_{>0}^{n_k}$. Then $\hat{\mathbf{x}}_k := \mathbf{x}_k^0 + \beta_k$ satisfies

$$\mathbf{t}_{P_m}^k \hat{\mathbf{x}}_k \geq -\log(\theta_m r_{P_m}^0) + \mathbf{t}_{P_m}^k \beta_k, \qquad \forall P_m \in \mathscr{P}_m, \forall v_m \in V_k, \tag{7.19}$$

and $\mathbf{t}_{P_m}^k \beta_k > 0$ following our assumption that $\mathbf{t}_{P_m}^k$ has at least one nonzero entry. We now define

$$X_k := \{\mathbf{x}_k \in \mathbb{R}_{\geq 0}^{n_k} | f_k(\mathbf{x}_k) \leq f_k(\hat{\mathbf{x}}_k)\}, \tag{7.20}$$

where $f_k(\mathbf{x}_k)$ is the cost of defense allocation $\mathbf{x}_k$ defined in (7.6). From the definition, it is easy to see that X_k is nonempty, convex (f_k is convex and X_k is its sublevel set), and compact (f_k is strictly increasing). In particular, $\mathbf{x}_k^0$ and $\hat{\mathbf{x}}_k$ belong to the set X_k because (i) $f_k(\mathbf{x}_k^0) < f_k(\hat{\mathbf{x}}_k)$ by the optimality of $\mathbf{x}_k^0$ and (ii) f_k is strictly increasing.

Now consider the set of constraints (7.7) for D_k that depend on the defense allocation of others. Formally, these constraints can be represented as a correspondence

$$C_k(\mathbf{x}_{-k}) := \{\mathbf{x}_k \in \mathbb{R}_{\geq 0}^{n_k} | \quad \mathbf{t}_{P_m}^k \mathbf{x}_k \geq -\log(\theta_m r_{P_m}^0) - \sum_{D_l \in \mathscr{D}, l \neq k} \mathbf{t}_{P_m}^l \mathbf{x}_l, \forall P_m \in \mathscr{P}_m, v_m \in V_k\}. \tag{7.21}$$

First observe that for any $\mathbf{x}_{-k} \in \mathbb{R}_{\geq 0}^{\sum_{l \neq k} n_l}$, $\mathbf{x}_k^0 \in C_k(\mathbf{x}_{-k})$ since entries in $\mathbf{T}_k$ are nonnegative for every $D_k \in \mathscr{D}$. Therefore, the optimal solution of the problem Equations (7.6) and (7.7), denoted by $\mathbf{x}_k^*(\mathbf{x}_{-k})$, has cost $f_k(\mathbf{x}_k^*(\mathbf{x}_{-k})) \leq f_k(\mathbf{x}_k^0)$, and accordingly $\mathbf{x}_k^*(\mathbf{x}_{-k}) \in X_k$. Therefore, without loss of generality, we can consider X_k to be the set of feasible defense allocation and redefine the constraint correspondence as $\hat{C}_k(\mathbf{x}_{-k}) := C_k(\mathbf{x}_{-k}) \cap X_k \subseteq X_k$.

Now, suppose $\mathbf{x}_{-k} \in \mathbb{R}_{\geq 0}^{\sum_{l \neq k} n_l}$. Then $\hat{C}_k(\mathbf{x}_{-k})$ is nonempty (contains $\hat{\mathbf{x}}_k$ following (7.19) and (7.21)), closed, and convex (intersection of closed and convex sets X_k and $C_k(\mathbf{x}_{-k})$). In addition, the constraint correspondence $\hat{C}_k$ is stated in terms of a set of inequalities where the associated functions (7.21) are continuous and affine (thereby, convex). Furthermore, from the definition of $\hat{\mathbf{x}}_k$ in (7.19), it satisfies all of the affine inequalities in (7.21) with strict inequality. Therefore, from Theorem 2 (in the chapter appendix), $\hat{C}_k$ is both upper and lower semicontinuous in X_{-k}.

Finally, the cost function f_k is independent of $\mathbf{x}_{-k}$ and is continuous and convex in X_k. Therefore, a straightforward application of Theorem 1 establishes the existence of a GNE. □

7.4.2 *Computing the Best Response of a Defender*

Recall that the cost function $f_k(\mathbf{x}_k)$ in (7.6) is independent of the strategies of other defenders and is convex. The set of constraints in (7.18) are affine. Note that (7.18) defines one constraint for every path P_m from s to a given node v_m. Thus, the number of such constraints can be exponentially large in the worst case. We therefore propose the following equivalent problem where the number of constraints is equal to the sum of the number of nodes and edges in the interdependency graph.

Consider the following problem:

$$\underset{\mathbf{y}\in\mathbb{R}^{|V|},\mathbf{x}_k\in\mathbb{R}^{n_k}_{\geq 0}}{\text{minimize}} \quad \sum_{i=1}^{n_k} g_i^k(x_i^k) \tag{7.22}$$

$$\text{subject to} \quad \mathbf{B}\mathbf{y} - \mathbf{T}_k\mathbf{x}_k \leq \mathbf{L}(\mathbf{x}_{-k}), \tag{7.23}$$

$$y_s = 0, \tag{7.24}$$

$$y_m \geq -\log(\theta_m), \qquad \forall m \in V_k, \tag{7.25}$$

where $\mathbf{B}$ is the incidence matrix and $\mathbf{L}(\mathbf{x}_{-k})$ is the vector of edge lengths under the defense allocation $\mathbf{x}_{-k}$ by defenders other than D_k. We now prove the following equivalence result.

Proposition 4. *A defense strategy* $\mathbf{x}_k^* \in \mathbb{R}^{n_k}_{\geq 0}$ *is the optimal solution of the problem defined in Equations* (7.22) *to* (7.25) *if and only if it is the minimizer of the problem defined in Equations* (7.6) *and* (7.7).

Proof. Let $\mathbf{x}_{-k}$ be the defense allocation by other defenders. Let $(\mathbf{x}_k, \mathbf{y})$ be feasible for the problem defined in Equations (7.22) to (7.25). We show that $\mathbf{x}_k$ is feasible for the problem defined in Equations (7.6) and (7.7). In particular, consider a path $P_m \in \mathscr{P}_m$ from s to $v_m \in V_k$. For every $(v_j, v_i) \in P_m$, (7.23) is equivalent to

$$y_j - y_i - \sum_{D_l \in \mathscr{D}} \mathbf{t}^l_{j,i}\mathbf{x}_l \leq -\log(p^0_{j,i})$$

$$\iff y_m - \sum_{D_l \in \mathscr{D}} \mathbf{t}^l_{P_m}\mathbf{x}_l \leq -\log\Big(\prod_{(v_j,v_i)\in P_m} p^0_{j,i}\Big) \qquad \text{(adding over all } (v_j, v_i) \in P_m\text{)}$$

$$\iff \sum_{D_l \in \mathscr{D}} \mathbf{t}^l_{P_m}\mathbf{x}_l \geq -\log(\theta_m) + \log\Big(\prod_{(v_j,v_i)\in P_m} p^0_{j,i}\Big)$$

$$\iff \sum_{D_l \in \mathscr{D}} \mathbf{t}^l_{P_m}\mathbf{x}_l \geq -\log(\theta_m r^0_{P_m}),$$

which satisfies (7.18). Therefore, $\mathbf{x}_k$ is feasible for the problem defined in Equations (7.6) and (7.7), and the optimal cost of the problem Equations (7.6) and (7.7) is at most that of the problem Equations (7.22) to (7.25).

Now, let $\mathbf{x}_k$ be feasible for the problem Equations (7.6) and (7.7). Define

$$y_m := \min_{P_m \in \mathscr{P}_m} \Big[\sum_{D_l \in \mathscr{D}} \mathbf{t}^l_{P_m}\mathbf{x}_l - \log\Big(\prod_{(v_j,v_i)\in P_m} p^0_{j,i}\Big)\Big],$$

and $y_s = 0$. In other words, y_m is the length of the shortest path from s to v_m under the joint strategy profile $(\mathbf{x}_k, \mathbf{x}_{-k})$. Thus, it satisfies (7.23). In addition, it follows from (7.18) that for every $P_m \in \mathscr{P}_m$,

$$\sum_{D_l \in \mathscr{D}} \mathbf{t}^l_{P_m} \mathbf{x}_l - \log\Big(\prod_{(v_j,v_i)\in P_m} p^0_{j,i}\Big) \geq -\log(\theta_m)$$
$$\Longrightarrow y_m \geq -\log(\theta_m).$$

Thus, $(\mathbf{x}_k, \mathbf{y})$ is feasible for the problem Equations (7.22) to (7.25). Therefore, the optimal cost of the problem Equations (7.22) to (7.25) is at most that of the problem Equations (7.6) and (7.7).

Combining both observations, we have the desired result. □

Centralized Defense Allocation to Minimize Defense Cost: The defense cost minimization problem for a central authority is to find a defense allocation $\mathbf{x}^{\text{OPT}} \in \mathbb{R}^{\sum_{D_k \in \mathscr{D}} n_k}_{\geq 0}$ which minimizes $\sum_{D_k \in \mathscr{D}} f_k(\mathbf{x})$ subject to risk tolerance constraints for every $v_m \in V_k, D_k \in \mathscr{D}$. This problem can be solved via an analogous reformulation as Equations (7.22) to (7.25); the equivalence result from Proposition 4 applies in this case.

In our case studies, we compute GNE defense allocations by employing the sequential best response algorithm as discussed earlier and compare the security risks under both centralized and game-theoretic defense allocations.

In the following section, we show how to compute optimal deployment of MTD by applying the framework developed thus far.

7.5 Moving Target Defense

As discussed in the introduction, one of our goals is to consider MTD techniques that eliminate the advantage that strategic adversaries have against a static defended system. This advantage arises from the fact that the adversary can seek to breach such a static system repeatedly, with different (and likely continually learning) attack techniques. In order to capture this mathematically, we consider the notion of *time-to-compromise* of an asset [30]. Specifically, the time to successfully compromise an asset v_i, via an attack launched from v_j, is a random variable denoted $Q_{j,i}$ with an associated distribution function $F_{j,i}$. We assume that the support of $Q_{j,i}$ is $[0,\infty)$ for every $(v_j, v_i) \in E$. As before, we denote the *baseline* attack probability on v_i, launched from v_j, by $p^0_{j,i} \in (0,1]$; this represents the probability of successful attack when (i) there is no defense allocation on the edge (v_j, v_i) and (ii) the attacker has an infinite amount of time-to-compromise v_i.

While deploying MTD, a key variable that determines its effectiveness as well as the deployment cost is *how fast the configuration is changed dynamically*. For instance, consider the class of Dynamic Network defense techniques that relies on randomizing network IP addresses that have been shown to be effective against many types of attacks [3, 22]. If the network addresses are changed more slowly (for instance, once every few months), it gives the attacker sufficient time to learn about system vulnerabilities and execute its attack. On the other hand, if the addresses are changed more frequently, then it deters certain types of attacks more effectively. However, this also increases the overhead cost, such as the number of IP addresses that the defender must own, as well as the cost to legitimate clients, e.g., due to disconnections of network sessions. We now formalize this idea.

For ease of exposition, we only discuss an edge-based defense strategy where each edge receives an independent MTD deployment. We denote $\tau_{j,i} \in [0,\infty)$ as the time period between two successive changes of configuration of the edge (v_j, v_i) under a certain MTD deployment. A smaller $\tau_{j,i}$ represents a higher frequency of configuration changes. While evaluating the effectiveness of MTD, we only consider attacks that succeed within a given configuration in this section. A change of configuration while the attack is in progress (i) prevents the attack from succeeding and (ii) enables the defender to detect the attack and take corrective measures. In other words, for the attack on v_i to succeed, we must have $Q_{j,i} \leq \tau_{j,i}$. Accordingly, the probability of a successful attack on v_i is given by

$$p_{j,i}(\tau_{j,i}) = p^0_{j,i} F_{j,i}(\tau_{j,i}). \tag{7.26}$$

More generally, we refer to $\tau_{j,i}$ as the defense allocation on the edge (v_j, v_i) and τ_E as the vector of defense allocations on all edges. As before, we assume that the success of this attack is independent of the success of attacks propagating through other edges in the graph.

The defender incurs a cost $g^m_{j,i}(\tau_{j,i})$ for its choice of MTD allocation $\tau_{j,i}$ on the edge (v_j, v_i). We make the following assumptions on the function $g^m_{j,i}$.

Assumption 1 *The functions $g^m_{j,i}$ have the following properties.*

1. *$g^m_{j,i}$ is strictly decreasing and convex.*
2. *$g^m_{j,i}(0) = \infty$ and $g^m_{j,i}(\tau) > 0$ for any finite $\tau \in [0,\infty)$.*

In other words, the defender incurs a higher cost for more frequent configuration updates, and this cost is infinite for updating continuously. For finite choice of period τ, the defender incurs a nonzero cost. As an example, the functions $g^m_{j,i}(\tau) = e^{-\alpha\tau}, \alpha > 0$ and $g^m_{j,i}(\tau) = \frac{1}{\tau}$ satisfy the above assumption.

In the context of MTD deployment, we will consider both security risk minimization and defense cost minimization problems stated in Section 7.2. Formally, the security risk minimization problem for a single defender is to

$$\text{minimize} \quad \sum_{v_m \in V} J_m \cdot \left(\max_{P \in \mathscr{P}_m} \prod_{(v_j, v_i) \in P} p_{j,i}(\tau_{j,i}) \right) \tag{7.27}$$

$$\text{subject to} \quad \tau_{j,i} \geq \gamma_{j,i}, \qquad (v_j, v_i) \in E, \tag{7.28}$$

$$\sum_{i=1}^{n_k} g^m_{j,i}(\tau_{j,i}) \leq b, \tag{7.29}$$

where $\gamma_{j,i}$ is a bound on how fast the configuration can be updated, possibly due to physical constraints, and b is the budget. The game-theoretic setting and the defense cost minimization problem can be defined in an analogous manner and are omitted.

7.5.1 Convexity Under Exponential Distributions

Many probability distribution functions (for instance, exponential and Laplace) are log-concave [4]. Log-concavity does not necessarily imply that the function is convex. Nonetheless, for exponentially distributed $Q_{j,i}$'s, we obtain sufficient conditions under which the problem defined in Equations (7.27) to (7.29) is in fact convex.

Let $F_{j,i}$ be any continuous strictly monotone distribution function, such as the distribution function of an exponential random variable. Similar to Section 7.3.2, we define the *length* of an edge (v_i, v_j) under defense allocation $\tau_{j,i}$ as

$$\begin{aligned} l_{j,i}(\tau_{j,i}) &:= -\log(p_{j,i}(\tau_{j,i})) = -\log(p^0_{j,i}) - \log(F_{j,i}(\tau_{j,i})) \\ &:= l^0_{j,i} + x_{j,i}(\tau_{j,i})). \end{aligned} \tag{7.30}$$

In other words,

$$x_{j,i}(\tau_{j,i})) := -\log(F_{j,i}(\tau_{j,i})) \tag{7.31}$$

$$\Longleftrightarrow e^{-x_{j,i}} = F_{j,i}(\tau_{j,i}) \Longleftrightarrow \tau_{j,i} = F^{-1}_{j,i}(e^{-x_{j,i}}). \tag{7.32}$$

Note that $l^0_{j,i}$ is the length of the edge without any defense allocation and the quantity $x_{j,i}$ (a function of $\tau_{j,i}$) increases the length linearly. Let $\mathbf{L}^0$ be the vector of lengths without any defense allocation, and let $\mathbf{x}$ be the vector of $x_{j,i}$ variables. We now state the following assumptions on cost functions $g_{j,i}$ and exponentially distributed time-to-compromise random variable $Q_{j,i}$.

Assumption 2 *For every edge (v_j, v_i), i) $g_{j,i}(\tau) = e^{-\alpha_{j,i}\tau}, \alpha_{j,i} > 0$, ii) $F_{j,i}(\tau) = 1 - e^{-\beta_{j,i}\tau}, \beta_{j,i} > 0$, and iii) $\beta_{j,i} < \alpha_{j,i}$.*

Now consider the following optimization problem.

$$\begin{aligned} \underset{\mathbf{y}\in\mathbb{R}^{|V|}_{\geq 0}, \mathbf{x}\in\mathbb{R}^{|E|}_{\geq 0}}{\text{minimize}} \quad & \sum_{v_m\in V} J_m e^{-y_m} && (7.33) \\ \text{subject to} \quad & \mathbf{B}\mathbf{y} - \mathbf{x} \leq \mathbf{L}^0, && (7.34) \\ & y_s = 0, && (7.35) \\ & \sum_{(v_j,v_i)\in E} (1 - e^{-x_{j,i}})^{\frac{\alpha_{j,i}}{\beta_{j,i}}} \leq b, && (7.36) \\ & 0 \leq x_{j,i} \leq \log(\frac{\alpha_{j,i}}{\beta_{j,i}}), && (7.37) \end{aligned}$$

where $\mathbf{B}$ is the incidence matrix of the interdependency graph. Our main result in this subsection shows that the above problem is convex and solves the optimization problem stated in Equations (7.27) to (7.29) when $\gamma_{j,i} = \frac{-1}{\beta_{j,i}} \log\left(1 - \frac{\beta_{j,i}}{\alpha_{j,i}}\right)$. We start with the following lemmas. We drop the subscript i, j in the following analysis.

Lemma 1. *Under Assumption 2, the function* $g(F^{-1}(e^{-x})) := (1-e^{-x})^{\frac{\alpha}{\beta}}$ *is convex in* x *over the domain* $x \in [0, \log(\frac{\alpha}{\beta})]$.

Proof. For the exponential distribution function, we have

$$F(\tau) = 1 - e^{-\beta\tau} \implies e^{-\beta\tau} = 1 - F(\tau) \implies \tau = \frac{-1}{\beta}\log(1 - F(\tau))$$
$$\implies F^{-1}(w) := \frac{-1}{\beta}\log(1-w),$$

where $w := F(\tau)$. Then, the cost function can be expressed as

$$g(F^{-1}(e^{-x})) = g\left(\frac{-1}{\beta}\log(1-e^{-x})\right) = \exp\left(\frac{\alpha}{\beta}\log(1-e^{-x})\right) = (1-e^{-x})^{\frac{\alpha}{\beta}}.$$

We now verify that the function $h(x) := g(F^{-1}(e^{-x})) = (1-e^{-x})^{\frac{\alpha}{\beta}}$ is increasing and convex for $x \in [0, \log(\frac{\alpha}{\beta})]$. We denote $\frac{\alpha}{\beta} = z$ and compute

$$\begin{aligned}
h'(x) &= z(1-e^{-x})^{(z-1)}(e^{-x}) \\
h''(x) &= z(z-1)(1-e^{-x})^{(z-2)}e^{-2x} + z(1-e^{-x})^{(z-1)}(-e^{-x}) \\
&= z(1-e^{-x})^{(z-2)}e^{-x}[(z-1)e^{-x} - (1-e^{-x})] \\
&= z(1-e^{-x})^{(z-2)}e^{-x}[ze^{-x} - 1].
\end{aligned}$$

We need $z > e^x$ for $h''(x) > 0$, or equivalently, $x < \log(z) = \log(\frac{\alpha}{\beta})$. □

Lemma 2. *Let* $\beta < \alpha$, *and then,* $x \leq \log(\frac{\alpha}{\beta}) \iff \tau \geq \frac{-1}{\beta}\log(1-\frac{\beta}{\alpha}) = \gamma$.

Proof. Recall from (7.31) that $x = -\log(F(\tau)) = -\log(1-e^{-\beta\tau})$. Then,

$$x = -\log(F(\tau)) \leq \log(\frac{\alpha}{\beta}) \iff F(\tau) = 1 - e^{-\beta\tau} \geq \frac{\beta}{\alpha},$$
$$\iff e^{-\beta\tau} \leq 1 - \frac{\beta}{\alpha} \iff -\beta\tau \leq \log(1-\frac{\beta}{\alpha}) \iff \tau \geq \frac{-1}{\beta}\log\left(1-\frac{\beta}{\alpha}\right).$$

This concludes the proof. □

We now prove the following result.

Proposition 5. *Suppose Assumption 2 holds, and let* $\gamma_{j,i} = \frac{-1}{\beta_{j,i}}\log(1-\frac{\beta_{j,i}}{\alpha_{j,i}})$. *Then Equations* (7.33) *to* (7.37) *represent a convex optimization problem that is equivalent to the security risk minimization problem stated in Equations* (7.27) *to* (7.29).

Proof. From Lemma 1, we observe that the constraints (7.29) and (7.36) are equivalent and are convex. Similarly, from Lemma 2, we observe that the constraints (7.28) and (7.37) are equivalent. We reach the desired result following identical arguments as the proof of Proposition 2. □

In the following section, we compare centralized and PNE defense allocation in a case study on the IEEE 300 bus power grid network. In Section 7.7, we compute optimal MTD deployment for an e-commerce system for both security risk and defense cost minimization problems.

7.6 Case Study 1 - IEEE 300 Bus Power Network

A large-scale network, such as the power grid, contains thousands of cyber and physical entities. Therefore, many different types of attacks are possible against such a system. Our first case study illustrates how our framework is applicable in this context via the following stylized example. Note that the following choice of the interdependency graph, cost functions, and attack probabilities is only made for illustrative purposes. Depending on the setting, a practitioner must instantiate the model appropriately.

We consider the widely used benchmark IEEE 300 bus power grid network [7]. We define the network itself as the interdependency graph where each node represents a bus (i.e., the network has 300 nodes) and the physical interconnection between the buses represents the edges. Each bus has generators and/or load centers associated with it. The 300 bus network data divides the buses or nodes into 3 different regions containing 159, 78, and 63 nodes, respectively [7]. We assume that each region is managed by an independent entity or defender. The defenders want to protect the buses within their region that contain the generators; each generator is valued at its maximum generation capacity. The attacker can directly access three nodes (specifically, bus 39, 245, and 272).

All computations in this section are carried out in MATLAB using the convex optimization solver CVX [13].

We first consider the security risk minimization problem. We assume that the cost function is $g(x) = x$. Here x potentially represents the monetary amount spent on securing an asset, while our assumption in (7.1) ($p(x) = p^0 \exp(-x)$) captures how effective the monetary investment is in reducing the attack probability. We further assume that every edge has an initial probability of successful attack of magnitude 1. For a given total budget, we compute the centralized defense allocation that minimizes the total expected loss. We divide the total budget among the players proportional to the number of nodes they control and compute the PNE defense allocation by iteratively computing their respective best responses. We observe that both simultaneous and sequential best response dynamics converge to PNE within 25 iterations starting from random initial defense allocations. Figure 7.2a shows the total expected loss (in the logarithmic scale with base e) experienced by all three players at the PNE and under the centralized defense allocation for different total budgets. The total expected loss is larger at the PNE, and the relative change in the total expected loss at the PNE grows from 1.8% to over 7500% as the budget increasesfrom 1 to 100. When the total budget is 100, the total expected loss at the

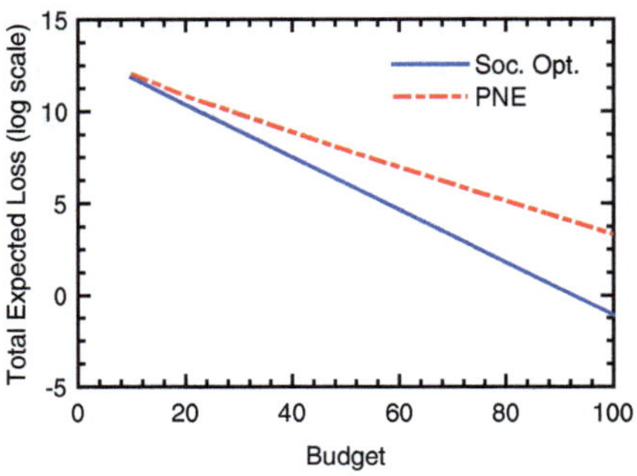

(a) Total expected loss

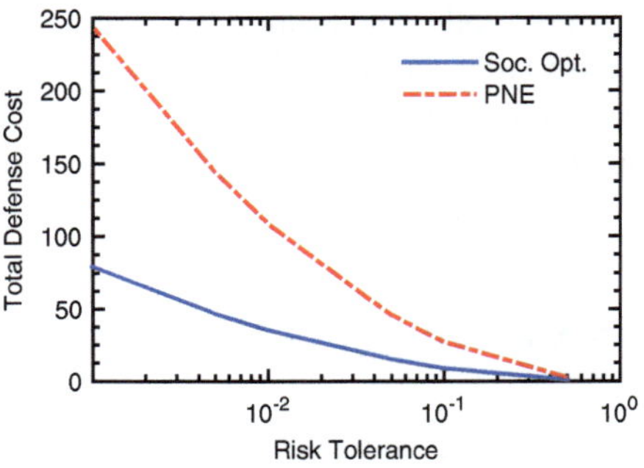

(b) Optimal and equilibrium cost

Fig. 7.2: Comparison of centralized and PNE defense allocations for the IEEE 300 bus power grid network. Figure 7.2a and Figure 7.2b correspond to the security risk minimization and defense cost minimization games, respectively. The costs are in an abstract unit

social optimum is 0.35, while it is 27.56 (or 3.3164 in the log scale as shown in the plot) at the PNE.

We then consider the defense cost minimization problem. We assume that the cost function is given by $g(x) = x^2$ for every defender and for every edge in the network. Our motivation behind this choice is the crash overdrive malware attack on the Ukraine power grid [14]. MTD techniques such as IP-address randomization are effective against reconnaissance scans which the above malware relies on; here x potentially represents the rate at which IP-addresses are updated. Following the discussion in Remark 2, we choose $g(x) = x^2$ which better captures nonlinear growth of certain types of overhead costs [6, 41]. Since $g(x)$ could be interpreted as both monetary as well as overhead costs, we assume that it is in an abstract unit. In Figure 7.2b, we compare the total defense cost required to enforce a given tolerance level (shown in the x-axis) at each generator node under centralized and PNE defense allocations. As the risk tolerance decreases, the defense cost at the PNE increases faster than the defense cost under the centralized defense allocation.

7.6.1 Interdependency Through Common Vendor

As we discussed earlier, strategic attackers have exploited vulnerabilities in assets prepared by a common vendor to increase the spread of their attacker in recent years. In this subsection, we show how our framework can be used by practitioners to quantify the (potentially higher) security risk they face when multiple assets are from a common vendor. This is a common occurrence in practice where the same hardware or software (or both) is in use at multiple sub-systems owned by different stakeholders and any vulnerability in it can affect multiple assets. We again consider

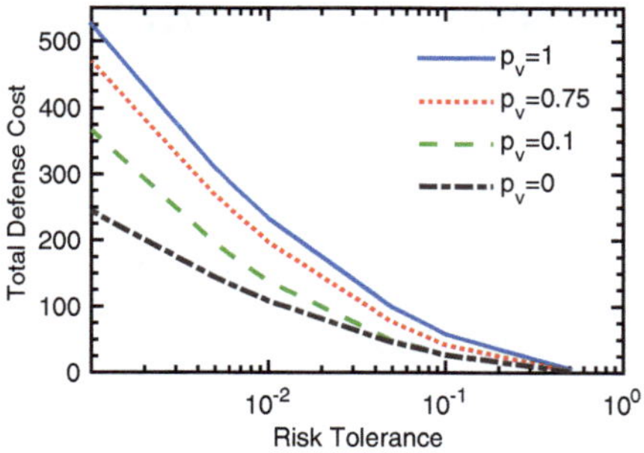

(a) Initial probability on edges: 1

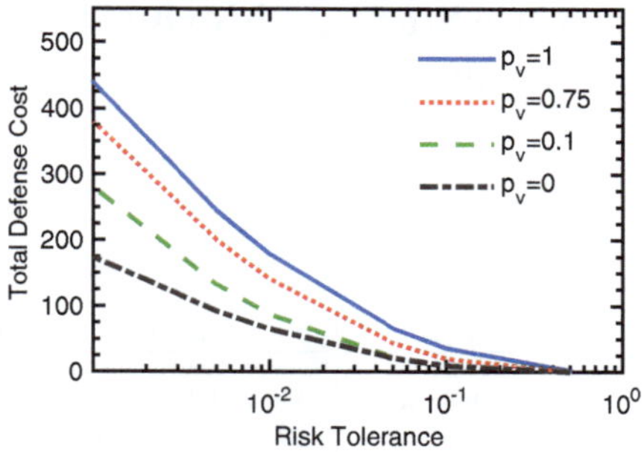

(b) Initial probability on edges: 0.75

Fig. 7.3: Total defense cost at a PNE for the IEEE bus power grid network with common vendor. The probability of successful attack on the vendor directly from the attacker is represented as p_v. Initial probabilities of successful attack on all edges are 1 and 0.75, respectively. The defense cost is in an abstract unit.

the IEEE 300 bus network with $g(x) = x^2$ (with an abstract unit). We represent the vendor by a new node and connect the vendor to eight different generator nodes (belonging to different players), i.e., if an attacker successfully compromises the vendor, it can launch attacks on the generators directly. The attacker can directly attack the vendor node.

We first assume that $p^0_{j,i} = 1$ on every edge in the network, except for the edge from the attacker node to the vendor node. In Figure 7.3a, we show the total defense cost at the PNE to meet a given risk tolerance; the quantity p_v represents the probability of successful direct attack on the vendor by the attacker. The case where the vendor is not present is denoted by $p_v = 0$ (which is the case from Figure 7.2b). As p_v increases, it becomes easier for the attacker to attack the generators via the vendor, and accordingly the budget required to meet a given tolerance increases. We find identical trends when the $p^0_{j,i} = 0.75$ on every edge in the network (except the edge from the attacker node to the vendor node) the results for which are shown in Figure 7.3b. When the risk tolerance is 0.5, the figure shows that the total defense costs are equal when $p_v = 0$, $p_v = 0.1$, and $p_v = 0.5$, which is expected because the attack probability via the vendor is smaller than 0.5.

The practical implication of this result is that quantifying the security risks due to assets from third-party vendors could lead to designing adequate countermeasures and financial incentives (such as adding appropriate security requirements in procurement and support contracts with the vendors), which will then potentially reduce the likelihood and spread of such attacks in the future. Our treatment enables any stakeholder to quantitatively calculate the risk of compromise of its asset due to shared vulnerability at a vendor.

7.7 Case Study 2 - Moving Target Defense of E-Commerce System

Nodes	Devices	Attack	Defense DN	DP	DE	DS	DD	CD
1,2	Web Server	Network Reconnaissance[a]	x					
3	Web Server	Vulnerability Exploit[b]		x	x	x		
4	Web Server	Credential attack[c]						x
5	Internal Network	Network Reconnaissance	x					
6	Web Server	Credential attack						x
7,8	DB or App. Server	Exploit a service		x	x	x		
9,11	DB or App. Server	Credential attack						x
10,12	Database Server	Read from Database					x	
T	Database Server	Read credit card data	No Defense					

(a) Description of attack steps and respective MTDs

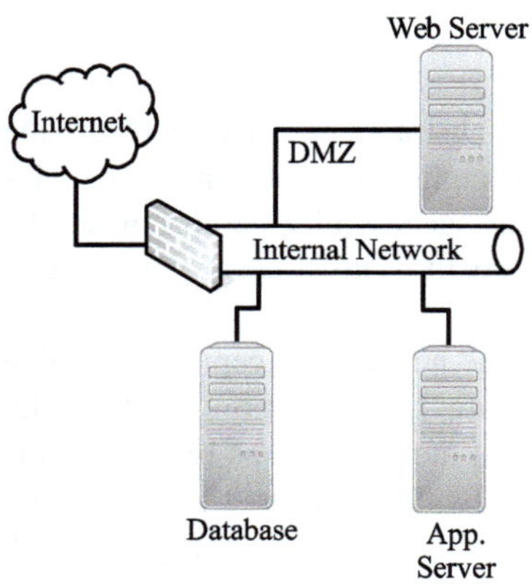

(b) E-commerce distributed system [29]

[a] Network reconnaissance actively or passively probes the network configuration to identify vulnerable systems.

[b] Exploitation relies on vulnerabilities in software which enable attackers to perform otherwise prohibited operations.

[c] Credential attacks involve obtaining valid credentials (e.g. passwords) to a system (for example, by brute force).

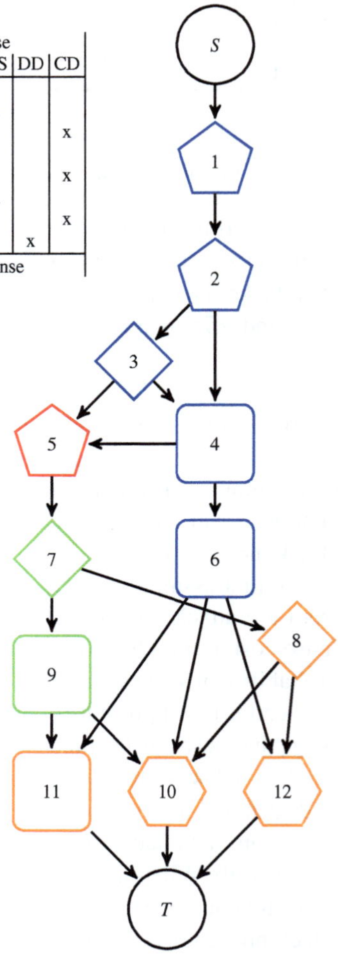

(c) Attack graph representation

Fig. 7.4: Representation of e-commerce network. In the attack graph, Figure 7.4c colors indicate the targeted device, and shapes indicate type of attack. For mapping see Figure 7.4a; defenses used in Figure 7.4a, are: dynamic networks (DN), dynamic platforms (DP), dynamic environments (DE), dynamic software (DS), dynamic data (DD), and credential defense (CD)

We consider an e-commerce distributed system studied in [29] to illustrate how our framework can be used to compute optimal MTD deployment. Figure 7.4b shows the devices, and Figure 7.4c shows the corresponding attack graph for the e-commerce system. The attacker aims to obtain the customer information, such as credit card numbers, from a database. The attacker needs to find a suitable subset of the twelve attack steps on four devices to achieve this goal. The devices are a web server (located in a DMZ), the internal network, a database server, and an application server; both servers are located on the internal network. For each node in the attack graph, we describe the type of attack that can compromise it, and the type of MTD from [32] that can be deployed in Figure 7.4a.

We treat the attack graph in Figure 7.4c as the interdependency graph. The attacker has a single entry point into the network at node S, and it targets node T. We assume that the initial probability of successful attack on every edge is 1. In practice, these initial probabilities can be defined in terms of their CVSS scores [33]. We consider a node-based defense strategy. At every node, the frequency at which the corresponding MTD is updated represents a decision variable.

We consider the setting in Section 7.5 where a higher frequency of updating the MTD reduces the attacker's advantage. We assume that the cost function and distribution of time required for successful compromise satisfy Assumption 2. Specifically, for every edge (v_j, v_i), we assume that the random variable $Q_{j,i}$ is exponentially distributed with distribution function $F_{j,i}(\tau) = 1 - e^{-\tau}$, i.e., the parameter $\beta_{j,i} = 1$. We also consider an identical cost function $g_{j,i}(\tau) = e^{-\alpha\tau}$ for every edge in an abstract unit and consider three different values of $\alpha \in \{3, 5, 10\}$ in our simulations. We consider both security risk minimization and defense cost minimization problems from the perspective of a single (centralized) defender. We used MATLAB's *fmincon* routine with *active-set* and *sqp* solvers to compute optimal defense allocation for both problems. Numerical results show that nodes $1, 2, 4, 6$, and 10 receive higher defense allocation than other nodes. This is expected because the initial attack probabilities are identical, and these nodes lie on a path with the smallest number of edges.

Figure 7.5a shows how the attack probability on the target node decreases under the optimal defense allocation with a given budget. First observe that at a given budget, when α is larger, the attack probability is smaller. For a given β, higher values of α imply that we can assign a larger defense allocation on the edges (i.e., $x_{j,i}$'s) without violating constraints (7.36) and (7.37). As a result, we obtain a smaller attack probability at the target node. Note further that, as the budget increases, the attack probability initially decreases, but it gets saturated beyond a certain budget. The reason for this is the constraints on the defense allocation (7.37) limits how fast the MTD configuration can be updated. While the constraint in (7.37) is imposed to preserve the convexity of the optimization problems, qualitatively similar behavior will emerge when the constraints are due to physical limitations on the frequency of configuration updates.

Figure 7.5b shows the cost of defense allocation to enforce that the probability of attack on the target is smaller than the risk tolerance. We observe that the relationship between the two is approximately piecewise linear in the logarithmic scale. As before, a higher value of α implies a smaller budget requirement for a given level of risk tolerance.

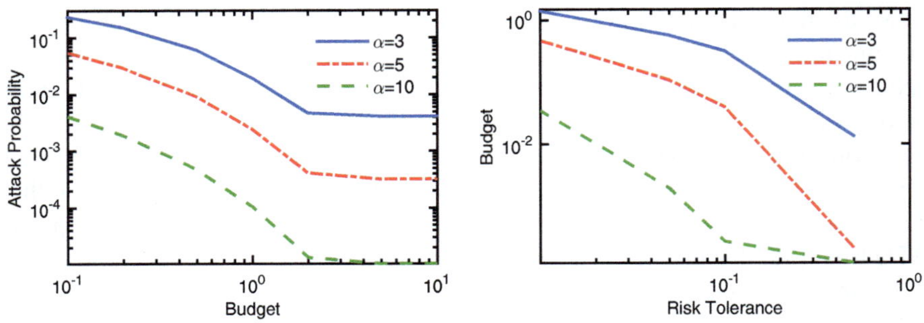

(a) Attack probability as a function of budget

(b) Optimal defense cost (budget) as a function of risk tolerance

Fig. 7.5: Optimal MTD deployment on e-commerce network. Figure 7.5a corresponds to the security risk minimization problem, and Figure 7.5b corresponds to the defense cost minimization problem. The budget is in an abstract unit

7.8 Conclusion

In this chapter, we presented two complementary game-theoretic models to study the security of networked systems. We considered multiple self-interested defenders, each of whom manages a set of assets represented by nodes in a directed graph. Attacks spread through the network via the interconnecting links in the graph. In the first class of games, each defender minimizes its expected loss subject to budget constraints on the defense investments, while in the second class of games, each defender minimizes its cost of defense investment subject to upper bounds on the probability of successful attack on its assets (or its risk tolerance). Under suitable assumptions on the effectiveness of defense investments in reducing attack probabilities, we showed the existence of (generalized) Nash equilibria in both settings and showed that each defender can compute its optimal defense allocation for a given allocation by other defenders by solving a convex optimization problem.

We demonstrated how our framework can be applied in diverse settings, including large-scale cyber-physical systems such as the power grid as well as enterprise networks. Motivated by recent cyber-attacks that exploit vulnerabilities in assets supplied by third-party vendors, we specifically studied the impact of such vendors on the Nash equilibrium defense allocation in a case study on the IEEE 300 bus power grid network. As the probability of successful attack on the vendor increases, the defenders need to invest more to meet a given risk tolerance constraints. In a second case study on an e-commerce network, we computed optimal deployment of moving target defense using our framework.

Our framework leaves several interesting avenues for future research. The impact of incentive mechanisms, such as imposing fines on defenders or vendors who do not take adequate security measures, can be studied within our framework. Another important future direction is to consider real-time interaction between attacker(s) and defender(s) in a dynamic game framework. Such interaction can proceed in multiple stages and can consider various levels of misinformation about the strategies of different parties.

Acknowledgements We thank Dr. Shaunak Bopardikar (United Technologies Research Center) and Dr. Pratyusha Manadhata (HP Labs) for fruitful discussions.

Chapter Appendix: Generalized Nash Equilibrium

In this section, we give a formal definition of a *generalized Nash equilibrium* (GNE) and state the required existence result that will be useful in our analysis.

Let there be N players. The strategy set of player i is denoted as $X_i \subseteq \mathbb{R}^{n_i}$. Let $X := \Pi_{i=1}^{N} X_i$, and $X_{-i} := \Pi_{j=1,j\neq i}^{N} X_j$. Let $C_i : X_{-i} \to 2^{X_i}$ be the set-valued map or correspondence that defines the feasible strategy set for player i at a given strategy profile of all other players. Let $f_i : X \to \mathbb{R}$ denote the cost function for player i. We denote this game as $\Gamma(N, \{X_i\}, \{C_i\}, \{f_i\})$.

Definition 1. A strategy profile $\mathbf{x}^* \in X$ is a GNE of $\Gamma(N, \{X_i\}, \{C_i\}, \{f_i\})$ if for every player i,

$$x_i^* \in \operatorname*{argmin}_{x_i \in C_i(x_{-i}^*)} f_i(x_i, x_{-i}^*). \tag{7.38}$$

Our proof of GNE existence in this chapter is based on the following general result.

Theorem 1. *Consider the game $\Gamma(N, \{X_i\}, \{C_i\}, \{f_i\})$. Assume for all players we have*

1. *X_i is a nonempty, convex, and compact subset of an Euclidean space,*
2. *C_i is both upper and lower semicontinuous,*
3. *$C_i(x_{-i})$ is nonempty, closed, and convex for every $x_{-i} \in X_{-i}$,*
4. *f_i is continuous on the graph of C_i, and*
5. *$f_i(x_i, x_{-i})$ is quasiconvex on $C_i(x_{-i})$ for every $x \in X$.*

Then there exists a GNE.

The proof of the above theorem relies on Kakutani fixed-point theorem and Berge's maximum theorem and is presented in [10, Theorem 3.1].

In many application, including for the defense cost minimization game studied in this chapter, we encounter a parametrized constraint set, i.e., $C_i(x_{-i}) = \{x_i \in X_i | g_{ij}(x_i, x_{-i}) \leq 0, j = \{1, 2, \ldots, m_i\}\}$. For this class of constraints, we have the following sufficient conditions for the upper and lower semicontinuity of C_i [16, Theorem 10,12].

Theorem 2. *Let* $C_i : X_{-i} \to 2^{X_i}$ *be given by* $C_i(x_{-i}) = \{x_i \in X_i | g_{ij}(x_i, x_{-i}) \leq 0, j = \{1,2,\ldots,m_i\}\}$.

1. *Let* $X_i \subseteq \mathbb{R}^{n_i}$ *be closed and all components* g_{ij}*'s be continuous on* X. *Then,* C_i *is upper semicontinuous on* X_{-i}.
2. *Let* g_{ij}*'s be continuous and convex in* x_i *for each* x_{-i}. *If there exists* $\bar{x}$ *such that* $g_{ij}(\bar{x}_i, \bar{x}_{-i}) < 0$ *for all* j*, then* C_i *is lower semicontinuous at* $\bar{x}_{-i}$ *and in some neighborhood of* $\bar{x}_{-i}$.

Remark 5. Some authors use the term hemicontinuity instead of semicontinuity [31]. The definitions coincide for closed and compact-valued correspondences, which is the case here.

References

1. Alpcan T, Başar T (2010) Network security: A decision and game-theoretic approach. Cambridge University Press
2. Amin S, Schwartz GA, Sastry SS (2013) Security of interdependent and identical networked control systems. Automatica 49(1):186–192
3. Antonatos S, Akritidis P, Markatos EP, Anagnostakis KG (2007) Defending against hitlist worms using network address space randomization. Computer Networks 51(12):3471–3490
4. Bagnoli M, Bergstrom T (2005) Log-concave probability and its applications. Economic Theory 26(2):445–469
5. Boyd S, Vandenberghe L (2004) Convex optimization. Cambridge university Press
6. Carroll TE, Crouse M, Fulp EW, Berenhaut KS (2014) Analysis of network address shuffling as a moving target defense. In: Communications (ICC), 2014 IEEE International Conference on, IEEE, pp 701–706
7. Christie R (1993) Power systems test case archives. URL https://goo.gl/1AOSXj, retrieved: 2017-06-07
8. Cook WJ, Cunningham WH, Pulleyblank WR, Schrijver A (1998) Combinatorial optimization, vol 605. Springer
9. Durkota K, Lisý V, Bošanský B, Kiekintveld C (2015) Approximate solutions for attack graph games with imperfect information. In: Decision and Game Theory for Security, Springer, pp 228–249
10. Dutang C (2013) Existence theorems for generalized Nash equilibrium problems. Journal of Nonlinear Analysis and Optimization: Theory & Applications 4(2):115–126
11. Fudenberg D, Levine DK (1998) The theory of learning in games, vol 2. MIT Press
12. Gordon LA, Loeb MP (2002) The economics of information security investment. ACM Transactions on Information and System Security (TISSEC) 5(4):438–457

13. Grant M, Boyd S, Ye Y (2008) CVX: Matlab software for disciplined convex programming
14. Greenberg A (2017) 'Crash Overdrive': The malware that took down a power grid. URL http://bit.ly/2raojOf, Wired Magazine, retrieved: 2017-09-20
15. Gupta A, Schwartz G, Langbort C, Sastry SS, Basar T (2014) A three-stage Colonel Blotto game with applications to cyberphysical security. In: American Control Conference (ACC), 2014, IEEE, pp 3820–3825
16. Hogan WW (1973) Point-to-set maps in mathematical programming. SIAM Review 15(3):591–603
17. Homer J, Zhang S, Ou X, Schmidt D, Du Y, Rajagopalan SR, Singhal A (2013) Aggregating vulnerability metrics in enterprise networks using attack graphs. Journal of Computer Security 21(4):561–597
18. Hong JB, Kim DS (2016) Assessing the effectiveness of moving target defenses using security models. IEEE Transactions on Dependable and Secure Computing 13(2):163–177
19. Hota A, Sundaram S (2016) Interdependent security games on networks under behavioral probability weighting. IEEE Transactions on Control of Network Systems 5(1):262–273
20. Hota AR, Clements AA, Sundaram S, Bagchi S (2016) Optimal and game-theoretic deployment of security investments in interdependent assets. In: International Conference on Decision and Game Theory for Security, Springer, pp 101–113
21. Israeli E, Wood RK (2002) Shortest-path network interdiction. Networks 40(2):97–111
22. Jafarian JH, Al-Shaer E, Duan Q (2012) Openflow random host mutation: Transparent moving target defense using software defined networking. In: Proceedings of the first workshop on Hot topics in software defined networks, ACM, pp 127–132
23. Jajodia S, Ghosh AK, Subrahmanian V, Swarup V, Wang C, Wang XS (2013) Moving target defense II. Application of Game Theory and Adversarial Modeling Series: Advances in Information Security 100:203
24. Jiang L, Anantharam V, Walrand J (2011) How bad are selfish investments in network security? Networking, IEEE/ACM Transactions on 19(2):549–560
25. Kunreuther H, Heal G (2003) Interdependent security. Journal of risk and uncertainty 26(2–3):231–249
26. Laszka A, Felegyhazi M, Buttyan L (2014) A survey of interdependent information security games. ACM Computing Surveys (CSUR) 47(2):23:1–23:38
27. Letchford J, Vorobeychik Y (2013) Optimal interdiction of attack plans. In: AAMAS, pp 199–206
28. Lou J, Smith AM, Vorobeychik Y (2017) Multidefender security games. IEEE Intelligent Systems 32(1):50–60
29. Modelo-Howard G, Bagchi S, Lebanon G (2008) Determining placement of intrusion detectors for a distributed application through Bayesian network modeling. In: International Workshop on Recent Advances in Intrusion Detection, Springer, pp 271–290

30. Nzoukou W, Wang L, Jajodia S, Singhal A (2013) A unified framework for measuring a network's mean time-to-compromise. In: Reliable Distributed Systems (SRDS), 2013 IEEE 32nd International Symposium on, IEEE, pp 215–224
31. Ok EA (2007) Real analysis with economic applications, vol 10. Princeton University Press
32. Okhravi H, Hobson T, Bigelow D, Streilein W (2014) Finding focus in the blur of moving-target techniques. IEEE Security & Privacy 12(2):16–26
33. Poolsappasit N, Dewri R, Ray I (2012) Dynamic security risk management using Bayesian attack graphs. IEEE Transactions on Dependable and Secure Computing 9(1):61–74
34. Roberson B (2006) The Colonel Blotto game. Economic Theory 29(1):1–24
35. Rosen JB (1965) Existence and uniqueness of equilibrium points for concave n-person games. Econometrica: Journal of the Econometric Society 33(3):520–534
36. Sanger DE, Perlroth N (2016) A new era of internet attacks powered by everyday devices. URL https://nyti.ms/2nsqrlT, The New York Times, retrieved: 2017-05-14
37. Schwartz G, Shetty N, Walrand J (2013) Why cyber-insurance contracts fail to reflect cyber-risks. In: Communication, Control, and Computing (Allerton), 2013 51st Annual Allerton Conference on, IEEE, pp 781–787
38. Sreekumaran H, Hota AR, Liu AL, Uhan NA, Sundaram S (2015) Multi-agent decentralized network interdiction games. arXiv preprint arxiv:150301100
39. Tambe M (2011) Security and game theory: Algorithms, deployed systems, lessons learned. Cambridge University Press
40. Van Dijk M, Juels A, Oprea A, Rivest RL (2013) Flipit: The game of "stealthy takeover". Journal of Cryptology 26(4):655–713
41. Van Leeuwen B, Stout WM, Urias V (2015) Operational cost of deploying moving target defenses defensive work factors. In: Military Communications Conference, MILCOM 2015-2015 IEEE, IEEE, pp 966–971
42. Wang L, Noel S, Jajodia S (2006) Minimum-cost network hardening using attack graphs. Computer Communications 29(18):3812–3824
43. Wang L, Jajodia S, Singhal A, Cheng P, Noel S (2014) K-zero day safety: A network security metric for measuring the risk of unknown vulnerabilities. IEEE Transactions on Dependable and Secure Computing 11(1):30–44
44. Zhang M, Wang L, Jajodia S, Singhal A, Albanese M (2016) Network diversity: A security metric for evaluating the resilience of networks against zero-day attacks. IEEE Transactions on Information Forensics and Security 11(5):1071–1086

Chapter 8
Random Damage in Interconnected Networks

Sandra König and Antonios Gouglidis

8.1 Introduction

Over the past few years, Industrial Control System (ICS) networks such as control networks supervising utility networks (e.g., SCADA systems) appear to have an increasing interconnection between different types of networks, i.e., between information technology (IT) and operational technology (OT) networks. While this development is useful in many ways, it also involves several risks. Specifically, it provides more opportunities with regards to deploying cyberattacks in these environments. Such attacks may range from simple types of malware and ransomware to more sophisticated attacks, such as advanced persistent threats (APTs). Recent incidents, as the WannaCry [24] and Petya [26] ransomwares, provide evident information that the number of attacks in critical infrastructures has increased and that we shall investigate approaches that may increase our level of awareness.

Besides cyberattacks, the increasing interconnection between different networks makes it even harder to analyze incidents. However, learning from past incidents and finding optimal ways to protect a system is a crucial task. In case of an incident, such a task will eventually help in finding the best way to reduce its expected damage. In interconnected networks, the consequences of an incident are not obvious, and their estimation involves a lot of uncertainty. Therefore, when looking at security incidents in ICS networks, it appears that the interplay between an attacker and a defender can be modeled using a game-theoretic approach. Setting up a game

S. König
AIT Austrian Institute of Technology GmbH, Centre for Digital Safety & Security, Giefinggasse 4, 1210 Vienna, Austria
e-mail: sandra.Koenig@ait.ac.at

A. Gouglidis (✉)
School of Computing and Communications, InfoLab21, Lancaster University, Lancaster LA1 4WA, UK
e-mail: a.gouglidis@lancaster.ac.uk

S. Rass, S. Schauer (eds.), *Game Theory for Security and Risk Management*,
Static & Dynamic Game Theory: Foundations & Applications,
https://doi.org/10.1007/978-3-319-75268-6_8

requires several steps, including the definition of attack/defense strategies, estimation of payoffs, etc. Specifically, during the preparation of a game, the estimation of payoffs (i.e., damage) for each possible scenario is one of its core tasks. However, damage estimation is not always a trivial task since it cannot be easily predicted, primarily due to incomplete information about the attack or due to external influences (e.g., weather conditions, etc.). Therefore, it is evident that describing the payoffs by means of a probability distribution may be an appropriate approach to deal with this uncertainty. These generalized payoffs contain more information than the crisp values used traditionally.

In this chapter, we elaborate on how the consequences of an infection of a network can be modeled during the process of risk analysis in the HyRiM framework (cf. Chapter 12) and therefore describe how to estimate the payoff distributions of such a game. The applied model captures two main characteristics of incidents with regard to its spreading on interconnected networks. Firstly, in case of a concrete problem, the behavior of the network is not always exactly predictable, which is why we choose a stochastic model to adequately describe the spreading. Allowing transmission of an error (or more general any kind of problem) yields a model that also captures external influences such as weather conditions. And secondly, an interconnected network can hardly be seen as one big homogeneous network since it contains several subnetworks that may have very different properties. We take this into account by classifying edges depending on their properties with respect to error transmission. Note that such a distinction between different connection types is indeed a common practice in risk analysis [3, 4, 19, 22, 27].

The remainder of this chapter is organized as follows: Section 8.2 elaborates on the random error spreading model used in the HyRiM framework. The model is applied to investigate how an infection (e.g., a malware) could propagate and how it spreads in networks. The model requires – among other – an estimation of the likelihood of transmission for each type of edges in a network (the network being represented as a graph). Ways of estimating the likelihood are described in Section 8.3. Specifically, the section provides information on how to assess the likelihood of transmission on the basis of existing threat models, experts' opinion, and different levels of trust. The various estimation approaches are required when modeling heterogeneous networks that depict not only the likelihood of technical threats to propagate on a device but also the likelihood for people to propagate a threat to each other through social interactions. Further, the choice of a specific estimation method to a large extent depends on the data being available. An implementation of the stochastic model that describes the error spreading is provided in Section 8.4, and simulations are carried out. Subsequently, it is shown in Section 8.5 how these simulations can be used for estimating random payoffs in security games, in combination with other means of assessing uncertain consequences of an incident and costs. Finally, we provide concluding remarks in Section 8.6.

8.2 Random Error Spreading on Interconnected Networks

Several epidemic models have been used to describe how an infection (such as a malware) can spread over a network [2, 14, 23, 28, 6, 29]. However, these models assume homogeneous spreading (just as the underlying epidemic models usually do), which is not suitable in most real-life situations. The reason why they are considered to be unsuitable is that the structure of networks is usually heterogeneous. A couple of approaches that take into account heterogeneity in the case of disease spreading are described in [12] and [15]. In this section, we briefly summarize a general model for random error spreading on an heterogeneous network that was introduced in [10]. Subsequently, we illustrate how it can be used in various situations (e.g., to describe consequences of a cyberattack) and discuss several issues that might arise when applying it.

8.2.1 Random Spreading on a Heterogeneous Network

Let an interconnected network be modeled as a graph $G(V,E)$ with a finite set V of nodes and a set E of directed edges between these nodes. Due to the diversity induced by several subnetworks, we distinguish edges depending on their properties concerning the spreading of an error. We therefore choose a finite number n of classes and assign a *type* t_k to every edge that belongs to class $k \in \{1,\ldots,n\}$. In order to describe the spreading of an error, we assign a characteristic likelihood p_k to each class that represents the probability that the error is transmitted over an edge of type k, i.e., the probability that an infection in one node causes an infection of a neighbor connected by an edge of type k. Having this description of a heterogeneous network, we can model the spreading of an infection by assuming that edges independently convey the infection with their characteristic likelihood. To formulate the results, we need the following

Notation: The topology of the network is described by the degree distribution $P(x_1,\ldots,x_n;y_1,\ldots,y_n)$ that gives the number of incoming and outgoing edges of each type, respectively. Let the function $\mathscr{G}$ be the generating function of the degree distribution of the network, and let $\mathscr{H}_i$ denote the generating function of the so-called excess degree distribution, i.e., the degree distribution that results when removing an edge of type t_i over which the infection reached a specific node. Further, let H_i be the generating function for the number of affected nodes due to failure of an edge of type t_i, and let the random variable S denote the actual damage (measured through the number of affected nodes).

This model yields the following main results (formal proofs can be found in [8]).

(i) In case the error spreads tree-like (i.e., does not loop back), the amount of infected nodes remains bounded in the long term. The expected number of infected nodes is

$$\mathbf{E}[S] = 1 + \sum_{j=1}^{n} \frac{\partial}{\partial y_j} \mathscr{G}(1,\ldots,1;1,\ldots,1) \cdot H_j'(1),$$

where $H_i'(1)$ is a solution of the linear equation system

$$H_i'(1) = 1 + \sum_{j=1}^{n} \frac{\partial}{\partial y_j} \mathscr{H}_i(1,\ldots,1;1,\ldots,1) \cdot H_j'(1).$$

(ii) In case the error spreads fast enough to loop back, the probability of an epidemic is

$$P_{ep} = 1 - \mathscr{G}(1,\ldots,1;H_1(1),\ldots,H_n(1)),$$

where for all i the value $H_i(1)$ is a solution of the system

$$H_i(1) = \frac{\partial}{\partial x_i} \mathscr{G}(1,\ldots,1;H_1(1),\ldots,H_n(1))/z_i,$$

where z_i is the average in- and out-degree of edges of type t_i. This system can be solved numerically.

These general results simplify for the case where the network can be modeled by the well-known Erdős-Rényi model [5].

Example 1 *Consider a network in which an edge of type i exists with probability q_i. Computation of the corresponding degree distribution yields the expected number of infected nodes*

$$\mathbf{E}[S] = \frac{1}{1 - np_1q_1 - \ldots - np_nq_n},$$

where p_i denotes the probability that an edge of type i fails. Specifically, we get the following simple criterion:
An epidemic will not *occur if*

$$1 - np_1q_1 - \ldots - np_nq_n > 0$$

is satisfied.

8.2.2 Components of Different Importance

While the basic model captures the intrinsic randomness of the error spreading process, it does not distinguish between different components of the network. However, the damage resulting from a fixed number of infected components of the network may vary depending on the importance of these affected components. For example, if we look at a ransomware that encrypts files on a computer, the type of data stored on the computer is a significant differentiator. Specifically, if that computer contains sensitive data of the organization or of its customers, the actual damage to the organization is higher compared to encrypting a computer that is mainly used for basic administrative operations. In order to capture this diversity, we assign a *value* to each component before investigating the spreading process. This enables a more precise estimation of the actual damage than the simple number of affected components.

Assigning an exact value to a single component may be a cumbersome process since consequences of a failure are often not clear before an incident happens. Nevertheless, since the infection of a specific node is a random event (due to the stochastic nature of the spreading process), the overall damage is random, and there is no need to specify a precise value. Moreover, it is often not feasible to assign a different value to each and every part of the network since this makes a plausibility check among the different values very difficult. Both issues can be solved by specifying a fixed number of categories, such as *"cheap," "normal,"* and *"expensive."* Thus, each component falls in one of these categories, and a comparison between the various parts of a network becomes more straightforward by comparing the corresponding categories.

Performing a simulation of the error spreading process will enable the estimation of the number of affected nodes for each category and thus yields a more accurate estimation of the damage to the network.

8.2.3 Time Until Infection

Besides the number of infected nodes due to an outbreak, it is also helpful to get some information about the time it takes until a specific node is infected. Since the transmission of an error is random, so is the remaining time during which the component still works correctly. We call this random variable *time-until-infection* and denote it by T_{ui}. The simulation of the error spreading model that we will describe in Section 8.4 yields an empirical distribution that allows direct estimation of the mean waiting time $\hat{T_{ui}} = \bar{x}$, where x_i is the recorded time until infection in the i-th of total N simulations, $i \in \{1,\ldots,N\}$. Alternatively, the mean can also be estimated by a trimmed mean or by the median.

While the time-until-infection can intuitively be seen as a waiting time, it does *not* follow a simple geometric distribution with one parameter since the infection may arise from various neighboring nodes. Thus, when describing the distribution of T_{ui} formally, we need a bit more notation. To this end, let T_i denote the time until infection of a fixed node due to infection of neighbor i and denote by F_i the corresponding distribution function. Then it holds

$$T_{ui} = \min\{T_1,\ldots,T_{nb}\},$$

where nb denotes the number of neighbors of that node. Then the cumulative distribution function F_{ui} satisfies

$$F_{ui}(t) \leq C(F_1(t_1),\ldots,F_{nb}(t_{nb})),$$

where $t = \min\{t_1, \dots, t_{nb}\}$ and C is the copula $C(u_1, \dots, u_n) = \min\{u_1, \dots, u_n\}$, i.e., the upper Fréchet–Hoeffding bound.

8.3 Assessing the Likelihood of Transmission

The biggest challenge when applying the model described in Section 8.2 is the estimation of the likelihoods of transmission for each type of edges. This estimation heavily depends on the structure of the network and its different components. While complexity of networks is increasing in a more and more connected world, data about such interconnections are rare (in fact, existing interconnections are not always recognized when analyzing a network). Type and quality of available data are very diverse, requiring various ways to handle them. We discuss the most common sources of information for interconnected networks and how to deal with them in this section. In particular, we consider the case where information about the devices of the network is available, e.g., when performing a vulnerability analysis. Another source of information is expertise of employees familiar with the organization. In this section, we show how to assess the likelihood of failure for different situations and various types of data.

In real-life situations, it is often difficult or even impossible to exactly assess the likelihood for an error to spread over a specific type of edges. Thus whenever possible we take into account all information available and estimate the likelihood based on *all* data available, as illustrated in Section 8.3.2.

8.3.1 Estimation Based on Threat Models

One of the facing challenges when applying threat modeling processes is how to estimate the *"likelihood"* of an adverse event to happen. It can be said that likelihood is closely connected with the easiness of obtaining knowledge that is required to successfully exploit a vulnerability [7]. When it comes to information technology (IT) systems, such information can be easily checked by automated tools, which are able to scan a system and discover the existence of already known vulnerabilities (e.g., vulnerability analysis). Another way to collect such information is by applying threat risk modeling approaches.

The Common Vulnerability Scoring System (CVSS) provides a way to calculate the risk of announced vulnerabilities and reflect their severity. CVSS defines three metric groups, i.e., base, temporal, and environmental. Each of the groups lists a number of specific metrics to capture the different characteristics of a vulnerability. Although all of them are required to calculate the overall score of a vulnerability, only the *"exploitability"* metric is required to be calculated to indirectly measure the *"likelihood"* of exploiting a vulnerability [7]. Specifically, exploitability in CVSS *"measures the current state of exploit techniques or code availability"* [11].

In addition to the above threat modeling system, the DREAD [13] threat risk model by Microsoft is also able to provide information with regard to the likelihood of an attack. DREAD stands for ***D**amage potential*, ***R**eproducibility*, ***E**xploitability*, ***A**ffected users*, and ***D**iscoverability*. Unlike CVSS, the likelihood of an attack in DREAD can be indirectly measured on the basis of several ratings, i.e., reproducibility, exploitability, and discoverability [7]. Specifically, reproducibility is related to the easiness of reproducing an attack; exploitability is related to the easiness of launching an attack; and discoverability is related to the easiness of finding a vulnerability.

From the above, it is obvious that – depending on the approach/model in use – same terms may refer to different things, e.g., exploitability appears to differ between CVSS and DREAD. However, in both approaches, the *"likelihood"* of an adverse event may be estimated, either using one or collating more metrics.

In the following, we briefly describe how to estimate the likelihood of exploiting a vulnerability using CVSS. In Figure 8.1, we depict several software and hardware components, such as a laptop that can be connected to various devices (e.g., programmable logic controllers (PLCs)) through specific software. In order to identify existing vulnerabilities on the software or hardware components, automated vulnerability scanners can be used, such as NESSUS [25], OpenVAS [18], etc. The results of a vulnerability scanner include detailed information about known vulnerabilities of a system. The identified vulnerabilities are commonly described by Common Vulnerabilities and Exposures (CVEs) [16]. Each CVE is uniquely identified by a number, a summary description of the vulnerability, and information about the CVSS severity of the vulnerability. The latter can be used to extract the value of the exploitability metric – information for individual CVSS metrics is provided by online vulnerability databases (e.g., NIST's National Vulnerability Database (NVD) [17]).

After conducting the vulnerability scanning and collecting all the prerequisite information, potential attack paths can be defined, and their likelihood can be estimated. Note that this likelihood is of informative nature rather than modeling the attackers' intension. In the HyRiM framework, the likelihood of the different attacks is a byproduct of the game-theoretic analysis (if the attacker acts rationally). However, it is not a major goal to calculate this since any behavior of the attacker that is not optimal only decreases the damage to the defender.

More information on how the likelihood is extracted from vulnerabilities in software or hardware components and used in a network is provided in Chapter 13.

8.3.2 Estimation Based on Expert Opinions

Most of the time, the only source of information about a network can be collected from experts. Since such expert opinions are subjective and sometimes vague, the resulting data is a collection of different and possibly contradicting assessments. It is common to deal with this diversity by agreeing on one single value (e.g., by taking the maximum). However, such approaches lose a lot of information that might bias the entire analysis of the network. The methods described here try to reduce such information loss.

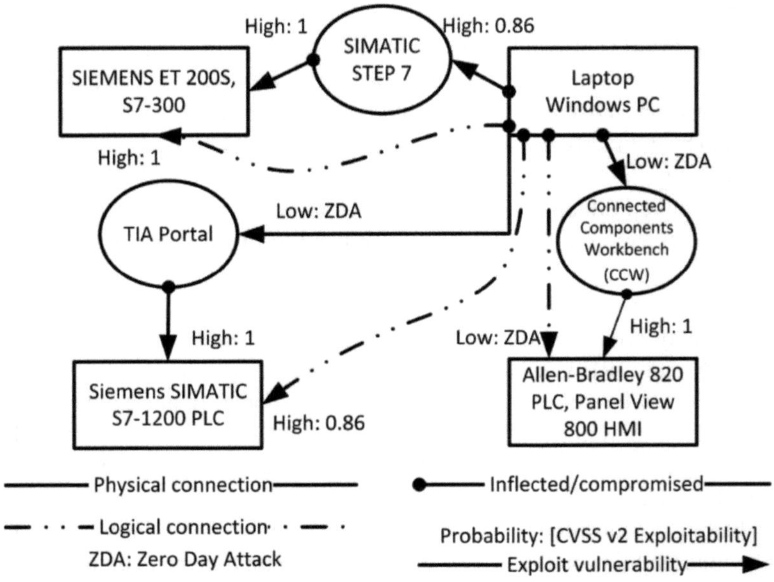

Fig. 8.1: Representation of likelihood using CVSS's exploitability

More precisely, we consider the transmission probability p_k to be a random variable X_k distributed according to the expert's opinions about edges of type t_k. As in [9], we will assume that the estimates of the transmission probability can be mapped from a qualitative scale (e.g., *"low," "medium,"* or *"high"*) to values in $[0,1]$. The expert opinions $p_{k,1},\dots,p_{k,n_k}$ are then nothing but samples from the (unknown) distribution F_k of X_k. A natural approach is to set

$$p_k := \mathbf{E}[X_k] \tag{8.1}$$

and look for suitable estimates. The most commonly used arithmetic mean has the disadvantage of not being robust, i.e., a single outlier has the potential to significantly influence the estimate. To avoid this effect, we choose

$$\hat{p}_k = \tilde{p}_k \tag{8.2}$$

where $\tilde{p}_k$ denotes the median of the observed values $p_{k,1},\dots,p_{k,n_k}$.

In case of a qualitative assignment, this estimate can be interpreted as illustrated in Figure 8.2 (cf. [9]). Several experts' opinions using a scale (i.e., *"negligible," "low," "medium," "high," "major"*) have been collected and represented by the (arbitrary) values $\{0, 0.25, 0.5, 0.75, 1\}$. The estimate (8.2) can then be interpreted as the sum of the gray bars that represent the likelihood of each of the values. Therefore, the likelihood of the opinion can be interpreted as being *"higher than medium"* and thus agrees with the intuition of a higher likelihood to transmit an error. Unlike the mean, a single outlier will not affect this estimate.

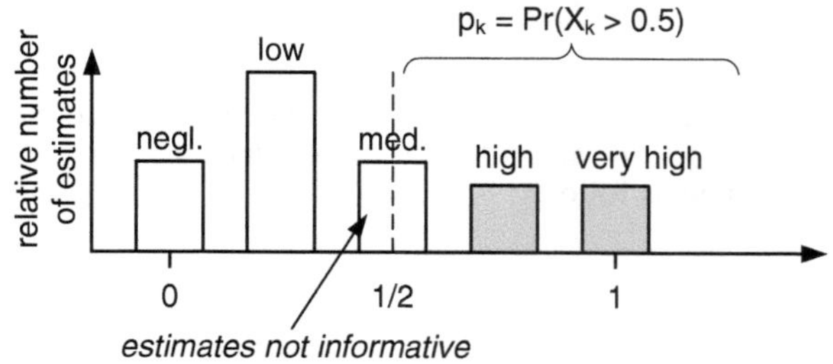

Fig. 8.2: Ambiguous experts' assessments

8.3.3 Estimation Based on Different Levels of Trust

Experts often find it difficult to characterize a type of connection by a single number. Rather, they often say *"it depends"* and give quite different reasons on which their assessment actually depends. In order to capture this fuzziness, we extend the basic model from Section 8.2 by introducing different *levels of trust*. Depending on this level, a different probability of error transmission is assessed to the specific type of edge. For example, the level may be *"low"* for non-reliable connections, *"medium"* for those that usually work, and *"high"* for very reliable ones.

Technically, the transmission probability p_k for an edge of type k from the basic model of Section 8.2 is replaced by a set $p_{k,1},\dots,p_{k,n_l}$ of probabilities where n_l gives the number of different trust levels. We assume the same number of levels for all classes since this refinement of likelihood assessment rather depends on the expert doing it than on the edge class. The corresponding probabilities of transmission are monotonic in the sense that edges with a lower level of trust are more likely to fail. Again, if expert opinions on these values vary, it is possible to work with an estimate as described in Section 8.3.2.

An illustration of this approach is given in Chapter 14.

8.4 Simulation of Random Error Spreading

The model of random error spreading on a network can straightforwardly be implemented in a software such as `R` [10]. This simulation can be used to estimate both the number of infected nodes and the time-until-infection. In particular, it allows estimation of random payoffs as we will demonstrate in Section 8.5.

The simulation algorithm for the basic model from Section 8.2 also works for the extension described in Section 8.3.3 since we still work with real-valued transmission probabilities. As soon as the transmission probabilities are random themselves, things may change.

8.4.1 Simulation of Random Transmissions

If the transmission probabilities depend on how much trust is put in a connection as described in Section 8.3.3, the algorithm has to be adapted slightly. In pseudo-code, this can be implemented as follows:

```
t ← 0
while t < T
  for each infected node v in V
    set N(v) ← {u ∈ V : (v,u) ∈ E};
    for each neighboring node u ∈ N(v)
      let k be the class in which the edge v → u falls into;
      let l be the level of trust of the edge v → u;
      with likelihood p_{k,l}, infect u;
      t ← t+1;
    endfor
  endfor
endwhile
```

The result of this simulation is a network containing infected and non-infected components. The situation can be visualized by marking non-infected nodes (sometimes called *"healthy"* nodes) of the representing graph as green and randomly coloring neighbors of infected components as red nodes [9].

8.4.2 Simulation for Components of Different Importance

If nodes are of different importance, as described in Section 8.2.2, the simulation described in the last section has to be adapted. In `R`, this can be done conveniently by using the `network` package that enables the assignment of different attributes to a network (both for nodes and edges). In the basic simulation, this package can be used to set the vertex attribute *"infected"* to either `"yes"` or `"no"` depending on whether the component is infected or not. As for the different importance of components, we define a `value` vector that holds the value of each node. Then we use the command

```
set.vertex.attribute(nw,"value",value)
```

to set these values for all nodes (vertices) of the network `nw`. Adapting the simulation in this way, it returns not only the total number of infected nodes but rather a list of how many nodes and of which importance are affected. For example, the nodes of a network of size 29 are classified as *"cheap"* (5 nodes), *"normal"* (13 nodes), and *"expensive"* (11 nodes). The adapted algorithm then returns a table as shown in Table 8.1.

Table 8.1: Number of infected nodes per class

Cheap	Normal	Expensive
5	12	11

In this case, only a single normal node has not been infected, while all other components are affected by the incident. Therefore, even without having explicit knowledge on which nodes are affected, we can acknowledge the importance of the incident and argue on the need for strategies toward reducing the damage. In the next section, we explain how this simulation may contribute to a game-theoretical analysis of cyberattacks with the aim to optimally protect the system under attack.

8.5 Estimating Payoffs of a Security Game

Security incidents are conveniently modeled as a game between an attacker and a defender (e.g., the security manager of a company). The advantage of such a model is its ability to find an optimal defense and at the same time to minimize the cost of implementing possible defense strategies. One of the biggest challenges when putting these models into practice is to assess the damage imposed in each possible scenario. While an exact assessment is hardly possible (or at least inaccurate), it is often useful to treat the payoffs as random variables and estimate their distribution. This yields a distribution-valued payoff matrix with entries as shown in Figure 8.3 for the discrete case. In this representation, each row corresponds to a defense strategy, while the columns represent the attack strategies, as in classical game theory. The game may then be solved with the generalized framework of distribution-valued games [21] (cf. Chapter 2). These games can be solved with the generalized fictitious play algorithm [20] (cf. Chapter 3).

In this section, we present different ways to estimate such random payoffs, including simulation of error spreading.

8.5.1 Simulation of Payoffs

When security incidents in an interconnected network are modeled as a game between an attacker and a defender, it is useful to work with distribution-valued payoffs. This is particularly true, if the aim is to reduce the caused damage. The latter is intrinsically random due to external influences or missing knowledge about consequences to new forms of attacks. For example, if a computer node in the IT network is infected with a malware, it is much more likely for other computer nodes in the same network to be infected, as opposed to computer nodes or components that re-

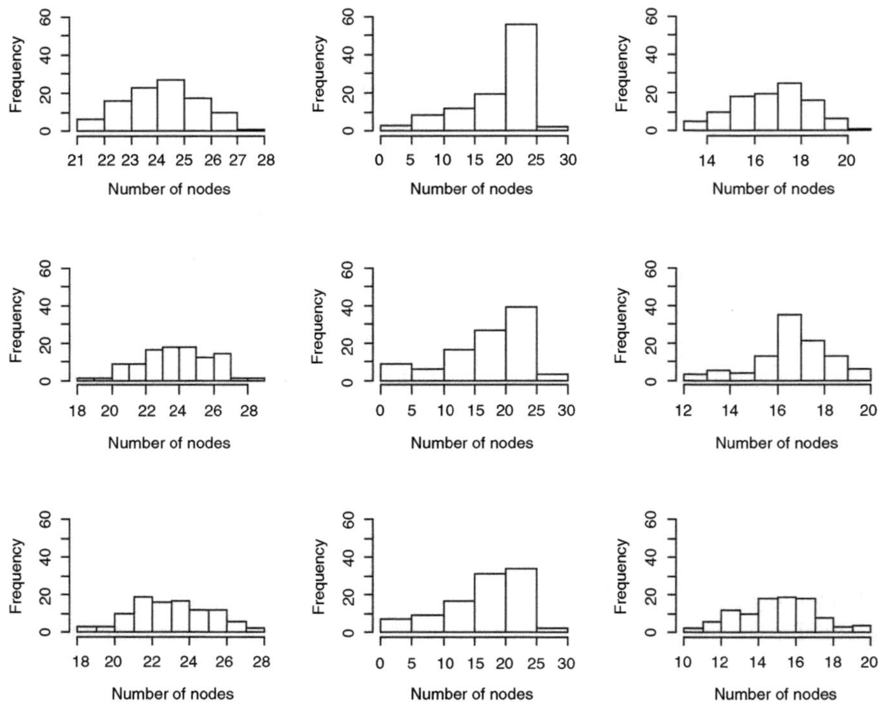

Fig. 8.3: Distribution-valued payoffs

side on a different network (e.g., in the operational technology (OT) network) of a utility network. The level of infection depends highly on the malware's capabilities.

On the other hand, the transmission probabilities are influenced by both an attacker and a defender. If we assume that the attack starts in a specific node (e.g., by plugging an infected USB stick to a PC), then the corresponding node in the graph representation is the first to be infected. On the other hand, any action taken by the defender aims to reduce the transmission probabilities for at least one type of edges. For example, the company decides to do training sessions with all employees to reduce the chance that an unknown USB stick is plugged to a computer and thus put the company network at risk. In our model, this can be represented by reducing the likelihood of transmission for logical connections as an employee is less likely to convey the threat to his or her laptop or PC. These two factors influence the simulation of the error spreading as illustrated in [10] by defining a starting point as well as a set of transmission probabilities. We illustrate the interplay between the game-theoretic model and this simulation in Chapters 13 and 14.

There are many other situations where payoffs are stochastic, especially if the actual damage to a utility provider is determined by the reaction of people, such as consumers, that influence each other. Additionally, external influences such as newspaper reports may have an impact on risk perception of people and thus influence their behavior.

8.5.2 Expertise

In many cases, simulation of the payoffs is not possible, e.g., due to missing information about the network. In this situation, the best we can do is ask for experts opinions on the expected damage. It is beneficial to use a nominal scale, say the 5-tier scale *"very low," "low," "medium," "high,"* and *"very high"*, that can be represented by a scoring from 1 (to represent *"very low"*) to 5 (to represent *"very high"*). Both experts and employees familiar with the everyday routine in the organization are asked to estimate the potential damage for each combination of attack and defense strategy (i.e., for each scenario). All these assessments are collected and represented in a histogram.

We explicitly refrain from aggregating the different opinions but rather use all available data. Assessments from people with different background and experiences may help in reducing a potential bias in the assessment and help in estimating the uncertainty about the assessment (i.e., the uncertainty is high if there is a big variance in the data). Further, our approach avoids the consensus problem that often appears in classical risk management where a single representative risk assessment is needed.

8.5.3 Additional Sources of Information

Many other ways exist to estimate payoff distributions of a security game. One model to describe the behavior of a population is the agent-based model applied in [1] for the analysis of security risks of critical utilities. It describes the risk beliefs of customers for different situations and thus helps the utility to analyze which option of risk communication is optimal.

8.5.4 Comparing Random Payoffs

In order to solve a security game with histogram payoffs, it is necessary to use the same scale for all goals. Only then, we are able to compute the weighted sums of all the different utility functions that occur during the generalized fictitious play algorithm. Besides, the use of the same scale for different assessments makes it easier for the experts to provide consistent information. However, the various ways used above to estimate payoffs yield very different data. While we can choose a suitable scale for expert assessments, the simulation algorithm returns a number of affected nodes, and things get even more involved if we allow those nodes to have different values. In this case, it is helpful to choose a common scale – i.e., a 5-tier scale that can be interpreted as *"very low," "low," "medium," "high,"* and *"very high"* – and then define a mapping from all possible results of the simulation to this chosen scale.

An example of such a mapping for a network consisting of 3 cheap, 18 normal, and 8 expensive nodes is shown in Table 8.2. This table is most conveniently read from right to left: if at least one of the conditions in the last column is satisfied, we assign a scale of 5 representing a very high damage. If none of these conditions is satisfied, we go to the column left to it and check any of these conditions are fulfilled. Again, if at least one condition is satisfied, we assign a 4 as a measure for the damage; otherwise we move on to the column left of the current one. Cells marked with "N/A" (not applicable) represent the situation that a failure of any number of nodes of this type never causes a damage in the corresponding category. For example, even if all cheap nodes fail the caused damage cannot be higher than 1. In a similar manner, if at least one expensive node fails the caused damage cannot be lower than 3.

Table 8.2: Mapping average number of infected nodes to a 5-tier scale

Cost of nodes / Scale	**1**	**2**	**3**	**4**	**5**
Cheap	up to 3	N/A	N/A	N/A	N/A
Moderate	up to 4	at least 5	at least 10	at least 12	at least 15
Expensive	N/A	N/A	at least 1	at least 3	at least 5

Following this procedure, we end up with a triple that describes the damage for each type of nodes on the 5-tier scale, e.g., with regard to the small example from Section 8.4.2, Table 8.2 returns a score of 1 for the cheap node and a score of 5 for both the normal and the expensive nodes. Since we want to measure the damage with a single number in each simulation, we apply the maximum principle here and set an overall score of 5 in this situation. Iterating the simulation of the error spreading process yields an empirical distribution of this score, corresponding to the situation where a precise prediction is not possible and the damage is random.

The resulting histograms of payoffs that use the same scale and can be compared with the stochastic ordering defined earlier (cf. Chapters 2 and 3). For example, considering experts' opinions for two different scenarios as shown in Figure 8.4, we can see that scenario 2 is *"preferred"* from scenario 1 since fewer experts expect a *"very high"* = 5 damage.

8.6 Conclusion

In this chapter, we have demonstrated how to apply a spreading model on interconnected and heterogeneous networks. Specifically, the described approach assumes that different edges in a network have different likelihoods to transmit a threat to a neighboring node. To this extend, we have described different ways of assessing the likelihood of transmitting threats, i.e., using information stemming from vulnerability scanners (i.e., extract the exploitability metric from CVSS) or from ex-

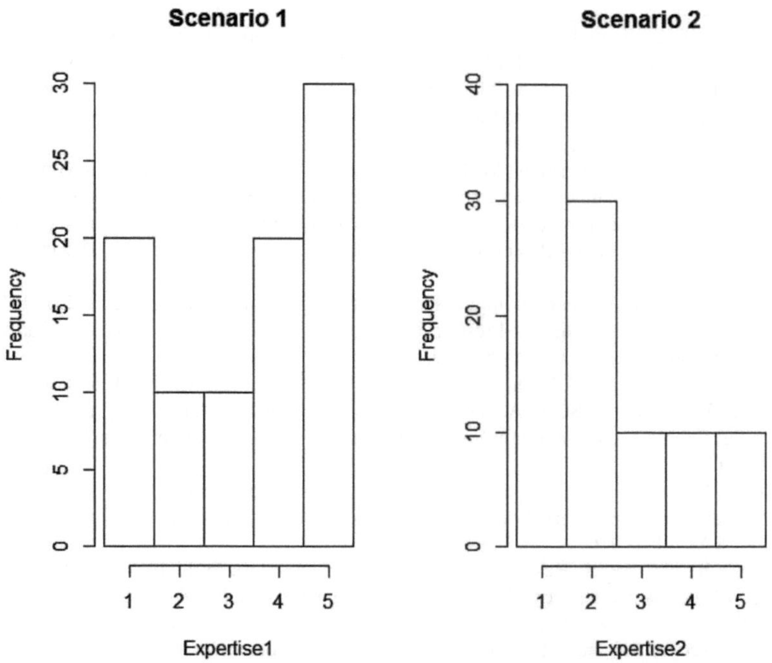

Fig. 8.4: Comparing two different experts' assessments

perts. Having this information, simulation of the infection is possible. Further, we demonstrated how our spreading model can be used in combination with a generalized game-theoretic model. This included the estimation of payoffs as distributions instead of single values and eventually depicted their use in calculating optimal defense strategies in the presence of specific attack vectors. With regard to the application areas of the described model, these may vary due to its abstract formulation. Examples of current application areas include the investigation of cyber security threats (cf. Chapters 13 and 14) and dependencies between critical infrastructures.

Acknowledgements The research leading to these results has received funding from the European Union Seventh Framework Programme under grant agreement no. 608090, Project HyRiM (Hybrid Risk Management for Utility Networks).

References

1. Busby, J., Gouglidis, A., Rass, S., König, S.: Modelling security risk in critical utilities: The system at risk as a three player game and agent society. In: Proceedings of 2016 IEEE International conference on System, Man, and Cybernetics (SMC) Budapest, October 9-12 (2016). https://doi.org/10.1109/SMC.2016.7844492

2. Chen, Z., Ji, C.: Spatial-temporal modeling of malware propagation in networks. IEEE Transactions on Neural networks **16**(5), 1291–1303 (2005)
3. Dudenhoeffer, D.D., Permann M.R. und Manic, M.: CIMS: A framework for infrastructure interdependency modeling and analysis. In: Proceedings of the 2006 Winter Simulation Conference. New Jersey (2006)
4. Dudenhoeffer, D.D., Permann M.R. und Boring, R.: Decision consequence in complex environments: Visualizing decision impact. In: Proceeding of Sharing Solutions for Emergencies and Hazardous Environments. American Nuclear Society Joint Topical Meeting: 9th Emergency Preparedness and Response/11th Robotics and Remote Systems for Hazardous Environments (2006)
5. Erdős, P., Rényi, A.: On random graphs. Publicationes Mathematicae **6**, 290–297 (1959)
6. Gao, C., Liu, J.: Modeling and restraining mobile virus propagation. IEEE Transactions on Mobile Computing **12**(3), 529–541 (2013)
7. Knapp, E.D., Langill, J.T.: Industrial Network Security: Securing critical infrastructure networks for smart grid, SCADA, and other Industrial Control Systems. Syngress (2014)
8. König, S.: Error propagation through a network with non-uniform failure. arXiv:1604.03558 (2016)
9. König, S., Rass, S., Schauer, S., Beck, A.: Risk propagation analysis and visualization using percolation theory. International Journal of Advanced Computer Science and Applications (IJACSA) **7**(1) (2016)
10. König, S., Schauer, S., Rass, S.: A Stochastic Framework for Prediction of Malware Spreading in Heterogeneous Networks, pp. 67–81. Springer, Cham (2016)
11. Mell, P., Scarfone, K., Romanosky, S.: A complete guide to the common vulnerability scoring system version 2.0. In: Published by FIRST-Forum of Incident Response and Security Teams, vol. 1, p. 23 (2007)
12. Meyers, L.A., Newman, M.E.J., Pourbohloul, B.: Predicting epidemics on directed contact networks. Journal of Theoretical Biology **240**(3), 400–418 (2006)
13. Microsoft: Chapter 3 threat modeling (2003). URL https://msdn.microsoft.com/en-us/library/ff648644.aspx,[retrieved:26/09/2017]
14. Miller, J.C.: Bounding the size and probability of epidemics on networks. Applied Probability Trust **45**, 498–512 (2008)
15. Miller, J.C., Volz, E.M.: Incorporating disease and population structure into models of SIR disease in contact networks. PLoS ONE **8**(8), 1–14 (2013)
16. MITRE: Common vulnerabilities and exposure. URL https://cve.mitre.org/,[retrieved:26/09/2017]
17. NIST: National vulnerability database. URL https://nvd.nist.gov/,[retrieved:26/09/2017]
18. OpenVAS: Open vulnerability assessment system. URL http://www.openvas.org/,[retrieved:26/09/2017]
19. Pederson, P., Dudenhoeffer, D.D., Hartley, S., Permann, M.R.: Critical infrastructure interdependency modeling: A survey of U.S. and international research. Tech. rep., Idaho National Laboratory (2006). INL/EXT-06-11464
20. Rass, S., König, S.: Package 'HyRiM': Multicriteria risk management using zero-sum games with vector-valued payoffs that are probability distributions. http://hyrim.net (2016). Version 1.0 (current stable release as of Sep.16; ongoing development)
21. Rass, S., König, S., Schauer, S.: Uncertainty in games: Using probability distributions as payoffs. In: M. Khouzani, E. Panaousis, G. Theodorakopoulos (eds.) Decision and Game Theory for Security, 6th International Conference, GameSec 2015, LNCS 9406. Springer (2015)
22. Rinaldi, S., Peerenboom, J., Kelly, T.: Identifying, understanding, and analyzing critical infrastructure interdependencies. IEEE Control Systems Magazine pp. 11–25 (2001)
23. Sellke, S.H., Shroff, N.B., Bagchi, S.: Modeling and automated containment of worms. IEEE Transactions on Dependable and Secure Computing **5**(2), 71–86 (2008)

24. Symantec: What you need to know about the wannacry ransomware (2017). URL https://www.symantec.com/connect/blogs/what-you-need-know-about-wannacry-ransomware, [retrieved:25/09/2017]
25. Tenable: Nessus vulnerability scanner. URL https://www.tenable.com/products/nessus-vulnerability-scanner,[retrieved:26/09/2017]
26. TrendMicro: Frequently asked questions: The petya ransomware outbreak (2017). URL https://www.trendmicro.com/vinfo/us/security/news/cybercrime-and-digital-threats/frequently-asked-questions-the-petya-ransomware-outbreak,[retrieved:25/09/2017]
27. US Government: Executive order, 13010. critical infrastructure protection (1996). Federal Register
28. Yan, G., Eidenbenz, S.: Modeling propagation dynamics of bluetooth worms (extended version). IEEE Transactions on Mobile Computing **8**(3), 353–368 (2009)
29. Yu, S., Gu, G., Barnawi, A., Guo, S., Stojmenovic, I.: Malware propagation in large-scale networks. IEEE Transactions on Knowledge and Data Engineering **27**(1) (2015)

Chapter 9
Optimal Dispatch of Electrical Transmission Systems Considering Interdependencies with Natural Gas Systems

Tianqi Hong, Francisco de León, and Quanyan Zhu

9.1 Introduction

Early in the twentieth century, the fundamental models for major civil infrastructures were developed and are now well-established. However, most of the managers still plan and operate their infrastructures individually even when those systems are physically interconnected. In the past decades, engineers in different areas devoted themselves to decouple the interdependencies between different infrastructures. Interdependencies have been reduced but not eliminated. Frequently, when one type of interdependency is reduced, other or several other interdependencies are introduced. Perhaps it is now, in the era of the smart grid (and smart everything), the time to stop eliminating the interdependencies between different infrastructures. Rather, one should be thinking about how to manage the response of the system of systems, including interconnected infrastructures in the analysis and optimization.

As the central link between different systems, electric power systems are customers and/or suppliers to other infrastructures. A robust electric power system is a prerequisite to improve the reliability of the combined system. Hence, enhancing the reliability of the electric power system by considering the interdependency impact (IDI) coming from other interconnected systems is necessary. In this chapter, we focus on the study of interconnected electric-natural gas (ENG) system. Figure 9.1 illustrates the structure of an example ENG system. Previous researchers have noticed the critical nature of the electric power system and investigated the interdependencies between the electric power and natural gas systems [1, 2, 3, 4]. In references [2, 3, 4], a unified model was established to model the coupled ENG system in normal operation by considering their interdependency. Instead of modeling the interdependencies, some researchers introduced the concept

T. Hong (✉) · F. de León · Q. Zhu
New York University, Five MetroTech Center, Brooklyn, NY, USA
e-mail: th1275@nyu.edu; fdeleon@nyu.edu; quanyan.zhu@nyu.edu

S. Rass, S. Schauer (eds.), *Game Theory for Security and Risk Management*,
Static & Dynamic Game Theory: Foundations & Applications,
https://doi.org/10.1007/978-3-319-75268-6_9

of energy hub [5, 6, 7]. They proposed several methods under different scenarios to allocate the energy hub and reduce the operational cost for the coupled system. Uncertainties, such as human activities, of the combined system are considered in [2, 8, 9, 10].

Instead of the interdependency impact (IDI) in normal operating condition, some groups have also put their efforts in modeling and analyzing the post-contingency interdependency impacts (PCIDIs) in the coupled networks. The authors of [10, 11, 12] have proposed various formulations to reduce the PCIDIs and improve the robustness of the joint systems.

All of the previous studies assume that their linked system is working under a co-dispatch operation scheme, meaning the utilities of the natural gas and electric power systems share their information and operate their systems together. In practice, this assumption may not be true for two reasons: First, the operational time scales of natural gas and electric power systems are quite different. Electric power systems can be dispatched every hour. In contrast, natural gas systems are normally dispatched on a period of several days. Second, only limited data can be shared between different utilities or departments for security reasons. To assess the IDIs with the limitations aforementioned, different models and solutions for enhancing the robustness of the electric power system are investigated in this chapter.

The original contributions of this chapter are as follows: (a) a detailed natural gas flow model is proposed for the evaluation of the IDIs from the natural gas transmission system – we model natural gas flow with long transmission pipelines (medium length for electrical lines); (b) a decentralized operating pattern with limited data

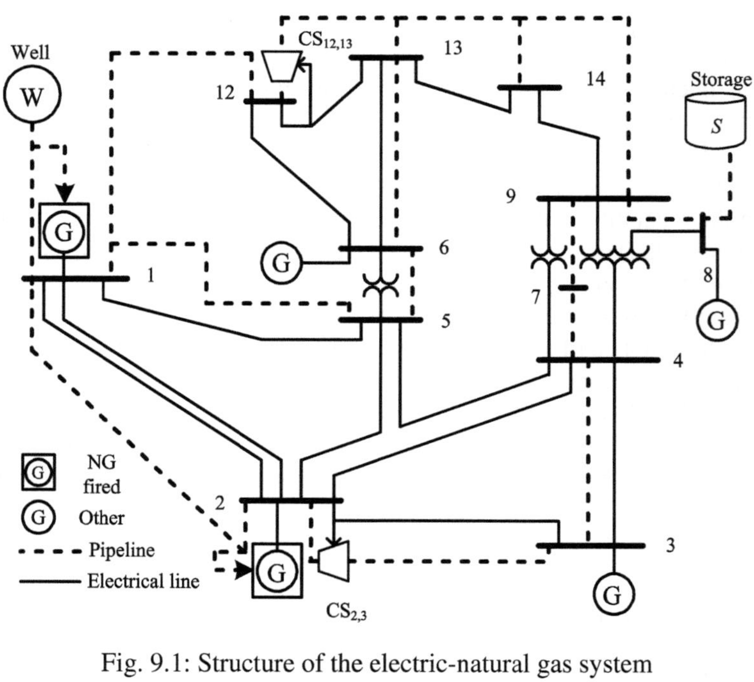

Fig. 9.1: Structure of the electric-natural gas system

exchanges between the natural gas and electrical transmission systems is analyzed; (c) several novel constraints for the integration of the PCIDIs into the electric power system are proposed and evaluated, for example, interdependency with compressor stations and gas-fired generation; and (d) the corresponding solution is provided.

9.2 Modeling of Natural Gas Transmission System

A natural gas system can be considered both a customer and a supplier of the interconnected electric power system. As a supplier, the natural gas system provides fuel to power plants to generate electricity. As a customer, the natural gas system consumes electricity that powers the pressure pumps. This interdependent relationship benefits each other and may cause instability on both systems. In this chapter, we only concentrate on reducing IDIs to the electric power system. To develop a physical model representing the IDIs generated by the natural gas system, we first establish a physical model of the natural gas system.

9.2.1 Physical Relationships of Natural Gas Transmission Systems

Different from [2] to [12], the Panhandle A equation is chosen to model the relationship between natural gas flow rate and pressure drops. The Panhandle A equation has higher accuracy than other equations for systems with large pressure drops. The natural gas flow rate from node k to node m (F_{km}) in steady state is given by [13]:

$$F_{km} = A_{km}^{NG} \left[\left| r_{km}^2 p_k^2 - r_{mk}^2 p_m^2 \right| \right]^{0.5394}, \tag{9.1}$$

where:

$$A_{km}^{NG} = \frac{4.5965 \times 10^{-3} \eta_{km}^{p} D_{km}^{2.6182}}{sign(p_k - p_m)(G^{0.854} T^f L_{km} Z_{km})^{0.5394}} \left(\frac{T^b}{p^b} \right). \tag{9.2}$$

F_{km} is the flow rate of pipeline km in m^3/day; r_{km} is the compression ratio of pipeline km at node k (note that r_{km} is different with r_{mk}, where r_{km} represents compression station installed at node k); p_k and p^b are the pressure at node k and base pressure in kPa; D_{km} is the inside diameter in mm; L_{km} is the length of the pipe in km; Z_{km} is the compressibility factors of pipeline km; η_{km}^{p} is the efficiency of the pipeline; T^f and T^b are the average gas flow temperature and base temperature in K; G is the specific gravity of the gas delivered by pipeline; and sign(x) is a function to extract the sign of variable "x."

Compressor stations are important elements for compensating the pressure drop in the natural gas transmission system. Centrifugal and positive displacement compressors are the most commonly used. By converting electric power into mechanical power, the compressor station is a critical link between the electrical and natural gas systems, especially at the transmission level. The relationship between the electric power and compression ratio is given in [13] as

$$P_{km}^{E} = \frac{Z_k + Z_m}{2\eta_{km}^{c}} \left(\frac{\gamma}{\gamma - 1} \right) \frac{4.0639 F_{km} T_k}{10^6} \left[(r_{km})^{\frac{\gamma-1}{\gamma}} - 1 \right] \tag{9.3}$$

where P_{km}^{E} is the electric power consumption of compressor at pipeline km in kW; Z_k is the compressibility factors at node k; η_{km}^{c} is the efficiency of the compressor station; T_k is the temperature at node k; and γ is the ratio of specific heats of gas according to [13], which is 1.4 for natural gas. Based on (9.3), the electric power consumption P_{km}^{E} has direct relationships with flow rate F_{km} and compression ratio r_{km}. The electric power consumption changes when the flow rate F_{km} or compression ratio r_{km} changes.

Based on (9.3), the electric power consumption P_{km}^{E} has direct relationships with flow rate F_{km} and compression ratio r_{km}. The electric power consumption changes when the flow rate F_{km} or compression ratio r_{km} change.

Natural gas-fired generators are other links between natural gas systems and electric power systems. Since all generators are equipped with an automatic generation controller (AGC), the flow rate consumption of the generators is constant.

Similar to the nodal constraints in the electric power system, the summation of the gas flow rate at each node must be zero at all times. Therefore, we arrive at:

$$F_k^{I} = F_k^{D} + \sum_{\substack{m=1 \\ m \neq k}}^{NN} F_{km}, \tag{9.4}$$

where F_k^{I} and F_k^{D} are the flow rates of the injection and demand of node k and NN is the number of nodes in the natural gas system.

9.2.2 Modeling the Natural Gas System Under Normal Operation

In normal operation, managers of a natural gas transmission company expect to operate their systems efficiently. To reduce the operational costs and improve the total efficiency of a natural gas transmission system, the gas dispatch becomes an optimization problem.

The dispatchable parameters in natural gas systems can be clustered in three types. The first variable is the pressure of the source. The natural gas source is normally considered as the gas plant or large gas storage where its pressure can be kept constant and its capacity is assumed to be infinite; see the natural gas well in Figure 9.1. The second variable is the flow rate of short-term storage. Since the manager can dispatch the flow rate of a short-term storage to improve the performance of the system, the terminal pressure of the storage varies according to the working status of the natural gas system. Naturally, the gas price of short-term storage is higher than that of a natural gas well. Thus, the flow rates from a natural gas source and short-term storage can be considered flow injections with different prices. The third variable is the compression ratio of each station. Compressor stations can regulate the flow rate or terminal pressures of a pipeline by adjusting the compression ratio. Note that the compressor station allocation problem is beyond the scope of our Chapter.

Considering the dispatchable parameters discussed above, an optimal natural gas flow (ONGF) problem can be formulated as:

$$\min_{\mathbf{p}^w,\mathbf{F}^I,\mathbf{r}} M^{NG} = \min_{\mathbf{p}^w,\mathbf{F}^I,\mathbf{r}} \sum_{k=1}^{NN} \left(C_k^{NG} F_k^I + C_k^E \sum_{m}^{NP} P_{km}^E \right), \tag{9.5}$$

where M^{NG} is the operational cost of the natural gas system, $\mathbf{F}^I$ is a column vector consisting of the flow rate injection at each node, $\mathbf{r}$ is a column vector consisting of compressor station ratios, $\mathbf{p}^w$ is a column vector consisting the pressure of the natural gas well, C_k^{NG} is the price of natural gas at node k in \$/Mm^3/s, C_k^E is the price of electricity at node k in \$/MW/s, and NP is the number of the pipelines.

The equality constraints of problem (9.5) are the flow rate nodal constraints presented in (9.4). The major inequality operation constraint of a natural gas transmission system is to keep pressures, flow rates, and injections within limits. These are expressed mathematically as:

$$\underline{p_k} \le p_k \le \overline{p_k}, \tag{9.6}$$

$$\underline{F_{km}} \le F_{km} \le \overline{F_{km}}, \tag{9.7}$$

$$\underline{F_k^I} \le F_k^I \le \overline{F_k^I}, \tag{9.8}$$

where $\overline{x}$ and $\underline{x}$ are the upper and lower bounds of variable "x".

The lower and upper bounds of the natural gas pipeline flow are calculated according to the erosional velocity of each pipeline. The erosional velocity $\overline{\mu_{km,k}}$ of the pipeline km with reference to the node k is (under the assumption that natural gas flow is an isothermal flow) [13]:

$$\overline{\mu_{km,k}} = 100\sqrt{\frac{Z_k R(T_k + 460)}{199.96 p_k G}}, \tag{9.9}$$

where $\overline{\mu_{km,k}}$ is in the unit of m/s and R is a constant that equals to 8.314 J/K/mol. The relationship between the maximum velocity and upper bound of the flow rate can be described as [13]:

$$\overline{F_{km}} = \min\left\{\overline{F_{km,k}}, \overline{F_{km,m}}\right\}, \tag{9.10}$$

where

$$\overline{F_{km,x}} = \left(\frac{D_{km}^2 \overline{\mu_{km,x}}}{14.7349}\right)\left(\frac{T_b}{p_b}\right)\left(\frac{\overline{p_x}}{Z_x}\right)\Bigg|_{x=k,m}. \tag{9.11}$$

To solve the non-convex optimization problem (9.5), some simplifications have been made as follows:

(1) The compressibility factor Z_x at each bus x is considered to be a constant value ($Z = 0.9$ for the natural gas [13, p. 67]);
(2) The inner temperature at each node is considered to be constant (T_k = 293 K);
(3) All natural gas pipelines are considered in a horizontal placement, therefore removing the gravity effect.

The simplifications listed above are used to reduce the computational complexity of solving natural gas flow problems, which remain very general and are believed to be more accurate than existing formulations.

Problem (9.5) can be solved using the Primal-Dual Interior Point Algorithm (PDIPA) [14] with truncated Newton step. The convergence of PDIPA is ensured and the first-order optimality conditions can be satisfied [15, 16, 17]. Normally, a current working status of natural gas transmission systems can be used as a feasible initial point for the ONGF problem. It is worth noting that the final solution may not be the global optimum due to the non-convexity of the original ONGF problem.

After obtaining the solution of the ONGF problem, the electricity consumption of each compressor station in normal operation can be obtained from (9.3). We can then construct a column vector $\mathbf{e}_{(0)}$ to represent the normal electricity consumption as:

$$\mathbf{e}_{(0)} = \left(P_1^E, \ldots, P_{NN}^E\right), \tag{9.12}$$

where

$$P_k^E = \sum_{\substack{m=1 \\ m \neq k}}^{NN} P_{km}^E. \tag{9.13}$$

9.2.3 Modeling the Natural Gas System Under Contingency

Since the combined systems are operated independently, we assume that the two systems cannot communicate with each other frequently. Additionally, the operational time scales of the two systems are different, and the electric power system cannot estimate when the PCIDI hits the electrical system. Hence, it is essential to consider the worst post-contingency impact of the natural gas system in the operation of the electrical system.

In pre-contingency scenarios, the natural gas system operates based on the ONGF results discussed in Section 9.2.2. When a contingency occurs, the pressures of each node and the flow rates of each pipeline may change. Consequently, the electrical consumptions of the compressors will change. Since generators are modeled as the customers of the natural gas system with constant flow rate demands, the natural gas system needs to satisfy the requirements from the connected generators unless the pressures of the nodes are reduced below certain limits. As a result, the impacts on the electrical system come from the electricity demand variations and the loss of generators regardless on how the natural gas system is dispatched. This is true for the original dispatch or re-dispatch after contingency. Hence, the worst PCIDI can be obtained through an exhaustive "what if" analysis.

In normal condition, the natural gas system is dispatched according to the solution of problem (9.5). For all the possible contingency scenarios, the pressure at each generator is checked to build the preindicator matrix $\mathbf{D}^{pr}$, where $d^{pr}_{(k)(j)}$ is the $(k,j)^{th}$ element in matrix $\mathbf{D}^{pr}$. Element $d^{pr}_{(k)(j)} = 1$ means the generator at node k needs to be disconnected from natural gas system when the jth contingency occurs. The consumption impact of each compressor station is recorded to form a pre-consumption impact matrix $\mathbf{E}^{pr}$ where $\mathbf{e}^{pr}_{(j)} - \mathbf{e}_{(0)}$ is the jth column of the matrix $\mathbf{E}^{pr}$. Vector $\mathbf{e}^{pr}_{(j)}$ can be built based on (9.12) under the jth contingency. Then, we assume that the natural gas system would be re-dispatched according to the solution of (9.5) with the natural gas system in post-contingency configuration. Similarly, a post-indicator matrix $\mathbf{D}^{po}$ and post-electrical consumption matrix $\mathbf{E}^{po}$ can be constructed. Figure 9.2 shows the flowchart of the proposed "what if" analysis. All the matrices can be built a day ahead or online according to load profiles.

Note that the electrical consumption of the compressor station changes regardless of whether the natural gas system is dispatched based on the security-constrained optimal natural gas flow [12] or just optimal natural gas flow, since the natural gas flow rate in (9.3) would change depending on the system operating status. Current literature has ignored this critical PCIDI between the natural gas system and the electric power system.

Matrices $\mathbf{D} = (\mathbf{D}^{pr}, \mathbf{D}^{po})$ and $\mathbf{E} = (\mathbf{E}^{pr}, \mathbf{E}^{po})$ are the interdependency impact model to the electric power system. To protect the network data of natural gas systems, the gas utilities can build all the matrices and only share numerical matrices with electric power utilities.

9.3 Electric Power System Modeling

After the IDI model has been obtained from the natural gas system, the IDIs on the electric power system needs to be solved. Hence, a set of interdependency constraints is formulated in this section. The interdependency constraints provide a tool to integrate the impact into the power system dispatch analysis. To illustrate this, we start from the traditional formulation of optimal power flow (OPF) problem.

9.3.1 Traditional Optimal Power Flow and Sensitivity Factors

To reduce the operational cost of the electrical transmission system, the power injections should be optimally dispatched. The classic operational cost of the electrical transmission system is defined as [16]:

$$M^E = \sum_{i=1}^{N_B} \left[a_i^E \left(P_i^G\right)^2 + b_i^E P_i^G + c_i^E \right] \tag{9.14}$$

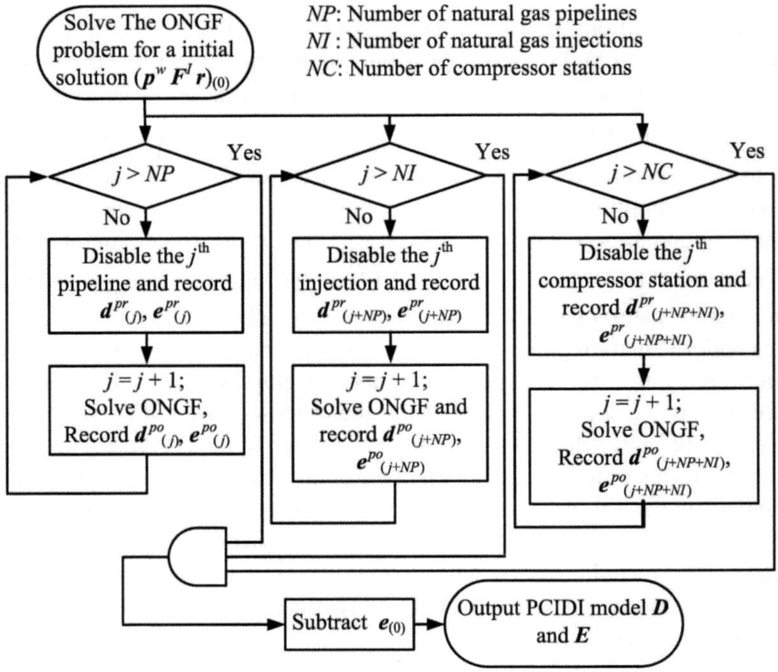

Fig. 9.2: Flowchart of "what if" analysis

where P_i^G is the active power of generator at bus i; a_i^E, b_i^E, and c_i^E are the cost coefficients of the thermal generator at bus i; and N_B is the number of the buses in the electric power system.

According to [16] and [17], the optimal power flow problem can be formulated as:

$$\min_{\mathbf{P}^G,\mathbf{Q}^G} M^E \tag{9.15}$$

subject to:

$$P_i^G - P_i^D = P_i \tag{9.16}$$

$$Q_i^G - Q_i^D = Q_i, \tag{9.17}$$

$$\underline{V_i} \leq V_i \leq \overline{V_i}, \tag{9.18}$$

$$\underline{I_{ij}} \leq I_{ij} \leq \overline{I_{ij}}, \tag{9.19}$$

$$\left(\underline{P_i^G}, \underline{Q_i^G}\right) \leq \left(P_i^G, Q_i^G\right) \leq \left(\overline{P_i^G}, \overline{Q_i^G}\right), \tag{9.20}$$

where Q_i^G is the reactive power of the generator at bus i; $\mathbf{P}^G$ and $\mathbf{Q}^G$ are the decision column vector generated by sequences $\left\{P_i^G\right\}_i^{N_B}$ and $\left\{Q_i^G\right\}_i^{N_B}$; P_i^D and Q_i^D are the active and reactive demand at bus i; P_i and Q_i are the active and reactive power flow at bus i; V_i is the voltage at bus i; and I_{ij} is the current of the line ij. It is worth to mention that the relations "$\leq$" and "$\geq$" are defined to hold component-wise in (9.20) and following discussion.

Apart from looking for an economic optimal dispatch solution to achieve minimal operational cost, most utilities also want to improve the reliability of their systems, i.e., to increase the robustness of their system and satisfy the $n-1$ contingency criterion (all loads can be restored if any single component fails). Hence, the security-constrained optimal power flow (SCOPF) is proposed [16, 18, 19].

The sensitivity factors are the major tools for solving the SCOPF problem. The fundamental sensitivity factors are the power transfer distribution factor (PTDF) [17, 20]. The PTDF is a sensitivity matrix. Element J_{ik}^P in the PTDF describes the change of the active power flow in transmission line i when there is a change of power injection at bus k. According to [17], the element J_{ik}^P is calculated with the dc power flow as:

$$J_{ik}^P = \frac{1}{X_i}\left(\frac{d\theta_{i:1}}{dP_k} - \frac{d\theta_{i:2}}{dP_k}\right), \tag{9.21}$$

where X_i is the reactance of transmission line i and $\theta_{i:1}$ and $\theta_{i:2}$ are the angles of the from-bus $i:1$ and to-bus $i:2$ of the line i, respectively. In matrix form, we have:

$$\Delta \mathbf{a}^G = \mathrm{PTDF} \cdot \Delta \mathbf{P}_G, \tag{9.22}$$

where $\Delta \mathbf{a}^G$ is a column vector representing the linearized change of active power flow of each transmission line induced by generator outage, $\Delta \mathbf{P}_G$ is a column vector representing the changing of active power injection at each bus.

The PTDF can be used to indicate the post-contingency status of generator outages in a given electric system. To indicate a line outage scenario, a line outage distribution factor matrix (LODF) is developed [21]. The element J_{ij}^L in the LODF describes the change of the active flow in transmission line i when there is a change of active power flow of transmission line j. The element J_{ij}^L is calculated as [21]:

$$J_{ij}^L = \begin{cases} \left.\dfrac{J_{i,j:2}^P}{1 - J_{j,j:2}^P}\right|_{i \neq j} \\ -1|_{i=j} \end{cases}, \tag{9.23}$$

where $J_{i,j:2}^P$ is the sensitivity of active power flow in line i with respect to the injection at the to-bus $j:2$ of line j, which can be found in the PTDF matrix. For each scenario of the electric system, a new constraint can be generated based on PTDF and LODF. By incorporating all scenario constraints into problem (9.15), a classic model of SCOPF problem can be obtained [19].

9.3.2 Integration of the Interdependency Model into the Power Dispatch Problem

According to (9.21), the PTDF matrix can be utilized for the estimation of system impacts from changes in power injections. The changes of a load can be seen as a negative power injection. Hence, substituting $\Delta\mathbf{P}_G$ by the impact $\Delta\mathbf{e}^{NG}_{(j)}$ obtained from Section 9.2.3, the linearized active power flow impact of all the electrical transmission lines can be computed from:

$$\mathbf{a}^{NG}_{(j)} = \mathbf{a}_{(0)} + \Delta\mathbf{a}^{NG}_{(j)} = \mathbf{a}_{(0)} + \text{PTDF}\cdot\Delta\mathbf{e}^{NG}_{(j)}, \tag{9.24}$$

where $\mathbf{a}^{NG}_{(j)}$ is a column vector representing the IDI of active power flows under the jth scenario in the natural gas system, $\mathbf{a}_{(0)}$ is a column vector of the active power flows in normal operation, and $\Delta\mathbf{e}^{NG}_{(j)}$ is the jth column vector in matrix $\mathbf{E}$.

The PCIDI of the currents in each transmission line can be modeled according to LODF factor:

$$\mathbf{I}^{NG}_{(j)} = \mathbf{I}_{(0)} + \Delta\mathbf{I}^{NG}_{(j)} = \mathbf{I}_{(0)} + \text{LODF}\cdot\Delta\mathbf{a}^{NG}_{(j)}, \tag{9.25}$$

where $\mathbf{I}^{NG}_{(j)}$ is the column vector corresponding to the impact of the current of the jth scenario in the natural gas system and $\mathbf{I}_{(0)}$ is a column vector of all the currents in normal operation. Comparing $\mathbf{I}^{NG}_{(j)}$ with the upper-bound limit of current $\bar{\mathbf{I}}$, the interdependency constraints of the line currents can be obtained from:

$$|I| \le \begin{cases} \bar{\mathbf{I}} + (\bar{\mathbf{I}} - \mathbf{I}^{NG}_{(j)}) & if\ \bar{\mathbf{I}} - \mathbf{I}^{NG}_{(j)} < 0 \\ \bar{\mathbf{I}} & else \end{cases} \quad \forall j, \tag{9.26}$$

where I is a column vector constructed from the I_{ij}. The repeated current constraints in (9.19), SCOPF, and (9.26) are removed, and the number of the constraints can be reduced. The line current constraints are ensured by the ac power flow.

The PCIDI of the natural gas-fired generators can be obtained similarly:

$$\mathbf{P}^{G}_{(j)} = -\mathbf{P}^{G}_{(0)}\cdot\mathbf{d}^{NG}_{(j)}, \tag{9.27}$$

where $\mathbf{P}^{G}_{(j)}$ is a column vector representing the PCIDI of all the generator outputs under j^{th} scenario in the natural gas system, $\mathbf{P}^{G}_{(0)}$ is column vector of all the generator outputs in normal operation, and $\mathbf{d}^{NG}_{(j)}$ is the jth column vector in matrix $\mathbf{D}$. Substituting $\Delta\mathbf{e}^{NG}_{(j)}$ by $\mathbf{P}^{G}_{(j)}$ in (9.24), and going through the same process, the upper and lower bounds of the line currents considering the natural gas-fired generator failure can be obtained.

The PCIDIs also affect the voltage profile of the electric system. To evaluate this type of impacts, the Jacobian matrix $\mathbf{J}^E$ of the Newton power flow is introduced. According to [22], we have:

$$\begin{bmatrix} \Delta\mathbf{P} \\ \Delta\mathbf{Q} \end{bmatrix} = \mathbf{J}^E \begin{bmatrix} \Delta\theta \\ \Delta\mathbf{V} \end{bmatrix}, \tag{9.28}$$

where $\Delta\mathbf{P}$ and $\Delta\mathbf{Q}$ are the change on active and reactive powers and $\Delta\theta$ and $\Delta\mathbf{V}$ are the linearized corresponding change on the phase angles and voltage amplitudes.

Based on matrix $\mathbf{E}$ built in Section 9.2.3, $\Delta\mathbf{Q}$ equals zero. Thus, the $\Delta\mathbf{V}$ can be solved from (9.28):

$$\Delta\mathbf{V} = \mathbf{H}^{VP}\Delta\mathbf{P} \tag{9.29}$$

where $\mathbf{H}^{VP}$ is a sub-matrix obtained from the inverse of Jacobian matrix $\mathbf{J}^E$. According to (9.29), the voltage constraints generated by a natural gas contingency can be described as:

$$\underline{\mathbf{V}} \le \mathbf{V}_{(0)} + \mathbf{H}^{VP}\Delta\mathbf{a}^{NG}_{(j)} \le \overline{\mathbf{V}}, \tag{9.30}$$

where $\mathbf{V}_{(0)}$ is the vector of voltage at each bus of OPF results. By integrating the interdependency constraints of the line current (9.26) and voltage profile (9.30) into either the OPF problem or SCOPF problem, a more robust dispatch solution can be obtained; see the flowchart in Figure 9.3. It is worth pointing out that one can easily implement the proposed impact constraints into other types of OPF problems. Instead of replacing the original method, the proposed tool provides a new view for the operator to dispatch the system.

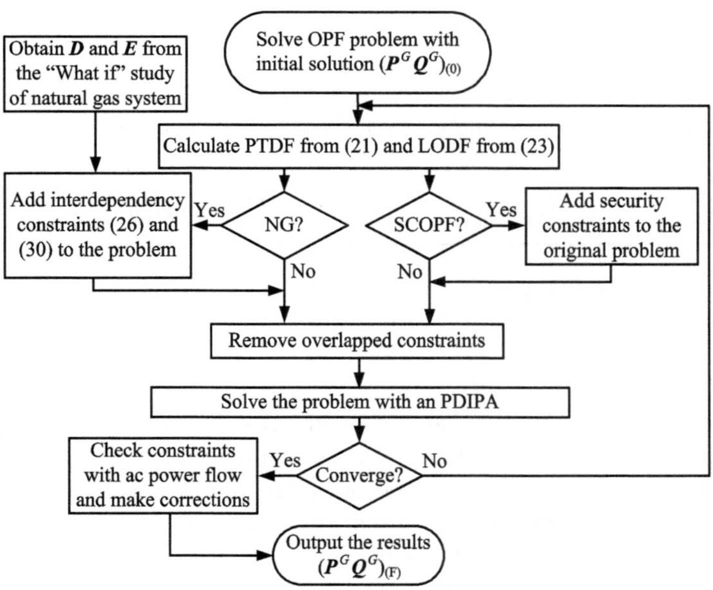

Fig. 9.3: Flowchart of interdependency based optimal power flow

9.4 Case Study

In this section, the IEEE 14-bus electric power system and an artificial 14-node natural gas transmission system are used for illustration purposes. Figure 9.1 illustrates the structure of the combined system. The detailed data of the 14-node natural gas transmission system are shown in the Appendix of this chapter.

Some modifications are applied to the IEEE 14-bus electric power system as follows: (1) the synchronous condensers at buses 3, 6, and 8 are replaced by distributed generators denoted G_3, G_6, and G_8; (2) the generators at buses 1 and 2 are assumed to be fired by natural gas; and (3) the voltage upper and lower bounds are 1.06 and 0.94 pu.

The cost coefficients of each generator are shown in Table 9.1. We assume that the costs of generators G_3, G_6, and G_8 are identical.

Table 9.1: Cost coefficients of generators

Generator	a_i^E [\$/MW^{2}h]	b_i^E [\$/MWh]	c_i^E [\$/h]
G_1	0.043	20	100
G_2	0.25	20	100
G_3, G_6, G_8	0.01	40	50

Table 9.2: "What if" contingency analysis for the outage of pipeline 1-5

Case: 1-5 outage	p_1^w [kpa]	F_8^I [Mm^3/day]	$r_{2,3}$	$r_{12,13}$	M^{NG} [\$/s]	$P_{2,3}^E$ [MW]	$P_{12,13}^E$ [MW]
Normal	5000	0.4	1.17	1.16	30.67	1.55	1.57
Pre-redispatch	5000	0.4	1.17	1.16	31.17	2.48	2.10
Post-redispatch	5114	4.13	1.28	1.20	32.21	2.67	1.98

The nodes of a natural gas system are numbered using the same order of the electrical transmission system; see Figure 9.1. The entire natural gas transmission system is a looped structure with a gas well and a short-term storage, which can ensure the possibility to satisfy the $n-1$ criterion. Two compressor stations are installed in the natural gas system to regulate the pressure at pipelines 2–3 and 12–13, respectively. The power consumed by those compressor stations is directly purchased from the electrical transmission system. The electricity price for all compressor stations is considered to be 0.05 \$/kW·h.

According to the detailed data of the natural gas system, one can solve the nonlinear flow problem by Newton's method. With the PDIPA, the optimal dispatch solution (normal operation) is shown in the first row of Table 9.2. To simulate the contingency scenario of the outage of pipelines 1–5, we follow the procedure proposed in Section 9.2.3. To illustrate the impact to the electric system generated by the natural gas system contingency, the system working statuses in the normal operation, pipeline outage contingency, and re-dispatch operation are shown in Figure 9.4. Table 9.2 provides the corresponding "what if" analysis results. When the fault occurs, the natural gas system works with operating violations (pressure and gas flow violations occur); see Figure 9.4. The power consumptions of the compressor stations increase due to the increased flow rates. After the manager re-dispatches the natural gas system, the flow rate of the short-term storage F_8^I and ratio of compressor stations are increased to force the system back to a safe operation. As a result, the electricity consumption of the compressor station $C_{2,3}$ further increases; see the electricity consumption variation in Table 9.2.

Impact matrices **D** and **E** can be computed following the procedure proposed in Section 9.2.3. The electrical consumptions during the natural gas contingency after re-dispatching are shown in the Appendix of this chapter. According to the "what if" studies, **D** is a zero matrix, meaning the generators can be connected to the natural gas system during the fault in natural gas system. The maximum power consumption of $CS_{2,3}$ is 2.67 MW, and the maximum consumption of $CS_{12,13}$ is 2.61 MW based on matrix **E**. According to 9.26 and 9.30, two extra sets of interdependency constraints can be obtained.

New dispatch results can be found in Table 9.3 after integrating the interdependency constraints into the OPF problem. The current upper bounds in the case are shown in Figure 9.5(b). The differences in the dispatch strategies and the operational cost of electric power system are very small. However, the robustness of the electric power system increases substantially; see Figure 9.5. When a contingency scenario occurs in the interconnected natural gas system, say pipeline 12-13 is out, the electricity consumptions of the compressor stations change. Correspondingly, two voltage violations occur in the electric power system; see the dotted arrows in Figure 9.5(a). Meanwhile, there is a line current violation at branch 19 (transmission line 12-13); see the dotted arrow in Figure 9.5(b). When considering the PCIDIs, the

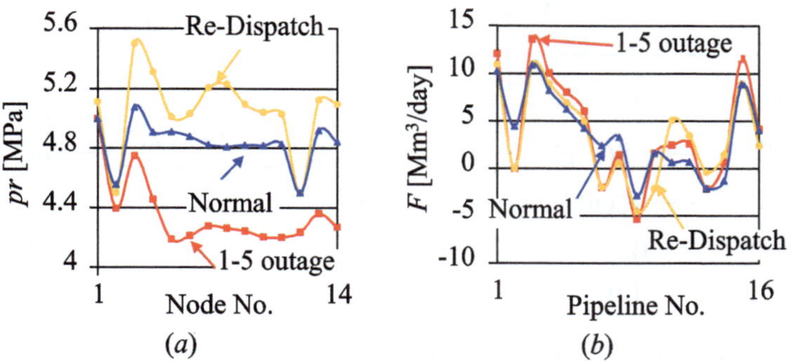

Fig. 9.4: Comparison between normal operation, post-contingency, and re-dispatch results

burden on transmission lines 7-8 and 6-13 reduced under normal condition. Consequently, the two voltage violations are eliminated as well as the line current violation when the outage occurs.

Table 9.3: Solutions of different formulations without the consideration of the contingency of the electric power system

		G_1	G_2	G_3	G_6	G_8	M^E [k$/h]
OPF	MW	194.3	36.7	28.7	0	8.5	8.0815
	Mvar	0	23.7	24.1	11.5	8.3	
OPF+NG	MW	194.0	36.7	28.4	0	9.2	8.0821
	Mvar	0	25.9	25.2	8.0	8.8	

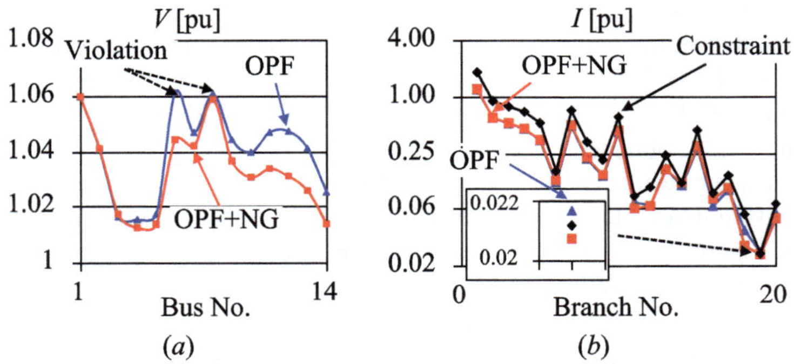

Fig. 9.5: Behavior of the electrical system after outage of pipeline 12-13 under OPF dispatch and proposed dispatch methods: (a) Voltage profile of the electrical system (b) Current profile of the electrical system

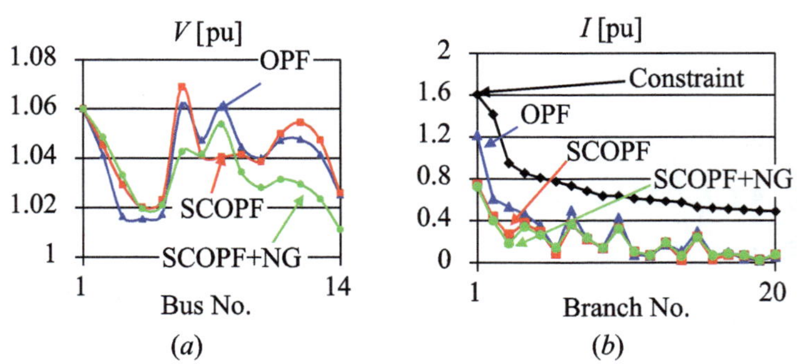

Fig. 9.6: Behavior of the electrical system following the outage of pipeline 12-13 under OPF dispatch, SCOPF dispatch, and proposed dispatch method: (a) Voltage profile of the electrical system (b) Current profile of the electrical system

Table 9.4: Solutions of different formulation with the consideration of the contingency of the electric power system

		G_1	G_2	G_3	G_6	G_8	M^E [k$/h]
OPF	MW	194.3	36.7	28.7	0	8.5	8.08
	Mvar	0	23.7	24.1	11.5	8.3	
SCOPF	MW	127.1	41.8	75.6	16.4	3.0	8.37
	Mvar	6.4	14.4	17.0	12.7	−0.4	(+3.6%)
SCOPF+NG	MW	120.7	27.7	90.0	25.5	0	8.49
	Mvar	10	28.0	15.4	−5.3	7.3	(+5%)

The dispatch results of the IEEE 14-bus system are obtained solving the SCOPF problem with the interdependency constraints. Since the system parameters in the previous example cannot satisfy the $n-1$ criterion, we increased the capacities of the electrical transmission lines. Hence, the line current upper bounds in this case are adjusted to the line shown in Figure 9.6(b). The operational costs and their corresponding dispatch solutions (OPF problem, SCOPF problem, and SCOPF problem with interdependency constraints) can be found in Table 9.4. The behavior of the electric power system under those different scenarios can be seen in Figures 9.6 and 9.7, respectively. Comparing the operational cost for different dispatch results, we can see that the operational cost of the OPF is the smallest. The operational costs of the SCOPF and proposed dispatch method increase 3.6% and 5%, respectively. However, the proposed dispatch method provides more robust systems than the OPF and SCOPF methods. When the outage of the pipelines 12–13 occurs, voltage violations happen when the electric power system is dispatched based on the OPF and SCOPF solutions; see Figure 9.6(a). In contrast, the proposed dispatch method solves the voltage violation problems (see Figure 9.6(b)). When the worst contingency scenario (the outage of transmission lines 1–2) in the electric power system

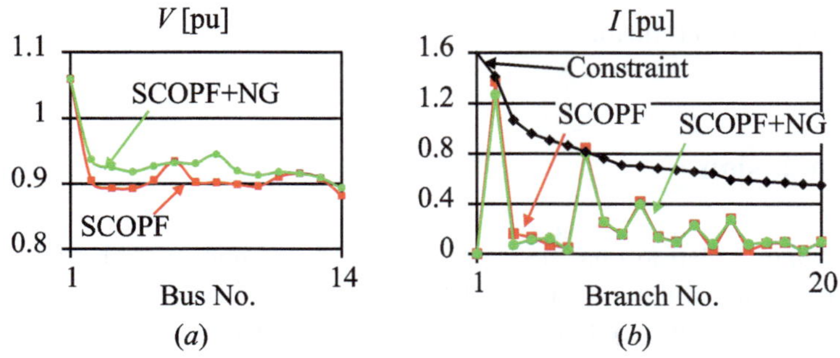

Fig. 9.7: Behavior of the electrical system after the outage of transmission line 1-2 under SCOPF dispatch and proposed dispatch methods: (a) Voltage profile of the electrical system (b) Current profile of the electrical system

occurs, the OPF dispatched system fails, and blackout occurs. The proposed dispatched system not only prevents the blackout but also resolves the voltage profile problem and the line current violations; see Figure 9.7(b). Through the proposed analysis method, the system operator can make an intelligent choice between the operational cost and robustness of the systems.

9.5 Discussion on Game Theory Formulation

In the game theory point of view, the proposed interdependency study can be formulated as a minimax problem. The fault in the electrical system and natural gas system acts as a maximizer who maximizes the operational cost of the electrical system. The electrical transmission system is a minimizer who wants to minimize its operational cost. Hence, the proposed problem can be formulated as:

$$\min_{\mathbf{P}^G,\mathbf{Q}^G} \max_{\mathbf{O}} M^E \tag{9.31}$$

subject to

$$\left(\left\{\underline{P_i^G}\right\}_i^{N_B}, \left\{\underline{Q_i^G}\right\}_i^{N_B}, \left\{\overline{P_i^G}\right\}_i^{N_B}, \left\{\overline{Q_i^G}\right\}_i^{N_B}\right) = \mathbf{f}(\mathbf{O}) \tag{9.32}$$

and (9.16)-(9.20), where $\mathbf{O}$ is a set that concludes all the types of faults, and function $\mathbf{f}()$ is a map which builds up the relationship between faults and the impacts to the strategy space of the electrical transmission system. An example of the function $\mathbf{f}()$ is Table 9.5 which links pipeline faults in the natural gas systems to the electricity load. Through (9.16), the electricity load links to the strategy space of the electrical transmission system.

Hence, the whole Section 9.2 can be concluded as the description of how to derive the fault mapping function $\mathbf{f}()$ from the faults in the natural gas system to the electrical system. It is worth to point out that the method used in Section 9.3 may not be the exact solution to solve the game problem due to the convexity and nonlinearity of the problem.

9.6 Conclusion

A co-simulation platform has been proposed, and two impact matrices have been introduced to model the interdependency impacts from interconnected natural gas system to electric power system. To enhance the stability of the target electric power system and eliminate the impacts, interdependency constraints have been proposed to provide new dispatch options to the system operator. The IEEE 14-bus system and a natural gas transmission system have been used to evaluate the impact of interdependencies and illustrate the advantages of the proposed interdependency constraints. The electric power system has been found to be more robust when considering the interdependencies. The discussion regarding the game formation of the interdependency problem is also proposed in this chapter.

It is worth pointing out that the solution of the interdependency analysis approach is more expensive than OPF and SCOPF. The proposed formulation integrates the existing models and provides a choice to utilities who aim to operate their systems more reliably in the presence of interdependencies with other systems.

Appendix

The electrical consumption data after re-dispatch are shown in Table 9.5. The information on the buses of the artificial natural gas system is presented in Table 9.6. The constant pressure node in the natural gas system is denoted as "CP," and the constant flow rate node is denoted as "CQ." The regular nodes are denoted as "L." All pipelines are assumed to have equal lengths of 80 km and equal inner diameters of 635 mm. The upper bounds of the pipeline are set to be 15 Mm^3/day. The data of the sources in the natural gas system are provided in Table 9.7.

Table 9.5: Electrical consumption data after re-dispatch

Contingency Scenario Pipeline index (from, to)	$P^E_{2,3}$ [MW]	$P^E_{12,13}$ [MW]	Contingency Scenario Pipeline index (from, to)	$P^E_{2,3}$ [MW]	$P^E_{12,13}$ [MW]
1. (1, 2)	fail	fail	09. (6, 13)	2.38	2.36
2. (1, 5)	2.67	1.98	10. (7, 8)	1.55	1.57
3. (1, 12)	fail	fail	11. (7, 9)	1.93	1.50
4. (2, 3)	0	1.29	12. (9, 10)	2.27	1.75
5. (3, 4)	0.26	1.18	13. (9, 14)	1.94	1.80
6. (4, 7)	0.65	1.15	14. (10, 11)	2.00	1.64
7. (5, 6)	2.67	2.15	15. (12, 13)	1.74	0
8. (6, 11)	1.66	1.59	16. (13, 14)	1.59	1.98

Table 9.6: Artificial 14-bus natural gas system data sheet of buses

Bus no.	Bus type	Load [Mm^3/day]	$(\underline{pr}, \overline{pr})$ [kPa]
1	CP	2	(4500, 6000)
2–7	L	2	(4500, 6000)
8	CQ	2	(4500, 6000)
9–14	L	2	(4500, 6000)

Table 9.7: Artificial 14-bus natural gas system data sheet of sources

Bus No.	Source Type	$(\underline{F},\overline{F})$ [Mm^3/day]	Price [\$/Btu]
1	CP	-	2.70
8	CQ	(0.4, 10)	3.71

List of Abbreviations and Symbols

Abbreviations

IDI	Interdependency impact
ENG system	Electric natural gas system
PCIDI	Post-contingency interdependency impact
ONGF	Optimal natural gas flow
PDIPA	Primal-dual interior point algorithm
OPF	Optimal power flow
SCOPF	Security-constrained optimal power flow
PTDF	Power transfer distribution factor
LODF	Line outage distribution factor

Symbols

F_k^I, F_k^D, F_{km}	Flow rates of injection, demand of node k, and pipeline km in m^3/day
p^b, p_k, p_k^w	Base pressure, pressure at node k, and well pressure at node k in kPa
r_{km}	Compression ratio at node k of pipeline km
G	Specific gravity of the gas delivered by pipeline, unitless
R	Ideal gas constant equals to 8.314 J/K/mol
γ	Ratio of specific heats of gas
D_{km}, L_{km}	Pipe inside diameter in mm, length of pipe in km
Z_k, Z_{km}	Compressibility factors of nodes k and pipeline km
η_{km}^p, η_{km}^c	Efficiencies of pipeline and compressor station of pipeline km
T^b, T^f, T_k	Base, average gas flow temperature; temperature at node k in K
NN, NC, NP	Number of nodes, number of compressors, and number of pipelines
N_S	Number of total contingency scenarios in natural gas system
M^{NG}	Operational cost of natural gas system
CS_{km}	Compressor station at pipeline km

P^E_{km}	Electric power consumption of compressor at pipeline km in kW
P^G_i, Q^G_i	Active and reactive powers of generator at bus i
P^D_i, Q^D_i	Active and reactive power demands at bus i
P_i, Q_i	Active and reactive power flows at bus i
V_i, I_{ij}	Per unit voltage of bus i and per unit current of line ij
a^E_i, b^E_i, c^E_i	Coefficients of the thermal generator at bus i
N_B	Number of buses
M^E	Operational cost of electrical system

Functions and Operators

$\text{sign}(x)$	Function to extract the sign of variable x
$\underline{x}, \overline{x}$	Lower and upper limits of variable x
$\leq, \geq$	Component-wise operators

References

1. Mohammad Shahidehpour, Yong Fu, and Thomas Wiedman. Impact of natural gas infrastructure on electric power systems. *Proceedings of the IEEE*, 93(5):1042–1056, 2005.
2. Alberto Martinez-Mares and Claudio R Fuerte-Esquivel. A unified gas and power flow analysis in natural gas and electricity coupled networks. *IEEE Transactions on Power Systems*, 27(4):2156–2166, 2012.
3. Zhinong Wei, Sheng Chen, Guoqiang Sun, Dan Wang, Yonghui Sun, and Haixiang Zang. Probabilistic available transfer capability calculation considering static security constraints and uncertainties of electricity–gas integrated energy systems. *Applied Energy*, 167:305–316, 2016.
4. Tao Li, Mircea Eremia, and Mohammad Shahidehpour. Interdependency of natural gas network and power system security. *IEEE Transactions on Power Systems*, 23(4):1817–1824, 2008.
5. Moein Moeini-Aghtaie, Ali Abbaspour, Mahmud Fotuhi-Firuzabad, and Ehsan Hajipour. A decomposed solution to multiple-energy carriers optimal power flow. *IEEE Transactions on Power Systems*, 29(2):707–716, 2014.
6. Michele Arnold, Rudy R Negenborn, Goran Andersson, and Bart De Schutter. Model-based predictive control applied to multi-carrier energy systems. In *Power & Energy Society General Meeting, 2009. PES'09. IEEE*, pages 1–8. IEEE, 2009.
7. Xianjun Zhang, George G Karady, and Samuel T Ariaratnam. Optimal allocation of chp-based distributed generation on urban energy distribution networks. *IEEE Transactions on Sustainable Energy*, 5(1):246–253, 2014.
8. Nilufar Neyestani, Maziar Yazdani-Damavandi, Miadreza Shafie-Khah, Gianfranco Chicco, and João PS Catalão. Stochastic modeling of multienergy carriers dependencies in smart local networks with distributed energy resources. *IEEE Transactions on Smart Grid*, 6(4):1748–1762, 2015.

9. Xiaping Zhang, Mohammad Shahidehpour, Ahmed Alabdulwahab, and Abdullah Abusorrah. Hourly electricity demand response in the stochastic day-ahead scheduling of coordinated electricity and natural gas networks. *IEEE Transactions on Power Systems*, 31(1):592–601, 2016.
10. Ahmed Alabdulwahab, Abdullah Abusorrah, Xiaping Zhang, and Mohammad Shahidehpour. Stochastic security-constrained scheduling of coordinated electricity and natural gas infrastructures. *IEEE Systems Journal*, 2015.
11. Cong Liu, Mohammad Shahidehpour, Yong Fu, and Zuyi Li. Security-constrained unit commitment with natural gas transmission constraints. *IEEE Transactions on Power Systems*, 24(3):1523–1536, 2009.
12. Carlos M Correa-Posada and Pedro Sanchez-Martin. Security-constrained optimal power and natural-gas flow. *IEEE Transactions on Power Systems*, 29(4):1780–1787, 2014.
13. E Shashi Menon. *Gas pipeline hydraulics*. CRC Press, 2005.
14. Stephen J Wright. *Primal-dual interior-point methods*. SIAM, 1997.
15. Hongye Wang, Carlos E Murillo-Sanchez, Ray D Zimmerman, and Robert J Thomas. On computational issues of market-based optimal power flow. *IEEE Transactions on Power Systems*, 22(3):1185–1193, 2007.
16. Allen J Wood and Bruce F Wollenberg. *Power generation, operation, and control*. John Wiley & Sons, 2012.
17. Ray Daniel Zimmerman, Carlos Edmundo Murillo-Sánchez, and Robert John Thomas. Matpower: Steady-state operations, planning, and analysis tools for power systems research and education. *IEEE Transactions on power systems*, 26(1):12–19, 2011.
18. Brian Stott and Eric Hobson. Power system security control calculations using linear programming, part i. *IEEE Transactions on Power Apparatus and Systems*, (5):1713–1720, 1978.
19. James Jamal Thomas and Santiago Grijalva. Flexible security-constrained optimal power flow. *IEEE Transactions on Power Systems*, 30(3):1195–1202, 2015.
20. Wai Y Ng. Generalized generation distribution factors for power system security evaluations. *IEEE Transactions on Power Apparatus and Systems*, (3):1001–1005, 1981.
21. Teoman Guler, George Gross, and Minghai Liu. Generalized line outage distribution factors. *IEEE Transactions on Power Systems*, 22(2):879–881, 2007.
22. Tianqi Hong, Ashhar Raza, and Francisco de León. Optimal power dispatch under load uncertainty using a stochastic approximation method. *IEEE Transactions on Power Systems*, 31(6):4495–4503, 2016.

Chapter 10
Managing Security Risks Interdependencies Between ICT and Electric Infrastructures: A Game Theoretical Analysis

Ziad Ismail, Jean Leneutre, David Bateman, and Lin Chen

10.1 Introduction

The power grid stands as one of the most important critical infrastructures on which depends an array of services. It uses a supervisory control and data acquisition (SCADA) system to monitor and control electric equipment. To that purpose, a SCADA system uses several telecommunication infrastructures such as telephone lines, cellular networks, etc. This renders the power system dependent on the reliability and security of the telecommunication system. An attack on a communication equipment used to control an industrial process can have severe impact on the power grid. Such attack recently targeted the Ukraine power grid on December 2015 [1]. Using a new variant of the malware BlackEnergy [2], attackers successfully compromised the information systems of several energy distribution companies and disrupted electricity supply to approximately 225,000 customers up to 6 hours. Reciprocally, an electrical node responsible of providing power to a set of communication equipment is important to the communication infrastructure: if the power source of these equipment is compromised, the communication nodes will not be able to achieve their objectives. Citing the Italian blackout of 2003, which impacted 56 million people, Rosato et al. [3] showed that a blackout can result from a cascade of failures between the power grid and the communication system.

Z. Ismail (✉) · J. Leneutre
Télécom ParisTech, Université Paris-Saclay, 46 rue Barrault, 75013 Paris, France
e-mail: ismail.ziad@telecom-paristech.fr; jean.leneutre@telecom-paristech.fr

D. Bateman
EDF, 1 Place Pleyel, 93282 Saint-Denis, France
e-mail: david.bateman@edf.fr

L. Chen
University of Paris-Sud 11, 15 Rue Georges Clemenceau, 91400 Orsay, France
e-mail: lin.chen@lri.fr

S. Rass, S. Schauer (eds.), *Game Theory for Security and Risk Management*, Static & Dynamic Game Theory: Foundations & Applications,
https://doi.org/10.1007/978-3-319-75268-6_10

In particular, some switches in the communication system can lose their power and fail because of failures in the power grid. As a consequence, due to a lack of control, some nodes in the power grid will fail.

Traditionally, the reliability of the power grid and the security of the ICT infrastructure are assessed independently using different risk methodologies, for instance, [4, 5] and [6, 7], respectively, for electric and ICT infrastructures.

In the last decade, a growing body of research has been dedicated to the analysis of interdependencies between critical infrastructures, focusing in particular on communication and electric systems. An overview of such work is provided in [8]. In what follows, we briefly recall related works. In one of the first work on this topic, Laprie et al. [9] proposed a qualitative model to address cascading, escalating, and common cause failures due to interdependencies between electric and communication infrastructures. In the case of quantitative models, we can distinguish two main categories: analytical-based and simulation-based models. In the first category of models, we find the work of Buldyrev et al. [10] in which a theoretical framework was developed to study the process of cascading failures in interdependent networks caused by random initial failures of nodes. Another work presented in [11] studies the minimum number of node failures needed to cause a total blackout: this problem is shown to be NP-hard in case of unidirectional interdependency between the networks but can be solved in polynomial time in case of bidirectional interdependency. In simulation-based models, the main techniques used include agent-based [12], Petri nets [13], Stochastic Activity Networks (SANs) [14, 15], and co-simulation [16]. Numerous works also focused on defining metrics to characterize the level of interdependencies between critical infrastructures and quantify their robustness (see for instance [17, 18, 19]).

In complex interdependent systems, the interactions between the attacker and the defender play an important role in defining the optimal defense strategy. In this context, game theory offers a mathematical framework to study interactions between different players with the same or conflicting interests. For example, Law et al. [20] investigate false data injection attacks on the power grid and formulate the problem as a stochastic security game between an attacker and a defender. Amin et al. [21] present a framework to assess risks to cyber-physical systems when interdependencies between information and physical systems may result in correlated failures.

In this chapter, we address the issue of the security risk management of interdependent communication and electric infrastructures in the smart grid by proposing an analytical model for hardening security on critical communication equipment used to control the power grid. Using noncooperative game theory, we analyze the behavior of an attacker and a defender. The attacker tries to compromise communication equipment to cause the maximum impact on the power grid. On the other hand, the defender tries to protect the power system by hardening the security on communication equipment, while taking into account the existence of backup control equipment in the communication infrastructure. In [22] and [23], we proposed an analytical model based on game theory for optimizing the distribution of defense resources on communication equipment taking into account the interdependencies between electric and communication infrastructures and defined a methodology to assess some of the parameters of the model. In this chapter, we make a number of extensions to this model in an attempt to answer the following questions: Is security by obscurity a good strategy for the defender? Under which conditions can a player guarantee a certain payoff? How can we strategically assess the initial security risk on communication equipment? Is deception required from the part of the defender to better protect the system? As we will see, while some of these questions can be analyzed analytically in the general case, some answers to these questions are system dependent and will therefore be analyzed in the case study. Throughout this chapter, the communication system refers to the telecommunication infrastructure responsible of controlling and monitoring the electric system.

The chapter is organized as follows. We start by presenting the interdependency and the risk diffusion models in Sections 10.2 and 10.3, respectively. In Section 10.4, we present the game theoretical model and analyze different equilibrium concepts. We propose an approach to evaluate the values of a number of parameters used in the analytical model in Section 10.5. In Section 10.6, we validate our model via a case study based on the Polish electric power transmission system. Finally, we conclude the chapter in Section 10.7.

10.2 Interdependency Model

We refer by initial risk, the risk on a node before the impact of an accident or an attack propagates between system nodes. We will denote by $r_i^e(0)$ and $r_j^c(0)$ the initial risk on electrical node i and communication equipment j, respectively. In the rest of this section, we assume that the initial risk on a system node is a nonnegative real number and has been evaluated using risk assessment methods. However, in Section 10.5.3, we propose a security game between the attacker and the defender to assess the initial risk on communication equipment. Therefore, instead of relying on subjective probability assessments, the evaluation of the probability of attacking a particular equipment is formally derived.

We use the framework proposed in [24] as a basis to represent the risk dependencies using a graph-theoretic approach. We model the interdependency between the electric and the communication infrastructures as a weighted directed interdependency graph $\mathscr{D} = (V, E, f)$, where $V = \{v_1, v_2, \dots, v_N\}$ is a finite set of vertices representing the set of electrical and communication nodes, E is a particular subset of V^2 and referred to as the edges of $\mathscr{D}$, and $f : E \rightarrow \mathbb{R}^+$ is a function where $f(e_{ij})$ refers to the weight associated with the edge e_{ij}.

Let $V = \{T^e, T^c\}$ where $T^e = \{v_1, v_2, \dots, v_{N_e}\}$ represents the set of electrical nodes in the grid and $T^c = \{v_{N_e+1}, v_{N_e+2}, \dots, v_{N_e+N_c}\}$ represents the set of communication nodes. Let $\mathscr{D}$ be represented by the weighted adjacency matrix $\mathbf{M} = [m_{ij}]_{N\times N}$ defined as follows:

$$\mathbf{M} = \begin{pmatrix} \mathbf{B} & \mathbf{D} \\ \mathbf{F} & \mathbf{S} \end{pmatrix}$$

where $\mathbf{B} = [b_{ij}]_{N_e \times N_e}$, $\mathbf{D} = [d_{ij}]_{N_e \times N_c}$, $\mathbf{F} = [f_{ij}]_{N_c \times N_e}$, and $\mathbf{S} = [s_{ij}]_{N_c \times N_c}$. Matrix $\mathbf{M}$ represents the effects of nodes on each other and is a block matrix composed of matrices $\mathbf{B}$, $\mathbf{D}$, $\mathbf{F}$, and $\mathbf{S}$. Elements of these matrices are nonnegative real numbers. Without loss of generality, we assume that these matrices are left stochastic matrices. Therefore, for each node k, we evaluate the weight of other nodes to impact node k. Finally, matrices $\mathbf{B}$ and $\mathbf{S}$ represent the dependency between electrical nodes and communication nodes, respectively.

10.3 Risk Diffusion and Equilibrium

We consider that the first cascading effects of an attack on communication equipment take place in the communication infrastructure itself. In the communication system, we consider that a set of Intrusion Detection Systems (IDSs) exists. We assume that devices that assure a security function, such as IDSs, have security mechanisms protecting the availability of their function. The attacker tries to compromise a set of communication nodes in order to control or disrupt the power system. The probability of being detected increases each time the attacker attempts to compromise a new equipment. Therefore, we consider that the payoff of future attacks decreases at each attack step. Let γ_c be a nonnegative real number that represents the weight of the impact payoff of future attacks s.t $\gamma_c \in [0,1]$. γ_c is a function of the probability of detection of the IDS and attacker's profile.

We introduce a metric t_c in the communication system that refers to the average time for the impact of an attack on communication equipment to propagate in the communication infrastructure. In this model, as opposed to our model in [22], we do not consider the average time t_e in the electric system that refers to the average time elapsed between the failure of a set of electric equipment and the response time of safety measures or operators manual intervention to contain the failures and prevent them from propagating to the entire grid.

Let $\mathbf{r^e(t)} = [r_i^e(t)]_{N_e \times 1}$ and $\mathbf{r^c(t)} = [r_i^c(t)]_{N_c \times 1}$ be the electrical and communication nodes risk vectors at time t, respectively. We take discrete time steps to describe the evolution of the system. Let $\mathbf{S^l} = [s_{ij}^l]_{N_c \times N_c}$ be the l-th power of the matrix $\mathbf{S}$. At attack step r, the payoff is decreased by a factor of γ_c^r. In fact, we consider that each action of the attacker in the system increases the probability of him being detected. Let the matrix $\mathbf{S^{max}} = [s_{ij}^{max}]_{N_c \times N_c}$ represent the maximum impact of an attack on communication equipment to reach communication nodes during time t_c, where $s_{ij}^{max} = \max_{l=1,\ldots,\lfloor t_c \rfloor} \gamma_c^l s_{ij}^l$. Let $\mathbf{S_n^{max}}$ be the normalized matrices of $\mathbf{S^{max}}$ with respect to their rows s.t. $\forall j,\ \sum_i {s_n^{max}}_{ij} = 1$.

Therefore, the system of equations for inter- and intra-infrastructure risk diffusion is given by:

$$\begin{cases} \mathbf{r^c(t+1) = S_n^{max} r^c(t)} \\ \mathbf{r^c(t+1) = F r^e(t)} \\ \mathbf{r^e(t+1) = B r^e(t)} \\ \mathbf{r^e(t+1) = D r^c(t)} \end{cases} \tag{10.1}$$

Solving the system of equations in (10.1), we will have:
$\mathbf{r^c(t+4) = S_n^{max} FBD r^c(t) = H r^c(t)}$ where $\mathbf{H} = [h_{ij}]_{N_c \times N_c} = \mathbf{S_n^{max} FBD}$.

Lemma 10.1. *Matrix* $\mathbf{H} = \mathbf{S_n^{max} FBD}$ *is a left stochastic matrix.*

Proof. Let $\mathbf{Z} = [z_{ij}]_{m\times n}$ and $\mathbf{Y} = [y_{ij}]_{n\times m}$ s.t $\forall j,\ \sum_i z_{ij} = 1$ and $\sum_i y_{ij} = 1$. Let $\mathbf{X} = [x_{ij}]_{m\times m} = \mathbf{ZY}$. Therefore:

$$\sum_i x_{ij} = \sum_i \sum_m z_{im} y_{mj} = (\sum_m y_{mj})(\sum_i z_{im}) = \sum_m y_{mj} = 1$$

Similarly, we can prove that matrix $\mathbf{H}$, which is the product of matrices $\mathbf{S_n^{max}}$, $\mathbf{F}$, $\mathbf{B}$, and $\mathbf{D}$, is a left stochastic matrix. □

We take a similar approach to [24] by balancing the immediate risk and the future induced one. Let β and τ refer to the weight of the initial risk on communication nodes and the weight of the diffused risk from electrical nodes to communication nodes at time $t = 0$, respectively, and δ the weight of future cascading risk with respect to the value of the total risk on communication nodes. The value of risk on communication equipment at a given time is given by $\mathbf{r^c(t+4)} = \delta \mathbf{H r^c(t)} + \beta \mathbf{r^c(0)} + \theta \mathbf{D^T r^e(0)}$, where β, τ, and δ are nonnegative real numbers and $\beta + \tau + \delta = 1$.

Theorem 10.1. *The iterative system of the cascading risk converges. An equilibrium solution exists whenever* $\delta < 1$ *and is given by:*

$$\mathbf{r^{c*}} = (\mathbf{I} - \delta \mathbf{H})^{-1}(\beta \mathbf{r^c(0)} + \theta \mathbf{D^T r^e(0)}) \quad \textit{where } \mathbf{H} = \mathbf{S_n^{max} FBD} \tag{10.2}$$

Proof. The spectral radius of any matrix is less than or equal to the norm of the matrix. The 1-norm of the matrix $\mathbf{H} = [h_{ij}]_{N_c\times N_c}$ is defined as $\|\mathbf{H}\|_1 = \max_{0\le j\le N_c} \{\sum_{i=1}^{N_c} |h_{ij}|\}$. From Lemma 10.1, we know that $\mathbf{H}$ is a left stochastic matrix. Therefore, $\|\mathbf{H}\|_1 = 1$ and the spectral radius $\rho(\mathbf{H}) \le 1$. The matrix $\mathbf{H}$ has at least one eigenvalue equal to 1 since $(1, \mathbf{e})$ is an eigenpair of $\mathbf{H^T}$ (where $\mathbf{e} = [1\dots 1]^T$). Since the matrix $\mathbf{H}$ is multiplied by $\delta < 1$, so are the eigenvalues of $\mathbf{H}$. Therefore, the sequence converges. The equation of the cascading risk $\mathbf{r^c(t+4)} = \delta \mathbf{H r^c(t)} + \beta \mathbf{r^c(0)} + \theta \mathbf{D^T r^e(0)}$ converges to the value $\mathbf{r^{c*}}$ given by $\mathbf{r^{c*}} = \delta \mathbf{H r^{c*}} + \beta \mathbf{r^c(0)} + \theta \mathbf{D^T r^e(0)}$.

The solution of the problem is given by $\lim_{t\to+\infty} \mathbf{r^c(t)} = (\mathbf{I} - \delta \mathbf{H})^{-1}(\beta \mathbf{r^c(0)} + \theta \mathbf{D^T r^e(0)})$. The existence of the solution depends on the existence of the inverse of the matrix $(\mathbf{I} - \delta \mathbf{H})$. However, we can notice that $|\ 1 - \delta h_{ii}\ | > |\ \delta \sum_{i\neq j} h_{ij}\ | = |\ \delta - \delta h_{ii}\ |$ $\forall i$ is true whenever $\delta < 1$. In this case, the matrix $(\mathbf{I} - \delta \mathbf{H})$ is a strictly column diagonally dominant matrix and therefore nonsingular. As a result, $(\mathbf{I} - \delta \mathbf{H})^{-1}$ exists.
□

From Theorem 10.1, we can predict how the risk on communication equipment diffuses between nodes of the communication and electric systems. If an attacker has access to $\mathbf{H}$, he can choose his targets in the communication system intelligently to maximize the impact of his attacks on the power system. In the next section, we propose a security game between an attacker and a defender and analyze the behavior of both players in this scenario.

10.4 Security Game

We formulate the problem as a noncooperative game between the attacker and the defender. The attacker's/defender's objective is to distribute attack/defense resources on the communication nodes in order to maximize/minimize the impact of attacks on the power system.

We associate, for each communication equipment, a load l_i that represents the amount of computational work the equipment performs. Let $\mathbf{L} = diag(l_i)_{N_c \times N_c}$ be the load matrix. The existence of redundant equipment in the communication system increases the resilience of the power grid against cyberattacks. Let $\mathbf{W} = [w_{ij}]_{N_c \times N_c}$ be the redundancy matrix where $\forall i$, $w_{ii} = -1$ and $\sum_{j, j \neq i} w_{ij} \leq 1$. If $i \neq j$, w_{ij} represents the fraction of the load of node i node j will be responsible of processing when node i is compromised.

10.4.1 Game with Symmetric Information

In this section, we consider the worst-case scenario where both players have complete knowledge of the architecture of the system. We will assume that the players' utilities are composed of three parts: the payoff of an attack taking into account both players' actions and the cascading impact of the attack in the communication and electric systems, the cost of attacking/defending, and the impact of redundant equipment in ensuring the control of the power system when a set of communication nodes is compromised.

Let $\mathbf{p} = [p_i]_{1 \times N_c}$ refer to the attacker's strategy where $0 \leq p_i \leq 1$ is the attack resource allocated to target $i \in T^c$, and let $\mathbf{q} = [q_j]_{1 \times N_c}$ refer to the defender's strategy where $0 \leq q_j \leq 1$ is the defense resource allocated to target $j \in T^c$. Let $\mathbf{R_D^c(0)}$, $\mathbf{R_D^{c*}}$, $\mathbf{C^a}$, and $\mathbf{C^d}$ be diagonal matrices. The diagonal elements of matrices $\mathbf{R_D^c(0)}$ and $\mathbf{R_D^{c*}}$ refer to the initial risk $r_i^c(0)$ and the equilibrium solution of the cascading risk r_i^{c*} on each communication node i, respectively. $\mathbf{C^a}$ and $\mathbf{C^d}$ refer to the cost of attacking and defending communication nodes, respectively. Let $\mathbf{I}$ refer to the identity matrix and let $\mathbf{e} = (1, \ldots, 1)_{1 \times N_c}$.

The utilities u_a and u_d of the attacker and the defender, respectively, can be defined as follows:

$$u_a(\mathbf{p}, \mathbf{q}) = \mathbf{p}\mathbf{R_D^{c*}}(\mathbf{e^T} - \mathbf{q^T}) - \mathbf{p}\mathbf{R_D^c(0)}\mathbf{C^a}\mathbf{p^T} - \psi\mathbf{p}\mathbf{L}(\mathbf{W}\mathbf{q^T} - \mathbf{I}(\mathbf{e^T} - 2\mathbf{q^T}))$$
$$u_d(\mathbf{p}, \mathbf{q}) = -\mathbf{p}\mathbf{R_D^{c*}}(\mathbf{e^T} - \mathbf{q^T}) - \mathbf{q}\mathbf{R_D^c(0)}\mathbf{C^d}\mathbf{q^T} + \psi\mathbf{p}\mathbf{L}(\mathbf{W}\mathbf{q^T} - \mathbf{I}(\mathbf{e^T} - 2\mathbf{q^T}))$$

The parameter $\psi \in [0,1]$ represents the impact of compromising the load of communication equipment and the existence of backup equipment in computing players' utility functions. ψ can be a function of the probability that backup equipment are able to take charge of the load of compromised communication equipment.

First, we analyze the interactions between the attacker and the defender as a one-shot game [25] in which players take their decisions at the same time. Let Γ^o refer to this game. We are interested in particular in the concept of Nash equilibrium (NE), in which none of the players has an incentive to deviate from unilaterally [25].

Let $\mathbf{p}^*$ and $\mathbf{q}^*$ denote the attacker and defender strategies at the Nash equilibrium, respectively. We have the following theorem:

Theorem 10.2. *If $r_i^c(0)$, c_i^a, and c_i^d are real positive numbers $\forall i$, a unique NE of the game Γ^o exists and is given by:*

$$\mathbf{q}^* = \frac{1}{2}\mathbf{e}(\mathbf{R_D^{c*}} + \psi\mathbf{L})(\mathbf{R_D^c(0)C^a})^{-1}\mathbf{M}[\frac{1}{2}\mathbf{M^T}(\mathbf{R_D^c(0)C^a})^{-1}\mathbf{M} + 2\mathbf{R_D^c(0)C^d}]^{-1}$$

$$\mathbf{p}^* = \mathbf{e}(\mathbf{R_D^{c*}} + \psi\mathbf{L})[\frac{1}{2}\mathbf{M}(\mathbf{R_D^c(0)C^d})^{-1}\mathbf{M^T} + 2\mathbf{R_D^c(0)C^a}]^{-1}$$

$$\textit{where } \mathbf{M} = \mathbf{R_D^{c*}} + \psi\mathbf{L}(\mathbf{W} + 2\mathbf{I})$$

Proof. Let $\overline{\nabla}$ be the pseudogradient operator of $U = u_a(\mathbf{u}) + u_d(\mathbf{u})$ where $\mathbf{u} = [\mathbf{p}\ \mathbf{q}]$. Let $g(\mathbf{u}) = \overline{\nabla}U = \begin{bmatrix} \nabla_{\mathbf{p}} u_a(\mathbf{u}) \\ \nabla_{\mathbf{q}} u_d(\mathbf{u}) \end{bmatrix}$ and let $\mathbf{G}(\mathbf{u})$ be the Jacobian of $g(\mathbf{u})$.

$$\mathbf{G}(\mathbf{u}) = \left(\begin{array}{c|c} -diag(2r_i^c(0)c_i^a) & -\mathbf{R_D^{c*}} - \psi(\mathbf{W^T} + 2\mathbf{I})\mathbf{L} \\ \hline \mathbf{R_D^{c*}} + \psi\mathbf{L}(\mathbf{W} + 2\mathbf{I}) & -diag(2r_i^c(0)c_i^d) \end{array} \right)$$

Since $r_i^c(0)$, c_i^a, and c_i^d are real positive numbers $\forall i$, $(\mathbf{G(u)} + \mathbf{G(u)^T})$ is a negative definite matrix. As a result, U is diagonally strictly concave. Based on [26], an equilibrium of the game in pure strategy exists and is unique.

To characterize the equilibrium, we need to find vectors $\mathbf{p}^*$ and $\mathbf{q}^*$ in which the gradients ∇u_a and ∇u_d are equal to 0. Solving these equations, we find $\mathbf{q}^*$ and $\mathbf{p}^*$ given in Theorem 10.2.

Let $\mathbf{M} = \mathbf{R_D^{c*}} + \psi\mathbf{L}(\mathbf{W} + 2\mathbf{I})$. The existence of $\mathbf{p}^*$ and $\mathbf{q}^*$ depends on the existence of the inverses of matrices $\boldsymbol{\xi}$ and $\boldsymbol{\zeta}$, where:

$$\boldsymbol{\xi} = \tfrac{1}{2}[\mathbf{M}(\mathbf{R_D^c(0)C^d})^{-1}\mathbf{M^T} + 4\mathbf{R_D^c(0)C^a}]$$

$$\textit{and } \boldsymbol{\zeta} = \tfrac{1}{2}[\mathbf{M^T}(\mathbf{R_D^c(0)C^a})^{-1}\mathbf{M} + 4\mathbf{R_D^c(0)C^d}]$$

The diagonal matrix $4\mathbf{R_D^c(0)C^a}$ is a positive definite matrix (diagonal matrix with strictly positive elements). To prove that $\mathbf{M}(\mathbf{R_D^c(0)C^d})^{-1}\mathbf{M^T}$ is a positive definite matrix, we need to show that:

$$\forall \mathbf{x} \neq 0,\ \mathbf{x^T}\mathbf{M}(\mathbf{R_D^c(0)C^d})^{-1}\mathbf{M^T}\mathbf{x} > 0$$

Let $\mathbf{y} = \mathbf{M^T}\mathbf{x}$. Therefore, we need to prove that:

$$\forall \mathbf{y} \neq 0,\ \mathbf{y^T}(\mathbf{R_D^c(0)C^d})^{-1}\mathbf{y} > 0 \tag{10.3}$$

However, $(\mathbf{R_D^c(0)C^d})^{-1}$ is a positive definite matrix, and equation (10.3) holds. Therefore, the matrix $\mathbf{M}(\mathbf{R_D^c(0)C^d})^{-1}\mathbf{M^T}$ is a positive definite matrix. Finally, the matrix $\boldsymbol{\xi}$ is a positive definite matrix because it is the sum of two positive definite matrices. Since $\boldsymbol{\xi}$ is a positive definite matrix, the inverse $\boldsymbol{\xi}^{-1}$ exists. Similarly, we prove that $\boldsymbol{\zeta}^{-1}$ exists. □

The analytical solution has multiple advantages. From a scalability point of view, the complexity resides in evaluating the input parameters of the model. In fact, by proving the existence and uniqueness of the Nash equilibrium and characterizing the solution analytically, we avoided the complexity of searching the set of all possible strategies to find the NE. Using an analytical solution, we can compute the optimal strategies of both players directly and be able to assess the sensitivity of players' strategies to estimation errors on the values of parameters used in the model.

In the rest of this section, we analyze the interactions between players as a Stackelberg game [25]. In this type of games, a leader chooses his strategy first. Then, the follower, informed by the leader's choice, chooses his strategy. The leader tries to anticipate the follower's response. In our case, the defender is the leader who tries to secure communication equipment in order to best protect the power system. Let Γ^s refer to this game.

Stackelberg games are generally solved by backward induction. The solution is known as Stackelberg equilibrium (SE). We start by computing the best response strategy of the follower as a function of the leader's strategy. Then, according to the follower's best response, we derive the optimal strategy of the leader.

The attacker solves the following optimization problem:

$$\mathbf{p}(\mathbf{q}) = \underset{\mathbf{p}\in[0,1]^{N_c}}{\operatorname{argmax}}\, u_a(\mathbf{p},\mathbf{q})$$

On the other hand, the defender solves the following optimization problem:

$$\mathbf{q}(\mathbf{p}) = \underset{\mathbf{q}\in[0,1]^{N_c}}{\operatorname{argmax}}\, u_d(\mathbf{p}(\mathbf{q}),\mathbf{q})$$

We have the following theorem:

Theorem 10.3. *The game Γ^s admits a unique Stackelberg equilibrium $(\mathbf{p^S},\mathbf{q^S})$ given by:*

$$\mathbf{q^S} = \mathbf{e}(\mathbf{R_D^{c*}} + \psi\mathbf{L})(\mathbf{R_D^c(0)C^a})^{-1}\mathbf{M}(\mathbf{Q}+2\mathbf{R_D^c(0)C^d})^{-1}$$

$$\mathbf{p^S} = \frac{1}{2}\mathbf{e}(\mathbf{R_D^{c*}} + \psi\mathbf{L})(\mathbf{R_D^c(0)C^a})^{-1}[\mathbf{I}-\mathbf{M}(\mathbf{Q}+2\mathbf{R_D^c(0)C^d})^{-1}\mathbf{M^T}(\mathbf{R_D^c(0)C^a})^{-1}]$$

$$\textit{where}\ \ \mathbf{Q} = \mathbf{M^T}(\mathbf{R_D^c(0)C^a})^{-1}\mathbf{M}$$

Proof. The solution can be found by solving the system by backward induction. We start by finding $\mathbf{p^S}$ by setting $\nabla u_a(\mathbf{p},\mathbf{q}) = 0$. Then we solve the equation $\nabla u_d(\mathbf{p^S},\mathbf{q}) = 0$ to find $\mathbf{q^S}$.

Similarly to the proof of Theorem 10.2, we can prove that $(\mathbf{Q}+2\mathbf{R_D^c(0)C^d})^{-1}$ exists. □

10.4.2 Game with Asymmetric Information

In the previous section, we made a strong assumption about the knowledge that both the attacker and the defender have about the architecture of the system. In particular, we assumed that the attacker has knowledge about the existence of backup equipment and can accurately assess the backup dependencies between the different communication equipment. In this section, under asymmetry of information between the attacker and the defender, we try to analyze whether unique Nash and Stackelberg equilibriums of the games exist. An answer to this question will have a direct impact on whether it is optimal for the defender to mislead the attacker about dependencies relationships between communication equipment. While it is challenging to answer this question in the general case analytically, we will analyze in the case study two variations of the game where that scenario arises.

Information asymmetry between the attacker and the defender can involve information about the electric system or the architecture of the communication system. In what follows, we analyze the general case where any of the following parameters are different for each player: the initial risk matrix $\mathbf{R_D^c(0)}$, the cascading risk matrix $\mathbf{R_D^{c*}}$, the load matrix $\mathbf{L}$, and the redundancy matrix $\mathbf{W}$. This assumption follows from the observation that, contrary to the defender, the attacker can have incomplete information about the system, which yields incorrect assessment of these matrices. While the defender can also have incomplete information in practice, in this section, we are interested in the case where only the attacker has incomplete information about the system (as if assuming a best-case scenario for the defender's knowledge about the system). The objective is to analyze the impact of the asymmetric nature of information about the system on the strategy of both players. In this game, there's an assumption that the defender knows about the attacker's inaccurate beliefs about the architecture of the system. It is as if the defender has deliberately published inaccurate information about the system publicly before the start of the game with the objective of misleading the attacker.

Let $\mathbf{R_D^{c\dagger}(0)}$, $\mathbf{R_D^{c*\dagger}}$, the load matrix $\mathbf{L}^\dagger$, and the redundancy matrix $\mathbf{W}^\dagger$ refer to the attacker's evaluation of matrices $\mathbf{R_D^c(0)}$, $\mathbf{R_D^{c*}}$, $\mathbf{L}$, and $\mathbf{W}$, respectively. Let Γ_{IA}^o and Γ_{IA}^s refer to the one-shot and Stackelberg games where information asymmetry exists between the attacker and the defender. We note that Γ_{IA}^s is a game in which the defender is the leader and the attacker is the follower.

We have the following theorem:

Theorem 10.4. *The game Γ_{IA}^o (resp. Γ_{IA}^s) can have 0, 1, or an infinite number of Nash (resp. Stackelberg) equilibriums.*

Proof. The utility of the attacker and the defender are given by:

$$u_a^{IA}(\mathbf{p},\mathbf{q}) = \mathbf{p}\mathbf{R_D^{c*\dagger}}(\mathbf{e^T}-\mathbf{q^T}) - \mathbf{p}\mathbf{R_D^{c\dagger}(0)}\mathbf{C^a}\mathbf{p^T} - \psi\mathbf{p}\mathbf{L}^\dagger(\mathbf{W}^\dagger\mathbf{q^T} - \mathbf{I}(\mathbf{e^T}-2\mathbf{q^T}))$$
$$u_d^{IA}(\mathbf{p},\mathbf{q}) = -\mathbf{p}\mathbf{R_D^{c*}}(\mathbf{e^T}-\mathbf{q^T}) - \mathbf{q}\mathbf{R_D^c(0)}\mathbf{C^d}\mathbf{q^T} + \psi\mathbf{p}\mathbf{L}(\mathbf{W}\mathbf{q^T} - \mathbf{I}(\mathbf{e^T}-2\mathbf{q^T}))$$

Setting $\nabla u_a^{IA} = 0$, we find $2\mathbf{p}\mathbf{R_D^{c\dagger}(0)}\mathbf{C^a} + \mathbf{q}\mathbf{M}^{\dagger\mathbf{T}} = \mathbf{e}(\mathbf{R_D^{c*\dagger}} + \psi\mathbf{L}^\dagger)$, where $\mathbf{M}^\dagger = \mathbf{R_D^{c*\dagger}} + \psi\mathbf{L}^\dagger(\mathbf{W}^\dagger + 2\mathbf{I})$. Similarly, setting $\nabla u_d^{IA} = 0$, we find $\mathbf{p}\mathbf{M} - 2\mathbf{q}\mathbf{R_D^c(0)}\mathbf{C^d} = 0$. Therefore, we get:

$$\mathbf{p}[2\mathbf{R_D^{c\dagger}}(\mathbf{0})\mathbf{C^a} + \tfrac{1}{2}\mathbf{M}(\mathbf{R_D^c}(\mathbf{0})\mathbf{C^d})^{-1}\mathbf{M^{\dagger T}}] = \mathbf{e}(\mathbf{R_D^{c*\dagger}} + \psi \mathbf{L}^\dagger)$$
$$\mathbf{q}[\tfrac{1}{2}\mathbf{M^{\dagger T}}(\mathbf{R_D^{c\dagger}}(\mathbf{0})\mathbf{C^a})^{-1}\mathbf{M} + 2\mathbf{R_D^c}(\mathbf{0})\mathbf{C^d}] = \tfrac{1}{2}\mathbf{e}(\mathbf{R_D^{c*\dagger}} + \psi \mathbf{L}^\dagger)(\mathbf{R_D^{c\dagger}}(\mathbf{0})\mathbf{C^a})^{-1}\mathbf{M}$$

For a NE to exist and be unique, the matrices $[2\mathbf{R_D^{c\dagger}}(\mathbf{0})\mathbf{C^a} + \frac{1}{2}\mathbf{M}(\mathbf{R_D^c}(\mathbf{0})\mathbf{C^d})^{-1}\mathbf{M^{\dagger T}}]$ and $[\frac{1}{2}\mathbf{M^{\dagger T}}(\mathbf{R_D^{c\dagger}}(\mathbf{0})\mathbf{C^a})^{-1}\mathbf{M} + 2\mathbf{R_D^c}(\mathbf{0})\mathbf{C^d}]$ must be nonsingular. Otherwise, either 0 or an infinite number of NE exist. Similarly, we can prove that for a Stackelberg equilibrium to exist and be unique, the matrix $[\mathbf{M^{\dagger T}}(\mathbf{R_D^{c\dagger}}(\mathbf{0})\mathbf{C^a})^{-1}\mathbf{M} + \mathbf{M^T}(\mathbf{R_D^{c\dagger}}(\mathbf{0})\mathbf{C^a})^{-1}\mathbf{M}^\dagger + 4\mathbf{R_D^c}(\mathbf{0})\mathbf{C^d}]$ must be nonsingular. □

In some cases where an infinite number of NE exists, it is possible that one of the players chooses the same strategy at each NE. This case is particularly disadvantageous for the defender if a unique strategy for the defender exists while an infinite number of strategies for the attacker at the NE exist. In this case, playing the NE strategy for the defender will not give him any insight on the potential strategy that will be played by the attacker, and therefore, predicting his payoff from playing a NE strategy will be challenging.

From the proof of Theorem 10.4, we can state the following lemma:

Lemma 10.2. *If* $\mathbf{M}(\mathbf{R_D^c}(\mathbf{0})\mathbf{C^d})^{-1}\mathbf{M^{\dagger T}}$ *and* $\mathbf{M^{\dagger T}}(\mathbf{R_D^{c\dagger}}(\mathbf{0})\mathbf{C^a})^{-1}\mathbf{M}$ *are positive definite matrices, then unique Nash and Stackelberg equilibriums exist for the games* Γ_{IA}^o *and* Γ_{IA}^s*, respectively.*

We note that the games in Section 10.4 are special cases of Γ_{IA}^o and Γ_{IA}^s and where Lemma 10.2 holds.

10.4.3 Maximin Strategies

In this section, we analyze the *maximin* strategies of both the attacker and the defender in the game where both players have complete knowledge about the architecture of the system. The *maximin* strategy of a player is the maximum payoff that he is guaranteed to get in the worst-case scenario for any strategy played by the other player.

In the rest of this section, we suppose that $r_i^c(0)$, c_i^a, and c_i^d are real positive numbers $\forall i$. We start by analyzing the *maximin* strategy of the attacker $\mathbf{p} = \operatorname{argmax}_{\mathbf{p}'} \min_{\mathbf{q}} u_a$. Let $\kappa_i = (r_i^{c*} + \psi l_i)p_i + \psi \sum_{j \neq i} l_j w_{ji} p_j \ \forall i$.

Theorem 10.5. *When there are no constraints on the defender budget* $\sum_i q_i$*, the attacker cannot guarantee any positive payoff from attacking the system. Otherwise, if* $\sum_i q_i = K_d \leq N_c$*, there exists a sensible target set* X_d *that will be of interest to the defender.*

Proof. The utility of the attacker can be written as $u_a = \sum_i \{p_i(r_i^{c*} - r_i^c(0)c_i^a p_i + \psi l_i) - \kappa_i q_i\}$. When there are no constraints on the defense budget $\sum_i q_i$, since $\kappa_i \geq 0 \ \forall i$, the defender can minimize the utility of the attacker by setting $q_i = 1 \ \forall i$. In this case, $u_a \leq 0\ \forall \mathbf{p}$, and as a result, the attacker cannot guarantee getting any positive payoff from attacking the system.

In the case of a constrained defense budget $\sum_i q_i = K_d$, in order to minimize u_a, the defender will allocate his resources on the set of communication equipment with the highest values of $\kappa_i\ \forall i$. Let X_d refer to this set. Therefore, $\forall t \in T^c \backslash X_d$, we have $q_t = 0$. If $\sum_i q_i = K_d < 1$, we have $\kappa_i = \operatorname{argmax}_j \kappa_j\ \forall i \in X_d$. □

The sensible target set X_d can actually be controlled by the attacker. In particular, adjusting the attacker's strategy $\mathbf{p}$ will determine the set X_d that will be of interest to the defender. Therefore, we have the following lemma:

Lemma 10.3. *If* $\sum_i q_i = K_d < 1$, *the complexity of finding a* maximin *strategy for the attacker is* $O(2^{N_c})$.

Lemma 10.3 follows from the fact that the attacker needs to check all possible combinations of X_d to find his *maximin* strategy. For a given sensible target set X_d, we have the following theorem:

Theorem 10.6. *If* $K_d < 1$, *then for a given set* X_d, *there is at most one* maximin *strategy for the attacker.*

Proof. Let $K_d < 1$ and $m \in X_d$. Since $\forall i, j \in X_d$, we have $\kappa_i = \kappa_j$, u_a can be written as $u_a = \sum_i \{p_i(r_i^{c*} - r_i^c(0)c_i^a p_i + \psi l_i) - (r_m^{c*} p_m + \psi l_m p_m + \psi \sum_{j \neq m} l_j w_{jm} p_j) K_d\}$. Therefore, finding a *maximin* strategy for the attacker is equivalent to solving the following optimization problem:

$$
\begin{aligned}
\max_{\mathbf{p}}\ & u_a \\
\text{s.t.}\ \ & \kappa_t - \kappa_m \leq 0\ \forall t \in T^c \backslash X_d \\
& \kappa_i - \kappa_j = 0\ \forall \{i, j\} \in X_d \\
& p_i \in [0, 1]\ \forall i \in T^c
\end{aligned}
$$

The utility function u_a is strictly concave and the constraints form a convex set. Therefore, there exists at most one optimal solution to the problem. □

Let S be a large positive number. By analyzing the attacker's utility function u_a, we have the following lemma:

Lemma 10.4. *In the case of a constrained defense budget* $K_d < 1$, *finding a* maximin *strategy for the attacker is equivalent to solving the following mixed integer quadratic program (MIQP):*

$$
\begin{aligned}
\max_{\mathbf{p}, \mathbf{q}, \mathbf{y}, b}\ & u_a \\
s.t.\ \ & 0 \leq b - \kappa_i \leq (1 - y_i) S \\
& q_i \leq y_i S \\
& \sum_i q_i = K_d \\
& y_i \in \{0, 1\},\ p_i \in [0, 1],\ q_i \in [0, K_d],\ b \in \mathbb{R}
\end{aligned}
$$

We now analyze the *maximin* strategy of the defender $\max_{\mathbf{q}} \min_{\mathbf{p}} u_d$. Let $\eta_i = -(r_i^{c*} + \psi l_i)(1 - q_i) + \psi \sum_{j \neq i} l_i w_{ij} q_j \ \forall i$. We have the following theorem:

Theorem 10.7. *Independent of whether there are constraints on the defense budget* $\sum_i q_i$*, the complexity of finding a* maximin *strategy for the defender is* $O(2^{N_c})$.

Proof. The utility of the defender u_d can be written as $u_d = \sum_i \{p_i \eta_i - r_i^c(0) c_i^d q_i^2\}$. The attacker will allocate his resources on the set of communication equipment i where $\eta_i < 0$. Let X_a refer to this set. Since the set X_a depends on the strategy of the defender $\mathbf{q}$, the defender needs to check all possible subsets of the N_c communication equipment to find his *maximin* strategy. □

In the rest of this section, we will analyze the case where we a have a constrained defense budget $\sum_i q_i = K_d \leq N_c$.

By analyzing the defender's utility function u_d, we have the following lemma:

Lemma 10.5. *Finding a* maximin *strategy of the defender is equivalent to solving the following mixed integer quadratically constrained program (MIQCP):*

$$\begin{aligned}
&\max_{\mathbf{q},\mathbf{p}} \ u_d \\
&s.t. \quad 0 \leq 1 - p_i \leq \eta_i (1 - p_i) S \\
&\qquad\ \ 0 \leq p_i \leq -\eta_i p_i S \\
&\qquad\ \ \sum_i q_i = K_d \\
&\qquad\ \ p_i \in \{0,1\}, \ q_i \in [0, \min(1, K_d)]
\end{aligned}$$

Let Y be an integer in the range $[\![0, N_c]\!]$. Through the change of variables $z_{ij} = p_i q_j \ \forall \{i,j\} \in T^c$ in the MIQCP problem in Lemma 10.5, we have the following lemma:

Lemma 10.6. *Finding a* maximin *strategy of the defender is equivalent to finding the maximum value of N_c mixed integer quadratic programs, where for each $Y \in [\![0, N_c]\!]$, the MIQP is given by:*

$$\begin{aligned}
\max_{\mathbf{z},\mathbf{p}} \ & \sum_i \{-(r_i^{c*} + \psi l_i)(p_i - z_{ii}) + \psi \sum_{j \neq i} l_i w_{ij} z_{ij} - \frac{r_i^c(0) c_i^d}{Y^2} (\sum_j z_{ji})^2\} \\
s.t. \ \ & 0 \leq 1 - p_i \leq \big(-(r_i^{c*} + \psi l_i)(1 - \tfrac{1}{Y} \sum_j z_{ji}) + \tfrac{\psi}{Y} \sum_{j \neq i} l_i w_{ij} \sum_m z_{mj} \\
& \qquad\qquad + (r_i^{c*} + \psi l_i)(p_i - z_{ii}) - \psi \sum_{j \neq i} l_i w_{ij} z_{ij} \big) S \\
& 0 \leq p_i \leq \big((r_i^{c*} + \psi l_i)(p_i - z_{ii}) - \psi \sum_{j \neq i} l_i w_{ij} z_{ij} \big) S \\
& \sum_i \sum_j z_{ij} = Y K_d \\
& p_i \leq \frac{1}{K_d} \sum_j z_{ij} \leq 1 \\
& \sum_i z_{ij} \leq Y \min(1, K_d) \\
& \sum_i p_i = Y \\
& p_i \in \{0,1\}, \ z_{ij} \in [0, \min(1, K_d)]
\end{aligned}$$

We note that we can find a lower bound for $\max_{\mathbf{p}} \min_{\mathbf{q}} u_a$ and $\max_{\mathbf{q}} \min_{\mathbf{p}} u_d$ by replacing the quadratic cost functions in both u_a and u_d with linear cost functions $\sum_i p_i r_i^c(0) c_i^a$ and $\sum_i q_i r_i^c(0) c_i^d$, respectively, and solving the mixed integer linear programs derived by analyzing the resulting utility functions.

10.5 Parameters Evaluation

In this section, we present an approach to assess the impact of attacks in the electric and communication infrastructures and therefore evaluate matrices B and S, respectively. In addition, we propose a game between the attacker and the defender to assess the initial security risk on communication equipment. While the problem of the assessment of the other parameters of the model remains, we discuss at the end of this section potential avenues for their evaluation.

10.5.1 Evaluation of Matrix **B**

We assess the impact of cascading failures in the power grid by solving power flow equations using the DC power flow approximation [27]. Following a similar approach as in [28], we simulate individual failures and assess their impact on the power grid such as identifying generators with insufficient capacities to meet the demand and overloaded lines.

In our model, we analyze the impact of tripping transmission lines or losing generators on the power grid. The flowchart diagram in Figure 10.1 shows the cascading algorithm used in our model to analyze the impact of tripping transmission lines. In general, this could have a significant impact on the power grid and could lead to the formation of islands in the electric system. In our algorithm, we shut down islands where the demand (denoted as d in Figure 10.1) exceeds the maximum generation capacity in the island (denoted as $max(g)$ in Figure 10.1). We then solve the DC power flow problem in the electric transmission system using MATPOWER [29] and check the existence of overloaded lines. These lines are tripped, and the process is repeated until a balanced solution emerges. Similarly, we assess the impact of losing generators on the power grid.

In our approach, we consider the worst-case scenario where load shedding is not an option when we conduct our analysis of the impact of cascading failures on the power grid. Further work taking into account more fine-grained analysis of the behavior of the power grid will allow us to quantify more precisely the values of elements of matrix **B**.

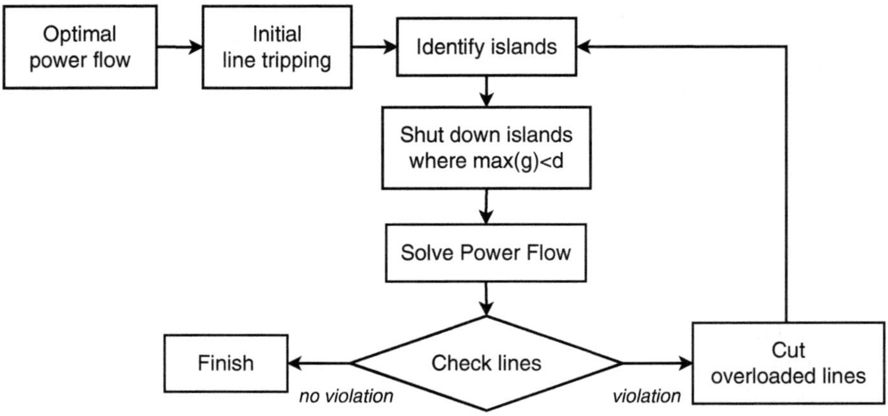

Fig. 10.1: Flowchart of the cascade algorithm in the case of tripped transmission lines

10.5.2 Evaluation of Matrix S

To address the challenge of evaluating the impact of cyberattacks on the communication infrastructure, attack graphs [30] are a promising solution to generate all possible attack steps to compromise a target node. These graphs could be used in conjunction with risk assessment methods to evaluate the impact of each attacker action on the communication infrastructure.

Let $\mathscr{G} = (\mathscr{X}, \mathscr{E})$ be an attack graph where $\mathscr{X}$ refers to the set of nodes in the graph and $\mathscr{E}$ refers to the set of edges. In our case, a node $x \in \mathscr{X}$ in the graph refers to a state of the attacker in the system, and an edge $e = (x_i, x_j) \in \mathscr{E}$ refers to an action executed by the attacker after which the state of the attacker in the system changes from x_i to x_j. A state of the attacker refers to his knowledge at a particular time of the topology and the configuration of the system, the set of access levels acquired on equipment, and the set of credentials at his disposal. $\mathscr{G}$ represents all attack paths that can be used by the attacker to compromise a set of equipment or services in the system. In [31], we defined such graph and implemented a proof of concept for constructing it.

Let θ^r_{lm} be the number of paths of length r an attacker can use to compromise communication equipment m from communication equipment l. Let $\Theta_{lm} = \sum\limits_r \gamma^r_c \theta^r_{lm}$ refer to the impact metric of a communication equipment l on a communication node m. Θ_{lm} is a measure of the cumulated impact on communication equipment m of an attack originating from equipment l. We consider that each action of the attacker in the system increases the probability of him being detected. Therefore, at attack step r, the payoff is decreased by a factor of γ^r_c representing the uncertainty for the attacker of getting the payoff of the rth future attack step. In this case, $s_{lm} = \frac{\Theta_{lm}}{\sum\limits_i \Theta_{im}}$, where $\mathbf{S} = [s_{lm}]_{N_c \times N_c}$.

10.5.3 Evaluation of the Initial Risk

In this section, we present a game between the attacker and the defender to assess the initial security risk on communication equipment before the impact of an attack propagates to the electric system. To simplify notations, we assume that in this game, both players have knowledge about the architecture of the communication system. The general case where the information about the architecture is asymmetrical or incomplete can be analyzed similarly.

The security risk is generally defined as the product of the probability of an attack occurring times its impact. Let g_i refer to the impact of compromising communication equipment i and $\mathbf{p^0}$ and $\mathbf{q^0}$ two vectors referring to the strategies of the attacker and the defender, respectively, where $0 \leq p_i^0, q_i^0 \leq 1\ \forall i$. p_i^0 and q_i^0 can be viewed as the probability of attacking and defending equipment i, respectively. The impact of compromising an equipment will eventually depend on the amount of defense resources allocated to defend it. Therefore, the security risk on equipment i can be defined as $r_i^c(0) = p_i^0(1-q_i^0)g_i$. The challenge for evaluating $r_i^c(0)$ resides in evaluating p_i^0 and q_i^0. From the point of view of the defender, estimating the risk is closely related to estimating the potential distribution of the attacker's resources on system equipment. Therefore, to best protect the system against a powerful and a strategic adversary, it is important to evaluate this distribution as a result of a strategic interaction between the attacker and the defender.

In general, attacking and defending a target i can depend not only on the impact g_i on i but also on the cost associated with attacking and defending i. In the case of interdependent communication equipment i and j, an attack on equipment i can have an impact on the cost of attacking equipment j. In what follows, we assume that this cost is proportional to s_{ij} where $\mathbf{S} = [s_{ij}]_{N_c \times N_c}$ represents the dependency between the communication equipment. Assessing $\mathbf{p^0}$ and $\mathbf{q^0}$ can therefore be the result of analyzing a strategic game between the attacker and the defender. In this game, the utility of the attacker can be defined as $u_a^0 = \sum\limits_i \big(p_i^0(1-q_i^0)g_i - (p_i^0)^2 c_i^{a,0} g_i + \phi \sum\limits_j s_{ji} p_j^0 (1-q_j^0)\big)$, where $0 \leq \phi \leq 1$ and $0 \leq c_i^{a,0} \leq 1\ \forall i$. $(p_i^0)^2 c_i^{a,0} g_i - \phi \sum\limits_j s_{ji} p_j^0 (1-q_j^0)$ represents the cost of attacking equipment i, where ϕ represents the extent in which compromising the dependent neighbors j of equipment i can have on reducing the cost of attacking equipment i. Similarly, the utility of the defender can be defined as $u_d^0 = \sum\limits_i \big(-p_i^0(1-q_i^0)g_i - (q_i^0)^2 c_i^{d,0} g_i - \phi \sum\limits_j s_{ji} p_j^0 (1-q_j^0)\big)$, where $0 \leq c_i^{d,0} \leq 1\ \forall i$. In this case, compromising the dependent neighbors of equipment i will increase the cost of defending equipment i.

One of the challenges for evaluating $\mathbf{p^0}$ and $\mathbf{q^0}$ will be the choice of the type of interactions that can take place between the attacker and the defender. If the attacker scans the system to assess its defenses, his interaction with the defender can be viewed as a Stackelberg game $\Gamma^{s,0}$ where the leader is the defender and the attacker is the follower. In this case, the interaction between the attacker and the defender can be viewed as taking place in two stages. $\mathbf{q^0}$ can be seen as a first allocation of defense resources on system equipment, while the result $\mathbf{q}$ of the game in Section 10.4 can be viewed as a security hardening measure to minimize the residual risk taking into account the impact of an attack on the electric system. The interaction can also take place in two stages when the knowledge of both players about the electric system is incomplete. Therefore, the game in this section is played before playing the game in Section 10.4 when players' knowledge about the electric system is updated.

The Stackelberg game $\Gamma^{s,0}$ can also refer to the case where the defender signals to the attacker his allocation of defense resources on communication equipment. When scanning the system or the defender's signaling does not take place, the game can be viewed as a one-shot game $\Gamma^{o,0}$. Let $\sum_i q_i^0$ refer to the defense budget and let $\sigma_i = (g_i + \phi \sum_j s_{ij})^2 + 4c_i^{a,0} c_i^{d,0} g_i^2$. We have the following theorem:

Theorem 10.8. *In the absence of constraints on the defense budget, a unique NE of the game $\Gamma^{o,0}$ exists and is given by:*

$$q_i^{0*} = \frac{(g_i + \phi \sum_j s_{ij})^2}{\sigma_i} \qquad p_i^{0*} = \frac{2(g_i + \phi \sum_j s_{ij}) c_i^{d,0} g_i}{\sigma_i} \qquad \forall i \in T^c$$

The result in Theorem 10.8 follows directly from setting $\nabla u_a = 0$ and $\nabla u_d = 0$. Similarly, by solving the system by backward induction, we have the following theorem:

Theorem 10.9. *In the absence of constraints on the defense budget, a unique Stackelberg equilibrium of the game $\Gamma^{s,0}$ exists and is given by:*

$$q_i^{0,s} = \frac{(g_i + \phi \sum_j s_{ij})^2}{\sigma_i - 2c_i^{a,0} c_i^{d,0} g_i^2} \qquad p_i^{0,s} = \frac{(g_i + \phi \sum_j s_{ij}) c_i^{d,0} g_i}{\sigma_i - 2c_i^{a,0} c_i^{d,0} g_i^2} \qquad \forall i \in T^c$$

In the case of a constrained defense budget $\sum_i q_i^0 = K_d^0$, we have the following theorem:

Theorem 10.10. *If $\sum_i q_i^0 = K_d^0 < 1$, the games $\Gamma^{o,0}$ and $\Gamma^{s,0}$ admit at most one Nash and one Stackelberg equilibrium, respectively.*

Proof. A necessary condition for a NE for $\Gamma^{o,0}$ to exist and be unique is having $-\min_k \left(\frac{(g_k+\phi \sum_j s_{kj})^2}{2c_k^{a,0} g_k} \right) \sum_i \frac{2c_i^{a,0} g_i}{\sigma_i} \leq \sum_i \frac{(g_i+\phi \sum_j s_{ij})^2}{\sigma_i} - K_d^0 \leq \min_k \left(2c_k^{d,0} g_k\right) \sum_i \frac{2c_i^{a,0} g_i}{\sigma_i}$. This result can be found by solving $\Gamma^{o,0}$ and verifying that $0 \leq q_i^0 \leq 1 \; \forall i$. Similarly for $\Gamma^{s,0}$, we can verify that a necessary condition for a Stackelberg equilibrium to exist and be unique is having $-\min_k \left(\frac{(g_k+\phi \sum_j s_{kj})^2}{2c_k^{a,0} g_k} \right) \leq \frac{K_d^0 - \sum_i \frac{(g_i+\phi \sum_j s_{ij})^2}{\sigma_i - 2c_i^{a,0} c_i^{d,0} g_i^2}}{\sum_i \frac{2c_i^{a,0} g_i}{\sigma_i - 2c_i^{a,0} c_i^{d,0} g_i^2}} \leq \min_k (c_k^{d,0} g_k)$. $\square$

10.5.4 Other Parameters

In our case study, we rely on experts' knowledge to evaluate matrices **D** and **F**, which represent the dependency relation on communication nodes by electrical nodes and vice versa, respectively. However, at the end of the case study in the next section, we conduct a sensitivity analysis to evaluate errors in the outputs of our model to estimation errors on the values of the elements of matrix **F**.

In our model, we introduced parameters β and τ, which represent the weight of the initial risk on communication nodes and the weight of the diffused risk from electric equipment to communication equipment at time $t = 0$, respectively, and δ which reflects the weight of future cascading risk with respect to the value of the total risk on communication equipment. These parameters can be evaluated as a result of the application of a risk assessment method coupled with quantitative metrics derived from the attack graph of the communication infrastructure. In fact, depending on the assessment of the efficiency of deployed defense mechanisms in thwarting threats, the value of β and τ with respect to δ can be adjusted. In particular, by analyzing the attack graph, we can evaluate the probability of compromising critical communication equipment given existing defense measures in the system.

10.6 Case Study

In this section, we validate our model on a case study based on the data set of the Polish electric transmission system at a peak load in the summer of 2004 provided in the MATPOWER computational package [29]. The data set consists of 420 generators and 3504 transmission lines. The analysis of an electric system at a peak load is important, as it allows us to assess the maximum impact on the power grid as a result of a cyberattack.

10.6.1 System Architecture

We made a number of assumptions on the architecture of the communication infrastructure that we use in our case study to assess the impact of attacks on the power grid. In addition, to simplify our analysis, we combined a set of communication equipment in a single communication node depending on their functions, thus reducing the number of nodes to be represented in each electric transmission system control center. Let $\mathscr{Y}$ represent the Polish electric transmission system. We assume that $\mathscr{Y}$ is controlled by 10 TSO (Transmission System Operator) control centers. Each center controls 42 generators and about 350 transmission lines in a specific area of the power grid. We assume that communication equipment in control centers are vulnerable to attacks, and the attacker has enough resources and both players know the architecture of the system. As we study the impact of attacks on the power grid in the worst-case scenario, this assumption holds. A unique TSO ICT control center is introduced to manage all communication equipment in TSO control centers. The communication architecture of the electric transmission system is represented in Figure 10.2.

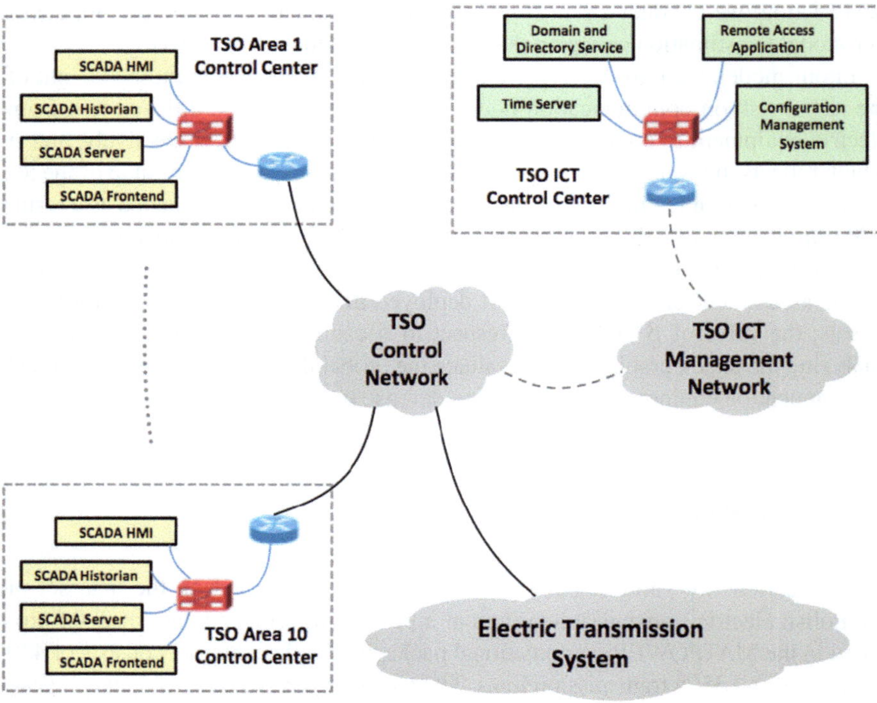

Fig. 10.2: Example of a control network of an electric transmission system

TSO ICT Control Center. In the TSO ICT control center, four types of communication equipment are represented. A Time Server synchronizes the clocks in all communication equipment. A Domain and Directory Service manages access controls on communication equipment. The Remote Access Application is used by ICT administrators to access equipment remotely via secured connections. Finally, the Configuration Management System is responsible of pushing OS and software updates to equipment. Updates can be installed automatically or require specific authorizations on equipment performing critical operations.

TSO Area Control Centers. We represent four types of communication equipment in each TSO area control center: a SCADA HMI, a SCADA server, a SCADA frontend, and a SCADA historian. The SCADA HMI is a human-machine interface that provides a graphics-based visualization of the controlled area of the power system. The SCADA server is responsible of processing data collected from sensors in the power grid and sending appropriate control commands back to electrical nodes. The SCADA frontend is an interface between the SCADA server and electrical nodes control equipment. It formats data in order to be sent through communication channels and to be interpreted when received by control equipment and vice versa. Finally, the SCADA historian is a database that records power state events.

Impact Matrix. We use the algorithm presented in Section 10.5.1 to assess the impact of stopping generators or tripping transmission lines on the electric transmission system and compute matrix B. We rely on experts' knowledge to evaluate matrices F and D. In the communication infrastructure, we consider that each equipment in a TSO control center is also the backup of an equipment in another TSO control center. Therefore, if a communication equipment i fails, another communication equipment j takes charge of processing the load of equipment i.

In this case study, we assume that the values of the initial risk on communication equipment have been computed, and for each communication equipment, the cost to defend is always greater than the cost to attack. We fix $\beta = 0.4$, $\tau = 0$, $\delta = 0.6$, and $\psi = 0.5$. Therefore, the future cascading risk has more weight than the initial risk with respect to the value of the total risk on communication equipment.

10.6.2 Results

Figure 10.3 shows the value of risk on communication equipment in each TSO area control center after the impact of attacks propagates in the interdependent communication and electric infrastructures. We can notice that the highest risk values in TSO control centers are on SCADA servers. In particular, risk values on SCADA servers in TSO 1 and TSO 2 control centers are significantly higher than risk values on SCADA servers in the other TSO control centers.

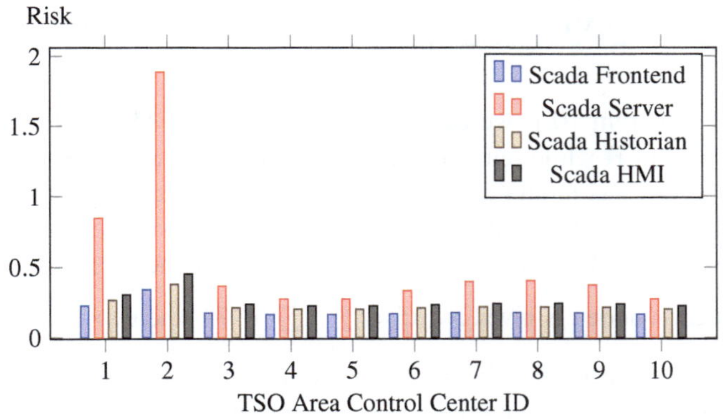

Fig. 10.3: Risk on communication equipment in TSO area control centers

In order to understand the values of risk on communication equipment, we introduce the *impact betweenness centrality* Λ_t^c of communication node t. Λ_t^c represents the weight of node t to propagate the impact of attacks originating from the communication infrastructure on other communication equipment. Before giving the expression of Λ_t^c, we introduce $\lambda_{ij} = \sum_{r=0}^{+\infty} \delta^r (H^r)_{ij}$ as the impact metric of a communication node i on a communication node j. λ_{ij} represents the possible impact of an attack on the communication node i to affect another communication node j taking into account the interdependent electric and communication infrastructures. λ_{ij} is a measure of the cumulated impact on communication node j of an attack originating from node i. This measure takes into account all possible cascading impact paths that could exist between nodes i and j. At each step r, the weight of the payoff of the future impact is multiplied by δ, which represents in a sense the uncertainty for the attacker of getting the payoff of the rth future step. Similarly to

Table 10.1: Nash and Stackelberg equilibriums for Γ^o and Γ^s

		r_i^{c*}	One-Shot game		Stackelberg game	
			p^*	q^*	p^S	q^S
TSO ICT	**Time Server**	2.547	0.287	0.972	0.146	0.986
	Domain Server	2.885	0.183	0.972	0.093	0.986
	Remote App.	2.089	0.202	0.966	0.103	0.9823
	Config. Manag.	3.073	0.21	0.985	0.106	0.992
TSO 1	**SCADA Fontend**	0.226	0.275	0.537	0.15	0.591
	SCADA Server	0.844	0.295	0.688	0.156	0.744
	SCADA Historian	0.266	0.315	0.515	0.177	0.584
	SCADA HMI	0.305	0.329	0.51	0.187	0.586
TSO 2	**SCADA Fontend**	0.339	0.302	0.648	0.162	0.697
	SCADA Server	1.888	0.213	0.895	0.108	0.909
	SCADA Historian	0.379	0.344	0.618	0.189	0.684
	SCADA HMI	0.451	0.358	0.631	0.197	0.7

the proof of Theorem 10.1, we can prove that $\lambda_{ij} = (\mathbf{I} - \delta\mathbf{H})^{-1}_{ij}$. Let $\mathbf{H}(\mathbf{l})$ be the matrix identical to $\mathbf{H}$ while removing elements relative to the edges of node l. The importance λ_{ilj} of a communication node l in diffusing the impact of an attack on communication node i to reach communication node j can be computed as follows $\lambda_{ilj} = (\mathbf{I} - \delta\mathbf{H})^{-1}_{ij} - (\mathbf{I} - \delta\mathbf{H}(\mathbf{l}))^{-1}_{ij}$. Therefore, the impact betweenness centrality of a communication node t is given by $\Lambda_t^c = \sum_{r \neq t} \sum_{s \neq \{t,r\}} \frac{\lambda_{rts}}{\lambda_{rs}}$ where $\{r,s,t\} \in T^c$. Λ_t^c can be thought of as a measure of the importance of node t in diffusing the impact of an attack on any node r to reach any node s with respect to all possible ways that the attack on r can reach s, where $\{r,s\} \in T^c$, $r \neq s$, and $\{r,s\} \neq t$.

In our analysis, the values of risk on a communication node i are highly correlated to $\sum_j h_{ij} \Lambda_j^c$ (correlation coefficient of 99.76% between $\mathbf{r^{c*}}$ and $H\Lambda^c$). In fact, the risk on communication node i depends on the identities of the nodes it will eventually impact following an attack. The more critical these nodes are in propagating the impact of attacks in the interdependent electric and communication infrastructures, the higher the risk value is on node i.

Table 10.1 presents the results of the one-shot and Stackelberg games between the attacker and the defender for the TSO ICT and TSO area 1 and area 2 control centers.

One-Shot game Γ^o. From Figure 10.3 and Table 10.1, we notice that the Time, Configuration, and Domain Servers have the highest risk values. These equipment are often connected to the internet which significantly increases their attack surface. In addition, given their functions, compromising these equipment could lead to important disruptions in the communication infrastructure. As a result, at equilibrium, the defender allocates a large amount of defense resources to protect these equipment. However, this does not prevent the attacker from allocating attack resources on these equipment considering their potential impact on the power grid in the case of a successful attack.

The utilities of the attacker and the defender in the one-shot game are $u_a = 0.941$ and $u_d = -6.151$, respectively. In addition to the risk on communication equipment, the cost to attack and defend and the existence of a backup play an important role in the strategy of both players. In our case study, we noticed that in the case where the values of risk on equipment in two different TSO control centers are similar, the attacker/defender allocate more resources to attack/defend backup equipment. Therefore, by attacking backup equipment, the attacker improves the efficiency of his attacks and increases the probability of succeeding in his attempts to disrupt the power system. On the other hand, the defender responds by allocating more defense resources to protect backup equipment.

Stackelberg game Γ^s. The utilities of the attacker and the defender in the Stackelberg game are $u_a^S = 0.307$ and $u_d^S = -5.746$, respectively. Compared to the one-shot game, the defender allocates more defense resources on each communication equipment, which forces the attacker to reduce his attack resources on these equipment. In fact, an additional security investment by the defender by 2.908 reduced the attacker's allocated resources by 6.082. As a result, from the point of view of the defender, the benefits of operating at the Stackelberg equilibrium outweigh the additional cost of increasing security investments on communication equipment.

Impact of redundancies. Figure 10.4 shows the variation of total attack and defense resources in Γ^o and Γ^s with respect to the weight of the existence of redundancies in players' utility functions ψ. We notice that ψ has a negative effect on the total amount of resources allocated by the attacker. This is consistent with the fact that increasing the weight of redundancies in player's utilities leaves the attacker with fewer choices to achieve a better payoff since the defender will increase the protection of backup equipment. In addition, we notice that when the value of ψ increases, the difference between the one-shot and Stackelberg games total defense resources allocation decreases. However, we do not notice any significant change in the difference of the total attack resource allocations between the two games. When ψ approaches 1, the total amount of defense resources in the one-shot game approaches those allocated in the Stackelberg game. In this case, the defender is better off playing a Stackelberg game, thus reducing the total amount of attack resources allocated on communication equipment.

Figure 10.5 shows the variation of the attacker and the defender strategies on two communication equipment in TSO area 2 control center with respect to variation of elements of the redundancy matrix **W**. We analyze the behavior of the attacker and the defender when varying elements w_{ij}, the fraction of the load of node i, node j will be responsible of processing when node i is compromised. We notice that the behavior of the attacker and the defender depends on the type of the communication equipment. For example, the behavior of both players does not change significantly with respect to **W** for critical equipment such as the SCADA server. However, this behavior is different for the other equipment in TSO area 2 control center. Finally, increasing w_{ij} leads both the attacker and the defender to decrease their attack and defense resources on communication equipment.

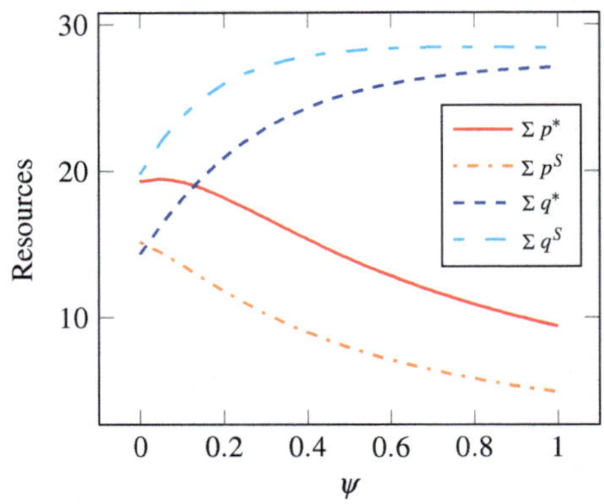

Fig. 10.4: Variation of attack and defense resources with respect to ψ in Γ^o and Γ^s

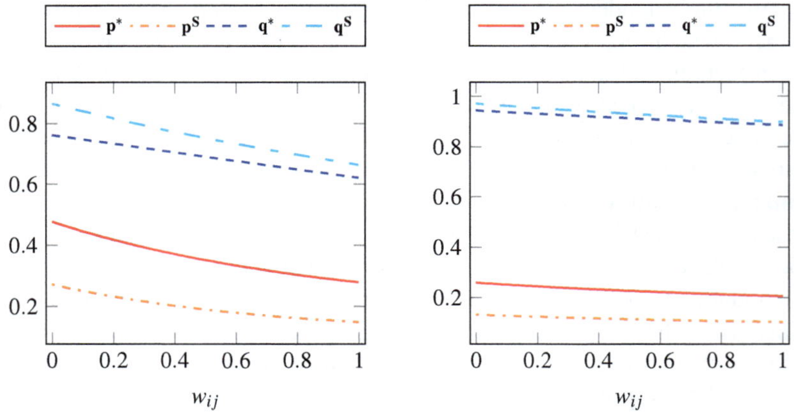

(a) SCADA frontend (b) SCADA server

Fig. 10.5: Variation of attack and defense resources on TSO 2 with respect to redundancy matrix **W** in Γ^o and Γ^s

Sensitivity analysis. We conducted a sensitivity analysis of the diffused risk $\mathbf{r}^{c*}$, the NE in Γ^o, and the Stackelberg equilibrium in Γ^s with respect to the values of the initial risk $\mathbf{r^c(0)}$ and the elements of matrices **S** and **F**. We averaged the results of 10000 iterations. At each iteration, we assume that a random number of elements of $\mathbf{r^c(0)}$ deviate from their correct values by $\pm 10\%$ (sign of the deviation is chosen randomly). We repeat the experiment taking into account errors in a random number of elements in matrices **S** and **F**.

Sensitivity to $\mathbf{r^c(0)}$. The maximum error on the values of $\mathbf{r^{c*}}$ was around 4%. The attacker strategy seems more sensitive than the defender strategy with respect to errors in $\mathbf{r^c(0)}$ at equilibrium. In Γ^o, the maximum error on the attacker strategy was about 4.1%, whereas the error on the defender strategy was about 2.1%. However, in Γ^s, we noticed that the maximum error on the attacker strategy has increased compared to Γ^o and was about 5.1%. In the case of the defender, the maximum error has decreased and was about 1.2%.

Sensitivity to matrices S and F. The maximum error on the values of $\mathbf{r^{c*}}$ was around 3.4%. We do not note a significant change in the maximum errors on the attacker and defender strategies in the case of the one-shot game Γ^o compared to the Stackelberg game Γ^s. The maximum error on the attacker and defender strategies was about 2.1% and 1.3%, respectively.

Game with asymmetric information. We analyze two variants of the game Γ_{IA} with asymmetric information. In the first variant Γ_{NB}, we assume that the attacker does not know or does not have access to the redundancy matrix W. Let Γ^o_{NB} and Γ^s_{NB} refer to the corresponding one-shot and Stackelberg games. The utility of the attacker can be written as $u_a^{NB}(\mathbf{p},\mathbf{q}) = \mathbf{pR_D^{c*}}(\mathbf{e^T} - \mathbf{q^T}) - \mathbf{pR_D^c(0)C^a p^T} + \psi \mathbf{pL}(\mathbf{e^T} - \mathbf{q^T})$. This utility represents what the attacker assumes he will get and not necessarily what he will eventually get by attacking the system if backups exist. Let $R^\Gamma(u_a)$ refer to the utility the attacker will eventually get after attacking the system in game Γ.

In the second variant Γ_{EB} of the game, the attacker's assessment of the matrix W is imprecise. The source of the errors on W can originate from the attacker's incomplete information about the architecture of the communication system or by erroneous information communicated by the defender. For example, before the game starts, the defender provides false information about the architecture of the communication system. Let Γ^o_{EB} and Γ^s_{EB} refer to the corresponding one-shot and Stackelberg games. In this case study, we assume that the attacker overestimates the backup ratios by 10%.

In both variants of the game Γ_{IA} in this case study, we can prove the existence and uniqueness of Nash and Stackelberg equilibriums. Let the L1-norm $\|\mathbf{x}\|_1^\Gamma$ of a strategy $\mathbf{x}$ of a player refer to the amount of resources deployed by the player in game Γ. Tables 10.2 and 10.3 present the total deployed resources and the utilities of players in the one-shot and Stackelberg games, respectively.

Table 10.2: Resources and utilities at the NE

	$\|\mathbf{q}\|_1^\Gamma$	$\|\mathbf{p}\|_1^\Gamma$	u_d	u_a	$R^\Gamma(u_a)$
Γ^o	25.281	14.025	-6.1506	0.94146	0.94146
Γ^o_{NB}	31.468	17.918	-6.4319	1.5463	-1.2637
Γ^o_{EB}	24.702	13.661	-6.0991	0.89283	1.0912

Table 10.3: Resources and utilities at the Stackelberg equilibrium

	$\|\mathbf{q}\|_1^\Gamma$	$\|\mathbf{p}\|_1^\Gamma$	u_d	u_a	$R^\Gamma(u_a)$
Γ^s	28.189	7.9435	-5.7462	0.30673	0.30673
Γ^s_{NB}	31.687	17.403	-6.3864	1.4743	-1.3503
Γ^s_{EB}	27.877	6.8006	-5.6054	0.22681	0.34018

In the case of the game Γ^o_{NB}, at the NE, an increase in the attacker' resource allocation by 3.893 units translates in an increase in the defender's resources allocation by 6.187 units with respect to Γ^o. However, while the attacker thinks he is getting a higher payoff with respect to Γ^o, he is actually getting a negative payoff. In this case, he is better off not attacking the system. The defender's utility at the NE also decreased in Γ^o_{NB} with respect to Γ^o. Therefore, an assumption about the nonexistence of backup equipment by the attacker leads to an increase in the amount of attack resources deployed, thus making both players worse off than the case where they both know the architecture of the system. As a result, in case of multiple interactions between the two players, if the defender signals the existence of backups before the start of the game, he needs to weigh the additional costs incurred by misleading the attacker with the potential benefits of this strategy in the future.

In the case of the game Γ^o_{EB}, the attacker overestimated the backup ratios in the communication system. As a result, at the NE, we notice that he tends to decrease his total resources on equipment with respect to Γ^o. This decrease leads the defender to decrease his total allocation of defense resources to maintain the NE. While the attacker thinks that his utility decreased with respect to Γ^o as a result, he is actually getting an improvement in his utility which is 3 times the improvement observed in the defender's utility.

We observe a different behavior for the attacker when analyzing the Stackelberg equilibriums in Table 10.3. In Γ^s_{NB}, the attacker significantly increased his total resources deployed to attack the system with respect to Γ^s. In addition, the payoff he will eventually get is negative. Therefore, he is better off not attacking at all. As a result, the defender can take advantage of the leading role and harden security on communication equipment before any attack attempt takes place. When such attack occurs, and in the absence of any knowledge about the existence of backup equipment, the attacker spends significant resources without being able to achieve a positive payoff.

A interesting scenario can occur when the attacker gets information about the existence of backups from different sources. For example, one of the sources provides him with a certain configuration of backup equipment in the system. The attacker cannot be sure in general whether he received the correct configuration. It may happen that, before the start of the game, the defender sends a public signal describing a configuration of backup equipment in the system. In this case, the attacker can either believe his first source or the public signal sent by the defender. Let us consider the case where the first source provided a configuration where backups do not exist in the system, while the defender communicated the correct configuration via the public signal. Let Υ be the belief of the attacker in the defender's public signal. Figure 10.6 shows the variation of the total allocated resources and utilities of both players at the Nash and Stackelberg equilibriums in the one-shot game and the Stackelberg game (where the defender is the leader and the attacker is the follower), respectively. We notice that the amount of allocated defense resources is significantly reduced when the attacker believes the defender's signal. Therefore, in this case, security by obscurity is not an optimal strategy for the defender who is better off being transparent about the configuration of backups in the system. This allows him to optimize the allocation of his resources without overspending to defend certain equipment due to the uncertainty related to the attacker's knowledge about the system. While this case study was restricted to one-time interactions between players, the attacker's trust in the defender's public signal can be more important in cases where players can have feedbacks about the impact of their actions after each interaction taking place between them.

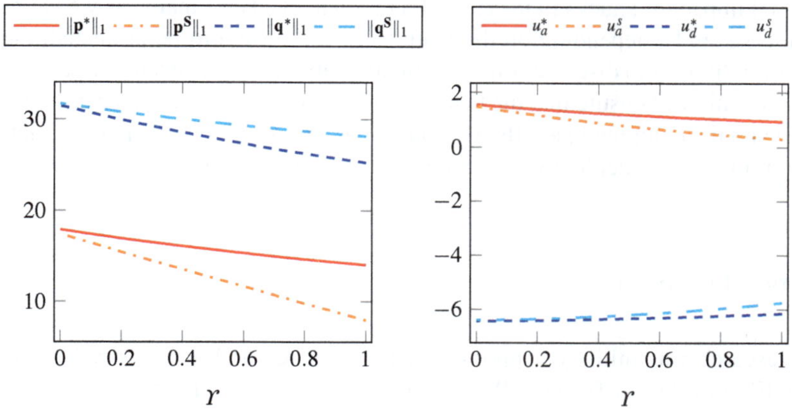

Fig. 10.6: Variation of resources and payoffs with respect to the belief Υ

10.7 Conclusion

In this chapter, we presented a quantitative model, based on game-theoretic analysis, to assess the risk associated with the interdependency between the cyber and physical components in the power grid. We proposed a methodology to evaluate the values of parameters used in our model to assess the impact of equipment failures in the power system and attacks in the communication infrastructure. The structure of player's utility functions, taking into account the existence of backups in the communication system, allows us to characterize analytically players' strategies at the NE. Therefore, we are able to evaluate potential changes in the behavior of players to estimation errors on the values of a set of model parameters. We validated our model via a case study based on the Polish electric transmission system.

The model presented in this chapter is an initial step to analyze the cyber-physical interdependencies in the power grid, and future work must take into account more fine-grained analysis of the behavior of the power grid. In addition, in this chapter, we studied different types of games and solution concepts. However, it is still challenging to provide strict guidelines on the type of game to use to analyze the interactions between the attacker and the defender and the choice of the best solution concept to compute. Many factors can affect these choices, which include the type of the adversary, his knowledge about the system, and the timing of his attacks. However, in practical scenarios, when hardening the system's defenses, analyzing the interactions between the players as a Stackelberg game seems more reasonable than a one-shot game. In particular, the defender is in general better off including the reaction of the attacker to his defense strategy in his analysis, thus bounding the potential impact of attacks in the case of a rational attacker. Finally, as we have seen in the case study, depending on the features of the system to be protected, the defender might need to disclose some information about the architecture of the system publicly. This will ensure that the behavior of a rational attacker can be correctly assessed and the computed payoffs will match the real payoffs of the players' actions, thus optimizing the deployment of valuable defense resources.

Acknowledgments

This research was initially supported by the joint research laboratory SEIDO between EDF R&D and Télécom ParisTech and later by the Cyber CNI Chair of Institut Mines-Télécom held by Télécom Bretagne and supported by Airbus Defence and Space, Amossys, BNP Paribas, EDF, Orange, La Poste, Nokia, Société Générale, and the Regional Council of Brittany acknowledged by the Center of Excellence in Cybersecurity.

References

1. R. Lee, M. Assante, and T. Conway, "Analysis of the cyberattack on the ukrainian power grid," E-ISAC & SANS ICS, Report, 2016.
2. R. Khan, P. Maynard, K. McLaughlin, D. Laverty, and S. Sezer, "Threat analysis of blackenergy malware for synchrophasor based real-time control and monitoring in smart grid," in *Proceedings of the 4th International Symposium for ICS & SCADA Cyber Security Research*, 2016, pp. 1–11.
3. V. Rosato, L. Issacharoff, F. Tiriticco, S. Meloni, S. De Porcellinis, and R. Setola, "Modelling interdependent infrastructures using interacting dynamical models," *International Journal of Critical Infrastructures*, vol. 4, pp. 63–79, 2008.

4. Wenyuan Li, *Risk Assessment Of Power Systems: Models, Methods, and Applications*. Wiley-IEEE Press, 2005.
5. A. Koonce, G. Apostolakis, and B. Cook, "Bulk power risk analysis: Ranking infrastructure elements according to their risk significance," *International Journal of Electrical Power & Energy Systems*, vol. 30, no. 3, pp. 169–183, 2008.
6. Agence Nationale de la sécurité des systèmes d'information, "EBIOS Risk Management Method," 2010, URL: https://www.ssi.gouv.fr/uploads/2011/10/EBIOS-1-GuideMethodologique-2010-01-25.pdf [retrieved: 13/09/2017].
7. ETSI TS 102 165-1 V4.2.3, "Telecommunications and internet converged services and protocols for advanced networking (tispan); methods and protocols; part 1: Method and proforma for threat, risk, vulnerability analysis," 2011.
8. S. Chiaradonna, F. Di Giandomenico, and P. Lollini, *Evaluation of Critical Infrastructures: Challenges and Viable Approaches*. Springer Berlin Heidelberg, 2008, pp. 52–77.
9. J. Laprie, K. Kanoun, and M. Kaniche, "Modeling interdependencies between the electricity and information infrastructures," in *SAFECOMP*, 2007, pp. 54–67.
10. S. Buldyrev, R. Parshani, G. Paul, H. E. Stanley, and S. Havlin, "Catastrophic cascade of failures in interdependent networks," *Nature*, vol. 464, pp. 1025–1028, 2010.
11. M. Parandehgheibi and E. Modiano, "Robustness of bidirectional interdependent networks: Analysis and design," *CoRR*, vol. abs/1605.01262, 2016.
12. E. Casalicchio, E. Galli, and S. Tucci, "Federated agent-based modeling and simulation approach to study interdependencies in it critical infrastructures," in *IEEE 11th International Symposium on Distributed Simulation and Real-Time Applications*, 2007, pp. 182–189.
13. T. Chen, J. Sanchez-Aarnoutse, and J. Buford, "Petri net modeling of cyber-physical attacks on smart grid," *IEEE Transactions on Smart Grid*, vol. 2, no. 4, pp. 741–749, 2011.
14. S. Chiaradonna, F. Di Giandomenico, and N. Nostro, "Modeling and analysis of the impact of failures in electric power systems organized in interconnected regions," in *Proceedings of the IEEE/IFIP International Conference on Dependable Systems and Networks (DSN)*, 2011.
15. M. Beccuti, S. Chiaradonna, F. Di Giandomenico, S. Donatelli, G. Dondossola, and G. Franceschinis, "Quantification of dependencies between electrical and information infrastructures," *International Journal of Critical Infrastructure Protection*, vol. 5, no. 1, pp. 14–27, 2012.
16. H. Lin, S. Veda, S. Shukla, L. Mili, and J. Thorp, "GECO: Global Event-Driven Co-Simulation framework for interconnected power system and communication network," *IEEE Transactions on Smart Grid*, vol. 3, no. 3, pp. 1444–1456, 2012.
17. R. Setola, "How to measure the degree of interdependencies among critical infrastructures," *International Journal of System of Systems Engineering*, vol. 2, no. 1, pp. 38–59, 2010.
18. S. Ruzzante, E. Castorini, E. Marchei, and V. Fioriti, "A metric for measuring the strength of inter-dependencies," in *SAFECOMP*, 2010.

19. E. Casalicchio and E. Galli, *Metrics For Quantifying Interdependencies.* Springer US, 2008, pp. 215–227.
20. Y. W. Law, T. Alpcan, and M. Palaniswami, "Security games for voltage control in smart grid," in *50th Annual Allerton Conference on Communication, Control, and Computing (Allerton)*, 2012, pp. 212–219.
21. S. Amin, G. Schwartz, and A. Hussain, "In quest of benchmarking security risks to cyber-physical systems," *IEEE Network*, vol. 27, no. 1, pp. 19–24, 2013.
22. Z. Ismail, J. Leneutre, D. Bateman, and L. Chen, "A game-theoretical model for security risk management of interdependent ict and electrical infrastructures," in *IEEE 16th International Symposium on High Assurance Systems Engineering (HASE)*, 2015, pp. 101–109.
23. ——, "A methodology to apply a game theoretic model of security risks interdependencies between ict and electric infrastructures," in *Proceedings of the 7th International Conference on Decision and Game Theory for Security (GameSec)*, 2016.
24. T. Alpcan and N. Bambos, "Modeling dependencies in security risk management," in *Proceedings of the 4th International Conference on Risks and Security of Internet and Systems (Crisis)*, 2009.
25. M. J. Osborne and A. Rubinstein, *A course in game theory.* MIT Press, 1994.
26. J. Rosen, "Existence and uniqueness of equilibrium points for concave n-person games," *Econometrica*, vol. 33, no. 3, pp. 520–534, 1965.
27. J. Zhu, *Optimization of Power System Operation.* Wiley-IEEE Press, 2009.
28. R. Pfitzner, K. Turitsyn, and M. Chertkov, "Statistical classification of cascading failures in power grids," in *2011 IEEE Power and Energy Society General Meeting*, 2011, pp. 1–8.
29. R. Zimmerman, C. Murillo-Sánchez, and R. Thomas, "Matpower: Steady-state operations, planning, and analysis tools for power systems research and education," *IEEE Transactions on Power Systems*, vol. 26, no. 1, pp. 12–19, 2011.
30. K. Ingols, R. Lippmann, and K. Piwowarski, "Practical attack graph generation for network defense," in *22nd Annual Computer Security Applications Conference (ACSAC)*, 2006, pp. 121–130.
31. Z. Ismail, J. Leneutre, and A. Fourati, "An attack execution model for industrial control systems security assessment," in *Proceedings of the First Conference on Cybersecurity of Industrial Control Systems (CyberICS)*, 2015.

Part III
Case Studies

Chapter 11
Security and Interdependency in a Public Cloud: A Game-Theoretic Approach

Charles A. Kamhoua, Luke Kwiat, Kevin A. Kwiat, Joon S. Park, Ming Zhao, and Manuel Rodriguez

11.1 Introduction

As cloud computing thrives, many organizations – both large and small – are taking advantage of the multiple benefits of joining a public cloud. Public cloud computing is cost effective: a cloud user can reduce spending on technology infrastructure and have easy access to their information without an up-front or long-term commitment of resources. Despite such benefits, concern over cyber security deters many large organizations with sensitive information to use a public cloud such

C. A. Kamhoua (✉)
Army Research Laboratory, Network Security Branch, Adelphi, MD, USA
e-mail: charles.a.kamhoua.civ@mail.mil

L. Kwiat
Vanderbilt University, Owen Graduate School of Management, Nashville, TN, USA
e-mail: kwiatluke@gmail.com

K. A. Kwiat
Haloed Sun TEK, LLC, Sarasota, FL, USA
e-mail: kwiatk@sunyit.edu

J. S. Park
Syracuse University, School of Information Studies (*i*School), Syracuse, NY, USA
e-mail: jspark@syr.edu

M. Zhao
Arizone State University, School of Computing and Information Sciences, Tempe, AZ, USA
e-mail: mingzhao@asu.edu

M. Rodriguez
Air Force Research Laboratory, Cyber Assurance Branch, Rome, NY, USA
e-mail: manuel.rodriguez-moreno.1.ctr@us.af.mil

S. Rass, S. Schauer (eds.), *Game Theory for Security and Risk Management*, Static & Dynamic Game Theory: Foundations & Applications,
https://doi.org/10.1007/978-3-319-75268-6_11

as the Department of Defense. This is because different public cloud users share a common platform such as the hypervisor. An attacker can compromise a virtual machine (VM) to launch an attack on the hypervisor which, if compromised, can instantly yield the compromising of all the VMs running on top of that hypervisor. In this chapter we evaluate the cloud user-attacker dynamic using game theory, which models competition among rational agents. This work will show that there are multiple Nash equilibria of the public cloud game. The Nash equilibrium profile that results will be shown to depend on several factors, including the probability that the hypervisor is compromised given a successful attack on a user and the total expense required to invest in security.

With software being one of the fastest-growing industries in the United States [1], the drive for growth in software technologies can easily outpace the information security needed to safely and reliably function. This can have far-reaching implications, from infrastructure protection to the home computer system. When information security is overlooked, the inattentiveness can be attributed to both the producer and consumer. Information security suffering due to under investment from both sides of the market can be counterintuitive since prevailing economic forces normally would indicate that both sides of the market have an inventive to invest. This can be explained by several factors, including perverse incentives, asymmetrical information, and interdependency (we will elaborate on the meanings of these terms from economics in the appropriate parts of our chapter). However, it will be seen that interdependency underpins all these causes and influences information and network security in general. The preliminary version of this chapter was published in [2].

Due to the fast-paced nature and rapid expansion of new technology in the software realm, first mover advantages can be enormous. This can create a software creator's philosophy where "they'll ship it on Tuesday and get it right by version 3" [3]. This philosophy clearly can cause suppliers to neglect many security aspects in their product. Frequently, the producers do not even know the true security of their own product [3]. This is especially true with emerging fields of technology, such as cloud computing [4]. Consumers, in turn, cannot truly know what they are purchasing, since many vulnerabilities created by the supplier can go undetected. Suppliers are not solely to blame, however. The idea of "get it out now and fix it later" is an unintended consequence that is created by the demanding aspects of consumers of the Internet economy. Although arguments can be made for who is more at fault for improperly secure products – the producers who produce it too quickly or the consumers who demand it too rapidly—there are real effects that result. Growth in up-and-coming sectors in the technology field, such as cloud computing, is severely hampered. It is indeed a sizable problem, as fears of leakage of sensitive or confidential data pose a "significant barrier to the adoption of cloud services" [5]. This fear prevents major industry entities from switching to cloud platform services, stifling its growth.

The concerns of individual organizations (hereby alternatively referred to as *users*) joining the cloud hold significant merit. What is notable of the cloud infrastructure compared to a regular network is that public clouds exhibit a unique type of interdependency between otherwise unassociated users. In a cloud network, an attacker has the ability to propagate his attack through shared resources on the cloud (i.e., attacking a hypervisor and then attacking all virtual machine on the shared hypervisor). This eliminates a very important aspect of regular network security in which an attacker would have to go through a multi-hop process in order to launch an indirect attack on multiple, unlinked users. Thus, a public cloud at its current stage leaves its users more susceptible to a "bad neighbor" effect where an unsecured cloud user might allow another to be indirectly attacked. Although our focus is on public clouds, the same research problems may also exist in private clouds, and our solution is also applicable. We focus on public clouds only because the problems are more pronounced in public clouds.

In an infrastructure of virtual machines (henceforth referred to as VMs) utilizing a common resource (usually a hypervisor), an attacker may launch an indirect attack on a User j by first compromising the VMs of User i and then attacking User j as a prime target. This creates a risk connection between the users of a cloud where a "large" user (one who has a high potential loss associated with successful compromise of his VM) may not be willing to use cloud services due to the risk imposed by a "small" user (low potential loss from a successful compromise). This threat is worsened when a small player will not invest in security measures since it could (correctly) rationalize that an attacker will attack the larger user anyway, so investing would be pointless. Definitely, a single user of a public cloud cannot protect itself if other users are not doing the same. This means that a user will be protected only if other users are also defending themselves. Given these scenarios, it is apparent that the security of one user is affected by another's actions on the cloud.

It is clear that the cloud platform unintentionally creates interactions between users due to the nature of shared resources. When there are two or more rational entities that face interdependent choices, we can use game theory to model their behaviors, as it is indeed "the study of mathematical models of conflict and cooperation between intelligent rational decision-makers" [6]. A survey of game theory applied to cyber security and privacy is available in [28].

There are several main contributions this chapter makes. Primarily, it aims to model the behaviors that govern the actions of different users in the cloud using game theoretical concepts. Along with modeling the choices of cloud users, it will be shown that the "small" user imposes a negative externality or a cost imposed unwittingly upon an otherwise uninvolved party—most notably the "larger" user. This will, in turn, spur the large user to invest more often than the small player since the large player is usually the prime target. The outcome is as follows: there is no Nash equilibrium in which all the players will fully invest in security. Lastly, we will prove that the probability that the hypervisor of a cloud is compromised given a successful attack on a VM will determine if we have a pure or mixed strategy Nash equilibrium.

11.2 Background

The background is divided into five sections. Section 11.2.1 looks at the interdependent nature of the critical infrastructure network in the United States and its connection to cyberspace. Section 11.2.2 discusses game theory and its connection to interdependency. Section 11.2.3 applies game theory to network security. Section 11.2.4 discusses interdependency in the context of cloud computing. Section 11.2.5 evaluates interdependency and cross-side channel attacks between VMs.

11.2.1 Critical Infrastructure Defense

Generally, the US government does not interfere in the affairs or operations of the Internet unless it pertains to national security. However, even when national security is at stake, the government is ill-prepared for a response, as Dave Clemente argues [7]. The main problem, he reasoned in his thesis, is that the infrastructures critical to the operations of the United States are mislabeled and overstated due to miscommunication at the local and national governmental levels. This causes many infrastructures that are not critical to be labeled critical (this is nicely stated in his aphorism: "When everything is critical, nothing is"). The problem is compounded by tying all these infrastructures together through a dense network of interconnectedness, making one network of infrastructure dependent on another. The backbone of this connected network is the Internet, which is becoming increasingly relied upon and only furthering the deep ties these sub-networks already have. Unfortunately, Clemente argues the Internet securitization process is not keeping pace with the current expansion of the Internet due to industry pressures to sacrifice long-term security needs for short- and mid term speed and efficiency needs. And until the critical infrastructure is taken out of private interests (which would cause much more harm than good), this problem will persist. And although no major solution was mentioned by Clemente—other than something must be done—a much more comprehensive solution was laid out by Kenneth Cukier [8].

The work done by Cukier and his colleagues addressed many of the issues raised by Clemente. The main issue was that there is an underinvestment of security within the critical information infrastructure of the United States. This problem was discussed at length and was cast as a symptom rather than a cause. The underinvestment was due to many underlying factors such as informational asymmetry (companies do not know the extent of their problem), conflict of interest (government interests vs. private), and interdependent security (this will be further analyzed in the context of game theory later). All these problems aggregate into a general deficiency of investment in cyber security. Although this seems like an economically counterintuitive outcome, it is a rational one given the constraints of various aforementioned forces. The solution offered by Cukier was essentially an insurance market for security risk, facilitated by a favorable environment created by the government.

Cukier goes on to state that many private companies do not know the extent of their risk because of a reluctance to share their vulnerabilities with others. Insurance companies will not insure the risk since they do not have access to the information to quantify it. This creates a cat-and-mouse game where neither the insurance market nor the companies in need of security will make the first move. This, according to Cukier, is where the government can step in and facilitate transactions of sensitive information as well as preserve anonymity. The creation of a beneficial environment through incentives and information exchange can create a market for risk, which by definition will reduce risk of infrastructure sectors (insurance premiums will discourage risky business and encourage security investing). Forrest Hare [9] reflects these sentiments as he argues that there is an underinvestment due to a conflict of interests. He contends that a public-private partnership should be formed to facilitate the transfer of information and to increase the incentives of private firms to invest in security. This will lead to noticeable positive externalities on the public (since they will be more secure) and everyone will be better off as a result.

Under the new Executive Order 13636—Improving Critical Infrastructure Cyber security [10]—the White House would like to provide incentive to private companies to voluntarily adopt a Cyber security Framework. The framework is a partnership with the owners and operators of critical infrastructure to improve cyber security information sharing and collaboratively develop and implement risk-based standards. The framework's goal is to share cyber security information such that the US government and the private sector may better protect and defend themselves against cyber threats and reduce cyber risk to critical infrastructure. In fact, a security breach on a government contractor (i.e., a private company) can compromise multiple government programs, which shows the interdependency between government and private sector security. The White House's Cyber security Framework is currently under development at the National Institute of Standards and Technology. The Cyber security Framework includes a set of standards and technological approaches to be adopted by each organization to minimize cyber risks.

11.2.2 Game Theory and Interdependency

Through globalization, firms are becoming increasingly dependent upon each other. Thus, it would be logical to assume that their choices would reflect the actions of their competitors and benefactors sharing a given set of information. Game theory accurately describes these conditions, as it is poised "the study of mathematical models of conflict and cooperation between intelligent rational decision-makers" [6]. This makes the case for interdependency among firms, as the actions of one affect the actions of many. The examples of interdependency observed here will include airline security, bankruptcy, and vaccinations.

Two of the papers from the National Bureau for Economic Research (NBER) carefully looked at multiple scenarios involving game theory and the subsequent interdependency of the players [11, 12]. The first paper looked at discrete and mostly static games [11]. It was shown that with airline security, one's own investment in baggage security was heavily dependent on the choices of the other airline in a simple two player game. In this analysis, one's own security is either compromised due to another airline's lack of security or complemented by the reinforcement of the rival's airline security. It was shown that the two Nash equilibria that exist in a simple two firm game occur when both airlines invest in security and when both airlines do not invest in security. As stated in the previous subsection, clearly only the outcome of both investing is desirable. However, economic costs and initial conditions can influence the firms to not invest. With government regulation or other methods to tip incentives toward investing, an economically-optimal situation can be achieved with certain modifications. Similar results were found with more than two firms, since the investing of one firm can cause multiple firms to change their decision to invest. This creates a 'cascade effect' in which one firm causes another to invest and so on. Within the same analysis [11], similar results were derived from firm bankruptcy. If each division of a large firm, such as bank, were to undergo risk reduction individually, the collective risk of a firm would be reduced. However, if one branch takes exceptional risks it can cause bankruptcy for the whole firm if the other divisions succumb to the cascading effect.

The second of the NBER papers demonstrated the cascading effect [12]. It was further shown that the incentive to invest is heavily dependent on the cost of investing compared to the benefit derived from both investing in security. The cost could be manipulated both by lowering the cost of investing as well as raising the cost of not investing.

Unlike an organization having exclusive use of computational resources, the resource sharing that occurs in the cloud enables unforeseen exploitation of weaknesses by attackers. Similarly, the commonality of computational resources without an equal commonality of user-instantiated security creates an avenue for launching an attack on other tenants i.e., a negative externality due to interdependency and resource sharing.

11.2.3 Applying Game Theory to Cyber Security

Sun et al. presented a model of investment security [13] where they simulated a security game between two companies deciding whether to invest or not invest in information security. The payoffs for each company were based on several inputs, such as cost of investing and the possible loss from a security compromise. The most important parameter discussed was a penalty parameter p for not investing. It was shown that the 3 Nash equilibrium strategies produced from the game were two pure and one mixed strategy (a pure strategy is one that is played with certainty whereas a mixed strategy is two or more different strategies played probabilistically).The parameter p was shown to have the ability to effect the mixed strategy outcome. This could skew the results from what could be considered 'normal' and demonstrated that an outside force such as the government could manipulate the penalty parameter in order to achieve a more favorable outcome.

Though the previous example would have used a central manager or network administrator to decide if investing was the correct choice, Kamhoua et al. applied game theory to autonomous nodes in networks [14]. They used similar constraints to yield similar results: 3 Nash equilibria, two pure and one mixed with the mixed strategy being an unstable equilibrium. The main distinction was instead of a penalty parameter such as in Sun et al. [13], there was a trust parameter in which the resulting strategies heavily depended on. The trust parameter depended on how much the deciding node believed that other node will participate in a security mechanism. The main conclusion to drawn from the simulations was that it is impossible to move from the low trust state to a high trust state through an evolutionary process. In the replicator dynamic model [15], the final state depended entirely on the initial condition. As will be seen, this result can have broad reaching implications, from network security to cloud computing.

In Tamer Basar's and Tansu Alpcan's book [16], they explain the devastating costs of failure to properly protect a network. They show how an attacker can infiltrate a network at one node, and spread to other nodes (or infrastructures) due to contagion. This can cause a spillover effect where one node affects another and so forth. Since one unprotected node causes risks at all the other nodes, the decision of one affects the outcomes of many. Game theory was used to minimize the effects of the interdependency inherent in a node network.

Basar and Tansu only applied network security in a traditional computer setting. The rise and expansion of cloud computing has led to many questions about its security. To raise concerns further, cloud computing's annual growth is rapidly outpacing regular computing methods by a significant margin [17]. In the next subsection we will outline details on its expansion, tradeoffs in switching to cloud platforms, and further research in cloud security.

11.2.4 Interdependency Analysis in Cloud Computing

According to the National Institute of Standards and Technology some of the 'essential characteristics' that come with the term 'cloud computing' include resource pooling, elasticity, resource optimization, network access and on-demand self-service [18]. Though these characteristics can overcome many constraints posed by traditional computing, the emerging field of cloud computing currently carries some profound tradeoffs. Pearson and Benameur outlined several important drawbacks in cloud technology such as privacy, security, and trust concerns [19]. However, these three problems are not unrelated to each other. Security within the cloud is based on trust associated with the provider, and privacy is based on the relevant security issues. Trust is in turn built on the relationship of security and privacy that the cloud operator provides.

Not all types of cloud technology has these aforementioned problems due to their diverse nature. Zissis et al. [20] differentiate between public and private cloud structures by stating that private cloud technology is for inter-organizational operations (which requires no third party provider) while public and community cloud computing utilizes a third party for a variety of service platforms. Such service platforms that are provided can include Infrastructure as a Service (IaaS), Software as a Service (SaaS), and Platform as a Service (PaaS).

An IaaS cloud provides a user access to virtualized hardware, presented by a hypervisor (e.g., VMware, Xen, KVM) and encapsulated in a VM, where the user is able to deploy and run arbitrary software including operating systems and applications on the underlying shared hardware. A PaaS cloud provides a user a language-specific platform (e.g., JVM, .Net) to deploy and run arbitrary applications developed using the given language on the underlying shared platform. A SaaS cloud provides a user access to a particular application (e.g., web-based email, document editor) where the user can use the functionality provided by the underlying shared application. Although these different levels of cloud services can be built separately, it is increasingly common to build a high-level cloud service using resources provided by a lower-level one (e.g., build a SaaS on resources from PaaS and a PaaS on resources from IaaS), so that the former can benefit from the elasticity and economics provided by the latter. Therefore, although our chapter focuses on VM-based hosting of mission-critical applications in an IaaS setting, its outcomes can also generate an impact to other models of cloud computing (further information can be seen in [20]). Although private clouds do share some of the benefits and drawbacks of public clouds, the issues of privacy, security, and trust arise from mainly public cloud platforms, as many of the users' computing capabilities are outsourced to a third party owner who leases the technology in a variety of ways. Therefore we focus on the public cloud; so in this chapter private cloud entities will not be discussed further. In fact, private clouds allow users from the same organization to run their internal applications on shared resources. Therefore, in a game-theoretic sense, there should be less conflict of interest among private cloud users since they belong to the same organization.

As stated before, these problems that involve the public cloud are not unrelated as they all underpin a unique relationship between the third party provider and the cloud user. This can give rise to interdependency between the user and the operator of the cloud. If we apply the behavior of network nodes as described in [14] to a cloud's VMs, then we can see that cloud computing yields very interdependent structure. Cloud computing gives way to two types of interdependent relationships: cloud host-to-client and cloud client-to-client.

Client-to-client interdependency is much less studied than to the above-mentioned cloud host-to-client relationship. Although, it can still carry the negative externalities provided by the first relationship since a security compromise is the same no matter where it has originated. A simple example of this involves the airline security problem found in [11] and [12] where a bomb infused baggage is sent through an unsecured airline, which in turn reaches a heavily secure airline because no inter-airline security screening is used (and it usually is not). Thus, an under-secure airline can impose negative externalities onto a seemingly secure airline. Similarities can be drawn to two clients operating in the same cloud environment. An attacker can compromise an unsecured client and make its way to the more secure and larger client through the hypervisor. However, unlike the airline interdependent security problem where a bomb can only destroy one airline, a virus in a public cloud or computer network can compromise many VMs including the VM in which the attack originated.

We have already seen that interdependency lays the foundation for game theory in previous subsections. Indeed, this scenario between two clients also involves two or more intelligent rational entities with conflicting incentives. Analogous to the previous example, a small firm with high overhead will see little point to invest in security since its cost to invest is most likely diminished by the fact it has lower possible loss from being compromised. However, a larger firm has a much higher potential loss from being compromised, especially if they carry sensitive information (This has been seen in [4] when large firms refuse to use cloud computing because of its risks). Thus, a rational attacker might attack a smaller firm, compromise the hypervisor, and then target the larger firm if the potential gain from a successful indirect attack outweighed the potential gain from a direct attack.

11.2.5 Interdependency and Cross-Side Channel Attacks Between VMs

The support for security isolations from existing cloud systems is limited. The different VMs sharing the same resources may belong to competing organizations as well as unknown attackers. From the perspective of a cloud user, there is no guarantee whether the underlying hypervisor or the co-resident VMs are trustworthy. The shared resource makes privacy and perfect isolation implausible. There is a risk that a covert side channel be used to extract another user's secret information [21, 22]. Cross-side channel attacks between VMs are possible in a public cloud when the VMs share the same hypervisor, CPU, memory, and storage and network devices. Some of the resources can be partitioned (e.g., CPU cycles, memory capacity, and I/O bandwidth). VMs also share resources that cannot be well partitioned such as last-level cache (LLC), memory bandwidth, and IO buffers. The shared resources can be exploited by attackers to launch cross-side channel attack. Although a multi-tenant public cloud-computing environment provides various advantages, it also introduces new challenges and concerns, especially on security issues. For instance, the security problems on a shared cloud resource (e.g., cloud storage devices, network services, software components, etc.), which are originally rooted from one of the tenants via internal vulnerabilities or external cyber-attacks, may eventually affect the service quality and security of all the tenants in the same cloud-computing environment. Unfortunately, we cannot simply assume that there would be a single authority who could comprehensively maintain all the possible issues, not only technical but also non-technical, across the tenants.

Moreover, existing cloud service providers do not provide sufficient security guarantees to their tenants. In fact, the service-level agreements (SLAs) of representative cloud providers (e.g., Amazon EC2/S3, Windows Azure, Google Compute Engine) specify only the provisions related to service up time, and there is no mentioning of security in these SLAs at all.

Many researchers have investigated the cache based side channel. Ristenpart et al. [21] show that a malicious user can analyze the cache to detect a co-resident VM's keystroke activities and map the internal cloud infrastructure and then launch a side-channel attack on a co-resident VM. Bates et al. [22] demonstrate the ability to initiate a covert channel of 4 bits per second, and confirm co-residency with a target VM instance in less than 10 seconds. Li et al. [23] proposed several techniques to protect VMs from untrusted management VM, which includes modifying the hypervisor to restrict access of the privileged domain to the memory mappings of the VM, encrypting all of the memory pages and vCPU registers before they are accessed by the privileged domain, and providing a hash value of the kernel image to be compared with the one residing on the VM. HyperSentry [24] enables stealthy in-context measurement of hypervisor integrity using a hardware channel to trigger the measurement and, using the system management mode, to protect the measurement agent's base code and critical data.

Given the danger of a cross-side channel attacks, some users may require physically isolated resources from the cloud provider. Zhan et al. [25] introduce Home-Alone - a defensive tool that helps users determine if their VMs have an exclusive use of a physical machine. HomeAlone can detect the activity of an intruder's co-resident VM by analyzing a portion of the L2 memory cache set aside by his VMs. The same technique can be used to detect adversarial VMs which try to extract information through the side channel due to their usual cache activity pattern. This solution, however, requires that all the user VMs to be co-resident which is often difficult to achieve and makes them more vulnerable to hardware and hypervisor failures.

Approaches that dedicate a physical machine to a specific user also greatly limit some of the benefit of a public cloud such as the on-demand dynamic resource allocation. This means that a user can no longer purchase exactly the capacity they require when they require it. Therefore this chapter only considers schemes in which the VMs from different users share the same resources. We can see that a cross-side channel attack between VMs is closely related to the problem of interdependency when many users share the same resource that they depend on. This chapter provides a comprehensive analysis of direct vs. indirect attack, collateral damage, and negative eternality in a public cloud. Finally, there are other prior work that have investigated cloud security based on game theory [29, 30, 31, 32], cloud security certifications [33, 34, 35], and Blockchain technology [36, 37, 38, 39].

11.3 System Model

Figure 11.1 illustrates our system model: A public cloud with n users that we denote User 1, User 2 ... User n. Each user runs several applications illustrated by Application 1 ... Application k in Figure 11.1. Technically, the users may run a different number of applications without any impact on this model. The different applications require an operating system to function and that operating system, in turn, manages a VM in the cloud. In practice, a single user may use several operating systems or numerous VMs.

However, we consider the architecture in Figure 11.1 to simplify the exposition. As it is a common practice in a public cloud, we consider that the different VMs from the different users share the same hypervisor and hardware as in Figure 11.1. The hypervisor can be of different types such as the Kernel-based Virtual Machine (KVM), Xen, and VMware. The common factor is that the VMs share one or more platforms that expose each VM user to collateral damage. We consider the possibility of a random hardware failure to be a rare event and neglect that possibility in our analysis. It is well known that the users' security heavily depends on the cloud provider. We are analyzing security interdependency among the users; therefore our model considers that the attacker compromises the hypervisor in two steps. The first step is to compromise a user's VM, or masquerade as legitimate user to obtain a VM in the public cloud. The second step is to use the compromised VM to attack the hypervisor.

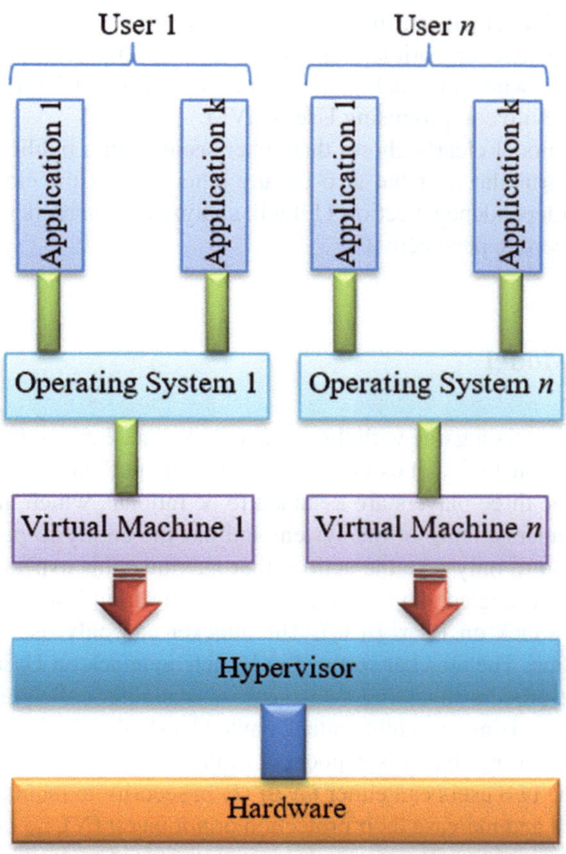

Fig. 11.1: System model illustration

This means that the public cloud provider takes all the necessary measures to prevent an attacker from directly compromising the hypervisor without using a compromised VM. This is to separate cloud client-to-client interdependency and cloud host-to-client interdependency. However, any model that analyzes cloud host-to-client interdependency can be superimposed to our model. We distinguish two types of attack depending on the extent of the consequence: a restricted attack and an unrestricted attack. A restricted attack on User i only compromises the applications, operating system and VM that belong to User i; the hypervisor is not affected after a restricted attack. An unrestricted attack has consequences that can cross a VM to reach the hypervisor, i.e. the hypervisor is compromised [21]. We consider that all the users suffer the consequences (damage) if the hypervisor is compromised. This is because an attacker that compromises the hypervisor can then freely compromise all the VMs on that public cloud.

We can see that an unrestricted attack causes collateral damage. A direct attack on User i can go through that user's VMs to compromise the hypervisor and ultimately affect the VM of another User j. We also refer to this as an indirect attack on j. Thus, each user in a public cloud can suffer from two types of attack. A direct attack on a User i is when an attacker's primary target is User i. Furthermore, an indirect attack on User i happens when an attack that is launched on another User j compromises the hypervisor before compromising User i's VM.

This system model clearly shows that cyber security in a public cloud depends not only on a particular user but also on any other user of the cloud. This is the problem of interdependency. Section 11.4 will analyze the interdependency problem from a game-theoretic perspective.

11.4 Game Model

This section considers a game with three players: An attacker and two users (User i and User j). Section 11.7 will extend this model to more than two users and multiple attackers. The three players are assumed to be rational, which means that each player has an understanding of the system and has the ability to perform the necessary calculation to only take the actions that maximize his expected payoff. The attacker has two strategies: launch an unrestricted attack on User i (A_i) and launch an unrestricted attack on User j (A_j). The attacker can only use one of the two strategies at a time. The attacker strategy to launch an attack on User i may consist of a multi stage process involving steps such as scanning, collecting information, credential compromising, executing attack payload, establishing backdoor, cleaning footholds, and avoiding firewalls. Choosing to invest is a binary decision for each user in which the two users can either *invest* (I) in security to maintain a minimum security standard and increase their protection or *not invest* (N), i.e., there is no partial investment in security. The strategy *invest* may consist of multiple actions such as system monitoring, reconfiguration, patching, updating software, and buying a new antivirus. Investment in security requires a total expense e. A strategy profile is a 3-tuple that indicates the action of each player. For instance, the strategy profile (N, I, A_j) indicates that User i does not invest (N), User j invests (I), and the attacker launches an attack on User $j(A_j)$.

The probability of a successful attack on a user, given that he has invested in security, is q_I, and the probability of a successful attack on a user, given that he has not invested, in security is q_N. We assume that

$$0 \leq q_I < q_N \leq 1. \tag{11.1}$$

We have $q_I < q_N$ because any rational user will only invest in security measures that diminish his chance to get compromised. It can also be seen that this model assumes that User i and j hold the same risk of being compromised if they have the same strategy (in other words, q_I is the same for both users and q_N is the same for both users). This was done in order to make the calculations that follow in the game analysis tractable, whereas the effect on the numerical results and interpretation will be negligible.

The probability that the hypervisor is compromised given a successful attack on a user is denoted π. Our model considers that at least some successful attack on a VM will reach the hypervisor or that $\pi > 0$. In fact $\pi = 0$ means that a successful attack on a VM would never reach the hypervisor which would be a strong assumption. We also consider that not all the successful attacks on a VM can compromise the hypervisor ($\pi < 1$). Thus we have

$$0 < \pi < 1 \tag{11.2}$$

We consider that there is a high profile User j and a low profile User i. In case of a security breach, the high profile user incurs more loss than the low profile user. The high profile User j's expected loss from a security breach is L_j, and the expected loss from User i is L_i. Then we consider that

$$0 < L_i < L_j \tag{11.3}$$

We will show that this imbalance affects the investment decision of each player and may yield positive and negative externalities. A positive (negative) externality is an action of a player that transfers a positive (negative) effect onto a third party. In fact, when (high profile) users in a public cloud invest in security to protect their applications, operating systems, and VMs, they also protect the hypervisor which in turn protects other users from an indirect attack or cross-side channel attack. This yields a positive externality to other users in a public cloud. On the contrary, if a (low profile) user chooses not to invest in security, then an easy attack path to the hypervisor is created and thus exposes all other users operating on the hypervisor to a cross-side channel attack. This yields a negative externality to other users in a public cloud.

The accuracy of our model depends on the correct estimation of the probabilities $q_I, q_N, \text{and} \pi$ and the loss L_i and L_j. We propose two different approaches to estimation. The first approach is the QuERIES approach [26]. The QuERIES approach estimates the probabilities and costs of successful attacks by first building an attack graph represented as a partially observable Markov decision process (POMDP). Then QuERIES uses a controlled red-team experiment and information market mechanisms to estimate the POMDP parameters. The outcome of an information market is a collective estimate of a quantity. The red-teams have real financial incentives for making correct predictions of the POMDP probabilities. Finally, the POMDP's optimum policy is calculated to derive the different probabilities and cost.

The second approach to estimate the relevant probabilities and cost associated with our model is based on historical data. In fact, in October 2011, the US Securities and Exchange Commission (SEC) issued a new guidance [27] requiring that companies disclose cyber incidents including a description of the costs, other consequences, and the relevant insurance coverage. Those data can now be aggregated to estimate the relevant probabilities and costs associated with our model.

Each user has a reward R from using the cloud computing services. The reward R can be calculated as a function of a user's multiple benefits of using the cloud such as reduced spending on technology infrastructure, easy access to their information without up-front or long-term commitment of resources, and dynamically growing and shrinking the resources provisioned to an application on demand. While we do not have to assume a constant R across different users as different users will have different benefits, we will take R as the same for user i and j for simplicity in future calculations.

We consider that a user can detect and identify a co-resident VM from another user in the cloud via side-channel analysis as in HomeAlone [25]. Further, a skillful attacker will first scan a public cloud to learn about the different users – gaining knowledge of their weaknesses and vulnerabilities before launching an attack. Also, each of the following can be made known or can be estimated about a player [26, 27]: the expected loss from a security breach and the related probability, the total expense required to invest in security, and the reward from using the cloud. Therefore, our model assumes that a player's identity, strategy, and payoff are common knowledge among the players.

Table 11.1: Game model in normal form

		Attack j	
		User j	
		I	N
User i	I	$\{R-e-q_I\pi L_i;$ $R-e-q_I L_j;$ $q_I\pi L_i+q_I L_j\}$	$\{R-e-q_N\pi L_i;$ $R-q_N L_j;$ $q_N\pi L_i+q_N L_j\}$
	N	$\{R-q_I\pi L_i;$ $R-e-q_I L_j;$ $q_I\pi L_i+q_I L_j\}$	$\{R-q_N\pi L_i;$ $R-q_N L_j;$ $q_N\pi L_i+q_N L_j\}$

		Attack i	
		User j	
		I	N
User i	I	$\{R-e-q_I L_i;$ $R-e-q_I\pi L_j;$ $q_I L_i+q_I\pi L_j\}$	$\{R-e-q_I L_i;$ $R-q_I\pi L_j;$ $q_I L_i+q_I\pi L_j\}$
	N	$\{R-q_N L_i;$ $R-e-q_N\pi L_j;$ $q_N L_i+q_N\pi L_j\}$	$\{R-q_N L_i;$ $R-q_N\pi L_j;$ $q_N L_i+q_N\pi L_j\}$

Table 11.1 shows the game model in normal form. We can see that Table 11.1 is a combination of two tables (top and bottom). The top table shows the game model when the attacker targets User i. On this part of the table, User j can only

be subject to collateral damage after a successful attack on User i and then the hypervisor (which can happen with probability $q_I\pi$ if User i invests or probability $q_N\pi$ if User i does not invest). Similarly, the bottom table shows the game model when the attacker targets User j. Likewise, User i can only be subject to collateral damage on this side of the table. The first row in each table shows when User i chooses to invest while the second row shows when User i chooses not to invest. The decision of User j to invest and not invest is represented in the first and second columns, respectively. The payoffs in each block are represented in three lines. The first line is User $i's$ payoff, the second line is User j's payoff, and the third line is the attacker's payoff.

The payoffs are calculated as follows: if the strategy profile is (I,I,A_i), then both users get the reward R and incur expense e because both of them have invested in security. Since the attacker targets User i, he will be compromised with probability q_I (because User i has invested) and will incur a loss L_i if compromised. This will result in an expected loss of $q_I L_i$. User j is not targeted but can incur a loss L_j if the attack on User i is successful (which happens with probability q_I) and the hypervisor is compromised (which happens with probability π). This is an expected loss of $q_I\pi L_j$ and represents collateral damage or loss from an indirect attack. The attacker's payoff is the sum of the expected loss of all the users: $q_I L_i + q_I\pi L_j$. The attacker's partial payoff $q_I L_i$ comes from a direct attack on User i while the second part of his payoff $q_I\pi L_j$ is the result of an indirect attack on User j through the hypervisor.

Taking another example, strategy profile (N,I,A_i), we can see User i has not invested (N), User j has invested (I), and the attacker targets User i (A_i) (top table, second row, first column). User i does not incur any expense e because the user has not invested in security. However, his likelihood of being compromised increases to q_N. Moreover, although User j has invested in security, his potential losses from collateral damage increases to $q_N\pi L_j$. The attacker's payoff resulting is $q_N L_i + q_N\pi L_j$. The inequality holds because of (11.1). The players' payoffs in the other six strategy profiles are calculated in a similar way. It is worthy to note that the difference $q_I\pi L_j - q_N\pi L_j = (q_I - q_N)\,\pi L_j$ is a negative externality that User i imposes on User j by not investing while User i is the prime target of the attacker.

11.5 Game Analysis

The main goal of this analysis is to derive the different Nash equilibria of the game in Table 11.1 and understand the consequences for both users. At a Nash equilibrium profile, no player's payoff can be increased by a unilateral deviation. As a result, at Nash equilibrium, each player is playing his best response to every other players' best strategies. Therefore, the Nash equilibrium can help predict the behavior of any rational player (i.e., that want to maximize their payoff in a game).

We observe that a user that is the prime target must be hurt before the other user suffers any collateral damage. Recall that the prime target's VM must be compromised before the hypervisor is compromised. Thus, we consider in the remainder of this analysis that each user prefers to invest instead of not investing when he believes that he is the attacker's prime target. For User i this translates to

$$R-e-q_I L_i \geq R-q_N L_i \Rightarrow$$
$$e \leq (q_N - q_I) L_i \quad (11.4)$$

Similarly, for User j this translates to

$$R-e-q_I L_j \geq R-q_N L_j \Rightarrow$$
$$e \leq (q_N - q_I) L_j \quad (11.5)$$

Also observe that investing in security is the best option for either User i or User j if and only if the user believes that he will be the attacker's prime target. The attacker targets the player that gives the higher total payoff (consisting of a direct and indirect payoff).

Theorem 1. If $\pi \leq \pi_0 = \frac{q_I L_j - q_N L_i}{q_N L_j - q_I L_i}$, then the game in Table 11.1 admits a pure strategy Nash equilibrium profile (N, I, A_j).

If $\pi > \pi_0$, there are three possible mixed strategy Nash equilibria depending on the required expense for security e.

Proof. We start by analyzing the eight different pure strategy profiles for possible Nash equilibrium.

Case 1: Both users invest,

$$U_a(I, I, A_j) - U_a(I, I, A_i) =$$
$$(q_I \pi L_i + q_I L_j) - (q_I L_i + q_I \pi L_j) = q_I (1-\pi)(L_j - L_i)$$

Then by considering (11.2) and (11.3), we have

$$U_a(I, I, A_j) - U_a(I, I, A_i) = q_I (1-\pi)(L_j - L_i) > 0. \quad (11.6)$$

Therefore, the attacker gets a higher payoff by targeting User j when both users invest. Thus the strategy profile (I, I, A_i) can never be a Nash equilibrium because the attacker can increase his payoff by changing his strategy to A_j. This gets us to the strategy profile (I, I, A_j) which cannot also be a Nash equilibrium because User i (not being the attacker's prime target) can increase his payoff by changing his strategy from I to N. This yields the strategy profile (N, I, A_j) that we study in Case 4.

Case 2: Both users do not invest,

$$U_a(N, N, A_j) - U_a(N, N, A_i) =$$
$$(q_N \pi L_i + q_N L_j) - (q_N L_i + q_N \pi L_j) = q_N (1-\pi)(L_j - L_i)$$

Then by considering (11.2) and (11.3), we have

$$U_a(N,N,A_j) - U_a(N,N,A_i) \\ = q_N(1-\pi)(L_j - L_i) > 0. \quad (11.7)$$

Thus, the attacker gets a higher payoff by targeting User j. The strategy profile (N,N,A_i) cannot be Nash equilibrium because the attacker can increase his payoff by changing his strategy to A_j. This gets us to the strategy profile (N,N,A_j) which cannot also be a Nash equilibrium because User j, being the attacker's prime target, can increase his payoff by changing his strategy from N to I (because of (11.5)). This yields again the strategy profile (N,I,A_j) that we study in Case 4.

Case 3: User i invests while User j does not.

We can see from Table 11.1 that

$$U_a(I,N,A_i) = U_a(I,I,A_i) = q_I L_i + q_I \pi L_j. \quad (11.8)$$

Moreover,

$$U_a(I,N,A_j) - U_a(I,I,A_j) = (q_N \pi L_i + q_N L_j) - (q_I \pi L_i + q_I L_j) \Rightarrow \\ U_a(I,N,A_j) - U_a(I,I,A_j) = q_N(L_j + \pi L_i) - q_I(L_j + \pi L_i) \\ = (q_N - q_I)(L_j + \pi L_i) > 0. \quad (11.9)$$

Note that the last inequality in (11.9) holds because of (11.1).

Combining (11.8) and (11.9), we have

$$U_a(I,N,A_i) = U_a(I,I,A_i)$$

and

$$U_a(I,N,A_j) > U_a(I,I,A_j) \Rightarrow \\ U_a(I,N,A_j) - U_a(I,N,A_i) > U_a(I,I,A_j) - U_a(I,I,A_i).$$

Taking (11.6) into consideration, we have

$$U_a(I,N,A_j) - U_a(I,N,A_i) > 0. \quad (11.10)$$

From (11.10), the attacker gets a higher payoff by targeting User j. Thus the strategy profile (I,N,A_i) cannot be Nash equilibrium because the attacker can increase his payoff by changing his strategy to A_j. This gets us to the strategy profile (I,N,A_j) which also cannot be a Nash equilibrium because User j (being the attacker's prime target) can increase his payoff by changing his strategy from N to I (because of (11.5)). We come back to the strategy profile (I,I,A_j) that we study in Case 1, which yields Case 4.

Case 4: User j invests while User i does not.

$$U_a(N,I,A_j) - U_a(N,I,A_i) = (q_I \pi L_i + q_I L_j) - (q_N L_i + q_N \pi L_j) \\ = (q_I L_i - q_N L_j)\pi + (q_I L_j - q_N L_i) = f(\pi).$$

We can see that $f(\pi)$ is a linear function with slope $(q_I L_i - q_N L_j)$ and initial value $(q_I L_j - q_N L_i)$. From (11.1) and (11.3), we have the slope $q_I L_i - q_N L_j < 0$. Thus, $f(\pi)$ is decreasing. Moreover, there is a unique value of π such that

$$f(\pi) = 0 \Rightarrow \pi = \pi_0 = \frac{q_I L_j - q_N L_i}{q_N L_j - q_I L_i}, \tag{11.11}$$

Furthermore, we have $f(\pi) > 0$ for $\pi < \pi_0$ and $f(\pi) < 0$ for $\pi > \pi_0$. Also,

$$f(1) = (q_I L_i - q_N L_j) + (q_I L_j - q_N L_i) \\ = (q_I - q_N)(L_i + L_j) < 0. \tag{11.12}$$

The last inequality holds because of (11.1).

In addition, the initial value is

$$f(0) = q_I L_j - q_N L_i, \tag{11.13}$$

which can be either negative or positive. Observe that because of (11.2) the condition $\pi \leq \pi_0$ can hold if $0 < \pi_0 < 1$, and by the intermediate value theorem and based on (11.12) and (11.13), it is only possible when $f(0) > 0 \Rightarrow q_N L_i < q_I L_j \Rightarrow$

$$L_i < \frac{q_I}{q_N} L_j. \tag{11.14}$$

Then we can distinguish two subcases (4a) and (4b).

Subcase (4a) If $\pi \leq \pi_0$, then we have $U_a(N,I,A_j) - U_a(N,I,A_i) \geq 0$. Thus the attacker prefers to attack User j than to attack User i. User j prefers to invest than not to invest (see (11.5)). User i, not being the attacker's prime target, prefers not to invest. Thus, the strategy profile (N,I,A_j) is the pure strategy Nash equilibrium of the game because no player can increase his payoff by a unilateral deviation.

Subcase (4b) If $\pi_0 < \pi$ (regardless of the sign of $f(0)$), we have $f(\pi) < 0$ and then $U_a(N,I,A_j) - U_a(N,I,A_i) < 0$. The attacker prefers to attack User i than to attack User j. Thus the strategy profile (N,I,A_j) cannot be Nash equilibrium because the attacker can increase his payoff by changing his strategy to A_i. This gets us to the strategy profile (N,I,A_i) which also cannot be a Nash equilibrium because User i, being the attacker's prime target, can increase his payoff by changing his strategy from N to I (see (11.4)). This brings us to Case 1 above, which you recall brings us to Case 4. Therefore, this circular reasoning tells us that there is no pure strategy Nash equilibrium. However, there will be a mixed strategy Nash equilibrium that we analyze next.

Mixed Strategy Nash Equilibrium

To find the mixed strategy Nash equilibrium, we set three variables $\alpha, \beta, \text{and} \lambda$ with

$$0 \leq \alpha, \beta, \lambda \leq 1. \tag{11.15}$$

α represents the probability by which the User i plays I. Since User i has only two strategies, User i plays N with probability $1-\alpha$. Similarly, User j plays I with probability β and plays N with probability $1-\beta$. Likewise the attacker attacks j with probability λ and attacks i with probability $1-\lambda$.

By definition, User i plays a mixed strategy if and only if his payoff $U_i(I)$ when playing I is equal to his payoff $U_i(N)$ when playing N. This translates to

$$\begin{aligned}U_i(I)=U_i(N) \Rightarrow\ & (1-\lambda)\beta(R-e-q_I L_i)+(1-\lambda)(1-\beta)(R-e-q_I L_i)\\ & +\lambda\beta(R-e-q_I\pi L_i)+\lambda(1-\beta)(R-e-q_N\pi L_i)=\\ & (1-\lambda)\beta(R-q_N L_i)+(1-\lambda)(1-\beta)(R-q_N L_i)\\ & +\lambda\beta(R-q_I\pi L_i)+\lambda(1-\beta)(R-q_N\pi L_i)\\ & \Rightarrow \lambda=\lambda_i=\frac{(q_N-q_I)L_i-e}{(q_N-q_I)L_i}.\end{aligned} \tag{11.16}$$

(11.4) shows that $0\le\lambda_i\le 1$. Also,

$$U_i(I)<U_i(N)\Rightarrow 0\le\lambda_i<\lambda\le 1, \tag{11.17}$$

and

$$U_i(I)>U_i(N)\Rightarrow 0\le\lambda<\lambda_i\le 1. \tag{11.18}$$

This means that, if the attacks on User j are more frequent than λ_i(and User i is attacked less often), then User i prefers to play N. User i plays I otherwise.

Similarly, User j plays a mixed strategy if and only if his payoff $U_j(I)$ when playing I is equal to his payoff $U_j(N)$ when playing N. This translates to

$$U_j(I)=U_j(N)\Rightarrow\lambda=\lambda_j=\frac{e}{(q_N-q_I)L_j}. \tag{11.19}$$

(11.5) shows that $0\le\lambda_j\le 1$. Also,

$$U_j(I)<U_j(N)\Rightarrow 0\le\lambda<\lambda_j\le 1, \tag{11.20}$$

and

$$U_j(I)>U_j(N)\Rightarrow 0\le\lambda_j<\lambda\le 1. \tag{11.21}$$

Further, the attacker plays a mixed strategy if and only if his payoff $U_a(A_i)$ when attacking User i is equal to his payoff $U_a(A_j)$ when attacking User j. This translates to

$$\begin{aligned}U_a(A_i)=U_a(A_j)\Rightarrow\ & \beta(L_j+\pi L_i)-\alpha(L_i+\pi L_j)\\ & =\left(\frac{q_N}{q_N-q_I}\right)[(L_j+\pi L_i)-(L_i+\pi L_j)].\end{aligned} \tag{11.22}$$

Given the condition in (11.16), (11.19), and (11.22), we can distinguish three cases that we denote M1, M2, and M3 depending if $\lambda_j=\lambda_i$, $\lambda_j<\lambda_i$, or $\lambda_j>\lambda_i$.

Furthermore, we will see that the total expense required to invest in security e determines which of the mixed strategy is used.

Case M1 If $\lambda_j = \lambda_i \Rightarrow$

$$e = e_0 = \frac{(q_N - q_I) L_i L_j}{L_i + L_j}, \tag{11.23}$$

then any strategy profile $\{\alpha I + (1-\alpha)N; \beta I + (1-\beta)N; \lambda_j A_j + (1-\lambda_j)A_i\}$, with α and β set according to (11.22) is a mixed strategy Nash equilibrium. Recall that (11.15) must hold.

We can see that when $\lambda_i \neq \lambda_j$, the conditions in (11.17), (11.18), and (11.20), (11.21) dictates that only one user plays a mixed strategy at a time while the other plays a pure strategy. Moreover, the attacker chooses the value of λ that corresponds to the user playing the mixed strategy. This consideration is critical to understand the next two cases.

Case M2 If $\lambda_j < \lambda_i \Rightarrow$

$$e < e_0 = \frac{(q_N - q_I) L_i L_j}{L_i + L_j}, \tag{11.24}$$

and $\lambda = \lambda_i$, then according to (11.21), User j plays the pure strategy I. This means that $\beta = 1$. Setting $\beta = 1$ in (11.22) yields

$$\alpha = \alpha_0 = \frac{q_N (L_i + \pi L_j) - q_I (L_j + \pi L_i)}{(q_N - q_I)(L_i + \pi L_j)}. \tag{11.25}$$

We can verify that $0 < \alpha_0 < 1$ when $\pi > \pi_0$ and (11.1), (11.2), and (11.3) hold. Therefore, the strategy profile $\{\alpha_0 I + (1-\alpha_0)N; I; \lambda_i A_j + (1-\lambda_i)A_i\}$ is a mixed strategy Nash equilibrium. Observe that the low profile User i is more likely to invest in this mixed strategy Nash equilibrium compared to the pure strategy Nash equilibrium (N, I, A_j). In this scenario, it is relatively cheap to invest in security as shown in (11.24).

However, if $\lambda_j < \lambda_i$ and $\lambda = \lambda_j$, then according to (11.18), User i plays the pure strategy I. This means that $\alpha = 1$. Setting $\alpha = 1$ in (11.22) yields

$$\beta = \frac{q_N (L_j + \pi L_i) - q_I (L_i + \pi L_j)}{(q_N - q_I)(L_i + \pi L_j)} > 1. \tag{11.26}$$

The last inequality in (11.26) holds when (11.1), (11.2), and (11.3) hold. This is a contradiction with (11.15).

Case M3 If $\lambda_j > \lambda_i \Rightarrow$

$$\frac{(q_N - q_I) L_i L_j}{L_i + L_j} < e < (q_N - q_I) L_i. \tag{11.27}$$

Note that the last inequality must hold because of (11.4). Thus according to (11.17), when $\lambda = \lambda_j$, User i plays the pure strategy N. This means that $\alpha = 0$.

Setting $\alpha = 0$ in (11.22) yields

$$\beta = \beta_0 = \frac{q_N\left[(L_j + \pi L_i) - (L_i + \pi L_j)\right]}{(q_N - q_I)(L_j + \pi L_i)}. \tag{11.28}$$

We can verify that $0 < \beta_0 < 1$ when $\pi > \pi_0$ and (11.1), (11.2), and (11.3) hold. Therefore, the strategy profile $\{N; \beta_0 I + (1-\beta_0)N; \lambda_j A_j + (1-\lambda_j)A_i\}$ is a mixed strategy Nash equilibrium. Observe that the high profile User j is less likely to invest in this mixed strategy Nash equilibrium compared to the pure strategy Nash equilibrium (N, I, A_j). In this scenario, it is relatively more expensive to invest in security as shown in (11.27).

However, if $\lambda_j > \lambda_i$ and $\lambda = \lambda_i$, then according to (11.20), User j plays the pure strategy N. This means that $\beta = 0$. Setting $\beta = 0$ in (11.22) yields

$$\alpha = -\frac{q_N\left[(L_j + \pi L_i) - (L_i + \pi L_j)\right]}{(q_N - q_I)(L_i + \pi L_j)} < 0 \tag{11.29}$$

The last inequality in (11.29) holds when (11.1), (11.2), and (11.3) hold. This is a contradiction with (11.15).

We have shown that the low profile User i imposes two different types of negative externalities on the high profile User j in the cloud. If L_i is low enough in such a way that (11.14) holds and $\pi \leq \pi_0$, then the pure strategy profile (N, I, A_j) shown in subcase (4a) results and the attacker targets the high profile user even though the high profile user (User j) invests in security while the low profile user (User i) does not invest. User j is the attacker's only target. This is the first type of negative externality. When L_i is high enough in such a way that (11.14) does not hold, then $\pi > \pi_0$, and the attacker is forced to play a mixed strategy. The specific mixed strategy is determined by the total expense required to invest in security e. However, User i produces the second type of negative externality by investing less often than User j in all those mixed strategies. In fact, there is no Nash equilibrium in which the low profile user (User i) plays the pure strategy I.

Furthermore, with low value of e (Case M2), it can be shown that User i's probability to invest α_0 (see (11.25)) increases with L_i to the benefit of User j. Recall that in Case M2, User j always invests. However, if the value of e is high (Case M3), it is easy to verify that User j's probability to invest in security β_0 (see (11.28)) decreases with L_i. Recall that in Case M3, User i does not invest (play N). A high value of e causes an under investment problem in cloud security.

In summary, it is important for users to be aware of who they are sharing the cloud with, because of the externalities inadvertently imposed onto them. It can be seen that some users can have a direct effect on the decisions and as a result the level of security of other users.

11.6 Numerical Results

Our game analysis has provided a detailed exposition of our game model and its equilibrium properties. The main variables used in calculating pure and mixed strategies equilibria were $R, q_I, q_N, L_i, L_j, \pi$, and e. For our numerical analysis, we selected eL_j, and π for further discussion. These variables were selected for our numerical analysis to most accurately convey the current attacker-user dynamic as seen in our game analysis. We will analyze how changes in their magnitude affect the payoffs and strategies of User j. As will be seen, a minor shift in the values of these variables will yield major results to interpret.

11.6.1 Changes in User j's Payoff with Probability π

In this first scenario, we will take the value of π to be variable while setting values for all the other parameters. We will take $q_N = 0.5; q_I = 0.1; R = 1.2; L_i = 1; \text{and} L_j = 10$. Those values are chosen to illustrate some of the non-intuitive implications of our game model. Using (11.11), we can see that $\pi_0 = 0.102$. Furthermore, with (11.23) we can see that $e_0 = 0.3636$. Moreover, (11.27) gives us $0.3636 < e < 0.4$. Recall that in case of a mixed strategy Nash equilibrium ($\pi > \pi_0 = 0.102$), the value of e determines which of the mixed strategy Nash equilibrium (Case M1, M2, or M3) is selected by the players. In Figure 11.2, we set $e = 0.3$ ($e < e_0$) so that once the critical value of π is reached, the mixed strategy Nash equilibrium will be as Case M2.

We immediately see that the payoff for User j in pure Nash equilibrium is negative. When the payoff of a rational user is negative, he prefers not to use the cloud. So, for all values of $\pi \leq 0.102$, the User j, which is assumed to be rational in our model, will not use the cloud because the risk of a security breach and negative externalities of using the cloud are greater than the multiple benefits that cloud computing provides. Recall that in the pure strategy Nash equilibrium, User j is at a disadvantage because he is the attacker's only target.

However, at $\pi = 0.102$, there is a strategy change from pure to mixed due to (11.11), and as at this point, the strategies shift. With a shift in Nash equilibrium and players' strategies, there is a concurring change in the function used as it is a new set of equations governing the strategies. This allows for a positive payoff for $0.102 < \pi \leq 0.47837$ and implies that User j will participate in the cloud for the aforementioned values of π. These results are seemingly counterintuitive since the hypervisor has a higher probability of being compromised when User j participates in cloud activities than when he does not. This is explained by the equilibrium shift to a mixed strategy where the attacker is not only attacking User j but also User i. This lowers User j's potential loss and thus shifts his payoff upward.

Examining Figure 11.2, the payoff becomes negative again as π crosses 0.47837, which shows that User j will again not participate in the cloud for all values of

$0.47837 < \pi \leq 1$ since the probability of being compromised from an indirect attack is now too high to justify cloud usage.

By setting $e = 0.38$ and upholding (11.27), Figure 11.3 shows the strategy shift from pure Nash equilibrium to the mixed Nash equilibrium in Case M3. Still, for values of $\pi \leq 0.102$, User j will not participate in the cloud because of his negative payoff. Although once π crosses 0.102, a change in payoff from negative to positive, as in Figure 11.2, makes the cloud a viable option. Interestingly, the payoff does not cross over again to become negative after this original movement of equilibriums. This means that for all values of $0.102 < \pi \leq 1$, User j will participate in the cloud if $0.3636 < e < 0.4$. Another surprising result is that User j's payoff is higher in Figure 11.3 compared to Figure 11.2 although the required expense in security e in Figure 11.3 is higher.

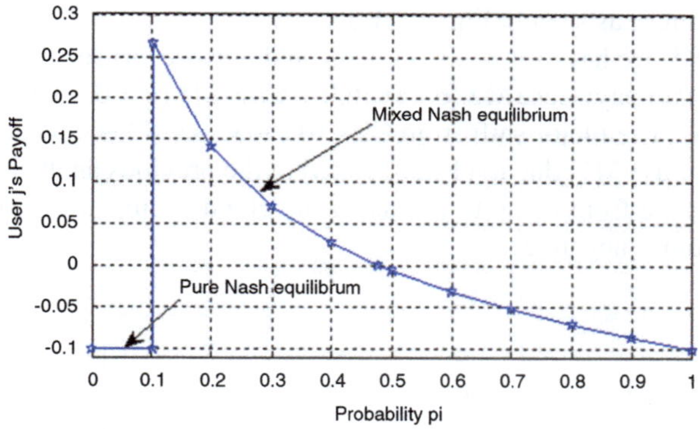

Fig. 11.2: Changes in User j's payoff with probability π with $e < e_0$

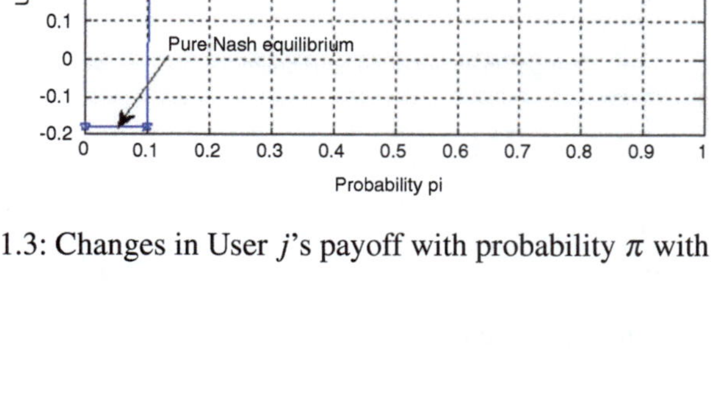

Fig. 11.3: Changes in User j's payoff with probability π with $e > e_0$

11.6.2 Changes of User j's Payoff with the Expense in Security e

We have already examined the case of pure Nash equilibrium and two cases of mixed strategy equilibrium dependent on the varying values of π. We will now make π a constant while varying the levels of e. As stated before, the value of $\pi_0 = 0.102$ is a focal point between mixed and pure strategy equilibrium. In this case of $\pi \leq 0.102$, User j has only one (pure) strategy, whose payoff of $R - e - q_I L_j$ yields the linear function in Figure 11.4.

The "x" intercept where the payoff is 0 (at $e = 0.2$) is yet another turning point where User j will no longer use the cloud. For values $0 \leq e \leq 0.2$, User j will participate in the cloud because of the low overhead of investing in security. However, for $e > 0.2$, the cost is too great to allow for a positive payoff, and User j will not use the cloud. For $.102 < \pi \leq 1$, the players' strategies are switched, and the entire payoff map changes as seen in Figure 11.5.

In Figure 11.5, we have set $\pi = 0.11 > \pi_0$, and thus we can see the three different cases of mixed strategy: Case M2 ($e < 0.3636$), Case M1 ($e = 0.3636$), and Case M3 ($3636 < e < .4$). The major shift from Case M2 to Case M3 occurs at the threshold of $e = 0.3636$ (Case M1) due to (11.23) stated in the previous analysis. For $0 \leq e <$ 0.3636, the change from using to not using the cloud occurs at $e = 0.08606$ when the payoff becomes negative.

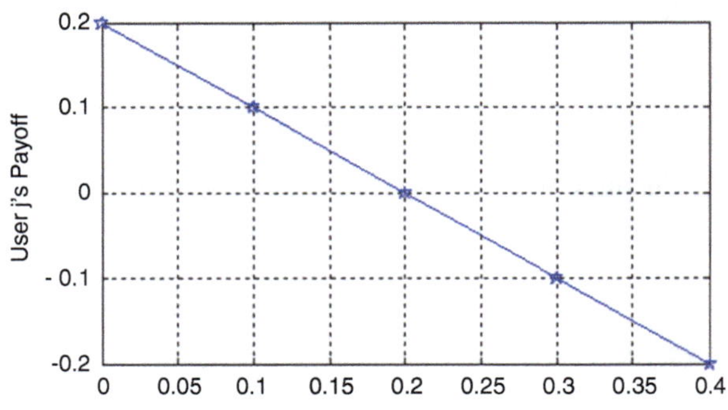

Fig. 11.4: Changes of User j's payoff with the expense in security e with $\pi < \pi_0$

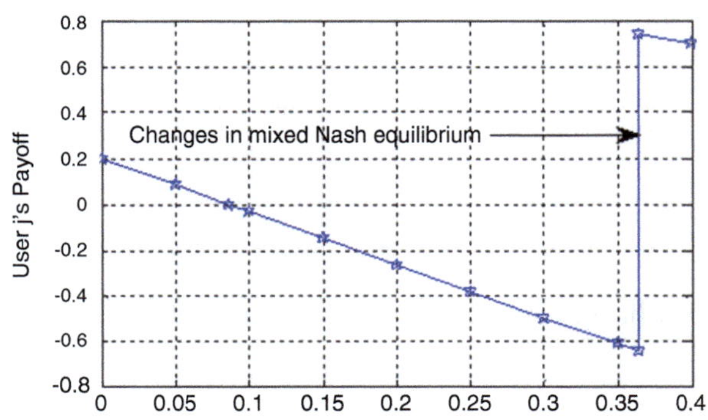

Fig. 11.5: Changes of User j's payoff with the expense in security e with $\pi > \pi_0$

When the expense e increases and $0.3636 < e < 0.4$, the shift in mixed Nash equilibrium from Case M2 to Case M3 causes the payoff to change and become positive. Thus it becomes possible for User j to profitably use cloud services. This is a counter intuitive result from this analysis. One may expect an increase of the expense e to never benefit User j. However, in this game-theoretic setting, User j's payoff depends not only of his own action but also on the action of User i and the attacker. The increase of the expense e changes User i's and the attacker's strategy in such a way that it has an overall positive effect on User j's payoff. In Case M3, User j invests with probability β_0 as opposed to 1 in Case M2. This yields some savings that increase User j's overall payoff. Recall that moving from Case M2 to M3 changes the mixed strategy Nash equilibrium from $\{\alpha_0 I + (1-\alpha_0)N; I; \lambda_i A_j + (1-\lambda_i)A_i\}$ to $\{N; \beta_0 I + (1-\beta_0)N; \lambda_j A_j + (1-\lambda_j)A_i\}$. Note also that for $e \geq 0.4$, Case M3 no longer applies as consistent with (11.4).

11.6.3 Changes in User j's Payoff with His Loss from Security Breach L_j

Now that the variability of π and e—and their resulting equilibrium shifts they cause—have been examined, we will examine Figure 11.6 and the equilibrium changes associated with varying values of L_j. Since L_j is a variable in both the equations that govern the values of π_0 (Equation (11.11)) and e_0 (Equation (11.23)), we must set specific values for π and e in order to avoid a problem of double variables. For the rest of the analysis of L_j, we will set $\pi = 0.1$ and $e = 0.3$. Recall that we have set $L_i = 1$. Therefore, L_j is a direct indication of how much time L_j is bigger than L_i.

Unlike the previous two problems in which a certain change in the discrete value of π with a varying e could cause an equilibrium shift, there is no such change here. Here the values of π and e are constant and L_j is the unique variable. As can be seen in Figure 11.6, any value of $L_j \geq 9.8$ will result in a pure Nash equilibrium due to (11.11). Further, (11.23) shows that when $3 < L_j < 9.8$ the mixed strategy Nash equilibrium profile of Case M2 will hold, Case M1 holds for $L_j = 3$, and if $1 < L_j < 3$, then Case M3 will be used.

These results show that Case M3 is the "best" of all the equilibriums because User j's potential loss L_j is so close to User i's loss L_i. An obvious result is that User j's payoff is maximized in Case M3 when L_j is close to $L_i = 1$. That is because there is no imbalance between L_i and L_j, and thus the negative externalities are minimized. The negative externality in a public cloud security can be mitigated by putting VMs that have similar potential loss from a security breach in the same physical machine. However, a surprising result is that User j's payoff jumps up concurrent with switching from the mixed Nash equilibrium (Case M2) to the pure Nash equilibrium despite the fact that L_j becomes substantially greater than L_i. For instance, User j's payoff when $L_j = 4L_i$ equals User j's payoff when $L_j = 10L_i$. This prediction is not possible without a thorough game-theoretic analysis.

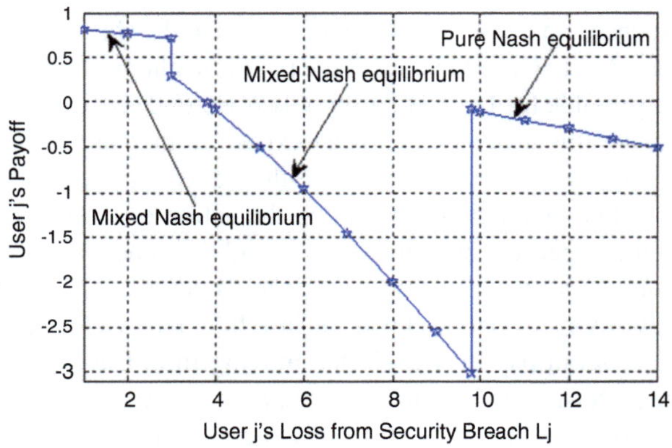

Fig. 11.6: Changes in User j's payoff with his loss from security breach L_j

11.6.4 Changes in User j's Payoff with His Reward from Using the Cloud

For the constant R, changing it will have a trivial effect on any of the given graphs shown. As seen in Figure 11.7, a change in the value of R will cause the graph to translate upward or downward depending on the new value of R selected. For this particular instance, if the reward for using the cloud is increased from 1.2 to 4.4, the entire payoff scheme from $1 \leq L_j \leq 14$ becomes positive since the increased level of reward increases the payoff.

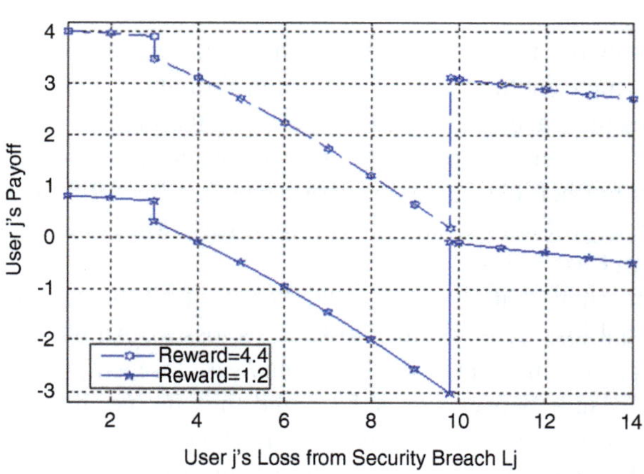

Fig. 11.7: Changes in User j's payoff with his reward from using the cloud

11.6.5 Interpretation of Results

Since π is the probability the hypervisor will be compromised and an indirect attack must be launched through the hypervisor, we use the value π to represent an estimation of potential negative externalities present in the cloud. Figures 11.2 and 11.3 both show that an increasing π causes a clear equilibrium shift that changes not only the user's Nash equilibrium strategy but also his propensity to use the cloud entirely. From a security and cloud provider standpoint, these are not beneficial results, as ideal conditions for a cloud user would command no equilibrium shifts across a wide range of externality conditions. This is because no equilibrium shifts would give a sense of stability and robustness in the user's investment strategy.

When a user has to account for multiple strategies and adjust accordingly, it can raise overhead (lower R), causing them to opt out of cloud services entirely. If this assurance of strategy consistency cannot be given to the user, it only hinders investment opportunities in the cloud that otherwise could have been possible. The numerical results only solidify the idea that negative externalities are imposed onto cloud users.

11.7 Model Extension and Discussion

The model we have presented so far has considered two users and one attacker. However, our model can be extended to more than two users and multiple attackers.

11.7.1 Model Extension to More Than Two Users and a Single Attacker

All the assumptions made in our game model in Section 11.4 remain valid except that we increase the number of users from *2* to n. The n users are denoted as User *1*, User *2*, ..., User $n-1$, User n. Their potential loss from a security breach is $L_1, L_2, \ldots, L_{n-1}, L_n$, respectively. We consider that $L_1 \leq L_2 \leq \ldots \leq L_{n-1} \leq L_n$. The attacker targets one of the n users. A similar analysis as above shows that the game admits a pure strategy Nash equilibrium if L_n is substantially greater than L_{n-1}. In this Nash equilibrium, User n is the attacker's only target. The attacker plays the strategy A_n, User n invests (plays I) while all the other users do not invest (play N). Regarding the threshold value of π below for which we have a pure strategy Nash equilibrium, (11.11) translates to

$$\pi_0^* = \frac{q_I L_n - q_N L_{n-1}}{q_N L_n - q_I L_{n-1}}. \tag{11.30}$$

As before, the game admits a multitude of mixed strategies if $\pi > \pi_0^*$. The expense e will determine the specific mixed strategy the players choose.

11.7.2 Model Extension to More Than Two Users and Multiple Attacker

In a game with multiple independent attackers, each attacker maximizes his own payoff. If $\pi < \pi_0^*$, each attacker plays the strategy A_n, and User n invests (plays I) while all the other users do not invest (play N). However, the game complexity increases if the attackers collude by coordinating their action and sharing the payoff. Nevertheless, an increase in the number of attackers increases the likelihood that a given user can be targeted by one attacker and eventually compromised. As the number of attackers increases, the cloud environment becomes more hostile, and more users will be forced to invest (because of (11.4) and (11.5)).

Another consideration is the users' payoff structure. There are applications in which a user incurs the same loss after being compromised by a single attacker or multiple attackers, e.g., information integrity can be lost when either a few bits or when many bits of a data item become useless.

11.8 Conclusion

The lack of an accurate evaluation of the negative externalities stemming from a high profile organization using the cloud could result in the refusal of such organizations from joining a public cloud in spite of the many advantages that cloud computing offers. The negative externalities of using a public cloud come from the fact that the users are not perfectly isolated from one another. They share common resources such as the hypervisor, the last-level cache (LLC), memory bandwidth, and IO buffers that cause interdependency.

This research has used game theory to provide a quantitative approach to perform a cost-benefit analysis of cloud services while taking into account the action of other cloud users and their different potential losses from a security breach. Our model takes into account the potential collateral damage from an indirect attack and cross-side channel attack. The game has multiple possible Nash equilibria that can be in pure or mixed strategy. Our research finds that an increase in the probability that the hypervisor is compromised, given a successful attack on a user's VM, may force the small cloud participant to protect their VM and thus increases the overall cloud security to yield better outcome to high profile users.

Additionally, this research has also shown that there is an intricate relationship between the total expenses required to invest in security and a high profile user's payoff. A change in security expense changes the game Nash equilibria that the players adopt with some of those equilibria being more desirable to high profile users. Most importantly, it was discovered that present cloud user-attacker dynamic causes equilibrium shifts to occur with only relatively minor changes in variables. At the raw implementation of this game model, the cloud is not conducive to users who seek stable results that are safe and secure across many factors.

According to Ross Anderson, information security is hard because defenders have to defend everywhere and attackers could attack anywhere [3]. This leads to many problems for network defenders, network users, software used in critical infrastructures, small businesses, or even the US government. Moreover, these security problems are exacerbated when using cloud computing. By utilizing game theory, we can more accurately describe the nature of the attacker and his motives. However, sometimes our best friend can be our worst enemy. Other players' behaviors can be seemingly erratic and even counterintuitive, which can be very dangerous when your decisions are based on the decisions of others. With game theory, we can quell some of this contradictory behavior that is characteristic of network security and bring clarity to this complex topic.

11.9 Acknowledgment

This research was performed while Dr. Joon Park and Dr. Manuel Rodriguez held a National Research Council (NRC) Research Associateship Award at the Air Force Research Laboratory (AFRL). This research was supported by the Air Force Office of Scientific Research (AFOSR). "Approved for Public Release; Distribution Unlimited: 88ABW-2013-5145."

References

1. Handbook, Handbook, Occupational Outlook, Bureau of labor statistics, United States Department of Labor, Spring (2008).
2. Charles Kamhoua, Luke Kwiat, Kevin Kwiat, Joon Park, Ming Zhao, Manuel Rodriguez, "Game Theoretic Modeling of Security and Interdependency in a Public Cloud" in the proceedings of IEEE International Conference on Cloud Computing, (IEEE CLOUD 2014) Anchorage, Alaska, June 2014.
3. R. Anderson, "Why Information Security is Hard – an Economic Perspective," Working paper, Computer Laboratory, Cambridge. 2001
4. C. Everett, "Cloud computing–A question of trust," Computer Fraud & Security 2009.6 (2009): 5–7.
5. J. Horrigan, "Use of cloud computing applications and services," Pew Internet & American Life project memo, September 2008.
6. R. Myerson (1991). "Game Theory: Analysis of Conflict," Harvard University Press, p. 1.

7. D. Clemente, "Cyber Security and Global Interdependence: What is Critical?", Chatham House, 2013.
8. K. Cukier, "Ensuring and Insuring Critical Information Infrastructure Protection: A Report of the 2005 Rueschlikon Conference on Information Policy," The Rueschlikon Conference, 2005.
9. F. Hare, "The Interdependent Nature of National Cyber Security: Motivating Private Action for a Public Good," PhD Dissertation, School of Public Policy, George Mason University, (2011).
10. Federal Register / Vol. 78, No. 33 / Tuesday, February 19, 2013 / Presidential Documents
11. G. Heal, H. Kunreuther. "You only die once: Managing discrete interdependent risks," No. w9885. National Bureau of Economic Research, 2003.
12. H. Kunreuther, H. Geoffrey "Interdependent Security: the Case of Identical Agents," Working paper, Columbia Business School and Wharton Risk Management and Decision Processes Center. Journal of Risk and Uncertainty, forthcoming, Special Issue on Terrorist Risks, 2002.
13. W. Sun, X. Kong, D. He, X. You. "Information security problem research based on game theory," International Symposium on Publication Electronic Commerce and Security, 2008.
14. C. Kamhoua, N. Pissinou, K. Makki. "Game theoretic modeling and evolution of trust in autonomous multi-hop networks: Application to network security and privacy," IEEE International Conference on Communications (ICC), 2011.
15. P. Tailor, L. Jonker "Evolutionary Stable Strategies and Game Dynamic," Mathematical Biosciences, 5:455–484, 1978.
16. T. Alpcan, T. Başar. "Network security: A decision and game-theoretic approach," Cambridge University Press, 2010.
17. N. Leavitt, "Is cloud computing really ready for prime time," Growth 27.5 (2009).
18. P. Mell, T. Grance. "The NIST definition of cloud computing (draft)," NIST special publication 800.145 (2011): 7.
19. S. Pearson, A. Benameur. "Privacy, security and trust issues arising from cloud computing," IEEE Second International Conference on Cloud Computing Technology and Science (CloudCom) 2010.
20. Zissis, Dimitrios, and Dimitrios Lekkas. "Addressing cloud computing security issues," Future Generation Computer Systems 28.3 (2012): 583–592.
21. T. Ristenpart, E. Tromer, H. Shacham, S. Savage. "Hey, You, Get Off of My Cloud: Exploring Information Leakage in Third-Party Compute Clouds," In the proceedings of the 16th ACM Conference on Computer and Communications Security, CCS'09, Chicago, IL, USA, October 2009.
22. A. Bates, B. Mood, J. Pletcher, H. Pruse, M. Valafar, K. Butler "Detecting Co-Residency with Active Traffic Analysis Techniques," in the proceedings of the 2012 ACM Cloud Computing Security Workshop (CCSW) in conjunction with the 19th ACM Conference on Computer and Communications Security, October 2012, Raleigh, North Carolina, USA.

23. C. Li, A. Raghunathan, N. Jha, "A Trusted Virtual Machine in an Untrusted Management Environment," IEEE Transactions on Services Computing, vol. 5, no. 4, pp. 472–483, Fourth Quarter 2012.
24. A. Azab, P. Ning, Z. Wang, X. Jiang, X. Zhang, N. Skalsky "HyperSentry: enabling stealthy in-context measurement of hypervisor integrity," In Proceedings of the 17th ACM conference on Computer and communications security (CCS '10). ACM, New York, NY, USA.
25. Y. Zhang, A. Juels, A. Oprea, M. Reiter "HomeAlone: Co-Residency Detection in the Cloud via Side-Channel Analysis," in the proceedings of IEEE Symposium on Security and Privacy, May 2011, Oakland, California, USA.
26. L. Carin, G. Cybenko, J. Hughes, "Cybersecurity Strategies: The QuERIES Methodology," Computer, vol.41, no.8, pp.20–26, Aug. 2008.
27. United States Securities and Exchange Commission, https://www.sec.gov/divisions/corpfin/guidance/cfguidance-topic2.htm [Retrieved: 28 September 2017]
28. Cuong T. Do, Nguyen H. Tran, Choongseon Hong, Charles A. Kamhoua, Kevin A. Kwiat, Erik Blasch, Shaolei Ren, Niki Pissinou, Sundaraja Sitharama Iyengar "Game Theory for Cyber Security and Privacy" ACM Computing Surveys (CSUR), Volume 50, Issue 2, Article No. 30, June 2017.
29. Luke Kwiat, Charles A. Kamhoua, Kevin Kwiat, Jian Tang, Andrew Martin "Security-aware Virtual Machine Allocation in the Cloud: A Game Theoretic Approach" in the proceedings of the IEEE International Conference on Cloud Computing, (IEEE CLOUD 2015), New York, June 2015.
30. Charles A. Kamhoua, Anbang Ruan, Andrew Martin, Kevin A. Kwiat "On the Feasibility of an Open-Implementation Cloud Infrastructure: A Game Theoretic Analysis" in the proceedings of the 2015 IEEE/ACM International Conference on Utility and Cloud Computing (UCC 2015), Limassol, Cyprus, December 2015.
31. Deepak K. Tosh, Shamik Sengupta, Charles A. Kamhoua, Kevin A. Kwiat "Game Theoretic Modeling to Enforce Security Information Sharing among Firms" in the proceedings of the IEEE International Conference on Cyber Security and Cloud Computing (CSCloud 2015), New York, November 2015.
32. Charles A. Kamhoua, Andrew Martin, Deepak Tosh, Kevin A. Kwiat, Chad Heitzenrater, Shamik Sengupta "Cyber-threats Information Sharing in Cloud Computing: A game Theoretic Approach" in the proceedings of the IEEE International Conference on Cyber Security and Cloud Computing (CSCloud 2015), New York, November 2015.
33. Carlo Di Giulio, Charles A. Kamhoua, Roy H. Campbell, Read Sprabery, Kevin Kwiat, Masooda N. Bashir "Cloud Standards in Comparison: Are New Security Frameworks Improving Cloud Security?" in the proceedings of the 2017 IEEE International Conference on Cloud Computing (CLOUD), Honolulu, Hawaii, June 2017.
34. Carlo Di Giulio, Charles A. Kamhoua, Roy H. Campbell, Read Sprabery, Kevin Kwiat, Masooda N. Bashir "IT Security and Privacy Standards in Comparison: Improving FedRAMP Authorization for Cloud Service Providers" in

the proceedings of the 2017 International Workshop on Assured Cloud Computing and QoS aware Big Data (WACC 2017), in conjunction with the 2017 IEEE/ACM International Symposium on Cluster, Cloud and Grid Computing (CCGrid), Madrid, Spain, May 2017.
35. Carlo Di Giulio, Read Sprabery, Charles A. Kamhoua, Kevin Kwiat, Roy H. Campbell, Masooda N. Bashir "Cloud Security Certifications: A Comparison to Improve Cloud Service Provider Security" in the proceedings of the International Conference on Internet of Things, Data and Cloud Computing (ICC 2017), Cambridge city, Churchill College, University of Cambridge, UK, March 2017.
36. Xueping Liang, Sachin Shetty, Deepak Tosh, Charles A. Kamhoua, Kevin Kwiat, Laurent Njilla, "ProvChain: A Blockchain-based Data Provenance Architecture in Cloud Environment with Enhanced Privacy and Availability" in the proceedings of the 2017 IEEE/ACM International Symposium on Cluster, Cloud and Grid Computing (CCGrid), Madrid, Spain, May 2017.
37. Deepak Tosh, Sachin Shetty, Xueping Liang, Charles A. Kamhoua, Kevin Kwiat, Laurent Njilla, "Security Implications of Blockchain Cloud with Analysis of Block Withholding Attack" in the proceedings of the 2017 IEEE/ACM International Symposium on Cluster, Cloud and Grid Computing (CCGrid), Madrid, Spain, May 2017.
38. Sachin Shetty, Val Red, Charles A. Kamhoua, Kevin Kwiat, Laurent Njilla "Data Provenance Assurance in Cloud using Blockchain" in the proceedings of the 2017 SPIE Disruptive Technologies in Sensors and Sensor Systems, Anaheim, California, April 2017.
39. Deepak Tosh, Sachin Shetty, Xueping Liang, Charles A. Kamhoua, Laurent Njilla "Consensus protocols for Blockchain based Cloud data provenance-Challenges and Opportunities" in the proceedings of the 2017 IEEE Annual Ubiquitous Computing, Electronics & Mobile Communication Conference (UEMCON), New York City, NY, October 2017.

Chapter 12
A Risk Management Approach for Highly Interconnected Networks

Stefan Schauer

12.1 Introduction

Today, critical infrastructures together with their utility networks are maintaining the backbone supply chains of modern society like the electrical power production and distribution, water and gas supply, as well as telecommunication networks, among others, operational. These critical infrastructures apply physical and cyber-based systems to monitor and control their underlying utility networks. Hence, these organizations are heavily relying on information and communication technology (ICT) as well as supervisory control and data acquisition (SCADA) systems for providing their services. Despite the fact that utility providers operate these interconnected networks, most of today's risk management tools only focus on one of these networks.

Over the last years, utility providers have become more and more the target of hackers, cyber criminals, and cyber terrorists, and the number of attacks on utility providers has increased [34]. In general, recent studies show that social engineering and malware are the most successful attack types [30]. Such attack vectors are used when deploying random attacks for opportunistic profit, e.g., the spread of ransomware, as it has been shown during the recent global ransomware infections with WannaCry [5, 24] and (Not)Petya [25, 8, 16] in 2017. Further, social engineering or phishing are often applied when planning targeted attacks, i.e., advanced persistent threats (APT) attacks, on organizations to get a foothold within their system.

S. Schauer (✉)
AIT Austrian Institute of Technology GmbH, Center for Digital Safety and Security, Klagenfurt, Austria
e-mail: stefan.schauer@ait.ac.at

S. Rass, S. Schauer (eds.), *Game Theory for Security and Risk Management*, Static & Dynamic Game Theory: Foundations & Applications,
https://doi.org/10.1007/978-3-319-75268-6_12

In particular, APT attacks on utility providers might have huge effects on the general operation of their utility networks, as recent events have shown in the Ukraine [48, 23, 10, 9] and Japan [38, 20] as well as in earlier incidents (cf. Chapter 13 for more details on APT attacks). Therefore, protecting and assuring the security of a utility provider's ICT and SCADA infrastructure is of the utmost importance for maintaining the availability of its service.

In the context of protecting utility networks and their connected ICT and SCADA systems, risk management has become a core duty in critical infrastructures. In the USA, the need for protection of critical infrastructure has been recognized, and the National Infrastructure Protection Plan [21], which is introduced to help critical infrastructure communities, develops technologies, tools, and processes that address near-term needs for the security and resilience. In Europe, the "Directive on security of network and information systems" (also known as the NIS Directive) [11] was adopted by the European Parliament. The objectives of the NIS Directive are to ensure a high level of network and information security, to improve the security of the Internet and the private networks, and to improve properness and cooperation between the member states of the EU.

12.1.1 Problem Overview

When looking at utility providers and the networks they are operating, we see that these networks rely on a high integration and a heavy interrelation among each other. This becomes more visible when considering the three main network layers within utility providers (cf. also Figure 12.1):

- the utility's physical network infrastructure, consisting of, e.g., gas pipes, water pipes, or power lines;
- the utility's control network including SCADA systems used to access and maintain specific nodes in the utility network;
- the ICT network, collecting data from the SCADA network and containing the organization's business logic

In more detail, the individual systems in the SCADA network are controlling the physical systems in the utility network, e.g., switching pumps on and off or administrating entire electrical substations. Further, SCADA systems are communicating with the ICT network, reporting status updates of the utility network or submitting monitored data required for billing. Due to these interconnections and communication pathways, an incident in one network might affect not only the network itself but might also have cascading effects on several other networks as well.

Certainly, most utility providers are aware of these interrelations between their networks and the potential risks caused by them. They are implementing security and risk measures according to state-of-the-art frameworks and guidelines to increase the security and be prepared for incidents. Nevertheless, current risk management frameworks like the ISO 31000 [27], the ISO/IEC 27005 [28], the NIST

SP800-30 [46], and the COBIT 5 for Risk [29] are mostly a matter of best practices. These standards and guidelines are compiled by experts and are tailored to classical business sectors and companies. Hence, they cover best practices but are not based on a well-defined basis.

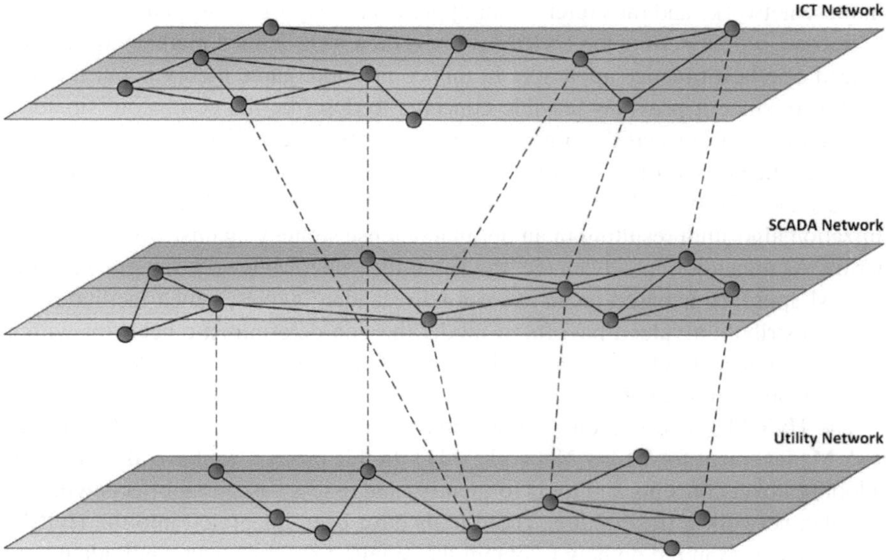

Fig. 12.1: Illustration of the interconnected networks operated by a utility provider

Additionally, the abovementioned standards and guidelines are often focusing only on one specific topic. For example, the ISO/IEC 27005 is specialized on the risk assessment for ICT systems and networks, the general focus of COBIT 5 for Risk lies on the business processes, and the ISO 31000 is deliberately generically designed to be applicable to organizations in a broad number of different fields. Moreover, there exist specific risk (or safety) policies for the utility networks, which are governing specialized issues within these networks. Therefore, the special requirements of utility providers (or critical infrastructures in general) with regard to the highly interconnected nature of their network infrastructure are not accounted for explicitly.

12.1.2 Hybrid Risk Management

With the Hybrid Risk Management (HyRiM) process we are describing in this chapter, we want to tackle in particular the two main shortcomings of standard risk management frameworks mentioned above. First, our approach is focusing specifically on the sensitive interconnection points between different networks operated by a utility provider. Moreover, besides the three network layers mentioned above (i.e., utility, SCADA, and ICT network), we are also including the human factor and the social interrelations (i.e., the social network) between employees in our considerations. Hence, by taking the technical as well as the social and organizational aspects within a utility provider into account, we are choosing a holistic or "hybrid" view on these networks and thus refer to our approach as "Hybrid Risk Management."

Second, we are setting the HyRiM process on a well-defined mathematical basis instead of a best practice approach. In this connection, game theory is our method of choice, since it provides feasible concepts to describe the combating situation between an attacker and the organization's security officer. Further, it particularly improves the process of risk mitigation, which is only dealt with marginally in most of the standards. We will describe how the HyRiM process implements a risk minimization algorithm resulting in an optimal defense strategy against worst case scenarios. As an additional advantage, the game-theoretic framework integrated in the HyRiM approach is based on an extension to standard games, which facilitates the use of distribution-valued payoffs to model the non-deterministic behavior within the interconnected networks [39, 41] (cf. also Chapters 2 and 3 for details on the game-theoretic framework).

The HyRiM Process is one major output of the FP7 project HyRiM ("Hybrid Risk Management for Utility Networks") [1]. In this project, we have also been developing novel concepts and tools to identify and assess cascading effects within a complex network infrastructure. These tools have been integrated into the HyRiM process as an example on how to obtain the required results for the individual process steps. We will describe where the respective tools might be applied and how their results can be used in the overall process.

12.1.3 Chapter Outline

In the following Section 12.2, the general structure of the HyRiM process is described. The seven steps, which build up the HyRiM process, are sketched in the Sections 12.2.1 to 12.2.7. Section 12.3 summarizes several concepts, algorithms, and tools, which are used in the individual steps of the HyRiM process. Therein, precise methods like simulation methodologies or the game-theoretic framework as well as "softer" techniques like ethnographic studies are illustrated. A conclusion wraps up the main strengths and opens gaps of the HyRiM process.

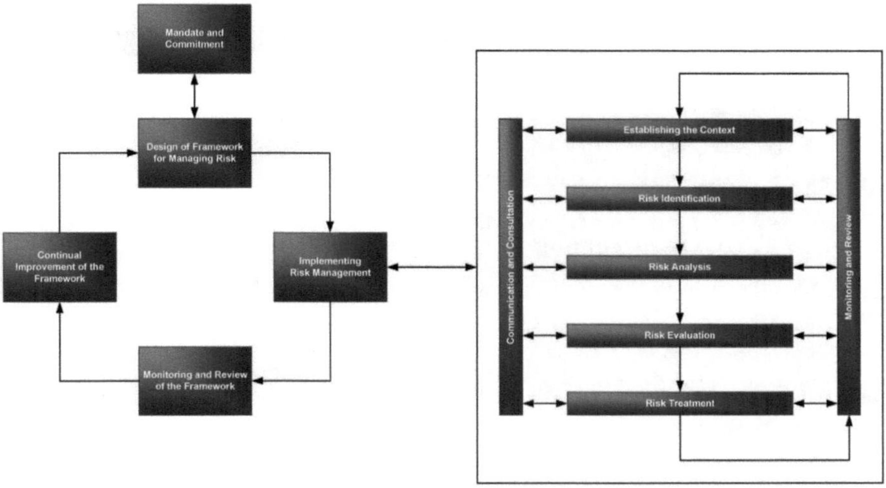

Fig. 12.2: Illustration of the two-tier structure of the ISO 31000 risk management process

12.2 The HyRiM Process

The Hybrid Risk Management process (in short HyRiM process) presented here is based on and compliant with the general ISO 31000 process for risk management [27] and most standards referencing the ISO 31000, for example, the ISO 27005 for risk management in ICT security [28] and the ISO 28001 for supply chain security [26]. The process is particularly suited for organizations operating highly interconnected networks at different levels, such as utility providers or critical infrastructure operators. In short, the generic risk management process of the ISO 31000 framework (as depicted on the right side in Figure 12.2) is adopted, and each step of the process is extended to address recurring challenges within interconnected networks (cf. Figure 12.3 below for an overview of the individual steps). Therefore, tools and concepts developed in the HyRiM project [1] are used, which cover different social and technical analysis techniques and simulation methodologies that facilitate the risk process (a mapping of these tools and concepts can be found in Figure 12.9 in the following Section 12.3). Due to the direct relation to international standards, the process can be integrated into existing risk management processes already running in the aforementioned organizations. Further, this makes it applicable in various operational areas.

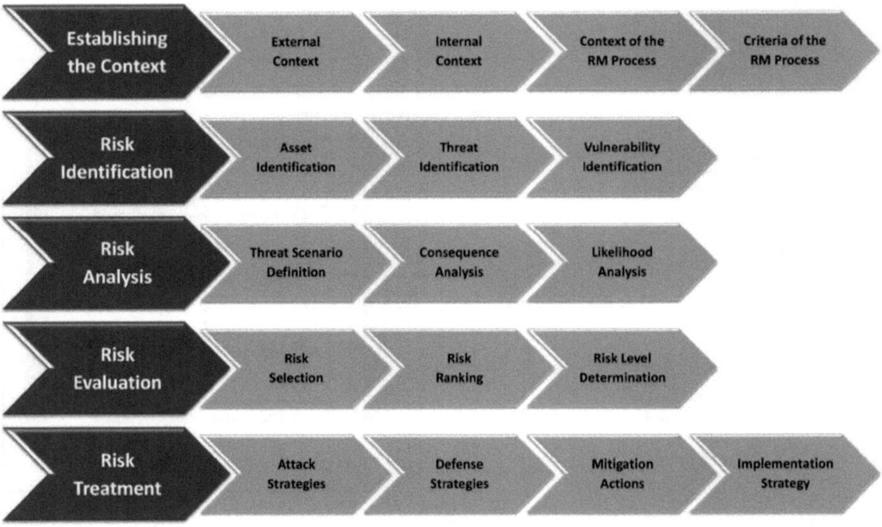

Fig. 12.3: Illustration of the HyRiM process including all sub-steps

The general framework applied in the HyRiM process to model the interplay between different networks is game theory. In this context, the process takes advantage of the sound mathematical foundation of game theory. By falling back onto a zero-sum game and a minimax approach [35], a precise model of an adversary's intentions and goals is not necessary. Moreover, the game-theoretic framework [40, 39] at the core of the HyRiM process allows modeling the intrinsic uncertainty and randomness encountered in the real-life application of interrelated (utility) networks. This is realized by collecting data describing consequences and likelihoods in distributions and working with these payoffs in the game [41, 42].

In the following subsections, we will describe in detail the specific steps of the HyRiM process. We will break up the five main steps of the process into more fine-grained sub-steps and highlight the core activities, inputs, and outputs of these sub-steps. More detailed information on potential tools supporting the activities in the respective steps are provided later on in Section 12.3.

12.2.1 Establishing the Context

The HyRiM process starts by defining the objectives which should be achieved and attempting to understand the external and internal factors that may influence the goal. In detail, the information about SCADA and IT communication networks (e.g., a network architecture diagram), components of the utility network (e.g., a detailed architecture of the physical utility network layer), industrial control functionalities, and information assets are taken into account. Besides the technical aspects, the

HyRiM process is designed to integrate also soft factors, e.g., the social and organizational aspects, into the risk management (cf. Figure 12.4).

Fig. 12.4: Illustration of the first step "Establishing the Context"

The first sub-step, "External Context," deals with all external influences and stakeholders, which impinge on an organization from the outside. These relations between the external stakeholders and the organization can be of different type. For example, an external company might provide some resources or services for the organization or a regulatory body is enforcing a legal framework, which defines how the organization's processes have to be designed.

Extending the information from the external context, all interrelations and interdependencies among the internal structures within an organization are described in the second sub-step "Internal Context." This includes technical, organizational, and social aspects and describes communication channels as well as dependencies between different technical and social networks.

The third sub-step, "Context of the Risk Management Process," inspects the relevant parts of the organization, which are covered by the risk management process. Whereas the previous sub-steps focus more on assets and their interrelations, this sub-step relates to the organization's management structure. These can include, for example, the organizational units under examination, the depth of the risk assessment process, or the resources and responsibilities required for the risk assessment process. The external and internal context together with the context of the risk management provides a detailed picture of the environment relevant for the risk management process.

Finally, some meta-information about the risk management process itself is required. This information is collected in the last sub-step "Criteria of the Risk Management Process" and includes the relevant criteria to evaluate the significance of a specific risk. These criteria can be based on the general characteristics describing the organization (e.g., size, revenue, legal form, etc.), as well as the organization's resources, objectives, and goals. This information will influence, for example, how the likelihood or the impact of an event is characterized or how the limits of the risk levels are defined during the later step "Risk Analysis" (cf. Section 12.2.3).

12.2.2 *Risk Identification*

Risk identification involves the application of systematic techniques to understand a range of scenarios describing what could happen, how, and why. Therefore, the information on organization's infrastructure relevant for the risk assessment process coming from the previous Step 1 "Establishing the Context" is required. This includes technical assets, organizational roles, and individual personnel as well as their interdependencies. Gathering this information is done in three sub-steps of the process: "Assets Identification," "Threat Identification," and "Vulnerability Identification." Based on that, potential vulnerabilities and threats can be identified (cf. Figure 12.5).

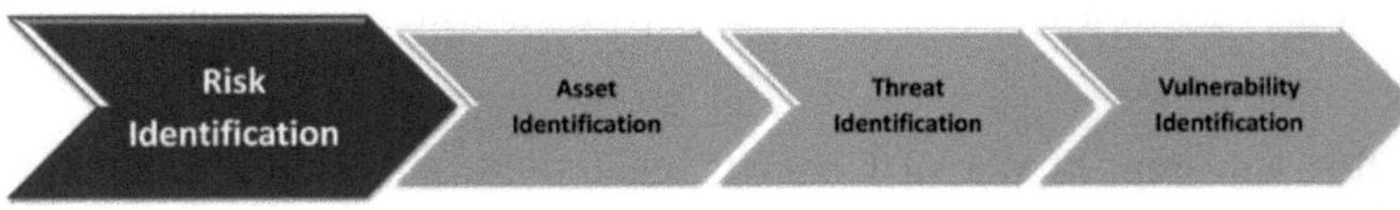

Fig. 12.5: Illustration of the second step "Risk Identification"

The first sub-step, "Asset Identification," looks at all the relevant assets of the organization's infrastructure based on the internal context discussed in the previous Step 1. In this context, an "asset" is anything of value to the organization (cf. also ISO 27005 [28]). Hence, an asset can be, for example, a part of the utility network, a cyber-physical system in the ICT or SCADA network, or an employee of the organization when considering at the social network. These assets and their different types of interconnections (e.g., technical, logical, social, etc.) need to be modeled using a proper representation, most conveniently, an interdependency graph.

Based on this list of relevant assets, a list of potential threats affecting the organization's infrastructure, i.e., the identified assets, is collected in the second sub-step "Threat Identification." The information on these threats can be gathered from external sources as well as internal sources. Such sources can be existing threat catalogs, for example, provided by the German BSI [6], online threat databases, or simply data collected on incidents which already happened to the organization in the past. An additional but maybe more subjective way to collect threat information is to use knowledge from (internal or external) experts.

In parallel to the collection of threat information, data on all vulnerabilities of the relevant assets is gathered in the third sub-step "Vulnerability Identification." Similarly to the threat information, data on vulnerabilities can also be found in online vulnerability databases like the National Vulnerability Database (NVD) [3] maintained by the National Institute of Standards and Technologies [2], closed (area-specific) discussion forums or expert knowledge. In general, most technical assets (e.g., software or hardware) have vulnerabilities due to a poor configuration or programming errors. This concept can also be extended to social assets (like employees) when thinking of social engineering attacks.

12.2.3 Risk Analysis

Risk analysis is concerned with developing an understanding of each risk, its consequences, and the likelihood of occurrence. In general, the level of risk is determined by taking into account the present state of the system, existing controls, and their level of effectiveness. Whereas in a classical risk analysis approach both the consequences and the likelihood of an incident are aggregated into a single value, in the HyRiM process, both are described by distributions or histograms including all the relevant information coming from different sources. In general, the more information is available to build up these distributions, the higher the quality of the results. Nevertheless, since most of the time only scarce information about potential threats and vulnerabilities is available within an organization, the HyRiM process is designed to work also with limited information (cf. Figure 12.6).

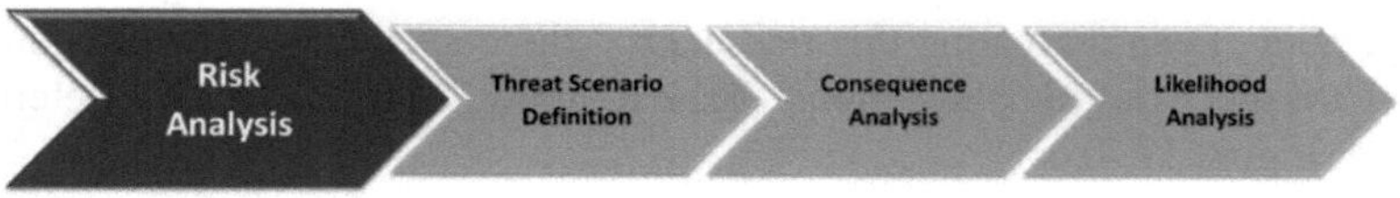

Fig. 12.6: Illustration of the third step "Risk Analysis"

Carrying on from the previous step, the connection between the identified assets, threats, and vulnerabilities is determined in the first sub-step "Threat Scenario Definition." This connection is understood as the following: a particular threat affecting a specific asset within the organization's infrastructure can only be effective if the asset has a vulnerability relevant for the specific threat (i.e., which the threat can exploit). Every such combination of threat, assets, and vulnerability is referred to as threat scenario. The first sub-step delivers a fine-grained list of potential threat scenarios as an output.

In the next step, "Consequence Analysis," the potential impacts of each single threat scenario are evaluated. This can be supported by quantitative results coming from a structured mathematical analysis, e.g., using percolation theory (cf. [32, 33]), a co-simulation approach (cf. [14]) or an intrusion simulation (cf. [4] and Chapter 6 as well). Additionally, the consequences can also be estimated full qualitatively by experts from within the organization or external advisors. To include all information gathered from the simulations as well as the experts, the consequences are represented as histograms or, more general, as distribution functions.

Similarly to the "Consequence Analysis," the likelihood for a specific threat scenario to occur is estimated in the last step "Likelihood Analysis." In comparison to the structured analysis of the consequences, determining the likelihood of an event is more vague since it is difficult, in particular in the context of ICT-related threat scenarios, to assign a specific number to that. Therefore, the HyRiM process is based on a fully qualitative estimation process carried out by experts from within the organization or external advisors. Nevertheless, information from external sources, e.g., reports containing statistical information on the likelihood of specific events, is used to support the decision-making. To be consistent with the consequence analysis, the likelihoods are represented as histograms or, more general, as distribution functions to include all information gathered from the experts.

12.2.4 Risk Evaluation

Risk evaluation involves making a decision about the level or priority of each risk by applying the criteria developed when the context was established. In classical risk management approaches, a cost-benefit analysis can be used to determine whether specific treatment is worthwhile for each of the selected risks. The game-theoretic model applied in the HyRiM process allows an optimization according to several tangible and intangible goals (i.e., not only costs but also employee satisfaction or social response). Nevertheless, the result can be visualized in a common representation (i.e., a risk matrix) to provide a high recognition value among the top-level management. Toward creating this risk matrix, three specific steps are required in the HyRiM process: "Risk Selection," "Risk Ranking," and "Risk Level Determination" (cf. Figure 12.7).

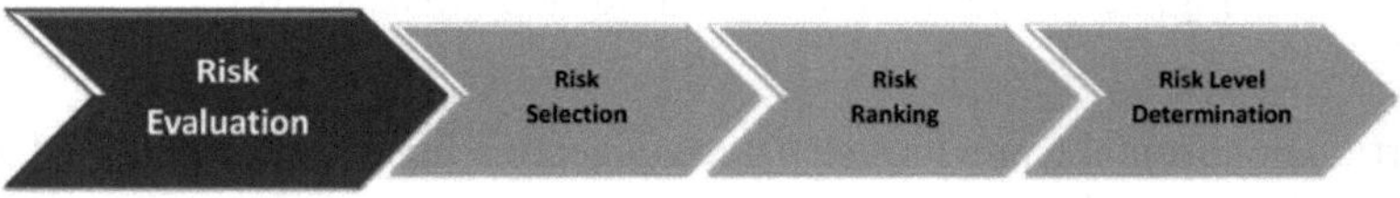

Fig. 12.7: Illustration of the fourth step "Risk Evaluation"

After all threat scenarios have been evaluated in the previous step "Risk Analysis," not all of them need to be considered in the overall risk management. Some of them might not be significant enough according on the organization's risk criteria (as described in "Establishing the Context"; cf. Section 12.2.1); others simply might be out of the scope of the process. The choice which of the threat scenarios are further evaluated is made in the sub-step "Risk Selection".

In the second sub-step "Risk Ranking," the remaining relevant threat scenarios are ordered according to their respective consequences and likelihood. In general risk management approaches, this is done based on the combination of their estimated consequences and likelihoods, following the commonly accepted formula risk = likelihood $\times$ consequences [37]. Nevertheless, in the HyRiM process, the consequences as well as the likelihood of each threat scenario are represented by distributions or histograms (cf. Section 12.2.3 above), and therefore finding an order is not trivial. To solve this problem, the preference relation $\preccurlyeq$ is applied [40, 39, 41, 42]. The sub-step outputs two ordered lists, one for the consequences and one for the likelihood, ranking the threat scenarios starting with the least severe consequences and highest likelihood, respectively.

Based on these two ordered lists, a risk matrix together with an overall priority list of the most important threat scenarios is created in the third sub-step "Risk Level

Determination." The threat scenarios are put into the risk matrix according to their relative position in the two lists. Hence, the top-ranked risks, i.e., the risks having the most severe consequences together with the highest likelihood, are located at the upper right corner of the matrix. Based on this risk matrix, a priority list of all identified risks can be created, starting from the upper right corner.

It has to be pointed out that in this approach all the identified threat scenarios are brought in relation to each other. However, the resulting risk matrix does not relate the individual threat scenarios to an absolute likelihood or impact scale, respectively, which might exist in the organization. Observe that the description of a threat by a probability distribution over its impact basically admits that a threat can have impacts of different magnitudes that occur with different likelihoods. The collection of all these possibilities makes up the loss distribution (the same goes for the probability distribution over the likelihood of a specific threat scenario to take place). The two-dimensional ranking of threats in risk matrices boils down to a "linear ordering" of loss distributions that describe each threat (we simply combine the likelihood and impact in one compound object that is the loss distribution).

However, it is not difficult to establish the more familiar (two-dimensional) risk matrices as well. For these, we merely need an individual (and independent) assessment of impacts and likelihoods (as usual) but compile data pairs of (impact, likelihood) values into two empirical distributions. We can then directly order these according to the stochastic order of Chapter 2 or confine ourselves to numbers by taking average impacts and likelihood values from the data. Either way, putting them in order on the vertical and horizontal axis and labeling the so-defined point in the 2D-plane with a threat, we end up with the common and familiar risk matrix representation. Nevertheless, we are not going further into detail on this topic because it is out of the scope of this chapter.

12.2.5 Risk Treatment

Risk treatment is the process in which existing controls are improved and new controls are implemented. In classical risk management approaches, the aim is to apply these new or improved controls to reduce either the likelihood of a specific threat to occur or the magnitude of the consequences. In classical risk management approaches, the decision about which controls to implement is often a subjective one carried out by the risk manager. On the contrary, the goal in the HyRiM process is to identify the optimal set of controls to reduce the maximum damage that can be caused by an attacker to a minimum. In this context, the optimality of the resulting controls is given due to the game-theoretic framework applied in the approach [40, 39]. Following this game-theoretic approach, the sub-steps "Attack Strategies" and "Defense Strategies" are carried out to obtain the payoff matrix. Further, the sub-steps "Mitigation Actions" and "Implementation Strategy" describe how the optimal solution is implemented in the organization (cf. Figure 12.8).

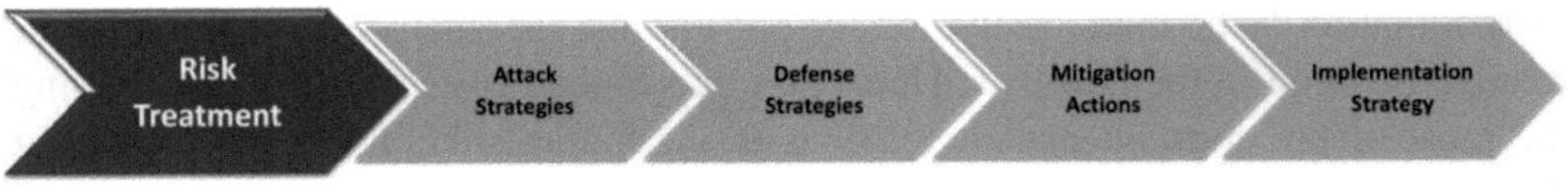

Fig. 12.8: Illustration of the fifth step "Risk Treatment"

The attack strategies identified in the first sub-step are based on the threat scenarios defined in the previous step "Threat Scenario Definition" (cf. Section 12.2.3). In more detail, any relevant combination of asset, threat, and vulnerability (as described above) represents a potential attack strategy for the game. If an adversary follows these attack strategies, the respective risks are manifested within the organization with the respective consequences identified in the step "Consequence Analysis" (cf. Section 12.2.3).

Accordingly to these attack strategies, related countermeasures are defined in the second sub-step "Defense Strategies." Therefore, a number of activities that can be carried out by the organization to mitigate the respective risks are collected. In general, the effect of these defense strategies can be diverse: such a strategy can reduce the damage done to a specific asset, cut down the (cascading) consequences of the risk, e.g., by lowering the probability to propagate through the networks or the number of connected (and thus affected) assets, or let a risk vanish completely, e.g., by closing specific vulnerabilities.

In the third sub-step "Mitigation Actions," the attack and defense strategies are used to build up the payoff matrix for the game. The payoff for each combination of attack and defense strategy is computed by rerunning the consequence analysis (cf. Section 12.2.3) for the organization's asset structure assuming that the specific defense strategy has been implemented. As discussed in the sub-step "Consequence Analysis" above, different methods (e.g., simulations based on percolation theory, co-simulation or physical intrusion, as well as expert interviews) can be used to get to the respective results. The derived payoff matrix is then fed into the game-theoretic framework, leading to a threefold output: the first result is an optimal security strategy for the defender, pointing at the optimal choice of defense strategies, the second is an optimal attack strategy for the attacker identifying the neuralgic assets within the organization, and the third is the maximum damage that can be caused by an adversary.

The final sub-step implements the optimal mitigation actions provided by the game-theoretic framework. In general, the optimal security strategy is a mixture of several of the identified defense strategies. This mixture indicates the frequency (or probability) at which these activities have to be performed. To reflect that in the organization's day-to-day business, these mitigation activities have to be carried out by the organization's employees precisely following the calculated frequencies (or probabilities). A deviation from the optimal security strategy might give the adversary an advantage and allow him to cause more damage than predicted by the game-theoretic framework.

12.2.6 Communication and Consulting

Concurrent with the five main steps of risk management (described in Sections 12.2.1 to 12.2.5) runs the Communication and Consultation step. Therein, the main and partial results of the process are communicated to the respective stakeholders in the underlying organization (as identified during the Step 1 "Establishing the Context"). This is a crucial part of the overall process, since it is of high importance that the stakeholders, in particular the organization's top-level management, are kept well-informed about the results from the process. Therefore, each output of this step needs to be tailored to its target group (e.g., the technical management, top-level management, etc.) such that the results and also their implications are understood by the recipients. It is important to maintain awareness for the risk management activities, since their continued support for the risk management process is crucial for the overall risk management framework (as described in beginning of Section 12.2).

The main results from each individual step of the risk management process represent the general inputs for this step. Among others, these include the list of potential threats and vulnerabilities determined in Step 2 of the process (i.e., "Risk Identification"; cf. Section 12.2.2), the threat scenarios together with their respective potential consequences, and likelihood of occurrence, which have been identified in Step 3 (i.e., "Risk Analysis"; cf. Section 12.2.3), the risk matrix and the prioritized list of all risks as created in Step 4 of the process (i.e., "Risk Evaluation"; cf. Section 12.2.4) as well as the list of potential attack strategies and the list of potential defense strategies, which are the inputs for the game in Step 5 (i.e., "Risk Treatment"; cf. Section 12.2.5).

The main outputs of the game, i.e., the list of critical nodes based on the optimal attack strategy, the sequence of mitigation actions implementing the optimal defense strategy, and the worst case risk level, are also the main outputs of this step. They are coming directly from the game evaluated in Step 5 (i.e., "Risk Treatment"; cf. Section 12.2.5).

12.2.7 Monitoring and Review

Besides the "Communication and Consultation" step described above, a second step running in parallel to the five main steps of risk management is "Monitoring and Review." Although the outputs of the game-theoretic model are optimal in the context of the equilibrium of the given attack and defense strategies, they are only as good as their inputs. Hence, the results from the various steps of the risk management process need to be evaluated after a certain amount of time. This allows the risk manager to verify whether the mitigation actions coming out of the risk treatment are still effective or not. In other words, this step implements a constant feedback loop into the HyRiM process.

Some modifications in the general organizational structure or the network interconnections, which are described in Step 1 (i.e., "Establishing the Context"; cf. Section 12.2.1), can be identified quite easily. Other information, for example, whether new threats or vulnerabilities are relevant for the inspected infrastructure, or if the likelihood and/or consequences for existing threat scenarios have changed, might influence the output of the game-theoretic model drastically. Therefore, all results need to be revised either by a simple checkup of the organization's infrastructure, by a new iteration of expert interviews, or by rerunning the simulations. These activities could be quite expensive, but even small differences can affect the equilibrium of the game. Without a review process, the security officer might not be aware that the optimal defense strategy has changed due to new initial conditions and an adversary might be able to cause additional damage.

This step produces a report on the effectiveness of the mitigation actions and changes in the overall scope of the risk management process as a main output. From a management perspective, this report serves as the basis for the next iteration of the risk management process (cf. Figure 12.9).

12.3 Supporting Tools and Concepts

12.3.1 Ethnographic Studies

To follow a structured and in-depth risk management process as the HyRiM process described here, it is important to have a precise overview on the systems, organizational units, and people involved in the areas which should be examined. The

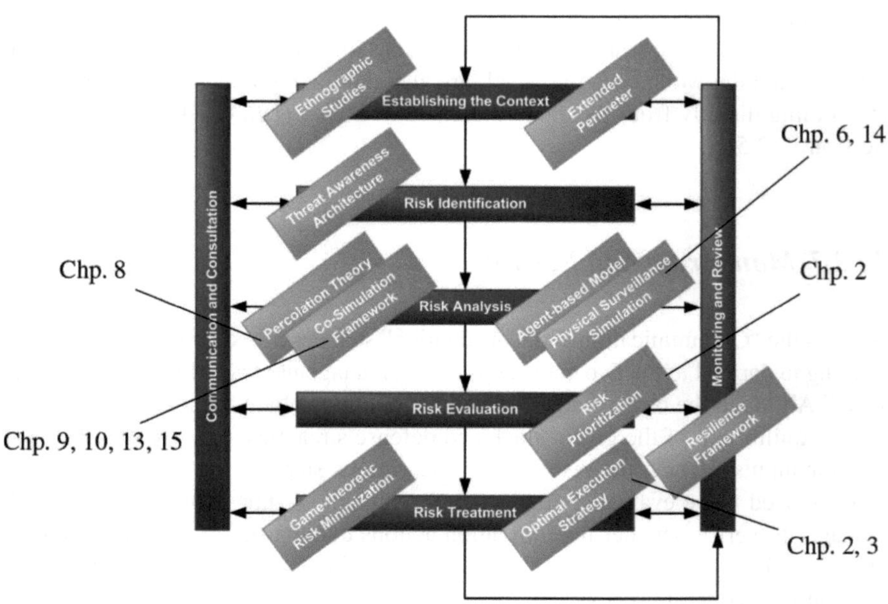

Fig. 12.9: Overview of the HyRiM process including relevant tools for each process step

technical aspects, like network diagrams, the ICT, ICS, or SCADA systems running within the organization as well as the software installed on these systems, are often more or less documented. Although in the ICT world such documentation (in particular if it is present in paper form) is outdated quickly, up-to-date information can often be gathered quickly using tools like network scanners, configuration managers, or vulnerability scanners.

A large part of an organization's infrastructure cannot be captured by such tools, i.e., the organizational structure together with the employees and their social interrelation. Although documentation on these parts exists (e.g., organization charts, policies for different areas, role descriptions, etc.), it is a complete different question, if these are practiced in the everyday life. Hence, for analyzing social aspects, it is suggested to use firsthand and more qualitative analysis techniques, like interviews or ethnography. This allows to identify the gap between the way policies and security measures are planned and should be implemented within the organization and how the organizational structure works in real life.

In particular in the field of ICT security, it has been shown that not the lack of technical security measures but human behavior represents a major risk factor [30] and is the central entry point for a multitude of attack scenarios. Therefore, human and organizational factors were vital parts of the investigations we performed in the HyRiM project. By carrying out ethnographic studies as part of our use cases, we gained a holistic and in-depth view on the relevant infrastructures of the involved end users. This included a visit from an ethnographers to the utility organizations, where discussions were conducted with employees and observations were made during the daily operations. Our studies provided information about the systems used in the utility organization, an identification of people and their roles, an understanding of organizational policies, the social relations among employees, and their behavior under specific circumstances or situations (for specific results from these studies, we refer to [22]).

These different pieces of information can further be used to extend and enhance the technical description of systems and networks and increase the insight into the respective infrastructures as well as all subsequent analyses.

12.3.2 Structured Threat Identification

Over recent years, not only the number but also the complexity of threats and attacks on cyber and physical systems have increased. Keeping up with the speed of this development is a core issue for organizations. Therefore, a structured approach to identify upcoming risks is required to avoid missing potential threats or vulnerabilities. Standardized and constantly updated vulnerability databases (e.g., the National Vulnerability Database) [3] have been established and are maintained by governmental, public, and private organizations.

In the context of critical infrastructures, we suggest to use an approach not only focusing on technical aspects but providing a broader perspective, having in mind the ethnographic studies mentioned in the previous Section 12.3.1. Hence, a threat awareness architecture was developed [18], which is based on organization, technology, and individual (OTI) viewpoints (cf. Figure 12.10). This architecture comprises a three-stage process, including situation recognition, situation comprehension, and situation projection. In this process, the OTI viewpoints serve as a basis and include not only the technical aspects (e.g., the organization's software and hardware systems together with the communication among them) but also cover policies and processes within an organization (i.e., the organization viewpoint) as well as how individual people behave under particular conditions (i.e., the individual viewpoint). The findings resulting from the OTI viewpoints are enriched by threat information coming from external sources, leading to a holistic view on an organization's threat landscape.

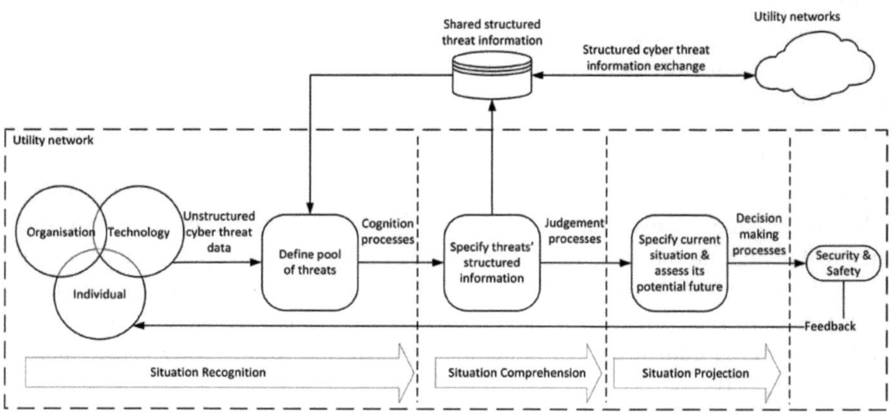

Fig. 12.10: Illustration of the Threat Awareness Architecture (cf. [18])

12.3.3 Simulation Approaches for Payoff Estimation

In general, there is a plethora of different methodologies for estimating the likelihood and consequences of a specific threat scenario. They range from simple questionnaires collecting expert opinions up to complex mathematical models. Especially in the context of utility networks, estimating the potential consequences of a threat often is quite complex due to the interconnected nature of the networks and the related cascading effects. Hence, for the HyRiM process, we suggest four specific simulation-based approaches, which are well suited for utility networks: percolation theory, co-simulation, physical surveillance simulation, and agent-based modeling.

Percolation theory is a common tool to describe, in general, how certain events trigger other events and, in detail, to analyze the spreading of a disease [45, 36, 31, 44]. Nevertheless, it has only been rarely used in the fields of security and risk management so far, although certain security incidents (e.g., the propagation of malware within an ICT network) have similar characteristics as a disease spreading. One reason for this might be that there is the common assumption in percolation theory that all nodes in the network are equally likely to trigger an event. While this might be true for most models of diseases, such an assumptions is too restrictive when looking at networks within a critical infrastructure.

In the course of the HyRiM project, the standard framework of percolation theory has been extended such that nodes and edges can be distinguished according to several characteristics [32, 33]. Based on these different types, a specific probability of failure is assigned to each type, and the propagation of an error is modeled according to these probabilities. This model allows us to compute the probability that an error affects a significant number of components, i.e., it causes an epidemic or even pandemic, as well as how many nodes are indeed affected in this case. A more detailed description of the percolation theory approach is given in the previous Chapter 8; specifics on the application in a real scenario use case can be found in Chapter 14.

In contrast to percolation theory, co-simulation is an approach for the joint simulation of models developed with different tools, where each tool is responsible for simulating a part of a complex system [15, 13, 12]. Each tool is representing one domain within a utility provider's infrastructure, e.g., the cyber and physical domain, and the co-simulation framework models and manages the communication between these tools, e.g., by exchanging variables, data, and status information. In this way, the separated simulations of the complex system are synchronized.

In the context of utility networks, the simulation message bus framework [13] has already been used to model the interactions between a smart grid and the ICT network [14]. In detail, the power grid is described by the DIgSILENT Power Factory [17], and the ICT network is modeled in OMNet++ [47] (cf. also Figure 12.11). With this setting, specific attack scenarios against both network layers can be simulated, evaluating their (cascading) effects onto their respective components.

In particular, when looking at the different networks operated by a utility provider, percolation theory [19, 32, 33] as well as co-simulation [13, 12, 14] can be used to describe the cascading effects spreading over the different networks. More precisely, percolation theory is more helpful when only high-level or sparse (i.e., qualitative) information is available [33]. If more details on the infrastructure and the communication between certain systems are present, a co-simulation approach can provide more accurate information about the spreading of a failure among these networks [14].

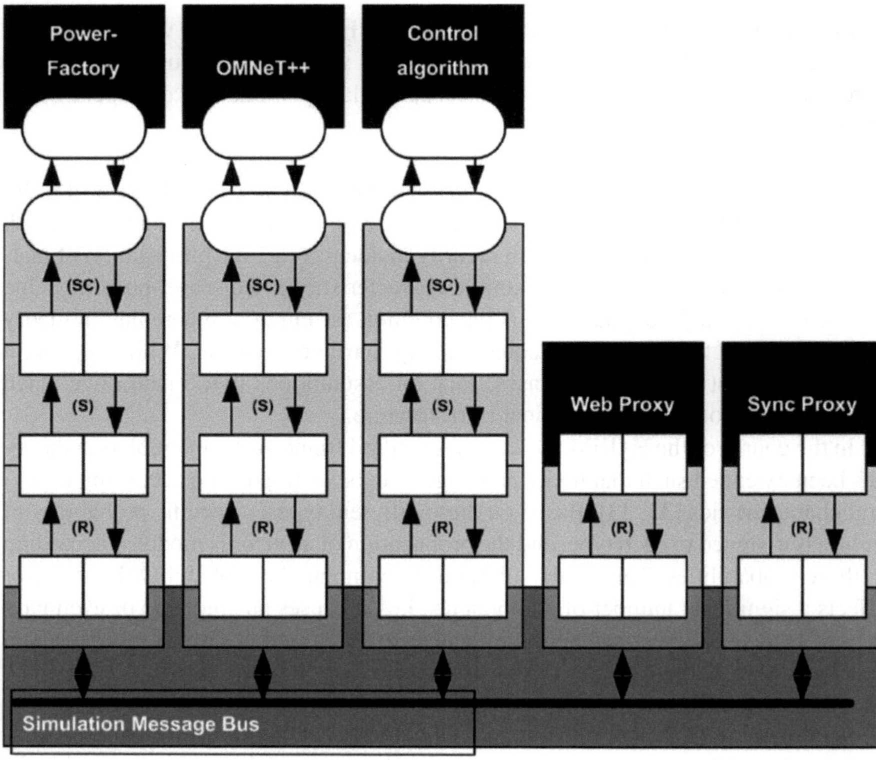

Fig. 12.11: Illustration of the SMB platform integrating the power grid and the ICT network simulation component

In case of threats against the physical infrastructure of a utility provider, e.g., the buildings, machinery, warehouses, tank depots, etc., a simulation framework for physical surveillance is more applicable. This takes the layout of the utility provider's premises, including the buildings and pathways connecting them, and allows to simulate the movements of an adversary entering the premises. In more detail, the adversary's capabilities, potential entry points, and targets can be modeled. Additionally, the security measures (cameras, id badges, etc.) together with the routes and routines of the security guards within the premises can be represented in the simulation. Such a framework has been developed in the HyRiM project [4] and allows reproducing and analyzing different attack scenarios together with the respective defensive actions. Using this framework, not only the potential physical damage caused by one or more intruders but also soft factors (like the effect of increased surveillance on the employees) can be estimated. Further details on the respective framework can be found in the previous Chapter 6.

Complementary to these methodologies, agent-based modeling is much more focused on the societal impact of specific actions taken by an organization. Since utility providers are, in general, critical infrastructures, incidents happening within utility providers as well as the respective security actions can directly affect societal structures in a certain region. As shown in the HyRiM project, an agent-based model can be used to simulate such social response and provide an overview on the potential implications on society [7].

12.3.4 Risk Prioritization and Risk Matrices

A general approach toward risk evaluation and the prioritization of risks in risk management frameworks (like the ISO 31000 [27], the ISO/IEC 27005 [28], and others) is to compute the risk as the product *risk* = *consequence* × *likelihood* [37]. In this case, the consequence as well as the likelihood are represented by a single number assigning a specific value to the resulting risk. This makes it easy to compare several risks and end up with a prioritization, i.e., a list ordered by the magnitude of the risk value.

As already mentioned in Section 12.2.4 above, in the HyRiM process, the consequences as well as the likelihoods are represented as histograms to prevent the loss of important information. The game-theoretic framework applied in the HyRiM process has been specifically designed to handle such histograms as payoffs. Nevertheless, it is nontrivial to find an ordering among these histograms to end up with a prioritized list of risks.

One direct solution for this is given by the $\preceq$-ordering, which has been introduced in [40, 39, 41, 42], and allows comparing two distributions. In the general case, this is done by mapping distributions onto hyperreal numbers, where the standard $\leq$-relation is defined. In the special case of histograms, a lexicographical ordering can be applied. Further technical details on $\preceq$-ordering are given in previous Chapter 2 as well as [40, 39, 41, 42]. By applying the $\preceq$-ordering to the unsorted lists of the threat scenarios' consequences and likelihoods, we end up with a ranking of all the threat scenarios (cf. also Figure 12.12).

In classical risk management frameworks, one core output is a risk matrix, where all the risks are depicted according to their respective consequences and likelihood. In the HyRiM process, one goal is to create a similar risk matrix resembling to the output of classical frameworks. Therefore, the rank of each threat scenario (according to the $\preceq$-ordering) with regard to consequence and likelihood is used. The more severe the consequences are, the further on the right side of the x-axis (i.e., the consequence axis) the threat scenario is placed. This works accordingly for the y-axis (i.e., the likelihood axis), and thus each threat scenario is placed in the 2D-coordinate system of the risk matrix (cf. Figure 12.12). Additional critical regions can be defined, e.g., the upper right corner is usually the most critical. The threat scenarios falling into this area need to be further addressed by the subsequent risk treatment (cf. Section 12.2.5 above).

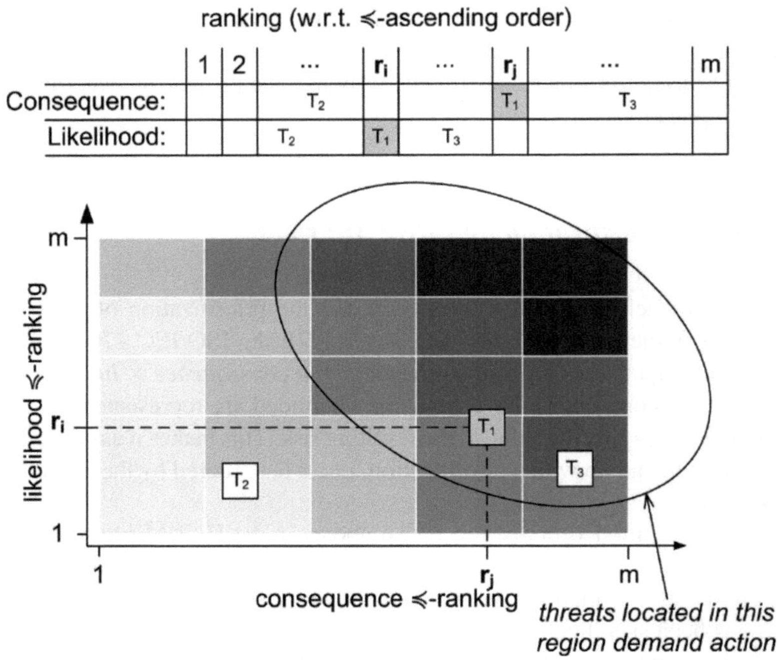

Fig. 12.12: Illustration of the resulting risk matrix based on the two ordered lists for the consequences and likelihoods

12.3.5 Game-Theoretic Risk Minimization

Most of the time, general approaches on risk management offer advice regarding the selection which controls to implement to counter the identified risks. Hence, this decision is often a subjective one, carried out by the risk manager. In the HyRiM process, the goal is to identify an optimal set of controls to reduce the damage that can be caused by an attacker to a minimum. To obtain such an optimal set of controls, game theory is the method of choice, since it provides several beneficial characteristics. The game-theoretic approach applied in the HyRiM process allows not only to identify the optimal choice of controls for a specific risk but also to cluster several risks with similar controls to identify the set of controls, which are most effective against all of the clustered risk. Additionally, the game-theoretic algorithm is capable of optimizing over different security goals, e.g., also taking the costs for implementing the controls into account.

As already mentioned briefly in the beginning of Section 12.2, the game is set up as a two-player, zero-sum game, applying a minimax approach [35]. With this setting, the combating situation between the organization's security officer (i.e., the defender) and an adversary (i.e., attacker) is modeled perfectly. To integrate also the uncertainty and the intrinsic randomness within a utility provider's interconnected network infrastructure, the game-theoretic approach supporting distribution-valued payoffs described in Chapter 2 as well as [40, 39, 41, 42] is applied. Using this approach, the consequences of a specific scenario can be described as distributions (or, as histograms, if necessary) and the game. Hence, the subjective opinions of multiple experts as well as the data coming from different simulations (cf. Section 12.3.3) can be integrated without the need for aggregation and thus without losing relevant information.

In the context of the HyRiM process, attack strategies are always connected to a specific node within the interconnected networks and thus are furthermore described by the identified threat scenarios (cf. Section 12.2.4). These nodes are not necessarily part of the technical network infrastructure, but can also be people handling physical devices. Due to the vulnerabilities (technical or social) of these nodes, they can be attacked. The defense strategies are given by the different security measures, which the organization is able to implement (cf. Section 12.2.5).

To compute the payoff matrix for the game, it needs to be evaluated how much a security measure influences the effects of an attack strategy. This can be done by rerunning the consequence analysis and the simulations or expert interviews carried out therein for each combination of attack and defense strategy. Certainly, some security measures may not have any effect on a given attack strategy, e.g., updating the virus scanner won't mitigate a physical attack on a system in any way. Therefore, the attack strategies as well as the defense strategies can be clustered to evaluate only the relevant combinations. In the end, each entry in the payoff matrix consists of a distribution (or histogram) including all the information gathered from the simulations or interviews (cf. Figure 12.13 as an example for such a payoff matrix).

It has to be noted that the game-theoretic approach is able to optimize over several quantitative or qualitative goals, e.g., the damage caused by an attacker, the costs for the mitigation actions, the effect on reputation of the organization, etc. In this context, the payoff matrix has to be calculated for each of these goals, individually.

In general, the output of the game-theoretic risk minimization algorithm is threefold. It yields

- an optimal security strategy for the defender. This can be a pure strategy (i.e., one single control) or a mixed strategy (i.e., a mixture of several controls) which has to be applied to reduce a specific risk (or a risk cluster) to a minimum.
- an optimal attack strategy for the attacker. This can also be a pure or mixed strategy indicating where an attacker can cause the highest damage to the system. This information can be used to identify weak spots and neuralgic points within the system. It must be noted, however, that this indication is ambiguous and the so-obtained worst-case scenarios are not the only ones possible. Other equilibria may be found by (hypothetically) mitigating the worst-case scenario and rerunning the game-theoretic analysis toward revealing other solutions.
- the maximum damage that can be caused by an adversary following the optimal attack strategy and a defender following the optimal security strategy. For multiple security goals (e.g., the costs for the implemented controls), the respective values are given.

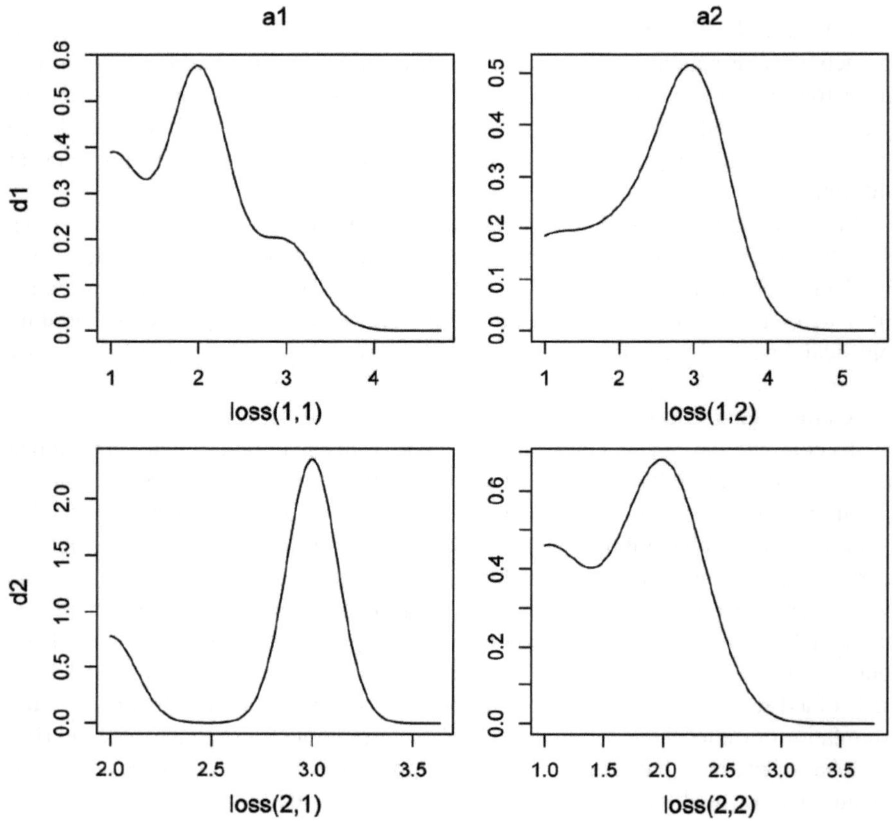

Fig. 12.13: Example of a payoff matrix consisting of distributions (taken from [43])

It must be emphasized that these guarantees hinge on the enforcement of the optimal defense actions. Some resilience against small deviations from the optimal behavior may be embodied in the randomness of the damage. Nevertheless, the distribution itself, consisting of the estimated likelihoods for damages of different magnitudes, is valid only if the optimal defense as delivered in the first of the above steps is implemented.

12.4 Conclusion

In this chapter, we presented a novel approach toward risk management, the HyRiM process for highly interconnected network infrastructures, for example, operated by utility providers. This approach is compliant with the international risk management standard ISO 31000 [27] and extends the steps specified therein by activities tailored to address the particular requirements utility providers are facing. In detail, the HyRiM process accounts for the "hybrid" nature of utility networks, i.e., the strong and complex interrelations between the different networks operated by utility providers. The compliance with the ISO 31000 allows also to directly integrate the HyRiM process into existing frameworks building on the ISO 31000, e.g., the ISO/IEC 27005 [28] or the ISO 28001 [26]. Further, it can also be easily adapted to other standard frameworks like the NIST 800-30 process [46] or the COBIT 5 for Risk framework [29].

The HyRiM process is unique in its concept of evaluating and handling risks; it builds upon a game-theoretic framework to improve mitigation actions and to identify an optimal risk minimization strategy. This game-theoretic framework allows to estimate the worst-case damage and to determine the corresponding optimal mitigation strategy for a given set of potential risks. This sound mathematical basis of the HyRiM process represents one of the main advantages over the abovementioned standard processes and frameworks, which are relying mostly on a best-practice approach. In this context, the HyRiM process also takes one step further by integrating distribution-valued payoffs into the game [41, 42]. This allows to describe real-life scenarios of utility providers and capture the intrinsic randomness within the highly interconnected networks, which is a another core advantage over other frameworks.

To illustrate how the individual steps of the HyRiM process can be carried out in praxis, we provided a detailed description of each steps. We highlighted techniques, concepts, and tools developed in the course of the HyRiM project [1] as a support for each of these steps. In this context, several simulation techniques (percolation theory, co-simulation, physical intrusion simulation, etc.) are sketched, which improve the analysis of the dynamics stemming from the interrelations in the network as well as their resulting cascading effects and serve as an input to the payoff matrix of the game. Moreover, the HyRiM process does not only focus on technical aspects but also takes organizational and human factors into account, including analysis techniques from the field of social and human studies. Therefore, the technical, individual, organizational, and social impact of risks is evaluated in the HyRiM process. Accordingly, the identification of an optimal risk minimization strategy can be done by optimizing over all these different impact types.

The process' practicality and applicability have been evaluated in real-life use case scenarios in the course of the HyRiM project. The following Chapters 13, 14, 15 and 16 show in detail how the steps of the HyRiM process are implemented for specific scenarios and how data is interchanged among them. Therein, the individual methodologies for gathering information as well as for simulating the cascading effects are applied. Further, a detailed formulation of the game according to the settings of the different use case scenarios can be found in these chapters.

By combining information from several different aspects (technical, organizational, social), the HyRiM process provides a holistic overview over an organization's risk situation. Due to the application of game theory and the ability of optimizing mitigation actions according to all these different aspects at the same time, the HyRiM process provides an improved support to the organization's risk manager.

Acknowledgements This work was supported by the European Commission's Project No. 608090, HyRiM (Hybrid Risk Management for Utility Networks) under the 7th Framework Programme (FP7-SEC-2013-1).

References

1. HyRiM | Hybrid Risk Management for Utility Providers. URL https://www.hyrim.net//
2. National Institute of Standards and Technology (NIST). URL https://www.nist.gov/
3. National Vulnerability Database (NVD). URL https://nvd.nist.gov/
4. Alshawish, A., Abid, M.A., Sui, Z., He, X., de Meer, H., Strobl, A., Opitz, A., Rass, S., Zambrano, A.: Deliverable 4.3 – Report on How to Enhance Perimeter Security Using New Surveillance Technologies. HyRiM Deliverable, Passau, Germany (2017). URL https://www.hyrim.net/project-deliverables/
5. Bill, B.: WannaCry: the ransomware worm that didn't arrive on a phishing hook. Tech. rep., Sophos Ltd (2017). URL https://nakedsecurity.sophos.com/2017/05/17/wannacry-the-ransomware-worm-that-didnt-arrive-on-a-phishing-hook/
6. Bundesamt für Sicherheit in der Informationstechnik: IT-Grundschutz-Kataloge. Bonn, Germany (2016). URL https://www.bsi.bund.de/DE/Themen/ITGrundschutz/itgrundschutz_node.html. English Version
7. Busby, J., Gouglidis, A., Rass, S., König, S.: Modelling security risk in critical utilities: the system at risk as a three player game and agent society. In: Systems, Man, and Cybernetics (SMC), 2016 IEEE International Conference on, pp. 1758–1763. IEEE, Budapest, Hungary (2016)
8. Cimpanu, C.: Petya Ransomware Outbreak Originated in Ukraine via Tainted Accounting Software (2017). URL https://www.bleepingcomputer.com/news/security/petya-ransomware-outbreak-originated-in-ukraine-via-tainted-accounting-software/
9. Condliffe, J.: Ukraine's Power Grid Gets Hacked Again, a Worrying Sign for Infrastructure Attacks (2016). URL https://www.technologyreview.com/s/603262/ukraines-power-grid-gets-hacked-again-a-worrying-sign-for- infrastructure-attacks/
10. E-ISAC: Analysis of the Cyber Attack on the Ukrainian Power Grid. Tech. rep., Washington, USA (2016). URL https://ics.sans.org/media/E-ISAC_SANS_Ukraine_DUC_5.pdf
11. European Comission: DIRECTIVE (EU) 2016/1148 OF THE EUROPEAN PARLIAMENT AND OF THE COUNCIL of 6 July 2016 concerning measures for a high common level of security of network and information systems across the Union. Official Journal of the European Union p. L 194/1 (2016). URL http://eur-lex.europa.eu/legal-content/EN/TXT/HTML/?uri=CELEX:32016L1148&from=EN

12. Faschang, M.: Loose Coupling Architecture for Co-Simulation of Heterogeneous Components. Ph.D. thesis, Vienna University of Technology, Vienna, Austria (2015)
13. Faschang, M., Kupzog, F., Mosshammer, R., Einfalt, A.: Rapid control prototyping platform for networked smart grid systems. In: Proceedings IECON 2013 - 39th Annual Conference of the IEEE Industrial Electronics Society, pp. 8172–8176. IEEE, Vienna, Austria (2013)
14. Findrik, M., Smith, P., Kazmi, J.H., Faschang, M., Kupzog, F.: Towards secure and resilient networked power distribution grids: Process and tool adoption. In: Smart Grid Communications (SmartGridComm), 2016 IEEE International Conference on, pp. 435 – 440. IEEE Publishing, Sidney, Australia (2016)
15. Fitzgerald, J., Pierce, K.: Co-modelling and Co-simulation in Embedded Systems Design. In: Collaborative Design for Embedded Systems, pp. 15–25. Springer, Berlin, Heidelberg (2014). URL https://link.springer.com/chapter/10.1007/978-3-642-54118-6_2. https://doi.org/10.1007/978-3-642-54118-6_2
16. Fox-Brewster, T.: Petya Or NotPetya: Why The Latest Ransomware Is Deadlier Than WannaCry (2017). URL http://www.forbes.com/sites/thomasbrewster/2017/06/27/petya-notpetya-ransomware-is-more-powerful-than-wannacry/
17. Gonzalez-Longatt, F., Luis Rueda, J.: PowerFactory Applications for Power System. Power Systems. Springer International Publishing (2014). URL http://www.springer.com/de/book/9783319129570. https://doi.org/10.1007/978-3-319-12958-7
18. Gouglidis, A., Green, B., Busby, J., Rouncefield, M., Hutchison, D., Schauer, S.: Threat Awareness for Critical Infrastructures Resilience. In: Resilient Networks Design and Modeling (RNDM), 2016 8th International Workshop on Resilient Networks Design and Modeling, pp. 196 – 202. IEEE Publishing, Halmstad, Sweden (2016)
19. Grimmett, G.R.: Percolation Theory. Springer, Heidelberg, Germany (1989)
20. Gross, J., Cylance SPEAR Team: Operation Dust Storm (2016). URL https://www.cylance.com/content/dam/cylance/pdfs/other/Op_Dust_Storm_Report.pdf
21. Homeland Security: NIPP 2013: Partnering for Critical Infrastructure Security and Resilience (2013). URL https://www.dhs.gov/publication/nipp-2013-partnering-critical-infrastructure-security-and-resilience
22. Hutchison, D., Rouncefield, M., Busby, J., Gouglidis, A.: Deliverable 3.1 - Analysis of human and organizational factors in utility vulnerability and resilience. HyRiM Deliverable, Lancaster, UK (2015). URL https://www.hyrim.net/project-deliverables/
23. ICS-CERT: Cyber-Attack Against Ukrainian Critical Infrastructure (2016). URL https://ics-cert.us-cert.gov/alerts/IR-ALERT-H-16-056-01
24. ICS-CERT: Indicators Associated With WannaCry Ransomware (2017). URL https://ics-cert.us-cert.gov/alerts/ICS-ALERT-17-135-01I
25. ICS-CERT: Petya Malware Variant (2017). URL https://ics-cert.us-cert.gov/alerts/ICS-ALERT-17-181-01C

26. International Standardization Organization: ISO 28001: Security management systems for the supply chain - Best practices for implementing supply chain security, assessments and plans - Requirements and guidance. Geneva, Switzerland (2007). English version
27. International Standardization Organization: ISO 31000: Risk Management – Principles and Guidelines. Geneva, Switzerland (2009). English version
28. International Standardization Organization: ISO/IEC 27005: Information technology - Security techniques - Information security risk management. Geneva, Switzerland (2011). English version
29. ISACA: COBIT 5 for Risk. Rolling Meadows, USA (2013)
30. ISACA: State of Cyber Security. Implications for 2016. An ISACA and RSA Conference Survey (2016). URL http://m.isaca.org/cyber/Documents/state-of-cybersecurity_res_eng_0316.pdf
31. Kenah, E., Robins, J.M.: Second look at the spread of epidemics on networks. Physical Review. E, Statistical, Nonlinear, and Soft Matter Physics **76**(3 Pt 2), 036,113 (2007). https://doi.org/10.1103/PhysRevE.76.036113
32. König, S., Rass, S., Schauer, S.: A Stochastic Framework for Prediction of Malware Spreading in Heterogeneous Networks. In: B. Brumley, J. Röning (eds.) Secure IT Systems. 21st Nordic Conference, NordSec 2016, Oulu, Finland, November 2–4, 2016. Proceedings, pp. 67–81. Springer International Publishing, Cham (2016)
33. König, S., Rass, S., Schauer, S., Beck, A.: Risk Propagation Analysis and Visualization using Percolation Theory. International Journal of Advanced Computer Science and Applications(IJACSA) **7**(1), 694 – 701 (2016)
34. Kovacs, E.: Critical Infrastructure Incidents Increased in 2015: ICS-CERT (2016). URL http://www.securityweek.com/critical-infrastructure-incidents-increased-2015-ics-cert
35. Maschler, M., Solan, E., Zamir, S.: Game Theory. Cambridge University Press (2013)
36. Newman, M.E.J.: Spread of epidemic disease on networks. Physical Review E **66**(1), 016,128 (2002). https://doi.org/10.1103/PhysRevE.66.016128. URL https://link.aps.org/doi/10.1103/PhysRevE.66.016128
37. Oppliger, R.: Quantitative Risk Analysis in Information Security Management: A Modern Fairy Tale. IEEE Security Privacy **13**(6), 18–21 (2015). https://doi.org/10.1109/MSP.2015.118
38. Paganini, P.: Operation Dust Storm, Hackers Target Japanese Critical Infrastructure (2016). URL http://securityaffairs.co/wordpress/44749/cyber-crime/operation-dust-storm.html
39. Rass, S.: On Game-Theoretic Risk Management (Part One) – Towards a Theory of Games with Payoffs that are Probability-Distributions. ArXiv e-prints (2015)
40. Rass, S., König, S., Schauer, S.: Deliverable 1.2 - Report on Definition and Categorisation of Hybrid Risk Metrics. HyRiM Deliverable, Vienna, Austria (2015). URL https://www.hyrim.net/project-deliverables/

41. Rass, S., König, S., Schauer, S.: Uncertainty in Games: Using Probability-Distributions as Payoffs. In: Decision and Game Theory for Security, no. 9406 in Lecture Notes inComputer Science, pp. 346 – 357. Springer, London, UK (2015)
42. Rass, S., König, S., Schauer, S.: Decisions with Uncertain Consequences - A Total Ordering on Loss-Distributions. PLOS ONE **11**(12), e0168,583 (2016). https://doi.org/10.1371/journal.pone.0168583. URL http://journals.plos.org/plosone/article?id=10.1371/journal.pone.0168583
43. Rass, S., König, S., Schauer, S.: Defending Against Advanced Persistent Threats Using Game-Theory. PLOS ONE **12**(1), e0168,675 (2017). https://doi.org/10.1371/journal.pone.0168675. URL http://journals.plos.org/plosone/article?id=10.1371/journal.pone.0168675
44. Salathé, M., Jones, J.H.: Dynamics and Control of Diseases in Networks with Community Structure. PLOS Computational Biology **6**(4), e1000,736 (2010). https://doi.org/10.1371/journal.pcbi.1000736. URL http://journals.plos.org/ploscompbiol/article?id=10.1371/journal.pcbi.1000736
45. Sander, L.M., Warren, C.P., Sokolov, I.M., Simon, C., Koopman, J.: Percolation on heterogeneous networks as a model for epidemics. Mathematical Biosciences **180**(1), 293–305 (2002). https://doi.org/10.1016/S0025-5564(02)00117-7. URL http://www.sciencedirect.com/science/article/pii/S0025556402001177
46. Stoneburner, G., Goguen, A., Feringa, A.: NIST SP800-30 Risk Management Guide for Information Technology Systems. Gaithersburg, USA (2002). URL http://csrc.nist.gov/publications/nistpubs/800-30/sp800-30.pdf
47. Varga, A., Hornig, R.: An Overview of the OMNeT++ Simulation Environment. In: Proceedings of the 1st International Conference on Simulation Tools and Techniques for Communications, Networks and Systems & Workshops, Simutools '08, pp. 60:1–60:10. ICST (Institute for Computer Sciences, Social-Informatics and Telecommunications Engineering), ICST, Brussels, Belgium, Belgium (2008). URL http://dl.acm.org/citation.cfm?id=1416222.1416290
48. Zetter, K.: Everything We Know About Ukraine's Power Plant Hack | WIRED (2016). URL https://www.wired.com/2016/01/everything-we-know-about-ukraines-power-plant-hack/

Chapter 13
Protecting Water Utility Networks from Advanced Persistent Threats: A Case Study

Antonios Gouglidis, Sandra König, Benjamin Green, Karl Rossegger, and David Hutchison

13.1 Introduction

Advanced persistent threats (APTs) naturally respond to the increasing diversity of security precautions by mounting attacks in a stealthy and equally diverse fashion to remain *under the radar* for as long as required and until the target system has been compromised. They combine a variety of different attack vectors ranging from social engineering to technical exploits and are tailored to attacking specific organizations, information technology (IT) network infrastructures, and existing security measures within organizations [15]. In particular, the application of social engineering in the opening stages of an APT lets the attacker bypass many technical measures, such as intrusion detection and prevention systems, to efficiently

A. Gouglidis (✉) · B. Green · D. Hutchison
School of Computing and Communications, InfoLab21, Lancaster University, Lancaster LA1 4WA, UK
e-mail: a.gouglidis@lancaster.ac.uk; b.green2@lancaster.ac.uk; d.hutchison@lancaster.ac.uk

S. König
AIT Austrian Institute of Technology GmbH, Centre for Digital Safety and Security, Giefinggasse 4, 1210 Vienna, Austria
e-mail: sandra.Koenig@ait.ac.at

K. Rossegger
Linz AG Telekom, Wiener Straße 151, 4021 Linz, Austria
e-mail: k.rossegger@linzag.at

S. Rass, S. Schauer (eds.), *Game Theory for Security and Risk Management*, Static & Dynamic Game Theory: Foundations & Applications,
https://doi.org/10.1007/978-3-319-75268-6_13

(and economically) get through the outer protection (perimeter) of the IT network. Thus, countermeasures may come too late to be effective since the sufficient damage has already been caused by the time the attack is detected. The diversity and usual stealth of APTs render them a problem of vital importance in critical infrastructures. Moreover, information on the attack vector(s) (e.g., zero-day vulnerabilities) may be unavailable, and the incentives of the attacker(s) may often be vague or impenetrable.

The number and severity of APT attacks against critical infrastructures have increased significantly over time, with the Stuxnet malware being a precursor of many such attacks [2, 10, 8]. Stuxnet was discovered into Iran's nuclear plants, sabotaging the nuclear centrifuges. In the following years, other APT attacks, such as Operation Aurora, Shady Rat, Red October, and MiniDuke, have reached public view [13, 3, 16]. Additionally, the Mandiant Report [12] explicitly stated how APTs are used on a global scale for industrial espionage, as well as that attackers are often closely connected to governmental organizations.

With regard to propagation techniques, APTs are not only focusing on a single vulnerability of a system (which could be detected and eliminated easily) but also using a chain of vulnerabilities in different systems to reach high-security areas within a company network. Adversaries often take advantage of the fact that most security controls are applied on the perimeter. However, once access to the internal network is achieved, a threat actor might have a good chance to be unnoticed, and a series of internal attacks may proceed with little or no resistance. Although several guidelines and recommendations exist to secure an internal network, e.g., using a demilitarized zone (DMZ), the intensity of monitoring them is not always adequate. Even more, intrusion detection or intrusion prevention systems might require a large amount of administration and human resources to monitor the output of these systems and cope with potential true/false negative and true/false positive notifications.

In this chapter, we apply processes developed within the HyRiM project (see Chapter 12) to ensure certain goals are met under the threat of an APT. The HyRiM processes are preventive in the sense of estimating and minimizing the risk of a successful APT from the beginning. The game-theoretic framework applied in HyRiM (see Chapters 2 and 3) is used to optimize the defense against a stealthy invader, who attempts to penetrate the system on a set of known paths, while the defender does its best to protect all of these paths simultaneously.

The remainder of this chapter is organized as follows. In Section 13.2, we provide the description of a case study based on a European water utility organization. In Section 13.3, we define the main goals for the risk management framework and establish the context of the water utility organization. Risks are identified and analyzed in Sections 13.4 and 13.5, respectively. An optimal solution is presented in Section 13.6, and concluding remarks are provided in Section 13.7.

13.2 Case Study Description

In this case study, we examine an APT in the context of a European utility organization that provides its services to more than a hundred municipalities in its region. In the following, we provide further information with regard to its water department, which will be considered throughout the chapter. The water department is focused on the water quality and is responsible for the planning, building, and maintenance of the whole water network. To ensure a sustainable water quality, the organization has its own institute for water processing, sewage cleaning, and research. All of these functions are supported by an industrial control system (ICS).

After analyzing the network of the utility organization, we compiled the collected data and prepared a high-level network architecture of the organization's network and then summarized its main characteristics and security posture. A detailed network diagram is omitted due to privacy and safety concerns. Instead, a high-level dataflow diagram is provided for this purpose. The dataflow diagram in Figure 13.1 is the result of an analysis of the actual network architecture. Furthermore, it depicts at a very basic level the variations in dataflows across different field sites and central systems, which take place between the *office* (information technology (IT)) network and the *process* (operational technology (OT)) networks. The applied layered categorization refers to the Purdue Enterprise Reference Architecture [6], allocating devices to their appropriate system levels. The designation of one- or two-way arrows is used to identify the predominant flows of data. While all communications are two way, in some instances, a one-way arrow is applied due to the predominant usage of the link in question.

Considering the above infrastructure, we examine the deployment of an APT within it. To do this, we assume the existence of an adversary who initially collects information about the organization under attack. The collected information may provide useful indications for deploying targeted attacks, e.g., to initiate social attacks on selected individuals or even identify contractors that are external to the organization. Such information may increase the likelihood of a successful attack. Specifically, we assume the following end-to-end scenario, also known as a *kill chain*:

- Using open-source intelligence (OSINT), a threat actor (TA), i.e., an individual or a group posing a threat [7], can identify employees working for the organization, e-mail structures, partners, and external contractors.
- By deploying a spear-phishing campaign, the TA can target appropriate individuals in order either to capture login credentials or to compromise their devices (introducing the APT).
- The TA can visit the facilities of partners and external contractors and look for alternative entry points, if step two yields insufficient results.
- The TA may review the information collected through each of the abovementioned steps.

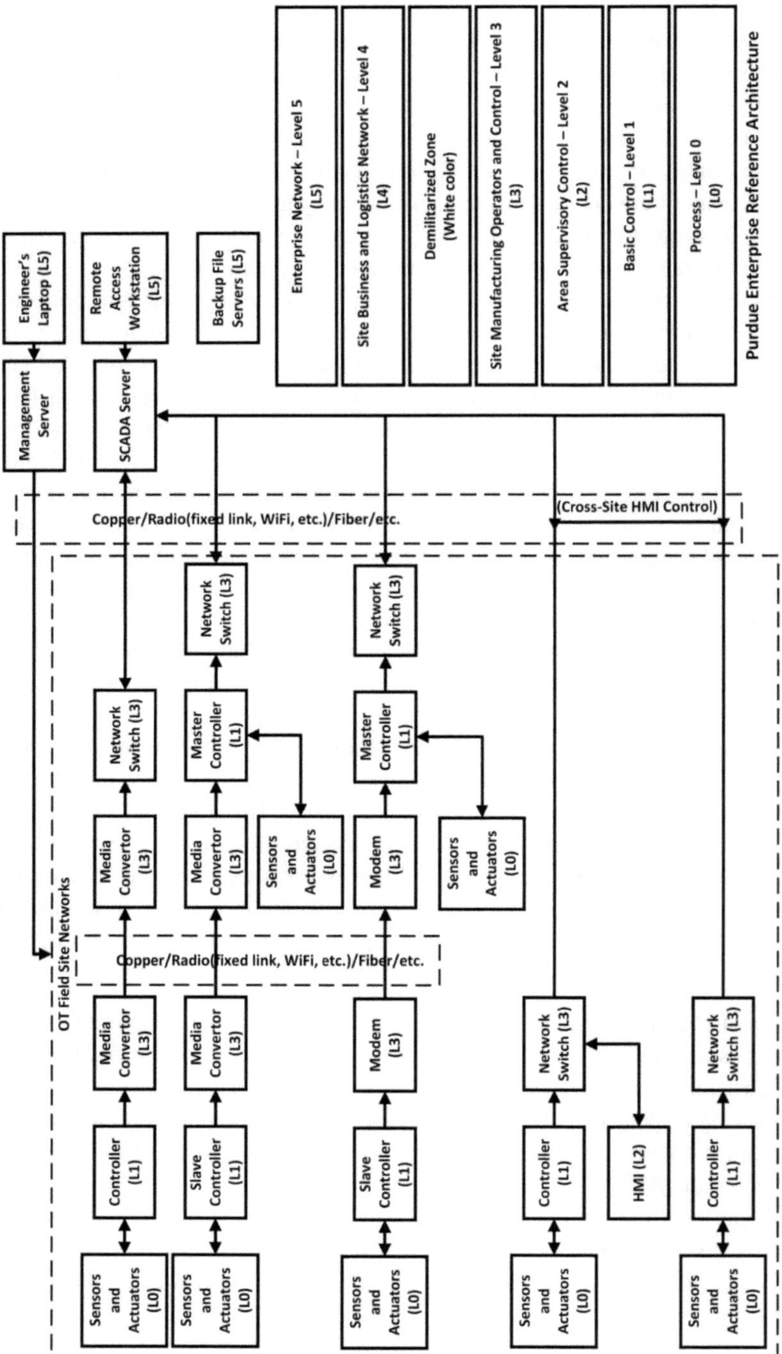

Fig. 13.1: High-level view of OT data-flows

- The TA/APT may explore and expand footholds gained on appropriate devices and networks to capture as much information about the target organization as possible.
- The TA may compromise potential WiFi devices on the OT network.
- Via the WiFi device, the TA may launch attacks to cause a failure on the controllers.
- The TA/APT may compromise the management server in the DMZ.
- Via the management server, the TA/APT may launch an attack to modify controller logic, send control requests, etc. achieving the desired change in process operations.
- Via the management server, the TA/APT may launch an attack to cause a failure of network switches.
- Via the management server, the TA/APT may launch an attack to cause a failure of the controllers.

Although the abovementioned end-to-end scenario (*kill chain*) is defined on the basis of Figure 13.1, it is further visualized in Figure 13.2 and clarified in the following sections.

13.3 Establishing the Context

Establishing the context is a first step of the HyRiM process (see Chapter 12) that starts by defining the objectives that should be achieved and attempts to understand the external and internal factors that may influence the goals. This section summarizes the external and internal environments of the organization.

13.3.1 Definition of Goals

After discussions with experts in the organization, three goals were agreed for further investigation. These are:

- Minimize the damage that can be caused by an attack to the provided service, the service being the provision of water.
- Minimize monetary damage caused by an attack, which may be of technical nature (e.g., substitute devices), of legal nature (e.g., fines), or of any other cost-related damage.
- Minimize reputation damage caused to the organization as a result of an attack. An organization may lose reputation when consumers start to disbelieve in it [1].

Assuming the above goals, several steps are considered in the subsequent risk management processes to ensure that they are achieved adequately through the application of a game-theoretic framework (see Chapter 2).

13.3.2 Data-Flow Based Analysis

In this step, the details of the teams responsible for operating and maintaining systems used to support core operational technology functions are covered. In line with concepts derived through our existing work [4], based on the well-documented Purdue Enterprise Reference Architecture [6], the scope of this exercise was all-inclusive. This means that it covered systems, a broad range of devices, and employees across all six levels, yet allowed for a high-level view of security in a more holistic sense. Due to the focus of HyRiM, more emphasis was spent on discussions around levels 0 to 4, essentially avoiding enterprise systems beyond the point of their interaction with the underlying ICS. For the purpose of this chapter, we will apply again the following terms, *operational technology* (OT) and *information technology* (IT). OT can be seen as any system used for the monitoring, control, and automation of operational processes, with IT systems covering aspects of computational resource applied within all other business areas. It is worth noting that while the participating organization can be seen as an independent body, it operates within a collective group of other organizations. Cross collaboration and shared service offerings within this group raise different issues when discussing internal and external resources.

13.3.3 Social Review Activity

Some points worth noting with regard to social factors are listed in the following:

- **Physical security monitoring:** Across each operational field site, physical security is applied. From a physical standpoint, this includes fencing, locks, etc. From an electrical and remote-monitoring perspective, this includes different types of security sensors, which feed into one alarm panel which in turn connects to an operational controller as a standard hard-wired sensor providing a binary state (intruder/no intruder).
- **ICT partner:** A partner organization is responsible for the IT equipment provided across all organizations within the group. While this resource does not necessarily require access to operational sites, some devices used across such sites are under their support framework.
- **Telecommunication partner:** A partner organization is responsible for inter-site communications, covering their support of some OT controllers at level 1; their responsibility for all communications and associated security controls is a salient point. The level of electronic access this partner has is organization wide (levels 1 to 5).

13.3.4 Business Process Analysis

As a next step, the main process and services carried out by the organization's employees are identified and analyzed. The findings listed in the following are the result of three different activities. The data flow analysis and social review, which have already been mentioned in the previous paragraphs, resulted in examining the organization on the basis of the Purdue model and helped in identifying the details on the teams responsible for operating and maintaining systems used to support core OT functions. Additionally, a technical review activity was carried out. A major result of this activity consists of the dataflow diagram (see Figure 13.1). In detail, this analysis resulted in the identification of the following critical, technical, networked components:

- **Controllers:** Different manufacturers' products are used to monitor, control, and automate the underlying operational process. For the purpose of standardization across the OT environment, a single (standard) protocol is applied by all equipment.
- **Switches:** Network switches provide a resource by which controllers can connect to the process network, offering two-way communications between field sites and centralized systems.
- **Modems:** Modems are used as a communication mechanism between master and slave controllers.
- **Media converters:** As with modems, media converters are used to facilitate communications between master and slave controllers. In addition, media converters are also used to connect controllers to process network switches.
- **Radio:** Several master radio stations concentrate data collection between master controllers and their associated slaves.
- **Compact Human Machine Interfaces (HMIs):** Compact HMIs are used on a number of field sites for local process monitoring and control.
- **WiFi:** Wireless technology is enabled in some areas of the network.
- **Servers/workstations:** Several devices within the OT network were identified as critical assets, e.g., SCADA servers and a management server used for remote management of controllers.
- **Laptops:** Laptops are used by the organization's employees and partners. The laptops are supported by a partner organization able to move between the IT and OT networks while also making use of mobile data technologies providing a level of remote access.
- **Communication mediums:** A partner company manages all communications, which include the use of fiber, copper, radio, etc. This management includes all currently applied network-based security controls.

13.4 Risk Identification

Risk identification involves the application of systematic techniques to understand a range of scenarios describing what could happen, how, and why. Therefore, the infrastructure within the scope of the risk management process needs to be defined, including technical assets, organizational roles and individual personnel, as well as their interdependencies. Based on that, potential threats to the main assets of the organization can be identified.

13.4.1 Threats to Main Assets

- **Radio jamming/data manipulation:** Attacking techniques, as described in [11], could be applied in order to achieve access to the process network. This level of access would be localized to the field site on which the WiFi is deployed. However, should vulnerabilities exist within the process network switches, or if it is possible for the attacker to masquerade as a trusted device, it may be possible to break out from the local network and access a wider variety of devices in alternative field sites and central systems. This level of access could be used to send illegitimate requests to devices across the process network with the ultimate goal being to manipulate process operations. Alternatively, the attacker could simply jam wireless signals, preventing devices communicating across the target link. While accessing the IP network via radio is unlikely, manipulation of data in transit may be possible, impacting the data sent to and from controllers at each end of the link. It is more likely, however, that an attacker would situate a radio jammer at each of the main radio masts, preventing a larger bank of controllers communicating back to the process network.
- **Becoming a HMI:** It is noted that one field site contains a HMI trusted to communicate with controllers locally and across the process network. Where exploitation of the core network switches, as previously described, may not be achievable, masquerading as a HMI with the described level of access could be applied to launch a wider attack. However, this attack would require physical access to the field site in question.
- **Becoming a master:** There exist several slave nodes communicating with one master. Similar to the HMI example described above, should an attacker gain physical access to a master, it may be possible to masquerade as the master and communicate with its associated slave nodes.
- **Network bridging:** Alternative to technical exploits could be the social engineering of trusted systems' users as an attempt to obtain legitimate credential sets. This attack would require physical access to the office network. The deployment of compact battery-powered devices (e.g., Raspberry Pi) could be used to provide remote access, meaning once deployed, an attacker could continue the attack while no longer on the organization's premises.

- **Target external resources:** An attacker could target external contractors, responsible for maintaining some of the organization's controllers and servers. Possibly seen as weaker in terms of their cybersecurity capabilities, the information they hold and trusted access they have would be of great value to an attacker. Compromising relevant resources could provide a level of access to field site networks. For example, should attackers gain access to an engineer's laptop, they could modify backups of controllers' logic/configuration, should those be downloaded. As a recall, controllers could begin to operate outside of their normal parameters as per the defined modifications.
- **Backup servers:** The previously discussed attack scenarios appear relatively simplistic, i.e., gain access and then attack. Achieving a level of impact, which would cause significant degradation of service, would likely be more challenging. It is for this reason an attacker may first need to gain access to configuration files, network diagrams, etc. Backups of critical resources are often stored on office network servers. Applying the same principle as described in *network bridging*, an attacker could gain access to the office network and seek to compromise backup servers to retrieve these resources. Once retrieved, the abovementioned attack scenarios become more achievable and meaningful.

13.4.2 Vulnerability Identification

Conducting penetration testing on devices or systems in a utility network may impose several risks to its operational state, e.g., cause service interruption. Therefore, we conduct penetration testing activities on devices and systems installed in Lancaster University's ICS test-bed [5] and consider a mapping between the two networks (i.e., the emulated and the actual network). This would eventually help us in identifying the likelihood of vulnerabilities of such devices/systems in an emulated environment and avoid any privacy and security concerns with regard to the actual infrastructure of the organization. The estimation of likelihood values for the vulnerabilities will be defined through the CVSS exploitability metric (see Chapter 8). The result of this analysis will be combined with additional semantic information, using the Purdue model. This process resulted in mapping likelihood values of vulnerabilities with elements of a network/system diagram, as depicted in Figure 13.2. Specifically, the main set of software/hardware components that are assessed for vulnerabilities in the emulated ICS test-bed are:

- **SCADA server:** The operating system of this host machine is Windows Server 2008 and has installed the ClearSCADA Server software. The latter is designed to work in water treatment facilities and provide features for the remote management of devices in the OT network and for keeping historical field data from controllers. Known vulnerabilities are CVE-2014-5411, CVE-2014-5412, and CVE-2014-5413.

- **Network switches:** These are ordinary network switches (i.e., not industrial switches). In the case of our test-bed, these are CISCO Catalyst 2950 switches. Known vulnerabilities are CVE-2001-0895 and CVE-2005-4258.
- **Controllers:** We have a diverse set of controllers installed in the emulated ICS test-bed. For the needs of this case study, we used the following: Siemens SIMATIC S7-1200 PLC, Siemens SIMATIC S7-300 PLC, ET200S PLC, and Allen-Bradley (AB) Micro 820 PLC. With regard to the individual hardware components:
  - **Siemens SIMATIC S7-300 PLC:** Vulnerabilities are CVE-2015-5698 and CVE-2016-3949.
  - **Siemens SIMATIC S7-1200 PLC:** Vulnerabilities are CVE-2013-2780 and CVE-2014-2250.
  - **Siemens ET 200S PLC:** Currently, no vulnerability in the CVE database;
  - **Siemens SCALANCE X208 WiFi switch:** Currently, no vulnerability in the CVE database.
  - **Allen-Brandley Micro 820 PLC:** Currently, no vulnerability in the CVE database.
- **Human Machine Interface (HMI):** The Allen-Bradley PanelView 800 HMI is used by a human operator to guide the control process on this field site, such as turning on/off automatic mode for the AB Micro 820 PLC, opening/closing valves that the PLC is connected to, and turning on/off manual mode of operation. Currently, no vulnerability exists in the CVE database.
- **Media convector:** The softing echolink S7-compact is a media converter for communication with Siemens S7 controllers. Currently, no vulnerability exists in the CVE database.
- **Management server:** In the emulated ICS test-bed, the role of the management server is taken over by the *engineer's laptop*, which can be connected in the network and configure controllers remotely by using the appropriate software packages per controller brand. Those are the SIMATIC STEP 7 SIMATIC Manager for connecting with SIEMENS S7-300 PLC, SIMATIC STEP 7 TIA Portal for connecting with SIEMENS S7-1200 PLC, and Connected Components Workbench for connecting with Allen-Bradley PanelView 800 HMI. With regard to the individual software components:
  - **SIMATIC STEP 7:** Vulnerabilities are described in CVE-2014-1594.
  - **Connected Components Workbench:** Currently, no vulnerability in the CVE database.
  - **TIA Portal:** Currently, no vulnerability in the CVE database.
  - **Allen-Brandley PanelView 800 HMI:** Currently, no vulnerability in the CVE database.

13.5 Risk Analysis

Risk analysis is concerned with developing an understanding of each risk, its consequences, and the likelihood of these consequences. In general, the level of risk is determined by taking into account the present state of the system, existing controls, and their level of effectiveness.

13.5.1 Likelihood Analysis

Since an APT attack is considered to be highly sophisticated, we can assume that the attacker can obtain information about the structure and the various devices of the network of the utility provider (see Section 13.4). Therefore, such an attack is tailored to the specific company and aims to exploit existing vulnerabilities (see Section 13.4.2).

However, it can be argued that the likelihood of such an attack is more difficult to estimate. Generally, the applied model can work with different types of data, e.g., with vulnerability assessments such as CVSS. Still, in case of an APT, these likelihoods are fraught with uncertainty since we only have limited knowledge about the attacker. Thus, the most feasible approach is to ask as many experts as possible, compile an empirical distribution, and then aggregate the received information to a single number [9].

Figure 13.2 depicts consolidated information, including the various components of the OT network, the main systems in the IT network that may provide access to the process network, as well as the likelihood of vulnerabilities in systems/devices to be exploited. This results in providing an understanding of the major risks and gives an indication of potential attack vectors and attack paths.

13.6 Risk Treatment

During risk treatment, existing controls are improved and new controls are implemented. In the HyRiM process, the goal is to identify the optimal set of controls to reduce the maximum damage that can be caused by an attacker to a minimum. In this context, the optimality of the resulting controls is given due to the game-theoretic algorithms applied in the approach. Risk can also be transferred (e.g., purchasing insurance) or retained.

In the following, we define a list of attack and defense strategies toward achieving the goals defined in Section 13.3.1.

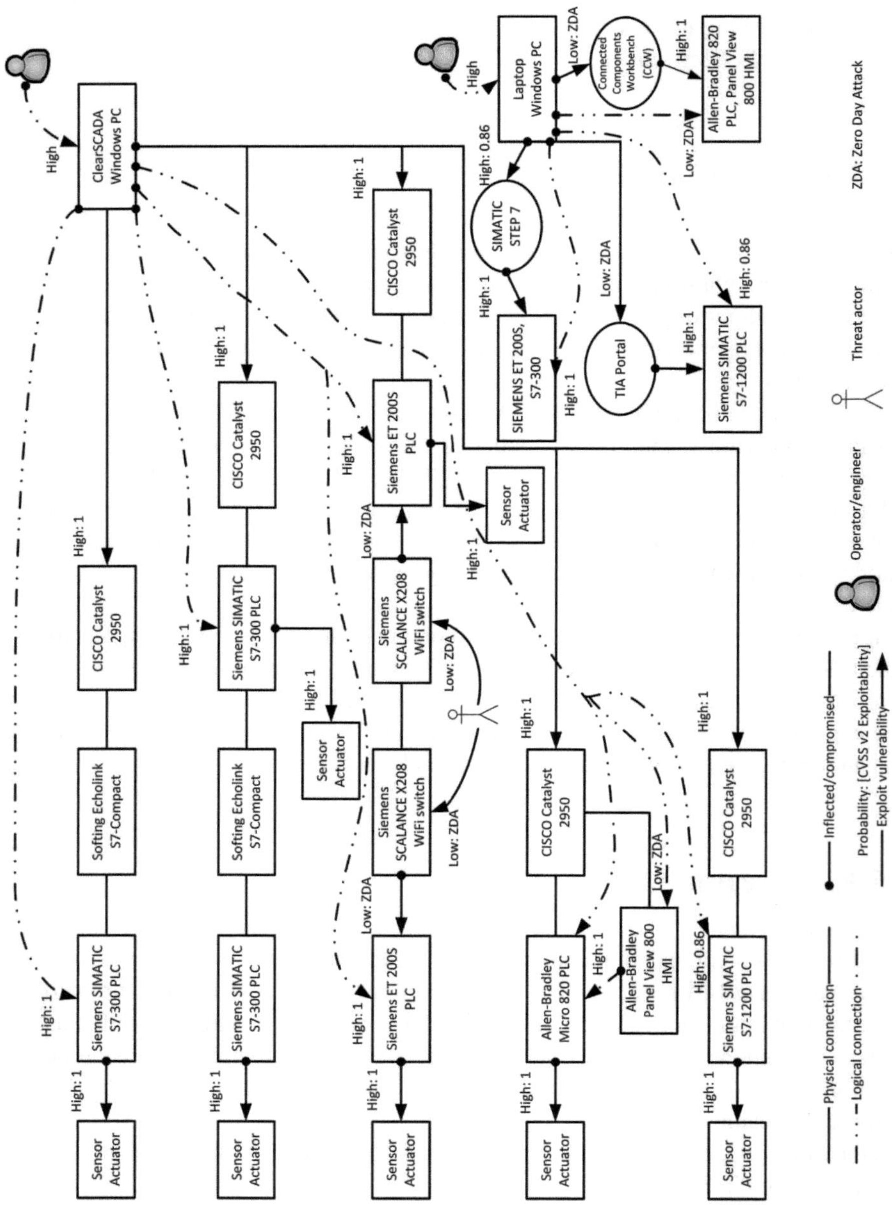

Fig. 13.2: Estimation of likelihood based on CVSS's exploitability metric

13.6.1 Attack Strategies

Considering the end-to-end scenario in Section 13.2 and the diagram in Figure 13.2, we define in the following a list of potential attack vectors and attack paths.

A first attack vector may include a social engineering attack on the operator of the SCADA server. This includes sending a spear-phishing e-mail to the operator that is read on the server or inserting an infected USB device on the server. In both cases, a malware will be executed automatically on the host machine and will try to scan the network for weak points, gain access to them, and escalate.

- **Attack vector 1:** An attack can be deployed from the SCADA server to all CISCO switches. This results in a denial-of-service (DoS) attack on each of the CISCO switches. Using this attack path, the attack cannot propagate further to the lower end of the OT network. The operation of the physical process is not affected. In the following, we list potential attack paths for this scenario:
  - Operator → ClearSCADA/Windows PC → CISCO Catalyst 2950 (x4).
- **Attack vector 2:** The SCADA server has a logical connection with all the underlying PLCs and the HMI on the OT network. The ClearSCADA software may collect data from master controllers, and known vulnerabilities on the software may allow unauthorized modification of data. Another attack vector assumes exploiting known or zero-day vulnerabilities on the logically connected devices. Such an attack may impact the normal operation of some devices and sensors due to existing vulnerabilities, e.g., vulnerabilities related with disruption of services and modification of the device's logic. In the following, we list potential attack paths for this scenario:
  - **Attack vector 2.1:** Operator → ClearSCADA/Windows PC → Siemens SIMATIC S7-300 PLC → Sensor/Actuator (x2).
  - **Attack vector 2.2:** Operator → ClearSCADA/Windows PC → Siemens SIMATIC ET 200S PLC → Sensor/Actuator (x2).

A social engineering attack can be conducted on an engineer of the organization or on an external partner who can visit the various field sites. In these cases, the laptop should be physically connected to a device to perform any required maintenance operations. The attack vector assumes the user of the laptop to be deceived to install a malware on it. Subsequently, when the laptop is connected on a device of the OT network, the APT may propagate and escalate to cause damage.

- **Attack vector 3:** A laptop having the appropriate interfaces and software to connect to devices may be infected by an APT. The APT can scan a device when the laptop is connected to it and exploit known (or zero-day) vulnerabilities. The vulnerabilities may be exploited either via the software that is used to maintain the devices or directly through the host operating system. In the following, we list potential attack paths for this scenario:

- **Attack vector 3.1**
  - Engineer/contractor → laptop/Windows PC → SIMATIC STEP 7 → SIEMENS ET 200S PLC → sensor/actuator (x2).
  - Engineer/contractor → laptop/Windows PC → SIEMENS ET 200S PLC → sensor/actuator (x2).
- **Attack vector 3.2**
  - Engineer/contractor → laptop/Windows PC → SIMATIC STEP 7 → SIEMENS S7-300 PLC → sensor/actuator (x3).
  - Engineer/contractor → laptop/Windows PC → SIEMENS S7-300 PLC → sensor/actuator (x3).

A threat actor can attack the WiFi switches located in the OT network of the organization. This would require either to exploit a zero-day vulnerability or try to decrypt the password through brute forcing or rainbow tables.

- **Attack vector 4:** If a threat actor can gain access to any of the WiFi switches, it would be possible to attack on devices that are logically connected to the switches. In the following, we list potential attack paths for this scenario:
  - Threat actor → Siemens SCALANCE X208 WiFi Switch → Siemens SIMATIC ET 200S PLC → sensor/actuator (x2).

13.6.2 Defense Strategies

In the following, we list countermeasures (i.e., defense strategies) for the abovementioned attack strategies:

- **Status quo (D1):** Do not change anything.
- **Training (D2-D4):** Employees should be aware of potential attack vectors aiming to them (e.g., social engineering attacks) and how to protect against them. The frequency of security awareness training courses needs to be examined:
  - Annually.
  - Once every 2 years.
  - Train only new personnel.
- **Password change (D5-D7):** Computing systems and devices may be password protected. The frequency of changing passwords needs to be examined:
  - Annually.
  - Every time a device is changed.
  - Every time an operator/engineer/contractor is changed.
- **Update (D8-D10):** Computing systems require updating the operating system, potential software offering protection against attacks (e.g., antivirus, antimalware), and software that interfaces with various PLC devices (e.g., SIMATIC STEP 7):

  - Enable automatic updates.
  - Update every year.
  - Apply only major updates.

- **Patch/replace (D11-D13):** Devices such as PLCs, HMIs, switches, etc. may need to be patched/replaced. The frequency of patching/replacement needs to be examined:
  - When a device fails to operate.
  - Annually.
  - When major vulnerabilities are known for a device.
- **Manual checking of water (D14-D16):** Additional checking of the water quality may be required to ensure the level of provided service to consumers. Although this information can be collected by the SCADA server, the latter may collect modified data in the case of an attack. The frequency of manually checking the water needs to be examined:
  - Daily.
  - Weekly.
  - Monthly.

13.6.3 Estimate Damage in Case of an Attack

For each scenario defined by a pair (d_i, a_j) of a defense strategy d_i and an attack strategy a_j, the damage is assessed by experts on a five-tier scale representing the categories *very low* (1), *low* (2), *medium* (3), *high* (4), and *very high* (5). Such an ordinal scale is often used in risk management, especially in cases where exact assessments are not possible. Each expert was asked to estimate the damage for a set of scenarios. This may result in either collecting or not an estimation from an expert, with the latter being the case for the expert to refuse to provide information. In our case study, four experts were asked to estimate the damage for three goals, i.e., minimize the downtime of a service, minimize monetary damage, and minimize reputation damage. In case there were no or not enough data for a scenario, we assumed a worst-case situation and considered the damage to be 5 (i.e., very high). The kernel density estimation used to estimate payoff distributions needs at least two data points to work properly.

The results from the experts' assessment on this five-tier scale are shown in Tables 13.1, 13.2, and 13.3 where AV1, ..., AV4 denote the attack strategies identified in Section 13.6.1 and D1, ..., D16 denote the defense strategies identified in Section 13.6.2. These assessments build up the random payoffs for the upcoming game-theoretic analysis.

Table 13.1: Experts' opinion on expected damage with regard to service downtime

	AV 1	AV 2.1	AV 2.2	AV 3.1	AV 3.2	AV 4
D1	3,5,2,3	5,5,4,4	4,4,4	5,4,3,3	4,3,3	4,3
D2	2,1,2,2	3,3,2,3	3,2,3	2,4,2,2	2,2,2	3,2
D3	2,3,2,2	4,4,3,3	4,3,3	3,4,3,2	3,3,2	4,3
D4	3,4,2,3	5,4,3,4	5,3,4	4,4,3,3	4,3,3	4,3
D5	2,5,2,2	3,5,3,3	3,3,3	3,2,2,2	3,2,2	3,3
D6	3,5,2,2	4,5,3,3	4,3,3	4,3,3,3	4,3,3	4,3
D7	3,5,2,3	4,5,3,4	4,3,4	4,3,2,3	4,2,3	4,3
D8	2,1,1,2	4,1,2,3	3,2,3	3,2,2	3,2,2	2,3
D9	1,3,1,2	3,3,2,3	3,2,3	3,1,2,3	3,2,3	3,3
D10	3,3,2,3	4,3,3,4	4,3,4	4,1,3,3	4,3,3	4,3
D11	3,5,2,3	4,1,2,3	3,2,3	4,2,2	4,2,2	4,2
D12	1,5,2,2	3,3,2,3	3,2,3	2,3,3	2,3,3	2,3
D13	3,5,2,2	4,3,2,3	4,2,3	3,2,3	3,2,3	3,3
D14	1,5,2,3	2,5,4,2	2,4,2	2,4,3	2,3	3
D15	3,5,2,3	3,5,4,2	3,4,2	3,4,3	3,3	3
D16	3,5,2,3	5,5,4,2	5,4,2	4,4,3	4,3	3

Table 13.2: Experts' opinion on expected cost damage

	AV 1	AV 2.1	AV 2.2	AV 3.1	AV 3.2	AV 4
D1	3,4,3,3	4,4,4,3	4,4,3	5,3,4,4	3,4,4	4,4
D2	3,1,2,2	3,1,2,2	3,2,2	2,3,2,2	2,2,2	2,2
D3	3,3,3,3	4,3,3,2	4,3,2	3,3,3,3	3,3,3	3,3
D4	3,4,3,3	4,4,3,3	4,3,3	4,3,3,3	4,3,3	4,3
D5	3,4,3,3	3,4,3,3	3,3,3	2,1,2,3	2,2,3	2,2
D6	3,4,3,2	4,4,3,2	4,3,2	4,2,3,3	4,3,3	3,3
D7	3,4,3,3	3,4,3,3	3,3,3	4,2,3,3	4,3,3	3,3
D8	1,1,2,2	4,1,2,2	4,2,2	2,2,3	2,2,3	2,2
D9	2,3,2,2	3,3,3,2	3,3,2	4,1,3,3	4,3,3	3,3
D10	3,3,3,2	4,3,4,2	4,4,2	5,1,4,3	5,4,3	4,4
D11	3,4,2,2	4,1,2,2	4,2,2	4,2,3	4,2,3	4,2
D12	2,4,3,2	3,2,2,2	3,2,2	3,3,3	3,3,3	3,3
D13	3,4,3,3	4,2,2,3	4,2,3	4,3,3	4,3,3	3,3
D14	4,3,3	4,4,4	4,4	3,4	4	4
D15	4,3,3	4,4,3	4,3	3,4	4	4
D16	4,3,3	4,4,2	4,2	3,4	4	4

Table 13.3: Experts' opinion on expected reputation damage

	AV 1	AV 2.1	AV 2.2	AV 3.1	AV 3.2	AV 4
D1	4,4,3,4	5,4,2,4	5,2,4	5,2,3,3	4,3,3	5,3,3
D2	2,1,2,2	3,1,2,2	3,2,2	2,2,2,2	2,2,2	3,2
D3	3,3,3,3	5,3,3,3	5,3,3	3,2,3,2	3,3,2	4,3
D4	4,3,2,4	5,3,3,3	5,3,3	4,2,3,3	4,3,3	4,3
D5	3,4,3,2	3,4,3,2	3,3,2	3,2,3,2	3,3,2	3,3
D6	3,4,3,3	4,4,3,3	4,3,3	4,3,3,2	4,3,2	4,3
D7	3,4,3,3	4,4,3,3	4,3,3	4,3,3,2	4,3,2	4,3
D8	2,1,2,3	4,1,3,2	4,3,2	3,2,2	3,2,2	2,2
D9	3,3,2,3	3,3,3,3	3,3,3	3,1,3,3	3,3,3	3,3
D10	4,3,3,3	4,3,3,3	4,3,3	4,1,3,3	4,3,3	4,3
D11	4,4,3,3	4,1,3,4	4,3,4	4,3,3	4,3,3	4,2
D12	2,4,3,3	3,2,3,3	3,3,3	3,3,3	3,3,3	3,3
D13	4,4,3,3	4,2,3,3	4,3,3	4,3,3	4,3,3	4,2
D14	2,4,3,3	2,4,3,2	2,3,2	2,2,3	2,3	3
D15	3,4,3,4	3,4,3,3	3,3,3	3,2,3	3,3	3
D16	4,4,3,4	4,4,3,4	4,3,4	4,2,3	4,3	3

13.6.4 Game-Theoretic Optimization of Defense Actions

Based on the strategies defined in Sections 13.6.1 and 13.6.2 and the payoffs resulting from the data presented in Section 13.6.3, we now set up a game to find the optimal defense strategy as well as the worst-case damage. To this extent, we apply a multi-objective security game (MOSG) between an attacker and a defender (e.g., a utility provider) with random payoffs. This game is assumed to be a zero-sum game where the defender tries to minimize his loss (payoff) while the attacker tries to maximize it.

Computation of an equilibrium is done by means of the generalized fictitious play algorithm (see Chapter 3). The set of strategies and the payoffs provide adequate input for computing the equilibrium. However, it is possible to prioritize the different goals. From discussions with experts, the different goals were assigned the following weights that represent their importance to the company: service 23/40, cost 6/40, and reputation 11/40.

Applying the adapted fictitious play algorithm contained in the `R` package `HyRiM` [15] with $T = 1000$ iterations, we find the following optimal defense strategy:

Table 13.4: Optimal defense strategy according to Nash equilibrium

	Training yearly	Update major	Patching failure	Patching major
Frequency	0.092	0.710	0.083	0.115

The defense strategies not listed in Table 13.4 were assigned with a zero frequency in the Nash equilibrium and thus shall not be chosen at all. This identified solution is optimal in the sense that it minimizes the expected damage, i.e., deviating from this strategy yields a loss that is never smaller than in this case (in fact, if only one equilibrium exists, the damage will be even higher). The optimal defense strategy is illustrated in the top row of Figure 13.3.

Depending on the goal, an APT may have different optimal attack strategies. These are illustrated in the three lower rows of Figure 13.3 (each labeled with the corresponding goal) together with the likelihood of damage to the defender in the case that this optimal strategy is applied (and the defender also follows his optimal strategy). When thinking in terms of service disruption, it may cause maximal damage by mainly choosing attack vector 2.1 (approximately 37% of time) and attack vector 4 (approximately 51% of time), occasionally applying attack vector 1 (approx. 2% of the time), and rarely playing attack vector 2.2 (approx. 10% of the time). With regard to the cost caused to the defender, the APT may cause highest damage when deploying attack vector 3.2 (which got a weight of 99.7% in the mixed equilibrium), while for the reputation, the APT is aiming to mix between attack vector 2.2 (67% of the time) and attack vector 4 (33% of the time). Since the APT may deploy several attack vectors in parallel, it might be able to choose its strategies according to all three equilibria so that it will not deviate from the optimal behavior, which in turn causes the worst-case damage for each goal to the defender.

On the defender's side, the organization should apply the optimal defense strategies described in Table 13.4 to protect against an attack strategy by an APT. Specifically, the defender shall apply only the four strategies listed in Table 13.4 in the corresponding relative frequency identified per se. To this extent, in 9.2% of the time, all employees should attend an annual training course. In 71.0% of the time, major updates of computer systems shall be applied. With regard to patching devices such as PLCs or HMIs, this is done upon failure in 8.3% of the time, while in 11.5% of the time, patches shall be applied on devices with known major vulnerabilities. All the relative frequencies are illustrated on the very top of Figure 13.3 per se.

As long as the overall frequencies correspond to the optimal solution, the defender can randomly choose the order in which these strategies are enforced. This means that the solution has a certain degree of freedom in the sense that if one strategy cannot be applied at some point in time, e.g., due to the absence of a key person, it can be postponed, and another defense mechanism can be used instead.

13.7 Conclusion

In this chapter, we reported on a water utility case study and demonstrated how the HyRiM process can be of benefit when applied in this sort of utility organization. Throughout the chapter, each of the applied processes is described. With regard to widely applied assurance activities (e.g., vulnerability scanning, interviews), only their integration with the process is demonstrated, i.e., presentation of output data. The application of the HyRiM process on the water utility organization resulted

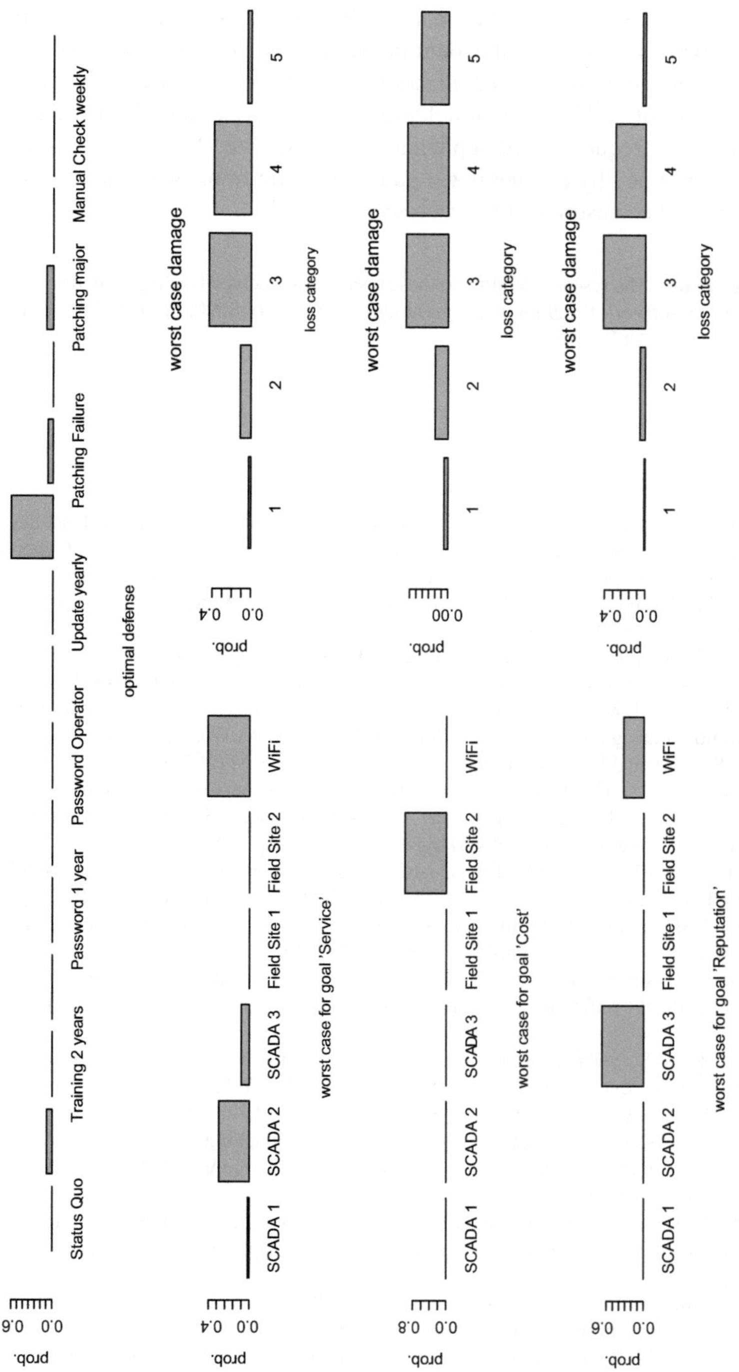

Fig. 13.3: Nash equilibrium and corresponding payoff

in defining optimal protection strategies against an APT and eventually improving its security posture. Specifically, the analysis showed that — based on the data provided by the experts — many of the identified defense strategies do not contribute to reducing the damage within the organization given the identified set of attacks. For example, only 4 out of 16 potential defense strategies are required for the equilibrium. The relative frequencies of application of the selected four defense strategies have been determined by a generalized game-theoretic framework, and the worst-case damage has been estimated for each security goal.

Acknowledgements The research leading to these results has received funding from the European Union Seventh Framework Program under grant agreement no. 608090, Project HyRiM (Hybrid Risk Management for Utility Networks).

References

1. Busby, J.S., Gouglidis, A., Rass, S., König, S.: Modelling security risk in critical utilities: The system at risk as a three player game and agent society. In: Systems, Man, and Cybernetics (SMC), 2016 IEEE International Conference on, pp. 001,758–001,763. IEEE (2016)
2. Falliere, N., Murchu, L.O., Chien, E.: W32. stuxnet dossier. White paper, Symantec Corp., Security Response **5**(6) (2011)
3. Friedberg, I., Skopik, F., Settanni, G., Fiedler, R.: Combating advanced persistent threats: From network event correlation to incident detection. Computers & Security **48**, 35–57 (2015)
4. Green, B., Krotofil, M., Hutchison, D.: Achieving ICS resilience and security through granular data flow management. In: Proceedings of the 2nd ACM Workshop on Cyber-Physical Systems Security and Privacy, October 2016, pp. 93–101. ACM (2016)
5. Green, B., Paske, B., Hutchison, D., Prince, D.: Design and construction of an industrial control system testbed. In: PG Net-The 15th Annual PostGraduate Symposium on the Convergence of Telecommunications, Networking and Broadcasting (2014)
6. ISA99: ISA-62443-1-1: Security for industrial automation and control systems - models and concepts. URL http://isa99.isa.org/ISA99%20Wiki/WP-1-1.aspx,[retrieved:11/09/2017]
7. Johnson, C., Badger, L., Waltermire, D., Snyder, J., Skorupka, C.: Guide to cyber threat information sharing. NIST Special Publication **800**, 150 (2016)
8. Karnouskos, S.: Stuxnet worm impact on industrial cyber-physical system security. In: IECON 2011-37th Annual Conference on IEEE Industrial Electronics Society, pp. 4490–4494. IEEE (2011)
9. König, S., Rass, S., Schauer, S., Beck, A.: Risk propagation analysis and visualization using percolation theory. Int. J. Adv. Comput. Sci. Appl.(IJACSA) **7**(1) (2016)
10. Kushner, D.: The real story of stuxnet. IEEE Spectrum **3**(50), 48–53 (2013)
11. Liu, Y., Jin, Z., Wang, Y.: Survey on security scheme and attacking methods of wpa/wpa2. In: Wireless Communications Networking and Mobile Computing (WiCOM), 2010 6th International Conference on, pp. 1–4. IEEE (2010)
12. MADIANT: APT1: Exposing one of china's cyber espionage units (2013). URL https://www.fireeye.com/content/dam/fireeye-www/services/pdfs/mandiant-apt1-report.pdf,[retrieved:28/09/2017]
13. Moon, D., Im, H., Lee, J.D., Park, J.H.: Mlds: multi-layer defense system for preventing advanced persistent threats. Symmetry **6**(4), 997–1010 (2014)
14. Rass, S., König, S.: R package 'hyrim': Multicriteria risk management using zero-sum games with vector-valued payoffs that are probability distributions (2017). URL https://hyrim.net/software/

15. Ross, R.S.: Managing information security risk: Organization, mission, and information system view. Special Publication (NIST SP)-800-39 (2011)
16. Tankard, C.: Advanced persistent threats and how to monitor and deter them. Network security **2011**(8), 16–19 (2011)

Chapter 14
Assessing the Impact of Malware Attacks in Utility Networks

Sandra König, Antonios Gouglidis, Benjamin Green, and Alma Solar

14.1 Introduction

Utility networks are nowadays often monitored and operated by industrial control systems (ICS). While these systems enhance the level of control over utility networks, they also enable new forms of attacks, such as cyberattacks. In the past few years, many cyberattacks appear to employ malwares, and in many cases, the impact of these attacks is considered to be considerably high, i.e., the damage caused to the systems under attack is high. The first step toward preventing such incidents is to increase our awareness and understanding on how a malware can propagate in a network. Malware spreading can be modeled as a stochastic process on a graph where edges transmit an infection with a specific probability. In practice, this probability depends on the type of the malware (e.g., ransomware, spyware, virus) as well as on the connection type between the nodes (e.g., physical or logical connection). In this chapter, we demonstrate how the abstract model can be put into practice through a case study. Specifically, the case study refers to a European electricity utility organization and takes into consideration the parts of the network that can be infected

S. König
AIT Austrian Institute of Technology GmbH, Centre for Digital Safety & Security, Giefinggasse 4, 1210 Vienna, Austria
e-mail: sandra.Koenig@ait.ac.at

A. Gouglidis (✉) · B. Green
School of Computing and Communications, InfoLab21, Lancaster University, Lancaster LA1 4WA, UK
e-mail: a.gouglidis@lancaster.ac.uk; b.green2@lancaster.ac.uk

A. Solar
Electrica d'Alginet, Alginet, Spain
e-mail: alma@electricadealginet.com

S. Rass, S. Schauer (eds.), *Game Theory for Security and Risk Management*, Static & Dynamic Game Theory: Foundations & Applications,
https://doi.org/10.1007/978-3-319-75268-6_14

by a ransomware. Although this does not consider the infection of devices on the operational network of the organization, it still considers them when estimating the level of damage caused to the utility network in the face of a ransomware threat.

In the past, there have been only a few models for malware propagation in a network, e.g., [1, 2, 3, 6, 14]. However, these models ignore an important property many attacked networks possess: they are not homogeneous in the sense that the malware is not equally likely to spread in all parts of the network. For example, if a utility network is equipped with sensors, a malware attack might yield incorrect measurements, and these incorrect information may later cause problems in the utility network itself. At the same time, the malware itself is not able to spread over this link if the sensor is used only for signaling and does not transmit any data.

One way to initiate a malware attack is by trying to infect a personal device of a member of a company (e.g., by manipulating a USB stick that would be eventually connected with an internal system) since this allows to circumvent security boundaries such as a firewall. Assuming a bring-your-own-device (BYOD) policy, we can illustrate how such an attack can work in practice: a USB stick infected with a malware may be used on a system in an organization (*"BYOD connection"*), and the malware may spread within the locally connected network through email exchange or even by physically connecting the USB stick to other systems. Such an incident was reported for the Iranian nuclear power plant, where Stuxnet was found to infect the systems, resulting in destroying the majority of nuclear centrifuges in the plant [4]. Although BYOD and protection mechanisms against malwares have received considerable attention in the literature [7, 10, 12], an epidemics-like treatment of malware infections (e.g., caused by the application of a BYOD policy) for risk management has just started to develop.

In this chapter, we apply the HyRiM risk management framework (cf. Chapter 12) and show how our model for random error spreading (cf. Chapter 8) can be applied to describe such an incident and, subsequently, how this supports the analysis of the security of the network with a game-theoretic model (cf. Chapters 2 and 3).

The rest of this chapter is structured as follows: in Section 14.2, we provide the description of the case study based on a European electricity provider. In Section 14.3, we define the main goals for the case study that will be investigated through the HyRiM risk management framework and identify internal and external factors that may influence the defined goals. Risks are identified and analyzed in Sections 14.4 and 14.5, respectively. An optimal solution is calculated in Section 14.6, and concluding remarks are presented in Section 14.7.

14.2 Case Study Description

In this case study, we focus on a ransomware attack on a European electricity provider since several incidents have been reported recently. Ransomware is a malware targeting the computing systems of a company with the aim to prevent users from accessing their data either by encrypting files or by simply locking the screens. Recent reported incidents refer to the WannaCry [11] and Petya [13] ransomwares. In such type of attacks, the user being attacked would be asked to pay a ransom to the attacker to regain access to her/his data on the infected system. For a utility provider, such an attack may not only affect data of the organization itself but also may affect customers' data. Besides the monetary loss, any incident involving consumer data may yield to a reputation damage for the company.

In the following sections, we evaluate such a (hypothetical) malware attack on an electricity provider and illustrate how our approach fits with a standard risk management process. To this extent, we consider an electricity distribution system that provides electricity for approximately 5000 users. All these end users are equipped with smart meters to measure the power consumption remotely and to control the grid efficiently. Such a network is controlled by SCADA systems.

For modeling the network, it is required to conduct initially an analysis of the network infrastructure of the electricity provider. Therefore, in Figure 14.1, we depict the – abstract – network diagram of an electricity provider. The diagram includes the main technical nodes and connection paths between the information technology (IT) and operational technology (OT) networks. On the left side of the diagram resides the main IT network of the organization. It is composed of the *"office network"* and the *"SCADA server"*. The main technical nodes of the office network are workstations, laptops, and mobile devices. The latter type of devices can connect to the office network through a wireless access point, which is installed in the office premises. The SCADA server is responsible for the supervision and data acquisition of the devices located in the OT network. Data collected at the office network is also forwarded to a cloud computing system, which is external to the organization and used for analytics and billing services. With regard to the OT network, the bottom of it includes an electricity substation and camera network, and the top of it represents the concentrators and distribution network of the electric power toward to its edge (i.e., consumers).

Based on the abovementioned information, we model the utility provider's network as a directed graph $G(V,E)$ with nodes in V and edges E between these nodes. Assume that all edges in E fall into different non-overlapping classes, where each class has distinct characteristics in how a problem propagates over the respective edges. These could be social connections that use email communication (class 1), technical connections in the actual network (class 2), and logical links between a person and her/his device (class 3) that actually may enable a BYOD attack scenario.

The goal of our case study is to estimate the risk of a malware infection due to the BYOD policy in place or due to a spear-phishing attack. Specifically, we assume the following end-to-end scenario:

- A threat actor prepares and releases a ransomware. This includes setting up a server to communicate with the installed ransomware and exchange keys, which are used for encrypting the storage space.

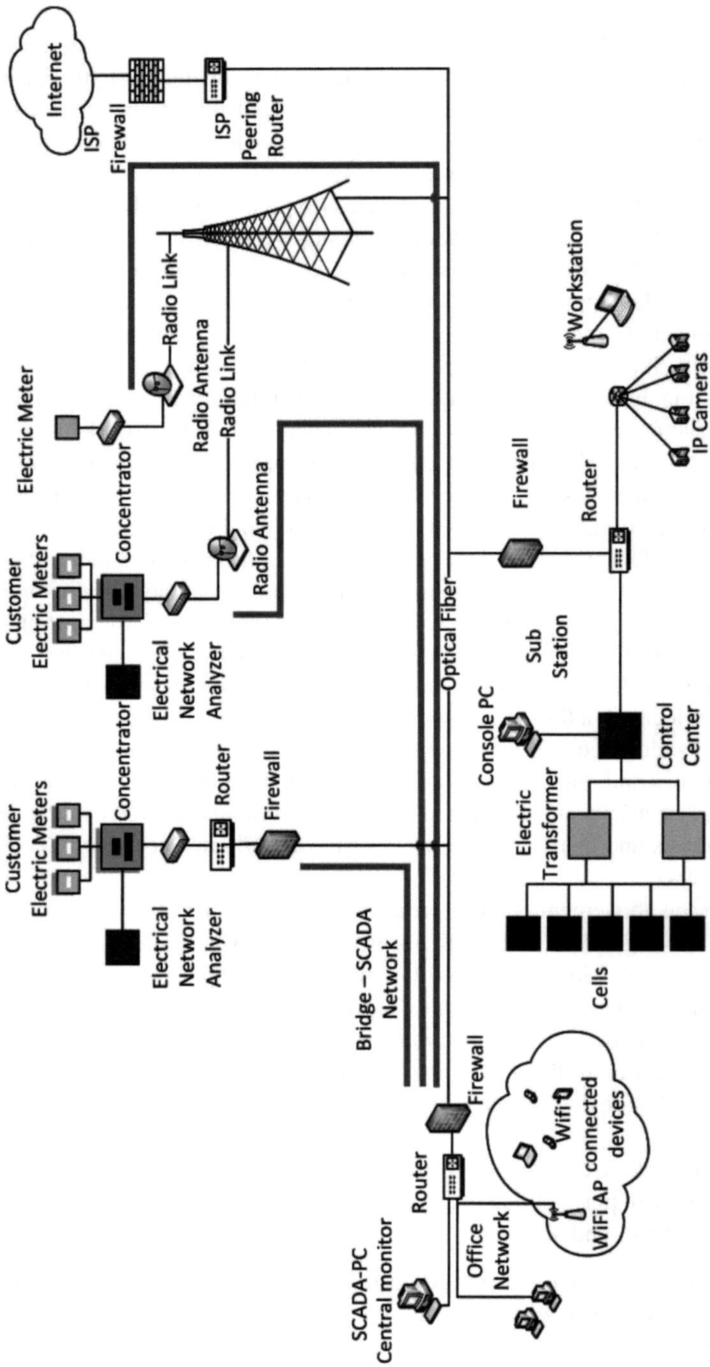

Fig. 14.1: Network diagram of the electricity utility organization

- Attack vectors:
  - A first attack vector can be its inclusion in file-sharing networks, e.g., in torrent files – this attack vector can be related with the BYOD policy in place;
  - A second attack vector can be a spear-phishing campaign (in this scenario, we do not consider the spear-phishing attack to be a targeted attack but instead an opportunistic one);
- The ransomware is installed on the IT network through any of the previous attack vectors;
- The ransomware propagates either by email exchange between the employees or by copying itself on the network file-sharing location;
- The ransomware connects with a remote server, exchanges cryptographic keys, and starts the encryption process;
- The ransomware displays a screen requesting for an amount of money to be paid in an untraceable cryptocurrency – the offer is time limited.

14.3 Establishing the Context

Establishing the context is the first step in the risk management process. It defines the objectives that shall be achieved and attempts to understand the external and internal factors that may influence the goals. Thus, this summarizes a description of the external and internal environment of the organization.

14.3.1 Definition of Goals

After discussions with experts in the organization, three goals were defined. These are:

- Minimize data loss caused by the ransomware. This includes the investigation of various systems given the fact that data of different importance are kept on different systems (e.g., laptops, servers);
- Minimize monetary damage caused by the attack. After the exploitation of a ransomware, various costs shall be estimated depending on how the organization responses to the incident and its mitigation actions;
- Minimize reputation damage caused to the organization as a result of an attack, e.g., due to stolen customer data. Potential reputation lost may be caused when consumers are informed for the incident from different sources of information.

Assuming these goals, several attack and defense strategies are considered in the subsequent processes. However, to ensure that the defined goals are achieved, we attempt to understand the internal and external factors that may influence the abovementioned goals.

14.3.2 Ethnographic Studies

Human and organizational factors were vital parts of our investigation, achieved by carrying out ethnographic studies as part of the case study. This included a visit from an ethnographer to the utility organization, where discussions were conducted and observations were made. The results included information about the systems used in the utility organization, identification of people and their roles, understanding of organizational policies, social relations among employees, and their behavior under specific circumstances or situations.

However, in order to preserve the confidentiality and anonymity of participants, and due to restrictions of the dissemination level of our findings, this information cannot be presented in this chapter in detail. Nevertheless, some general remarks can be provided. Briefly, the ethnographic studies along with the results of a penetration testing produced a technical understanding of some of its vulnerabilities, indicating certain vulnerabilities that the organization had formerly seemed unconcerned about. This, on face value, suggests that the organization members had imperfect mental models or interpretive schemes of the system under their management. But it is important to recognize that such models are adaptations to a wide variety of experiences of which technical experts have no knowledge. They cannot therefore be judged as deficient or otherwise in some general sense. What is more important is that there is an awareness within the system of how those models contribute and detract from its security.

14.3.3 Business Process Analysis

The main process and services carried out and provided by the organization's employees were identified and analyzed across their regular daily operations. This exercise serves as a basis to specify the main goals of the risk management process, discover possible expositions to risks of any kind, and establish possible mitigation actions preventing those risks in later steps. In detail, technical factors, organizational factors, and behavioral factors have been identified. Some examples are (the list is not complete but selected according to the relevance to the examined threat scenario):

- Technical factors:
  - The organization operates an information technology (IT) network (i.e., the office network) where several PCs (i.e., laptops and workstations) are operating in it.
  - A secure (i.e., password-protected) Wi-Fi connection is offered in the office premises of the organization;
  - The organization may operate surveillance technologies on the Operation Technology (OT) network to ensure physical security.

- Organizational factors:
  - Some operations (e.g., billing) are outsourced to third-party organization using cloud services;
  - Security controls are in place;
- Behavioral factors:
  - Usability is more important than prevention.

14.4 Risk Identification

Risk identification involves the application of systematic techniques to understand a range of scenarios describing what could happen, how, and why. Therefore, the infrastructure within the scope of the risk management process needs to be defined, including technical assets, organizational roles, and individual personnel as well as their interdependencies. Based on that, potential vulnerabilities and threats can be identified.

14.4.1 Ethnographic Studies

The results from the ethnographic studies introduced in the previous step can also be applied and extended here. For privacy and security concerns, comprehensive information is further omitted. In the following, we provide some of the main risks identified for the present case study:

- The organization employs a small number of people in their office location. Some staff (e.g., engineers) may visit local and remote field sites for maintenance purposes;
- The level of trust among the organization's employees can be considered medium to high;
- Cyberattacks are not seen as probable, but as very low risk;
- It is believed that the level of security they have is appropriate.

14.4.2 Interview-Driven Questionnaires

Questionnaires were passed to the organization's employees to understand better their reaction to specific attack vectors – social engineering resilience. Through conducting interview-driven questionnaires with individuals in the examined organization, we identified that 41% of users would action the malicious email (i.e., phishing and spear-phishing) content. This increased to 50% when the sender of the email was a colleague. In the case of a successful attack, the perimeter breach could be bypassed, providing direct internal system access and thus reducing even more the attack path.

14.4.3 Vulnerability Identification

This is the process applied for discovering vulnerabilities in computing systems or networks. Penetration testing was conducted on selected computing systems of the office network, resulting in identifying a list of known vulnerabilities for some of the examined systems. The tool used for performing this process is OpenVAS. Although several vulnerabilities were identified in the assessed network, we only refer here to risks that may be of interest to investigate under the assumption of a ransomware. Specifically, the information that would be relevant to the examined (ransomware) case study include the employees' behavior in responding to emails and examination software that can be exploited by ransomware, e.g., Adobe Flash and Microsoft Silverlight on Windows-based systems.

14.5 Risk Analysis

Risk analysis is concerned with developing an understanding of each risk, its consequences, and the likelihood of these consequences. In general, the level of risk is determined by taking into account the present state of the system, existing controls, and their level of effectiveness.

14.5.1 Likelihood Analysis: Technical

To estimate the likelihood of threats in this case study, we use the *"exploitability"* temporal metric from the Common Vulnerability Scoring System (CVSS), which is an open framework for communicating the characteristics and severity of software vulnerabilities. The use of such a metric to estimate a threats likelihood is recommended in the ICS-related literature [5]. This information is collected through the technical vulnerability assessment, which resulted in identifying several vulnerabilities. Common vulnerabilities and exposures (CVEs) were collected from the vulnerability reports, and subsequently the exploitability subscore was extracted from the vulnerability summaries for CVEs. Most of the laptops and workstations operate a version of Microsoft Windows. In the case of examining the infection of a system by a ransomware, we investigated the ease and technical means by which the vulnerability can be exploited by such a malware. As mentioned already, a common attack vector is to exploit an existing vulnerability in Adobe Flash or Microsoft

Silverlight. These applications are both found to be installed on most computing systems in the electricity organization. Thus, we have identified the likelihood for some of the technical nodes, in the examined network, to be infected by such a threat. The likelihood is extracted by examining the CVE describing the exploits and the CVSS assigned to it per se. However, due to privacy and security concerns, the details of these vulnerabilities are not disclosed.

14.5.2 Likelihood Analysis: Social

In addition to the technical analysis, the level of trust between the employees in the electricity organization was identified (cf. Chapter 8) to estimate the likelihood for an employee to respond to an email. The likelihood for such an action was identified via a set of discussions with employees, where relationships between employees were identified too. Specifically, employees were questioned on the level of trust they show in opening emails sent by their colleagues, as well as by external senders. The likelihood for such an action is depicted in Table 14.1, where applicable.

Table 14.1: Likelihood for responding to an email

Receiver ← sender	**Likelihood**	**Receiver ← sender**	**Likelihood**
User 1 ← User 2	Medium	User 5 ← User 4	Medium
User 1 ← User 3	Medium	User 5 ← User 6	Medium
User 2 ← User 1	High	User 5 ← User 8	Medium
User 2 ← User 3	High	User 5 ← User 7	Medium
User 2 ← User 5	High	User 6 ← User 1	Low
User 3 ← User 1	High	User 6 ← User 2	Low
User 3 ← User 2	High	User 6 ← User 3	Low
User 3 ← User 5	High	User 6 ← User 4	Low
User 4 ← User 5	Low	User 6 ← User 5	Low
User 4 ← User 6	Low	User 6 ← User 7	Low
User 4 ← User 7	Low	User 8 ← User 5	Medium
User 5 ← User 2	Medium	User 7 ← User 4	Medium
User 5 ← User 3	Medium	ALL ← External sender	Medium

14.6 Risk Treatment

During risk treatment, existing controls are improved and new controls are implemented. The aim of this step in the risk management process is to decide which controls are used to protect the organization as good as possible. We here apply a game-theoretic model to make this decision provably optimal.

In the following, we define attack and defense strategies toward achieving the goals defined in Section 14.3.1.

14.6.1 Attack Strategies

In our case study described in Section 14.2, we assume that employees are allowed to use their personal devices (e.g., laptops or mobile phones) at their working place since it is assumed that this increases efficiency of work. However, this also increases the potential for a malware infection since it enables attack vectors targeting employees directly.

Here, we consider the most typical ways to conduct a malware attack focusing on employees. The attacker is considered a random threat actor (and not a targeted attacker as in case of an APT attack) since she/he sends out these phishing emails at random. In particular, this type of attack highly depends on the behavior of people in regard to responding to malicious emails as well as any vulnerable software installed on their system. Another popular attack consists that of distributing infected USB sticks. Once plugged into a PC, the ransomware starts spreading as in the case of spear-phishing emails:

- **Attack vector 1:** A spear-phishing email is sent to one or more employees asking to open a link or download an attachment. If the employee follows this instruction, a ransomware is executed and his personal device is infected. Since the success of such an attack depends a lot on the employee (i.e., whether she/he clicks on the link), we distinguish different attack vectors here, depending on the type of employee and device:
  - **Attack vector 1.1:** Attacker (spear-phishing email) $\rightarrow$ highly educated employee $\rightarrow$ employee's PC
  - **Attack vector 1.2:** Attacker (spear-phishing email) $\rightarrow$ less educated employee $\rightarrow$ employee's PC
  - **Attack vector 1.3:** Attacker (spear-phishing email) $\rightarrow$ average educated employee $\rightarrow$ employee's PC

  Once this device is connected to the company network, the ransomware is behind the firewall and thus is able to encrypt sensitive files and spread further through the network by simple propagation mechanisms (e.g., by sending an email with a malicious link or by copying itself on a shared network directory).
- **Attack vector 2:** Another option is to infect the shared server, which is used for file exchange and keeping backups. The success of such an attack also depends on the employee that receives the spear-phishing email, but since the result is the same, we combine these different scenarios in one attack vector:
  - Attacker (spear-phishing email) $\rightarrow$ employee $\rightarrow$ employee's PC $\rightarrow$ shared server (files)

- **Attack vector 3:** Infection of the SCADA server, which collects information from the underlying OT networks (in particular the concentrators), might cause significant problems. Possible attack paths involve all employees with devices connected to the SCADA server:
  - Attacker (spear-phishing email) → employee 4, 5, or 8 → employee's PC 4, 5, or 8 → SCADA server
- **Attack vector 4:** Yet another option is to infect the camera server, which records information from different field sites. Again, we combine the different paths to one attack vector causing infection of the camera server:
  - Attacker (spear-phishing email) → employee → employee's PC → company network → camera server
- **Attack vector 5:** An opportunistic ransomware attack can alternatively be executed by placing an infected USB flash drive near to a company building. If an employee collects it and plugs it on her/his PC, the infection with ransomware starts as in the case of phishing emails. However, this yields an additional attack vector that might affect the operation of the provided service, namely, infection of the maintenance laptop. In case there is a problem with one of the concentrators and the infected laptop is used to solve that problem, the availability of the provided service may be interrupted. Potential attack paths are:
  - Attacker (infected USB flash drive) → engineer → maintenance laptop → concentrator

14.6.2 Available Defense Mechanisms

In order to find an optimal way to protect a system against a ransomware attack or at least to reduce the damage caused, the very first step is to identify all defense mechanisms available. Later, the game-theoretic analysis will then show how to optimally choose among them.

Here, we list countermeasures (i.e., defense strategies) against the abovementioned attacks:

- **Status quo:** Do not change anything;
- **Training:** Since we consider attack vectors targeting employees, an important defense mechanism is to train employees how to handle their devices (e.g., not to click on a link in an email that looks suspicious). This reduces the chance that a private device is infected and thus reduces the probability that a ransomware reaches the company network. Such trainings need to be repeated in order to be effective, and we consider different frequencies:
  - Annually;
  - Once every 2 years;
  - Train only new personnel.

- **Backup Policy:** Important data shall be backed up to still be available in case the computing system is encrypted. Different locations and frequencies are examined:
  - Weekly backup on local system (e.g., file server);
  - Monthly backup on local system;
  - Yearly backup on local system;
  - Weekly backup on a remote storage system (e.g., cloud storage);
  - Monthly backup on a remote storage system;
  - Yearly backup on a remote storage system;
  - Weekly backup on external media (e.g., CD, DVD, USB flash drive);
  - Monthly backup on external media;
  - Yearly backup on external media;
- **Update/patch:** The most important technical countermeasure is regular patching. Having less technical vulnerabilities in the network that can be exploited reduces the chance that a ransomware spreads inside the network. If this likelihood decreases, we might possibly avoid an epidemic outbreak where a significant part of the network is affected. We distinguish different frequencies of updating/patching:
  - Enable automatic updates;
  - Update every year;
  - Apply only major updates.

14.6.3 Estimate Damage in Case of an Attack

Each scenario of the game is characterized through a starting point of the infection given by the attack a_j and a probability matrix P representing the impact of defense d_i. The defense is useful if at least one probability is lower than in the matrix that originally defines the network. This point of view allows simulation of the payoffs from the percolation-based model for error spreading as described in the chapter on random damage (cf. Chapter 8).

The likelihood of transmitting an error over an edge depends on the type and level of trust put in this connection. For the current status of the network, these likelihoods are collected in a matrix:

$$P = \begin{pmatrix} 0.2 & 0.4 & 0.6 \\ 0.3 & 0.6 & 0.8 \\ 0.3 & 0.4 & 0.5 \end{pmatrix},$$

where p_{ij} gives the likelihood that an edge of type i and trust level j transmits the error. More explicitly, the first row gives the values for social links, the second row characterizes the technical links, and the third one characterizes the logical links, while the columns correspond to the levels low, medium, and high (from left to right). Note that P is *not* a stochastic matrix since its rows do not represent a distribution, but rather describe how the trust level of an edge influences the likelihood of error transmission. The probabilities are estimations based on interviews with experts and vulnerability analysis (i.e., based on the exploitability metric in CVSS). Depending on the defense strategy, these transmission probabilities may change.

In order to play a multi-objective security game (MOSG) that optimizes all identified goals simultaneously, all payoffs need to be described on the same scale. To this extent, we define a mapping from the number of infected nodes to a five-tier scale representing the categories *"very low"* (1), *"low"* (2), *"medium"* (3), *"high"* (4), and *"very high"* (5). In the following, we provide an example of such a mapping in Table 14.2 for the network under analysis that consists of 5 cheap, 13 normal, and 11 expensive nodes. Cells marked with `"N/A"` (not applicable) represent the situation that a failure of any number of nodes of this type never causes a damage in the corresponding category. For example, for the mapping in Table 14.2, even a failure of all cheap nodes will not cause a damage higher than 5, and similarly failure of an expensive node will not cause a damage lower than 3.

Table 14.2: Mapping from the number of infected nodes to a five-tier scale of damage in terms of data loss

Node value	Very low	Low	Medium	High	Very high
Cheap	Residual	≥ 3	≥ 5	N/A	N/A
Normal	Residual	≥ 4	≥ 7	≥ 9	≥ 12
Expensive	Residual	N/A	≥ 1	≥ 5	≥ 7

The cost of each defense strategy is assessed by experts on a five-tier scale (as used above) that represents categories from *"very low"* (1) to *"very high"* (5). The expert assessment on costs for this case study is shown in Table 14.3.

Table 14.3: Experts' opinion on expected cost of defense

Defense	D1	D2	D3	D4	D5	D6	D7	D8
Cost	1,1	2,3	2,2	1,2	3,3	2,3	1,2	4,4

Defense	D9	D10	D11	D12	D13	D14	D15	D16
Cost	3,4	3,3	3,3	2,3	2,2	3,3	2,3	1,2

The reputational damage seems to be the most difficult to assess. If possible, it should be assessed by experts as in the case of cost. However, this is not always a simple and straightforward task. When interviewing employees, they rather state that reputational damage is proportional to the data that is lost. Thus, we apply the same procedure as for the estimation of data loss, and we slightly adapt the mapping from the number of infected nodes to the five-tier scale. This is mainly due to the fact that not all data loss is reported and thus does not necessarily cause a high damage in the reputation of the organization. The mapping used for this goal is provided in Table 14.4.

Table 14.4: Mapping from the number of infected nodes to a five-tier scale of damage in terms of reputation

Node value	Very low	Low	Medium	High	Very high
Cheap	Residual	≥ 5	N/A	N/A	N/A
Normal	Residual	≥ 5	≥ 8	≥ 10	N/A
Expensive	Residual	N/A	≥ 3	≥ 6	≥ 9

14.6.4 Game-Theoretic Optimization of Defense Actions

Based on the strategies defined in Sections 14.6.1 and 14.6.2, the payoffs resulting from the simulation and the data presented in Section 14.6.3, we set up a game to find the optimal defense strategy as well as the worst case damage in case of an attack. To this extend, we set up a multi-objective security game (MOSG) between an attacker and a defender (e.g., a utility provider) with random payoffs as introduced in [8]. This game is assumed to be a zero-sum game where the defender tries to minimize his loss (payoff), while the attacker tries to maximize it.

The set of strategies and the payoffs provide adequate input for computing the equilibrium. However, it is possible to prioritize the different goals. From discussions with experts, the different goals were assigned with the following weights that represent their importance to the company: data loss 0.5, cost 0.3, and reputation 0.2.

Computation of an equilibrium is done by means of the generalized fictitious play algorithm that is implemented in the `R` package `HyRiM` [9]. With $T = 1000$ iterations, we find the optimal defense strategy shown in Table 14.5.

The defense strategies not listed in Table 14.5 were assigned with a zero frequency in the Nash equilibrium and thus shall not be chosen at all. This optimal defense strategy is illustrated in the top row of Figure 14.2.

The optimal attack strategy on the other hand usually changes with the goal. Therefore, Figure 14.2 lists both the optimal attack strategy (left) and the worst case damage (right) for each goal in one row. The worst case damage is the damage that occurs if both players choose their strategies according to the Nash equilibrium and is measured on the same five-tier scale as the payoffs. Since the attacker is assumed

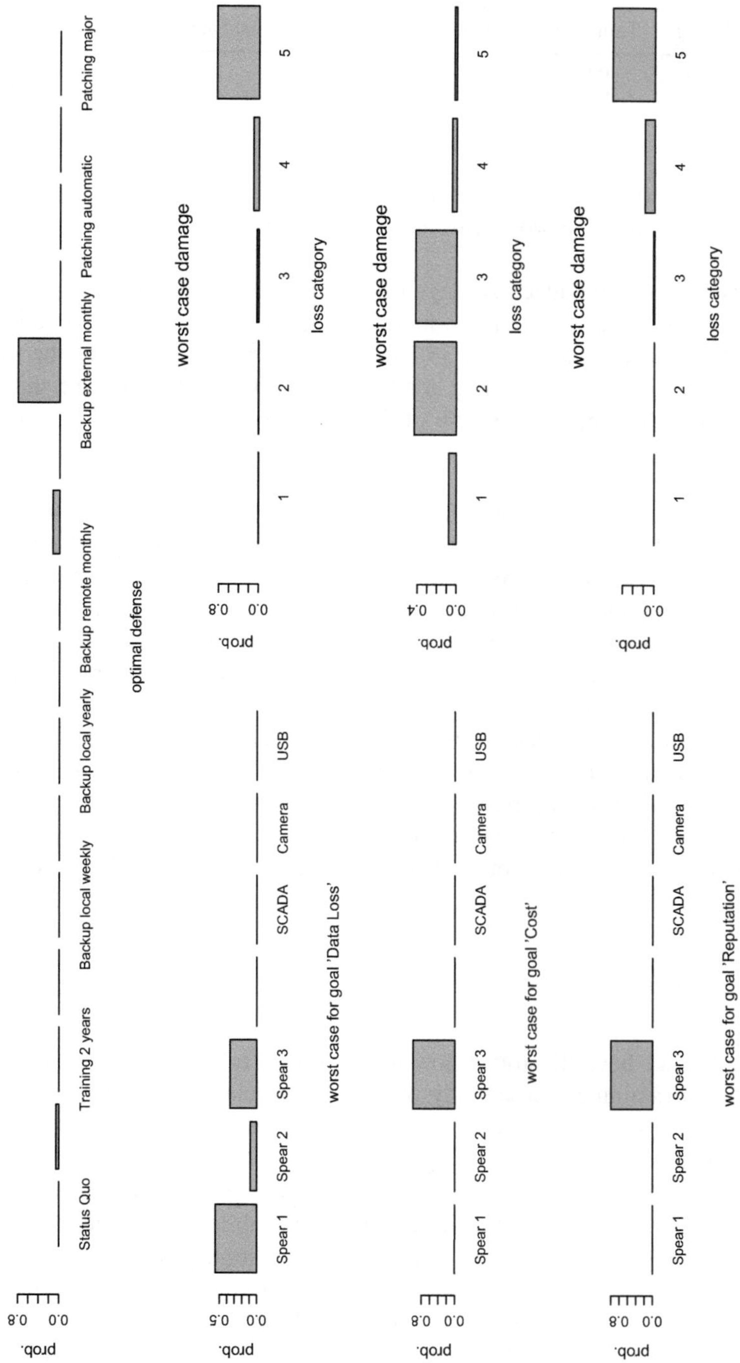

Fig. 14.2: Optimal solution from the game-theoretic analysis

Table 14.5: Optimal defense strategy according to Nash equilibrium

Strategy	Train yearly	Remote backup yearly	External backup monthly
Frequency	0.056	0.134	0.810

to be a single person, he usually is not able to play all optimal defense strategies simultaneously, so that the worst case damage is a lower bound to the occurring damage.

In this case study, the attack strategy of *"spear-phishing 3"* (i.e., targeting an employee with an average security awareness) is optimal in terms of cost and reputation, and it should be applied in 35.76% of all cases. In 55.65% of the time, the *"spear-phishing 1"* attack aiming at a highly security aware employee shall be deployed, and in the remaining 8.59% of the time the *"spear-phishing 2"*, attack shall be played, which will target a less security aware employee.

On the defender's side, the organization should apply the optimal defense strategies described in Table 14.5 to protect against a ransomware attack. Specifically, the defender shall apply only the three strategies listed in Table 14.5 in the corresponding relative frequency identified. To this extend, in 5.60% of the time, all employees shall attend one training course every year; in 81.00% of the time, a monthly external backup should be used; and in the remaining 13.40% of the time, a yearly remote backup shall be applied.

As long as the overall frequencies correspond to the optimal solution, the defender can randomly choose the order in which these strategies are enforced. In particular, the solution has a certain degree of freedom in the sense that if one strategy cannot be applied at some point in time, e.g., due to the absence of a key person, it can be postponed, and thus another defense mechanism can be used instead.

14.7 Conclusion

In this chapter, we have illustrated how to apply the HyRiM's risk management framework by examining a case study of a European electricity utility organization considering a single security threat, i.e., a ransomware attack. Although such threats are opportunistic, they pose a great concern to organizations, not least to utility networks. In this hybrid approach, we have examined the utility organization from multiple viewpoints and identified existing vulnerabilities to its infrastructure and potential attack vectors. A set of attack and defense strategies were defined, and experts' knowledge was collected to estimate the damage caused to the utility organization on the basis of three goals. The results indicate that using spear-phishing as an attack vector (i.e., attack strategy) would result in causing the most damage to the utility provider, when compared with other attack vectors. Finally, with regard to optimal defense strategies, our analysis indicate that performing remote and external backups on a regular basis as well as regular training of employees may serve as powerful defense mechanisms, when compared with the rest of the examined countermeasures.

Acknowledgements The research leading to these results has received funding from the European Union Seventh Framework Programme under grant agreement no. 608090, Project HyRiM (Hybrid Risk Management for Utility Networks).

References

1. Chen, Z., Ji, C.: Spatial-temporal modeling of malware propagation in networks. IEEE Transactions on Neural networks **16**(5), 1291–1303 (2005)
2. Cheng, S.M., Chon Ao, W., Chen, P.Y., Chen, K.C.: On modeling malware propagation in generalized social networks **15**(1), 25–27 (2011)
3. Ganesh, A., Massoulie, L., Towsley, D.: The effect of network topology on the spread of epidemics. In: Proc. INFOCOM05, vol. 2, pp. 1455–1466 (2005)
4. Karnouskos, S.: Stuxnet worm impact on industrial cyber-physical system security. In: IECON 2011 - 37th Annual Conference of the IEEE Industrial Electronics Society (IECON 2011), pp. 4490–4494. IEEE (2011)
5. Knapp, E.D., Langill, J.T.: Industrial Network Security: Securing critical infrastructure networks for smart grid, SCADA, and other Industrial Control Systems. Syngress (2014)
6. Moore, D., Shannon, C., Voelker, G.M., Savage, S.: Internet quarantine: Requirements for containing self-propagating code. In: Proc. INFOCOM03, vol. 3, pp. 1901–1910 (2003)
7. Morrow, B.: BYOD security challenges: control and protect your most sensitive data. Network Security pp. 5–8 (2012)
8. Rass, S.: On Game-Theoretic Risk Management (Part One) – Toward a Theory of Games with Payoffs that are Probability-Distributions. ArXiv e-prints (2015). http://arxiv.org/abs/1506.07368
9. Rass, S., König, S.: R package 'hyrim': Multicriteria risk management using zero-sum games with vector-valued payoffs that are probability distributions (2017). URL https://hyrim.net/software/
10. Scarfo, A.: New security perspectives around BYOD. In: Broadband, Wireless Computing, Communication and Applications (BWCCA), 2012 Seventh International Conference on, pp. 446–451 (2012)
11. Symantec: What you need to know about the wannacry ransomware (2017). URL https://www.symantec.com/connect/blogs/what-you-need-know-about-wannacry-ransomware, [retrieved:25/09/2017]
12. Thomson, G.: BYOD: enabling the chaos. Network Security pp. 5–8 (2012)
13. TrendMicro: Frequently asked questions: The petya ransomware outbreak (2017). URL https://www.trendmicro.com/vinfo/us/security/news/cybercrime-and-digital-threats/frequently-asked-questions-the-petya-ransomware-outbreak,[retrieved:25/09/2017]
14. Zou, C.C., Gong, W., Towsley, D.: code red worm propagation modeling and analysis. In: Proc. 9th ACM Conf. Computer and Communication Security, pp. 138–147 (2002)

Chapter 15
Game-Theoretic Optimization for Physical Surveillance of Critical Infrastructures: A Case Study

Ali Alshawish, Mohamed Amine Abid, and Hermann de Meer

15.1 Introduction

Critical infrastructures are physical or virtual assets that are essential for the proper functioning of any society and economy. The destruction of such systems and assets can adversely affect several vital sectors such as security, national economy, public health, and safety. Most countries identify the following critical infrastructures: telecommunications, electric power systems, natural gas and oil, banking and finance, transportation, water supply systems, government services, and emergency services [10]. Critical infrastructures interact at multiple levels to enhance their overall performance. These interactions often create complex relationships, dependencies, and interdependencies that cross infrastructures' boundaries. Therefore, these organizations constantly extend their physical perimeters to include other entities, such as vendors, business partners, service providers, or even customers, into their premises. Thus, access to the facilities is allowed to not only regular employees but also external entities including temporary workers, interns, independent contractors and subcontractors, or even visitors. Broadly speaking, all these entities need easy access to their workplaces. Therefore, surveillance and access control technologies are mostly deployed at the outer layer of the infrastructure system to ensure efficient mobility inside the facility.

A. Alshawish · M. A. Abid (✉) · H. de Meer
Faculty of Computer Science and Mathematics, Chair of Computer Networks and Computer Communications, University of Passau, Innstr. 43, 94032 Passau, Germany
e-mail: ali.alshawish@uni-passau.de; amine.abid@uni-passau.de; hermann.demeer@uni-passau.de

S. Rass, S. Schauer (eds.), *Game Theory for Security and Risk Management*, Static & Dynamic Game Theory: Foundations & Applications,
https://doi.org/10.1007/978-3-319-75268-6_15

At the entrance of the facility, the access control process is set according to a predetermined policies and procedures. Since the personnel structure changes considerably often, merely personal recognition of authorized persons through the security staff is not adequate. Therefore, security badges and identification cards are widely adopted for an easy and prompt access authorization. Although surveillance systems may be in place and operate within a critical system's premises, they are prone to technical as well as organizational failure. For example, security badges might be stolen without notification to the security personnel or without revoking them in a timely manner. Moreover, badges issued to temporary visitors and workers, or to employees that have already left the company, might not always be recovered before leaving the site. This gives adversaries the possibility to gain an easy access to the facility. As a consequence, the perimeter-centric physical security measures, such as traditional surveillance technologies (e.g., closed-circuit television (CCTV) systems or access control solutions) that use static surveillance devices mounted at specific locations, are not adequate to detect and prevent such potential intruders [11].

To summarize, surveillance technologies represent a standard practice for the protection of critical infrastructures such as utility networks. Although surveillance systems may be in place and operating within a utility provider's premises, they are prone to technical as well as organizational failures resulting in a fluctuating performance. Furthermore, several emergency and unforeseen events, such as human errors, can significantly impact the effectiveness of specific surveillance activities. Therefore, modeling surveillance needs to account for the characteristics and practicalities of surveillance systems, especially imperfect detection as well as fuzzy assessment of the performance.

To cope with this intrinsic dynamic nature of such critical infrastructures, and to achieve a decent level of situational awareness in such large-scale areas and taking into account the limited available resources (e.g., security personnel and badge check devices), badge verification activities have to be randomized to improve the effectiveness and detection probability. This can be achieved by mobilizing the involved security resources. Additionally, it is vital to implement some risk-based strategies that allocate and focus resources in highly sensitive areas and against real threats and therefore to effectively and efficiently mitigate risks of physical intrusion [8, 7]. Finding an optimal inspection layout involves simultaneously optimizing multiple objectives such as detection rate, privacy, damage, and incurred costs. This special scenario of conducting surveillance by human security staff has a natural reflection in game theory as the well-known "cops and robbers" game (a.k.a. graph searching). More details on application of game-theoretic concepts and algorithms in the general field of security and risk management are included in Chapter 6 as well as in our previous work [2].

In this chapter, we apply game-theoretic principles to solve zero-sum games with probability distribution-valued payoffs as a means to integrate the intrinsic uncertainty of surveillance systems. This model is an essential component of a comprehensive decision-making framework for physical surveillance games, called

"*G-DPS*-framework". The ultimate goal of this framework is to find the optimal configuration for physical surveillance system over multiple goals. As an evaluation scenario, we use an actual setup given within a critical infrastructure, henceforth referred to as "*the company*." For reasons of simplicity, we will focus solely on the use of security guards, who are controlling the area. Taking into account the details of the physical infrastructure (buildings, roads, etc.) as well as personnel requirements (working hours, available number of guards, etc.), we will make use of simulations to assess various real-life attack and defense scenarios with regards to different identified goals. Finally, the optimal solution obtained by the model will be implemented and empirically validated.

15.2 G-DPS: A Game-Theoretical Decision-Making Framework for Physical Surveillance Games – An Overview

Throughout this work, we apply *G-DPS* framework, which is a decision-making framework for physical surveillance games described in Chapter 6. This framework enables the involved security manager (i.e., defender) to identify and assess possible alternatives toward finding an optimal allocation of surveillance resources. The applied game-theoretical model facilitates finding an optimal solution for risk minimization through playing surveillance games with stochastic outcomes. This mainly addresses the uncertainty component, as the impact of surveillance systems cannot be fully expressed in a crisp numeric utility, but rather in fuzzy or probabilistic terms. Hence, the framework allows us to integrate the intrinsic randomness of the effects of surveillance action and subsequently provides a more realistic view of handling uncertainty in physical surveillance games. In addition, this framework allows us to optimize across different goals (e.g., damage caused by the adversary, costs for security measures, acceptance of the security measures by the employees, etc.). In a nutshell, the framework defines six main steps (cf. Chapter 6 for details):

- `Context Establishment` aims at understanding the environment of interest as well as the different objectives that should be achieved.
- `Identification of Strategies` involves identification and parameterization of possible configurations, layouts, and operational patterns for the surveilled infrastructure.
- `Identification of Goals` involves identification of relevant indicators related to the inspection activities.
- `Assessment of Strategy Effectiveness` aims at determining the effectiveness of identified strategies with regard to all identified goals. This can be achieved using different qualitative, quantitative, and semi-qualitative assessment methodologies (e.g., simulation, experts and stakeholders opinions, or social surveys).

- `Identification of Optimal Configuration` involves finding Nash equilibria in games with distribution-valued payoffs.
- `Implementation of Optimal Configuration` involves implementing and adjusting the surveillance configurations according to the optimal solution as well as analyzing and validating the feasibility of the obtained optimal strategy.

15.3 Scenario Description: Setup of the End User's Infrastructure

In this use case, we consider an industrial complex as a critical infrastructure that involves several industrial processing units and auxiliary facilities (e.g., petroleum refining units, a water purification center, a gas production plant, or an electricity production plant) and is illustrated by the map presented in Figure 15.1. Being sensitive (i.e., the business and the industrial processes), this infrastructure is a potential target of several attacks of different kinds.

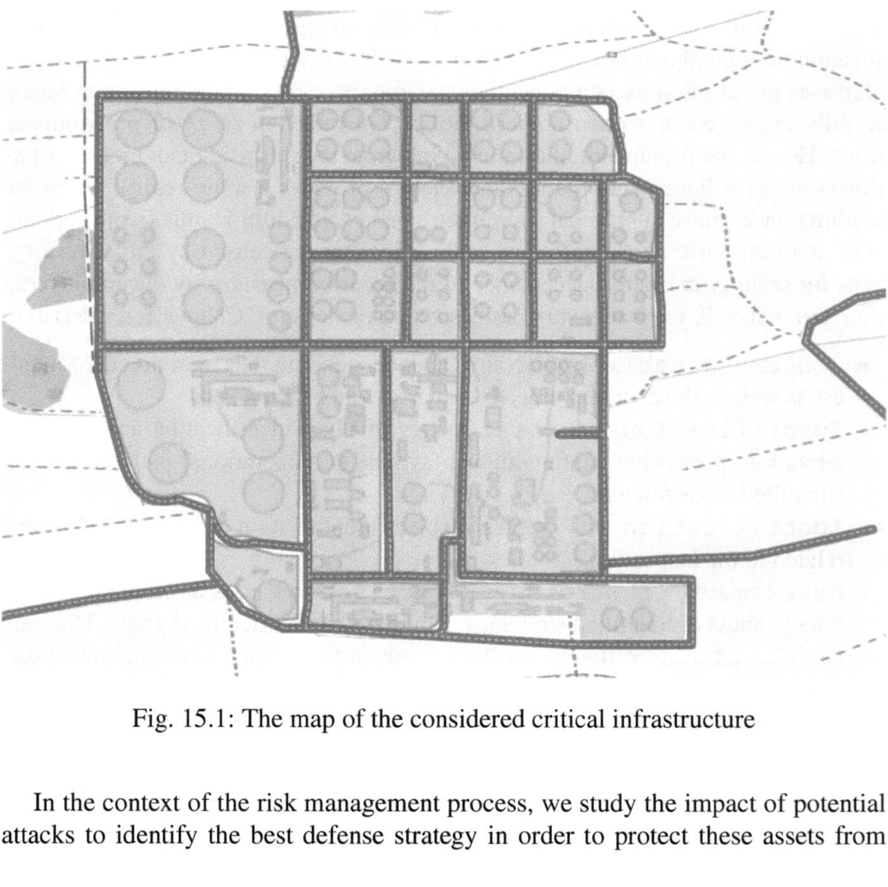

Fig. 15.1: The map of the considered critical infrastructure

In the context of the risk management process, we study the impact of potential attacks to identify the best defense strategy in order to protect these assets from potential external threats. For the sake of simplicity, we confine ourselves to the risk of physical intrusions. For that purpose, we will apply our aforementioned **G-DPS** framework. We may insist here that the use case we consider is derived from realistic environment settings and based on the knowledge of experts operating in critical infrastructures. Henceforth, we will refer to the investigated infrastructure as *the company*.

15.4 Application of G-DPS-Framework in This Use Case

This section describes the step-by-step application of the **G-DPS** framework, briefly introduced in Section 15.2, to infer the optimal configuration for the physical inspection activities of *the company*'s industrial complex.

15.4.1 Context Establishment

As a first step, a business process analysis was conducted to identify the main business processes in *the company*. It unveiled that *the company* carries several critical industrial processes including water purification as well as critical chemical engineering processes. These processes represent the basis for identifying possible expositions to risks, in order to find mitigation actions preventing those risks in later steps. Furthermore, ethnographic studies were carried out at *the company*. An ethnographer at the utility organization conducted several discussions and made observations, which included the organizational structure and the social interrelations between the employees. Finally, a study of *the company*'s layout and architecture was conducted to better assess its physical environment.

The results of these three steps are described in the following. We first provide some of the main risks identified for the present use case:

- Most of *the company*'s employees are field workers, who manipulate highly dangerous materials and have access to sensitive areas.
- Maintenance staff may visit sensitive fields and sites, even when they are already evacuated (during maintenance operations, the present personnel might be kept at its strict minimum, i.e., only maintenance staff).
- *The company*'s employees are considered as a potential source of leakage of sensitive information.
- Due to the prevalent believe of the security personnel that the deployed security solutions (i.e., CCTV cameras and the access control system at the entrance) are able to prevent any illegitimate access to *the company*, the risk of physical intrusions is seriously underestimated. This actually represents a great threat, as it means that the alert level at *the company* is very low. Thus, if such infiltration occurs, it would most likely not be detected early enough.

As such, *the company* can be a target of several attacks of different kinds: (i) attacks aiming to cause physical damage through the destruction of buildings/machines that contain highly dangerous materials, (ii) attacks causing damage to the nearby environment by tampering or disrupting some setup security measures and thus leading to polluting the area, or (iii) espionage attacks causing the leakage of sensitive information, either to a competitor, or simply to the public to breach *the company*'s reputation.

To deal with all these potential threats, *the company* set up the following security measures:

- Surveillance cameras: a network of fixed cameras is installed but rather acts as a reactive solution than a monitoring platform. It is mainly used to review taped events that could have happened as it is the case of many other critical infrastructures.
- Badges scanning and verification system at the entrance gates: to guarantee a basic level of security, each person entering the complex (i.e., employees of *the company*, temporary workers, and visitors) must have a security badge, which is automatically read at the entrance. These badges are used to grant them access to *the company* in the first place, and to working areas with special restrictions. On the security badge, the owner's identification number (ID), name, and photo are stored. All data can be retrieved upon scanning the badge with a specific verification device.

In this study, we will limit ourselves to potential physical intrusions. We will suppose that a single intruder or a team of intruders succeed in accessing *the company* using a stolen badge(s) of an employee(s) or a temporary worker(s) or simply by forging security badges.

The architecture of the area under surveillance (i.e., the buildings, road, perimeter, etc.) is known and depicted in Figure 15.1. Due to the complexity of the actual map and plant topography, and in order to achieve both simplicity and plausibility with respect to a complex real scenario, a simplified map layout has been realized for simulation purposes as shown in Figure 15.2.

The entire area is divided into 47 zones, each of which has a specific security level indicating its criticality. The security level of a specific zone depends on the assets located therein (e.g., areas where important machinery is operated or control room) or on the information stored in that zone (e.g., data centers, record storage rooms, etc.). Although these zones are equipped with surveillance systems such as video cameras or access control systems, the presence of security guards is also required. In particular, these zones need to be checked on a regular basis by a security guard to prevent unauthorized intrusion (which is partly covered by the technical solutions) in that zones. Table 15.1 indicates the assigned risk levels associated to each area presented in Figure 15.2. A risk level equal to 1 indicates a *low risk*. Obviously, the higher the risk level, the more critical the zone.

The company counts 180 employees in total. Every employee holds an individual security badge, or interchangeably an ID card, proving her/his identity and right to

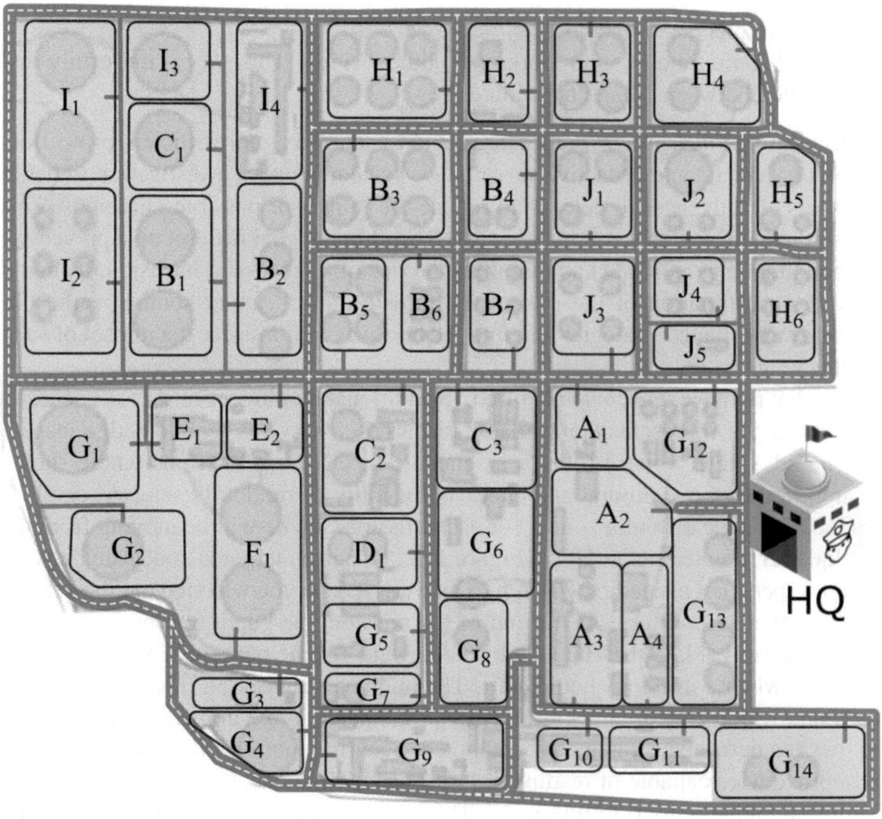

Fig. 15.2: *The company*'s simplified map

Table 15.1: Summary of the different subareas' risk levels

Subarea Label	Security Level
Subareas: $A_1,...,A_4$	Level *1*
Subareas: $B_1,...,B_7$	Level *2*
Subareas: $C_1,...,C_3$	Level *3*
Subarea: D_1	Level *4*
Subareas: $E_1,...,E_2$	Level *5*
Subarea: F_1	Level *7*
Subareas: $G_1,...,G_{14}$	Level *8*
Subareas: $H_1,...,H_6$	Level *9*
Subareas: $I_1,...,I_4$	Level *10*
Subareas: $J_1,...,J_5$	Level *12*

be in a given subarea or zone. They can move between subareas following the layout's paths connecting them. They are also free to move inside a subarea according to the following movement pattern:

1. select a random position inside the subarea,

2. move to this position, and
3. spend some time (a stay) working in this position. This stay is uniformly distributed between 10 and 60 minutes.

To reinforce these security measures, and to achieve a more efficient protection, a new on-demand and proactive surveillance system is under investigation to better assess its effectiveness to detect and reduce the impact of possible intrusions. This is also the main objective of this use case. The ultimate goal is to set up a preventive method that helps estimate and minimize the risk of a successful intrusion. It should provide an estimation of the risk of an intrusion attack under some specific on-demand surveillance strategies and find the best way to minimize the impact of such an attack.

For that purpose, *the company* hired 15 additional employees to serve as security guards (i.e., 15 represents the available resources to serve as mobile badge inspection guards). Every guard follows a schedule of checking missions where she/he is supposed to move around and check the identity of randomly selected employees located in the different zones. For each mission, she/he will be moving from the headquarter, pointed out as HQ in Figure 15.2, to a given targeted zone using a vehicle (at a speed not exceeding 20km/h). Once arrived, she/he will step out the vehicle to check the area on foot. At the end of the mission, the guard returns to the vehicle and drives back to the HQ where she/he waits for the next mission. A schedule indicates when a mission should start. The number of checking missions, when to start, and which area to target are further investigated in the upcoming stages. This defines the defensive strategy to be adopted. Security guards will be equipped with a mobile device capable of reading security badges and checking whether a person holding a badge is its rightful owner. This can be done by taking a picture of the person's face with the device and using face matching algorithms to compare it to the photo stored to the badge.

15.4.2 Identification of Strategies

Here, we need to identify the set of strategies of each player (i.e., defender and intruder). We will start with the intruder and then move to the defender side. As aforementioned, we will consider two case scenarios regarding the potential target of intruders:

- Espionage scenario: the intruders try to acquire critical information about *the company*. In this scenario, we will assume that intruders are roaming around and targeting zones randomly.
- Sabotage/vandalism scenario: the intruders target critical assets to manipulate them and cause physical damage. In this scenario, we will assume that the intruders have more knowledge about the zones of specific interest that they will try to target in the first place.

As such, we can identify two types of strategies according to the way areas are targeted: either randomly (R) or based on their criticality by targeting Higher Security Levels First (HSLF). Moreover, we will investigate attacks where a group of intruders tries to infiltrate *the company* and cause potential damage. The size of such a group can also be seen as a parameter defining different attacker strategies. In this study, we will consider attacks made by a group of 5, 10, and 15 intruders, respectively.

On the other hand, security guards will be carrying the mission of performing random checks in the different zones of *the company*. Certainly, it would not be advisable for the security guard to take tours on fixed routes and at fixed times, since such information can be obtained by an intruder, who then instantly knows where and when to sneak into the infrastructure. In this case, the intruder would be able to maximize the time she/he is able to cause harm until the security guard returns to that specific zone. Conversely, taking rounds on random routes or checking the specific zones for intruders at random times creates sort of a "moving obstacle defense" [5, 16], since the intruder is confronted with additional uncertainty. Additionally, since not all zones are of the same importance for *the company*, zones with a "high" security level have to be checked more frequently than zones with a "low" security level. In this context, the security guard may have better chances to catch an intruder the more often the guard checks a specific area (note that a guard is not able to check the badges of all persons within a specific zone at a specific time). This gives us once again two sets of possible strategies: random checks (R) or Higher Security Levels First checks (HSLF). Nevertheless, a certain number of security guards have a limited amount of time to check all zones within the complex. In fact, every security guard would have a limited number of checking missions assigned to him per day. The number of checking missions per security guard per working day (also defined as the frequency of checks per working day) can also give us a set of possible defensive strategies. For our case study, we will consider strategies with 2, 3, 5, and 8 checking missions per day per security guard.

To summarize, for the defender, a strategy is defined by:

- The number of missions per day (frequency of missions): *N_missions*
- How to target a given area: randomly (*R*) or Higher Security Level First (*HSLF*)

A defender strategy where we ask our security guards (in our case, their number is fixed to be 15) to target areas according to ***X*** (*R* or *HSLF*), ***Y*** times a day, is denoted D-NG15F***Y***T***X***.

On the other hand, an attacker strategy is defined by:

- The number of attackers
- How to target a given area: randomly (*R*) or Higher Security Level First (*HSLF*)

Similarly, an attacker strategy where ***N*** attackers enter the site and target areas according to ***X*** (*R* or *HSLF*) is denoted A-NI***N***T***X***.

In our case scenario, we enumerate in total the attacker/defender strategies depicted in Table 15.2.

Table 15.2: List of the strategies considered for defenders and attackers

#	Strategy Label	Description
- 8 Defender Strategies:		
1	D-NG15F*2*T*R*	freq = 2 & areas: targeted randomly
2	D-NG15F*3*T*R*	freq = 3 & areas: targeted randomly
3	D-NG15F*5*T*R*	freq = 5 & areas: targeted randomly
4	D-NG15F*8*T*R*	freq = 8 & areas: targeted randomly
5	D-NG15F*2*T*HSLF*	freq = 2 & areas: targeted Higher Sec. Lev. First
6	D-NG15F*3*T*HSLF*	freq = 3 & areas: targeted Higher Sec. Lev. First
7	D-NG15F*5*T*HSLF*	freq = 5 & areas: targeted Higher Sec. Lev. First
8	D-NG15F*8*T*HSLF*	freq = 8 & areas: targeted Higher Sec. Lev. First
- 6 Attacker Strategies:		
1	A-NI*5*T*R*	5 intruders & areas: targeted randomly
2	A-NI*5*T*HSLF*	5 intruders & areas: targeted Higher Sec. Lev. First
3	A-NI*10*T*R*	10 intruders & areas: targeted randomly
4	A-NI*10*T*HSLF*	10 intruders & areas: targeted Higher Sec. Lev. First
5	A-NI*15*T*R*	15 intruders & areas: targeted randomly
6	A-NI*15*T*HSLF*	15 intruders & areas: targeted Higher Sec. Lev. First

15.4.3 Identification of Goals

In the given scenario, we focus on more than one single goal, as we have to take several aspects into consideration to find an optimal solution for the game. In more detail, the overall game has four goals of interest: the caused damage, the privacy preservation, the comfort breach, and the detection rate. These goals can be quantified as follows:

- *Detection Rate* is the ratio of detected intruders to their total number (i.e., the number of detected intruders divided by the total number of intruders). Such a goal is to be maximized.
- *Caused Damage* is defined as the average time spent inside the targeted subareas per intruder, weighted with their respective security levels. Formally, if *NI* is the number of intruders, and *NA* is the number of subareas in *the company*, then the damage is understood as

$$CausedDamage = \frac{1}{NI} \times \sum_{i=1}^{NI} \sum_{j=1}^{NA} timeSpent(intruder_i, area_j) \times secLevel(area_j)$$

where $timeSpent(intruder_i, area_j)$ represents the total time spent by $intruder_i$ inside $area_j$; and $secLevel(area_j)$ gives the security level of $area_j$. Obviously, this goal is to be minimized.

- *Minimum Privacy Preservation* is inversely related to the maximum possible disclosure of employees' locations. Obviously, the more frequently manual ID checks are performed, the more effective the system can be. However, this comes at a price: a frequent ID checking may have an essential impact on location privacy of the employees, especially if such an information is leaked. Therefore, we are more interested in inspection strategies that maximize the minimum level of privacy preservation. In other words, we prefer strategies that keep the maximum privacy disclosure at its minimum.
- *Maximum Comfort Breach* is the maximum comfort breach experienced by the employees in *the company*. In fact, the more a worker is checked, the more uncomfortable she/he will feel. However, it is still a subjective issue after how many checks a person starts feeling uncomfortable and how much uncomfortable an employee would be after several checks. Thus, and to better assess such a measure, we may relay on the ethnographic studies that we conducted while establishing the context of our studied company. This particular point will be discussed in more details in Section 15.4.4. Given that one of the main objectives of *the company* is to satisfy its employees, this particular goal should be minimized.

We may insist here that we can define several additional key performance indicators (KPIs) and respective target goals such as resource cost, energy cost of the mobile checking devices, etc. However, we will limit ourselves to the four aforementioned measures: (i) detection rate, (ii) employees' comfort breach, (iii) caused damage, and (iv) minimum privacy preservation. We will have to take into account the multi-goal aspect while defining the optimal strategy. In fact, these four goals have to be minimized (e.g., damage) and maximized (e.g., detection rate) at the same time based on the different attack and defense strategies (cf. Table 15.2). In detail, more frequent security checks will increase the likelihood of detecting an intruder but will also cost more location disclosure of the employees and less comfort. Therefore, an optimization process is required, which necessarily relies on a measure of quality for the different defensive strategies we could apply. As already described in Section 15.2, our **G-DPS** framework is able to solve such a multiobjective game.

15.4.4 Assessment of Strategy Effectiveness: Simulation Setup and Results

For each known configuration, the effectiveness with regard to all aspects identified in the previous step needs to be determined. Since the response dynamics of the game, e.g., people's reactions, etc., may be uncertain (recall the comfort breach which remains a subjective feeling after all), we should be careful on how to assess our different strategies. For instance, we may rely on some experts opinion that will evaluate the different strategies in terms of our fixed goals. Another option would be to rely on simulation, which is the option we chose in this study. Following this particular method, we may easily quantify the outcome of certain goals: a more or less reliable risk estimation (e.g., given in terms of probabilities) may be achievable through simulation, but not necessarily so for all goals of interest. A "soft" indicator, such as the degree to which employees appreciate the surveillance or feel uncomfortable upon such monitoring, is one example of a goal that may not be estimated by simulation. In such cases, empirical data (e.g., coming from classical surveys or in our case from the ethnographic study we conducted) may be necessary before simulating the different configurations.

This subsection will then be devoted to present the simulation we developed to assess our different strategies. As a first step, let us describe our simulation model. We choose to use the INET 3.4 framework [3], on top of OMNeT++ 5.0 discrete event simulator [15] to integrate our model. Through this model, we need to be able to reproduce a faithful image of the physical environment of our monitored facility. We also have to reflect all the applied policies (zone restrictions, employees' profiles, badge checking policies, etc.) as well as actors' behaviors (security guard, field worker, or intruder).

15.4.4.1 The Physical Environment

In our developed simulation model, we reproduced the exact same (in terms of number of areas, their geographic repartition, their sizes, and the routes connecting them) simplified map layout given by Figure 15.2. In this figure, we can observe 47 zones plus a headquarter (HQ), reachable through a web of ways/paths to follow when moving from/toward any of these areas. These areas represent the smallest level of granularity of our site. Each of which has an attribute, called *security level*, indicating the criticality of the respective area (as described in Table 15.1). All this information, i.e., paths, fences, gates, and areas, is described in an XML file, parsed on the run time, to build and render the physical structure of our site.

15.4.4.2 Actors

In our case study, we can identify two main actor categories: employees and intruders. An employee can be either a worker or a security guard. They all hold security badges, meaning that they are known to the system. Unlike an employee, an intruder is someone from outside the facility. Hence, she/he doesn't hold a security badge, has a fake one, or has a stolen card that does not correspond to their biometrics (i.e., finger print or facial photo, etc.). In all these cases, she/he will not be recognized by the system as a regular employee. Thus, she/he should be caught at the first badge check, whenever it is done and wherever she/he is located inside the facility.

Depending on their job, employees are allowed to access certain areas of the facility but may be denied access to others. The restriction varies among employees. In our simulation model, we define a set of profiles, each of which indicates a subset of allowed areas. Using an XML file, we assign one of these profiles to each worker, indicating the areas she/he can access. This information is stored in her/his ID card. A *regular* worker is a person who does respect areas' restrictions. She/he will never access an area not figuring in his profile. Thus, upon a security check, her/his situation would always be fine. On the other hand, a *malicious* worker is an employee with a valid ID card but who intends to physically harm the facility. In our work, we suppose that such suspicious behavior manifests in targeting areas that she/he is denied to access. During a security check, a malicious worker can only be caught if she/he is behaving suspiciously at that time (i.e., he is in a restricted area when the check takes place). Such information can be acquired from the first step (i.e., context establishment). In our case all workers are allowed to be in all areas of *the company* (i.e., one single profile for all employees, all workers are regular).

Conversely, intruders are not authorized to be in any of the zones of our company. An intruder may choose to remain in the subarea where she/he is or move from one subarea to another following a given strategy (i.e., randomly or HSLF). At the cost of being possibly detected by a security guard, staying in the same zone means adopting a movement pattern similar to a regular employee.

On the other hand, security guards are allowed to access all areas of the facility. A special profile is then created just for them. A security guard owns two main devices: a navigation system and an ID checker (they are virtually two separate devices but could also be integrated into one single physical device). The navigation system serves as a mission scheduler. Checking missions are assigned to a security guard using this device. It first indicates which area a security guard needs to check, shows the way to reach this area, and decides the strategy to be adopted during the ID check. The ID checker is used to verify the identity of an employee. It starts with verifying the ID and the biometrics of the employee. If they match, it verifies whether this employee is allowed to be in the area where the check is performed.

15.4.4.3 Security Badge Checking Mission

In our simulation model, a mission consists of three phases: (i) select a target area, (ii) visit the targeted area and perform spot checking, and (iii) go back to the headquarter.

First phase of a mission: The first phase corresponds to selecting of a target area and guiding the security guard toward it. This selection is made according to a given strategy. We implemented the two strategy families identified in the previous section: random choice (R) and a choice based on the security level of the areas (HSLF). The navigation device, storing the map of the whole site (i.e., areas and paths), guides the security guard initially located at the headquarter (HQ), toward the gate of the targeted area. This is done by applying any shortest path algorithm on the graph representing the paths of our site, between the current position (the headquarter for the security guard) and the gate of the area to be checked. In this phase, our security guards are supposed to be equipped with vehicles and thus moving at a speed of 20km/h at most (recall the speed limit mentioned at the context establishment step).

Second phase of a mission: The second phase of a mission is checking the selected area. The security guard needs to walk (at a speed of 3.6 km/h, i.e., 1 m/sec, in average) all around and meet workers located in this area for an eventual ID check. Inside an area, we can apply any of the mobility models provided by the INET framework. Yet, we choose to use the well-known random waypoint mobility model [6, 4]. Basically, a mobile node inside a selected area uniformly generates a target position inside the polygon surface of our area, selects a speed, and then moves toward its target. At its arrival, the node waits at its position for a randomly generated time, before reproducing the same behavior once again. Notice here that all our actors are moving with respect to this same mobility model. The only difference might be the *move-wait* pattern. In fact, a worker would spend most of his time in the same place doing his work and then moves to another place to do some other work and so on and so forth. On the other hand, a security guard would spend most of his time moving from one position to another, with brief waits. A malicious subject, either an intruder or a worker, would be moving as a regular worker, spending as much time waiting as she/he is supposed to do some harmful work. In our simulation, we set these values to "10 to 60 min" for workers and intruders, and "few seconds" for a security guard. While moving, a security guard will meet persons who are in the checked area. For everyone in his direct vicinity, a security guard decides to check his ID with a given probability (by default, the probability is set to 0.5). This probability should be closely related to the security level of the area. Every selected subject, remains at his current place until the check is performed. If a malicious person (i.e., intruder or worker) is detected, a *handle situation* procedure is triggered. This procedure could be of any type like (i) calling a third party to drive the caught individual to an interrogation room; (ii) the security guard stopping the checking mission and driving the checked person back to the headquarter by himself; (iii) since we are running a simulation, remove the malicious node from the simulation and continue the checking mission (which is exactly what we are doing); or (iv) more drastically stopping the simulation. Besides, to avoid that a security guard repeatedly checks the same person again and again during one same spot checking mission, we added a memory module to the security guard. This module, being adjustable, will control the behavior of a security guard according to three basic features: how easily can she/he remember a new face, how long can she/he keep remembering it, and how many faces can she/he remember? The first feature, called *the memory quality*, is a probability-like parameter to be given as an input: it ranges between 0, meaning that she/he can't remember anything, and 1, meaning she/he remembers everything. The second feature, called *the memory time*, is a time duration to be given as an input. It can either be a fixed duration or a distribution (e.g., a uniform distribution) which indicates for how long a newly met face is remembered. Every new entry to the memory will be associated to a *memory time* value to decide when it is forgotten. The third feature represents the size of the memory and hence called *the memory size*. It is implemented as a circular buffer, so that if it is full, the oldest face (having the smallest *memory time* value) would be forgotten first. Based on the estimation of some experts, we set these values to 0.3, uniform (30min, 2h) (i.e., uniformly distributed between 30min and 2h), and 15, respectively. The end of the second phase can be determined in several ways: it can end after a time duration spent inside the area, after a number of checked persons is reached, or after the checking ratio goes beyond a given threshold (if the number of workers inside the area is a priori known). In any of these cases, the security guard announces the end of this phase using his navigator device. And the mission shifts to its third and final phase.

Third phase of a mission: It only involves guiding the security guard back to the headquarter using the reserve path stored in the navigation device. The security guard needs to empty his memory, because in the upcoming missions, she/he should be able to recheck a person as this person could move from one area to another at any time.

Moreover, the number of checking missions is equal to $N\_missions$ per security guard per day, uniformly spread over the 8 working hours. The different values given to this variable define the different defender strategies already described in Subsection 15.4.2. Every mission lasts for a duration between $MinDuration$ (set to 10) and $MaxDuration$ (set to 20) minutes. A checking operation may last between 1 and 3 minutes.

15.4.4.4 Implementation of Our Goals

In Section 15.4.3, we identified four goals to be measured: detection rate (to be maximized), caused damage (to be minimized), privacy preservation (to be maximized), and comfort breach (to be minimized). As already explained, some goals can easily be integrated to our simulation model as it is the case of the very first two goals. Others need more attention and a prior work before being able to integrate them to the simulation. Here after, we will try to explain how we managed to measure such goals.

Considering the comfort breach, we may rely on the ethnographic study we conducted in step 1, to extract the subjective feeling of workers regarding their comfort breach because of the repetitive security checks. Through a questionnaire, employees were asked about their feelings (scored between 0 and 1, where 0 is a total comfort preservation and 1 means a maximum comfort breach) if ever they get checked 1, 2, ..., 9 times (or more) a day. The collected data is summarized by the Average Comfort Breach as a function of the number of checks per day, which is depicted in Figure 15.3. Using this collected data, we created a multivariate Gaussian of dimension 9 (to represent the feeling upon 1, 2, 3, ..., and 9 or more checks per day) to be used in our simulator as a simple generator of degree of non-satisfaction, so that we can create as many workers as we want, with different subjective comfort breach measures, but following the same general shape as the one shown in Figure 15.3.

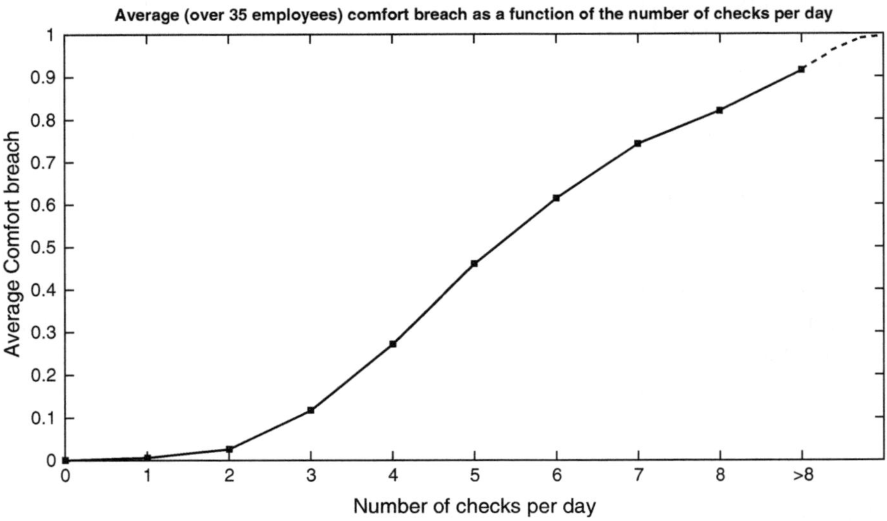

Fig. 15.3: Average Comfort Breach as a function of the number of checks per day

Besides, to assess the privacy preservation of a given applied defense strategy, a model based on the entropy theory and Markov chains was integrated to our simulator (more details on the model can be obtained from our previous work [1]). Briefly, when a worker is checked, his location is somehow revealed for a given period of time, and thus this model captures to what extent a checking policy could reveal employees' positions over time. Without going into the details of this model, we simply describe it as follows: it looks at location privacy as the capacity of an attacker, in the worst case scenario, to estimate/guess the employees' positions with high confidence at a given instant. It basically uses the Shannon entropy theory to estimate this maximum disclosure where an attacker may have more insights about how employees could move inside the facility leading to a more subsequent breach. To model movement of employees inside the facility, a continuous-time Markov chain (CTMC) is used. The latter helps in computing the probabilities over time of an employee to be inside or outside an area, given that she/he has been lately checked there. These probabilities, combined with the aforementioned entropy-based analysis, lead to a new metric that effectively measures location privacy as a function of time. Such a metric is very important as it allows capturing the decreasing significance of leaked location information over time. In our simulation we define our location privacy preservation as the minimum value of this metric that was reached during the whole run.

Finally, after presenting the simulation environment and the implemented key performance indicators measuring our four defined goals, we are able to run our simulations. Thus, we performed 100 runs (i.e., a big enough number of runs to obtain a satisfying distribution shape) for each possible configuration of attack and defense strategy. We may insist here that in a conventional approach, the results of these simulations would be averaged to get a single value for each of the four goals. Further, these averaged values would be used to solve the so-arising multiobjective game using standard methods. However, in our approach, we deviate from this route by *not aggregating* the data and calculating the average, but instead compiling the whole output of the simulation in terms of categorical distributions (effectively histograms or bar plots), over which the game payoff structures (one per goal) are defined. We may insist here that by processing our data into categories, we may obviously lose some information; however, we will be able to capture the overall shape of the distribution function. The main goal here is to have comparable distribution shapes regardless of the goal they may measure.

15.4.4.5 Simulation Results

As already explained in Section 15.4.2, the specific parameters are encoded in the string descriptors of each strategy. The resulting data was divided into a total of five categories that span the numeric range of all four goals, forming a histogram/bar plot for each result. The respective matrices displaying all this information, i.e., all the histograms/bar plots, are plotted in the Appendix of the chapter as Figures 15.17, 15.18, 15.19, and 15.20 with regard to all goals identified in Section 15.4.3. This collected data serves to identify the optimal configuration by applying our multiobjective game.

15.4.5 Identification of Optimal Configuration

To identify an optimal set of defensive actions, we will apply the game-theoretic model taking as input the results from the simulation of all the physical intrusion scenarios. As already mentioned, the game-theoretic framework is able to solve a multiobjective game. Thus, we can take several aspects into consideration and find an optimal solution for all of them at once. Our intrusion simulation identified the payoffs for the location privacy preservation of all employees, the maximum comfort/satisfaction breach (due to successive or continuous ID checks), the average time the intruder spends in a specific area/potential physical damage the intruder can cause (monetary or abstract), and the detection rate.

As a next step, the obtained results of our payoffs will be categorized into 5 fixed classes as presented in Table 15.3. This categorization is mandatory to be able to apply the game-theoretical framework capable of computing the optimal strategy of our multi-goal game [13]. This is simply done by dividing the resulting data into these five categories that span the numeric range of all goals, forming a histogram/bar plot for each result. In other words, the probability distribution including all available information must be constructed under the following constraint:

- All assessments are made in the same scale. This is required for the multi-criteria optimization to work (using scaling of the vector-valued payoffs). Numeric indicators are thus discretized onto a common categorical scale (that all categorical indicators use as well). Besides, categories of a *to-be-maximized* goal are simply inverted to transform our game into a pure minimization problem.
- The data source is reliable with regard to the intended risk assessment.

Table 15.3: Qualitative Risk Categories

1	2	3	4	5
very low	low	medium	high	very high

This categorized data can be seen as a distribution for each combination of (player 1: defender, player 2: attacker) strategies. These distributions are in fact the real payoffs of our two players in terms of the respective goals. After deciding on the priorities among all the different goals (i.e., even priorities in our case), the HyRiM tool [14] can be directly used to give recommendations regarding the best strategy, or more correctly the best mixed strategy, to be applied from the defender point of view, as well as the potential damage that can be caused by a worst case attack.

In general, the resulting solution is a mixture of all possible security strategies specified in the first step (cf., Subsection 15.4.2 and Table 15.2). In other words, each strategy of the solution has a specific probability to be carried out. Computing the equilibrium, the optimal *security strategies* for both players are as presented by Figures 15.4 and 15.5, respectively.

As expected, the result is a nontrivial mixed strategy. We have a mixture of three defense strategies, i.e., strategies #4, #7, and #8 (cf. Table 15.2). In more detail, strategies #4 ("D-NG1F*8*T***R***"), #7 ("D-NG15F*5*T***HSLF***"), and #8 ("D-NG15F*8*T***HSLF***") have to be applied with respective probabilities 0.1, 0.768, and 0.132. Hence, a practitioner could abandon the remaining strategies. This is also an interesting observation/lesson from the game: it also tells us which defense strategies are more relevant than others.

Besides, and as already stated above, the method for computing this equilibrium is essentially based on multi-criteria optimization for security (cf. [12, 9]). Provided that the defender plays his optimal strategy, the respective optimal loss distributions attained under any behavior of the attacker are given in Figures 15.6, 15.7, 15.8, and 15.9. Here, the optimal attack strategies are different between the four goals, meaning that the attacker can never cause maximal loss in all four goals at the same time. She/he can only cause a maximum loss in one or two goals using individually each identified attack strategy. Such an observation represents another good information to know, since it indicates that the computed loss distributions are only pessimistic and reality should look much better (expectedly).

Given these results, the next step should be the implementation of our computed optimal strategy. This is the object of the next subsection.

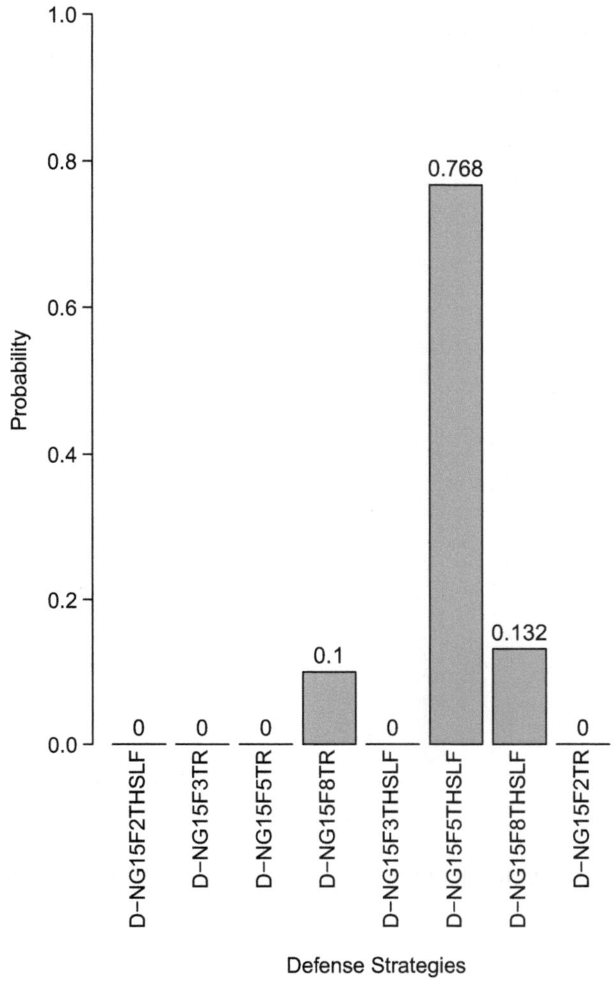

Fig. 15.4: Equilibrium for multiobjective security game (MOSG): optimal defense strategy

15.4.6 Implementation of Optimal Configuration

In general, the results from the game-theoretic optimization algorithm need to be implemented precisely since a deviation of the probabilities given in the equilibrium will increase the potential damage caused by an attacker under the worst case attack strategy. Therefore, the results from the algorithm can be fed into a random selection function to obtain the current advice according to the optimal randomized choice rule for the strategies. In other words, the manager of the security guards will use some kind of scheduling system, which provides him with an indication about which of the three security strategies (either #4, #7, or #8) to follow at any decision instant.

This needs to be done iteratively (e.g., each day or at the beginning of each shift) and can be pre-computed for several days or weeks to simplify personnel decisions (e.g., shift rotations, etc.).

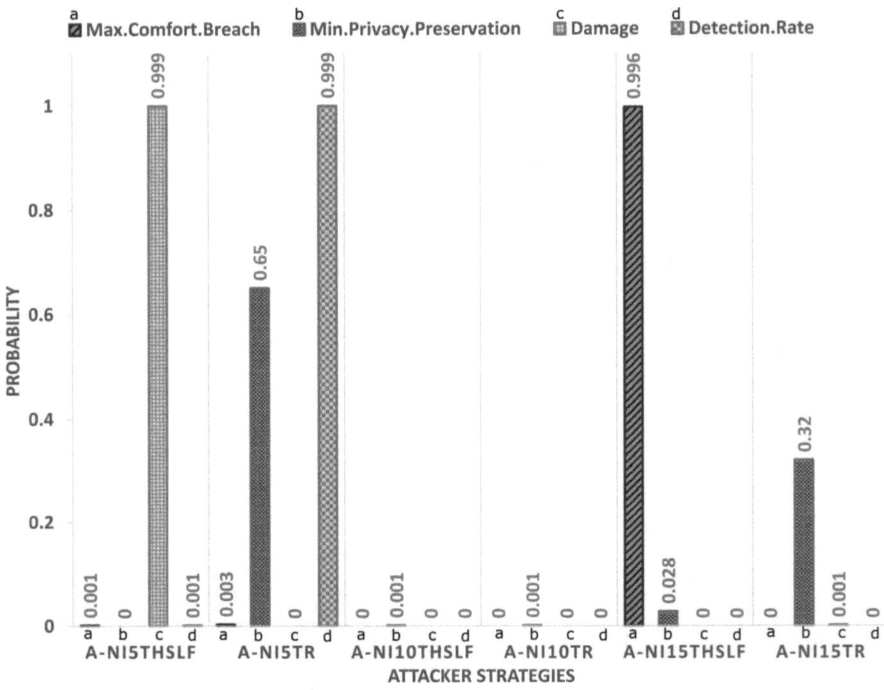

Fig. 15.5: Equilibrium for multiobjective security game (MOSG): worst case attack strategies per goal

The problem of such an approach is that our optimal defense strategy is achieved only over long period of time. One can think that it would be more interesting whether we are able apply our optimal strategy even for short terms. This would be exactly our goal that we will explain in the present section. First, let us recall that for all our pure strategies, we deployed the whole set of our resources (i.e., 15 security guards) during the whole day. In other words, each pure strategy was deployed at its respective full defense power all the time. Now, when we say that we need to apply a given strategy, say s, with a given probability, say p, we mean to apply s at its full power $(p * 100)\%$ of the time or, inversely, apply $(p * 100)\%$ of the full power of s all the time. Taking into account such an obvious observation, we only need to divide our resources (i.e., security guards) according to the optimal defense strategy. This gives us the mixed strategy (now seen as pure strategy) described by Table 15.4.

We insist here that we should respect as much as possible the results given by the equilibrium while dividing our resources among the detected useful strategies. This is the reason for which we decided to affect 12 security guards to strategy "D-NG15F*5*T***HSLF***" and one single security guard to "D-NG15F*8*T***R***."

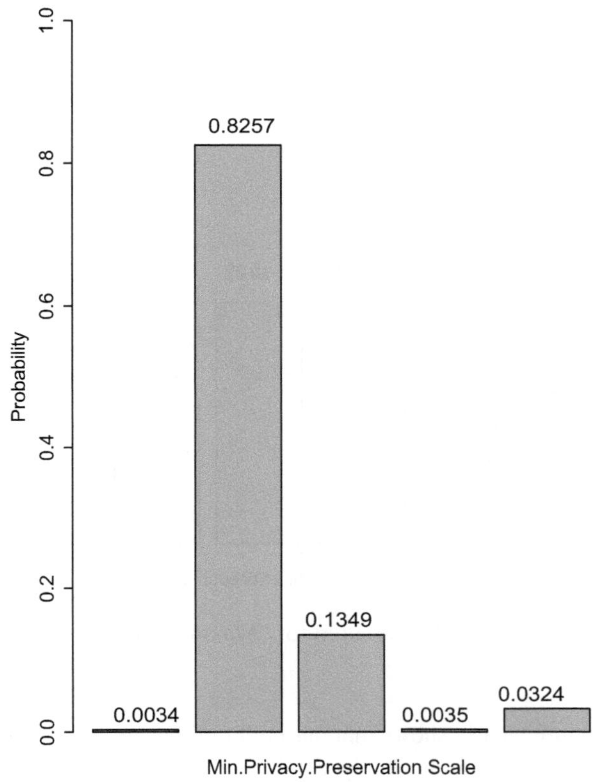

Fig. 15.6: Optimal loss assured for "Privacy Preservation"

15.5 Validation of Optimal Surveillance Configuration

This section is devoted to validate and verify the effectiveness of the implemented optimal strategy. For that purpose, and to assess the superiority of this strategy over all the others, we simulated it for 100 runs, extracted its relative distributions, and replayed the game with our newly added strategy (described in Table 15.4) and all the previously presented pure strategies (c.f. Table 15.2). As a result, we obtained the new equilibrium given by Figures 15.10 and 15.11.

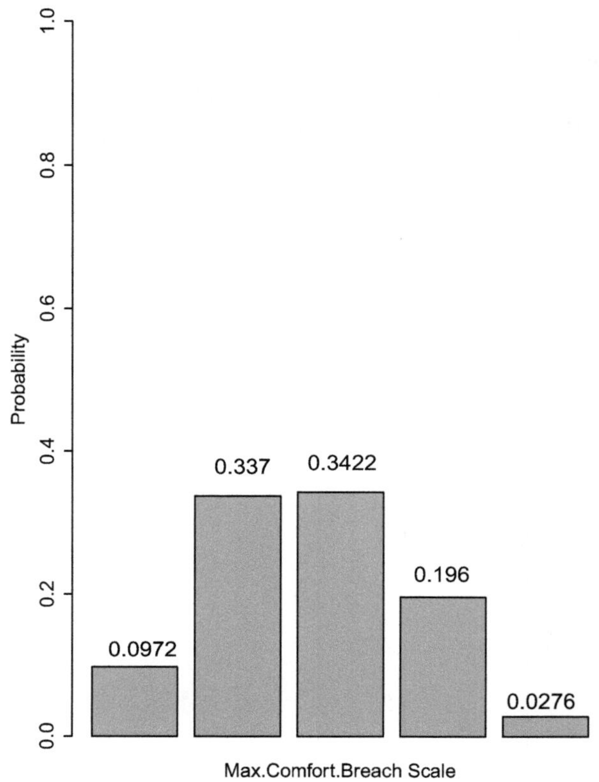

Fig. 15.7: Optimal loss assured for "Maximum Comfort Breach"

This equilibrium clearly shows that our built optimal strategy is indeed the most effective (99.5%). This result simply confirms that the defender's best choice to defend *the company* is by applying "D-NG15ImplMixed" almost all the time. This guarantees that the attacker's worst attack will never cause losses (in terms of our four considered goals) that exceed those presented in Figures 15.12, 15.13, 15.14, and 15.15.

A second way to validate our obtained results is to compare all the strategies to pick up the best one in terms of our considered four goals. For that, we describe each strategy by a 4-dimensional vector, where each dimension represents one of our 4 goals. By computing the average values of the respective goals of each of our strategies, we will be able to place each of them in the 4-dimension space of all possible strategies. Ideally, the fictive optimal strategy corresponding to the origin would represent the best possible strategy to apply (as it ensures a minimum value, 0, for all our four goals; bearing in mind the fact that all goals have been minimized as described in Section 15.4.5). All other strategies occupy the rest of the space, and assure a non-null value of at least one of the goals. Intuitively, we can say that the closer a strategy is to the origin, the better it would be (recall here that we considered

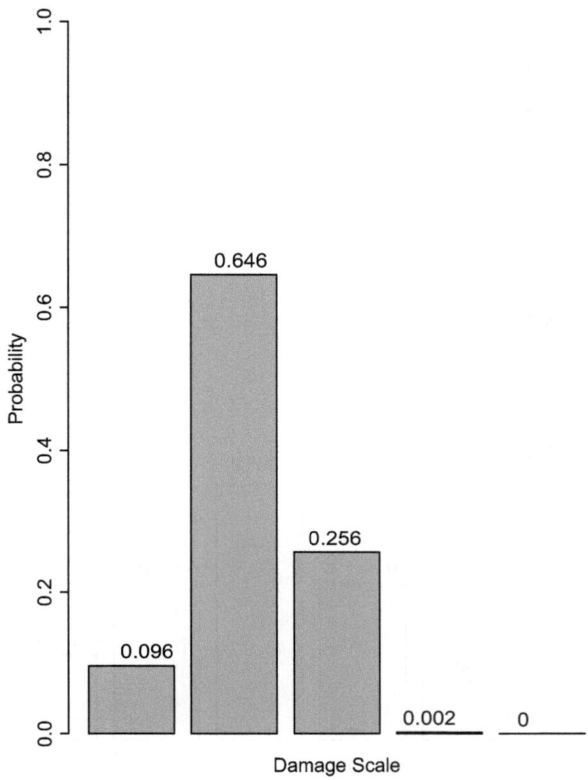

Fig. 15.8: Optimal loss assured for "Caused Damage"

even priorities among all goals). In other words, if we consider two strategies, say $s_1 = (g_{1,1}, g_{2,1}, g_{3,1}, g_{4,1})$ and $s_2 = (g_{1,2}, g_{2,2}, g_{3,2}, g_{4,2})$, we can say that s_1 is better than s_2, if the distance to the origin of s_1 is smaller than the distance to the origin of s_2:

$$dist(s_1) = \sqrt{g_{1,1}^2 + g_{2,1}^2 + g_{3,1}^2 + g_{4,1}^2} \leq dist(s_2) = \sqrt{g_{1,2}^2 + g_{2,2}^2 + g_{3,2}^2 + g_{4,2}^2}$$

where $g_{i,s}$ is the average value of goal i for strategy s. For our case study, we plotted the distance to the origin of all the considered pure strategies, in addition to our implemented mixed strategy "D-NG15ImplMixed," to better visualize where our 9 strategies are located in the space of all possible strategies. This plot is presented in Figure 15.16.

Once again, the obtained results clearly confirm that our computed mixed strategy is the best option that the defender should apply to insure a minimum average loss in terms of our four goals, as it is the closest to the origin compared to all the other pure strategies.

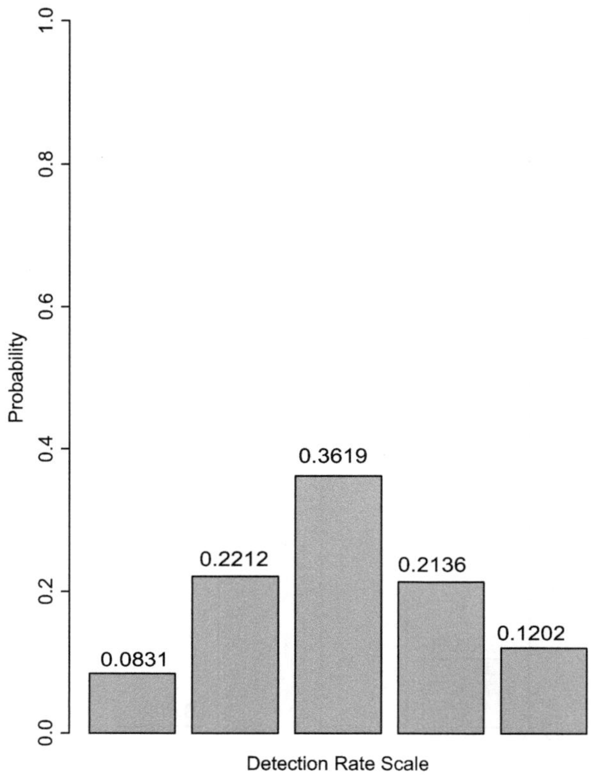

Fig. 15.9: Optimal loss assured for "Detection Rate"

Table 15.4: Implemented Optimal Strategy

Strategy Label	#Security Guards	Description
D-NG15F*8*T***R***	1.5 $1 securityguard$	≃ freq = 8 & areas: targeted randomly
D-NG15F*5*T***HSLF***	11.52 $12 securityguards$	≃ freq = 5 & areas: targeted Higher Sec. Lev. First
D-NG15F*8*T***HSLF***	1.98 $2 securityguards$	≃ freq = 8 & areas: targeted Higher Sec. Lev. First

15.6 Conclusion

In this chapter, we presented a realistic case study of a critical infrastructure that we intend to protect against physical intrusions. Our study is based on a game-theoretic approach for physical surveillance. Modeling surveillance as a theoretic game (i.e., cops and robbers) is quite common; however, in classical games, it is assumed perfect or crisp assessment of the payoffs to be used. Such an assumption turns out to

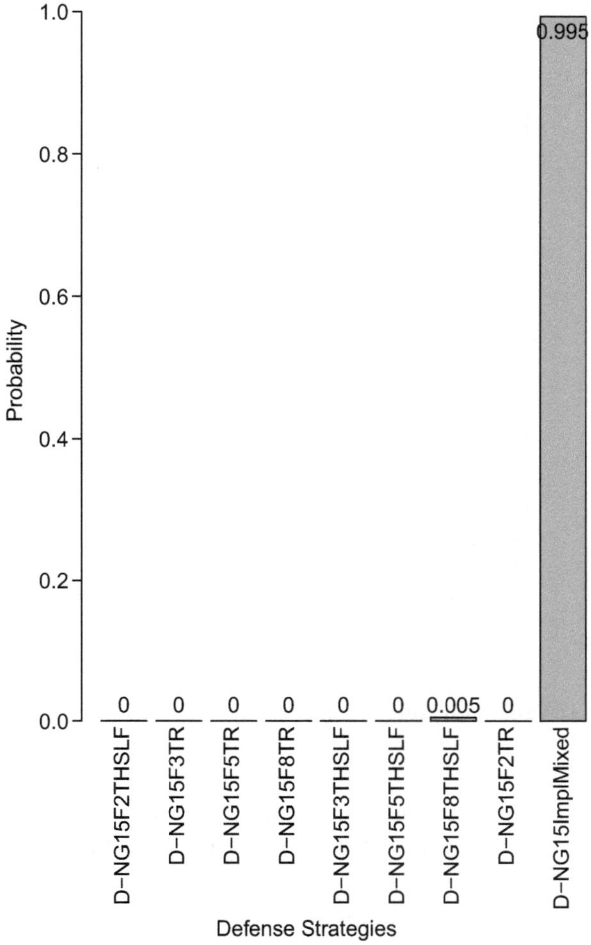

Fig. 15.10: Validation equilibrium: optimal defense strategy

be strong and do not really match reality. On the other hand, the HyRiM tool [14] appears to be a practically effective tool that accurately describes real surveillance scenarios, as it integrates the intrinsic uncertainty of surveillance systems and of the respective risk assessment into the game-theoretic model itself. Actually, applying this tool represents one of the six steps of the *G-DPS* framework, which aims at finding the optimal configuration for physical surveillance system over multiple goals.

In this chapter, we went through all the steps of this very decision-making framework, where we started by establishing the context of our targeted infrastructure. This first step helped us understand and better gauge risks and potential threats around our infrastructure. We then extracted the set of potential strategies that a risk manager would want to apply to implement surveillance (8 defense strategies were

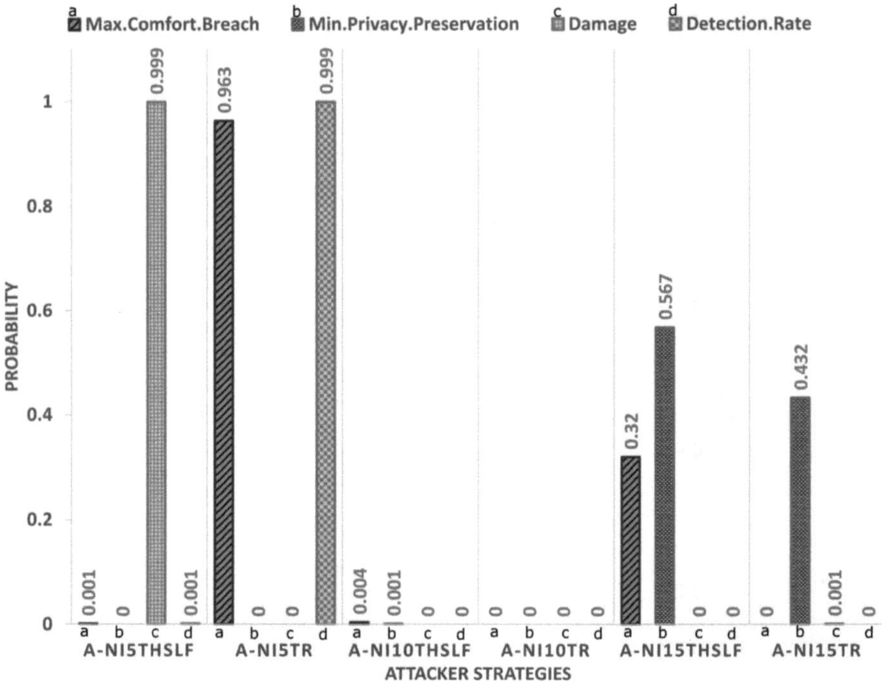

Fig. 15.11: Validation equilibrium: worst case attack strategies per goal

identified for our case study), as well as the potential attacker strategies (we identified 6 possible attack strategies) that can be adopted by an intruder to fulfill some of the risks identified in the first step. As a third step, we discussed four different security goals of capital interest to *the company*, i.e., caused damage, detection rate, location privacy preservation, and comfort of the employees, that we need to take into account while deciding on the optimal defense strategy to apply. The fourth step, i.e., strategy assessment, was deeply investigated in this chapter. In fact, we presented a simulation tool that was developed to easily and effectively measure our different payoffs. After collecting the different results, we moved to the fifth step which corresponds to applying the HyRiM tool to compute the optimal strategy, optimizing our four goals at once. The result was a nontrivial mixed strategy, and its implementation was described in the final and sixth step.

The last part of the chapter was devoted to validating the obtained results. At first, we simulated the implemented optimal mixed strategy and replayed our game once again. The obtained results showed that our newly found strategy is the most effective among all the considered strategies (i.e., 99.5% most effective). Furthermore, we showed that our strategy represents the closest one to a theoretical fictive 100%-optimal strategy (i.e., having a theoretical best outcome in terms of all the considered goals), confirming by the same way the results of applying our decision-making framework.

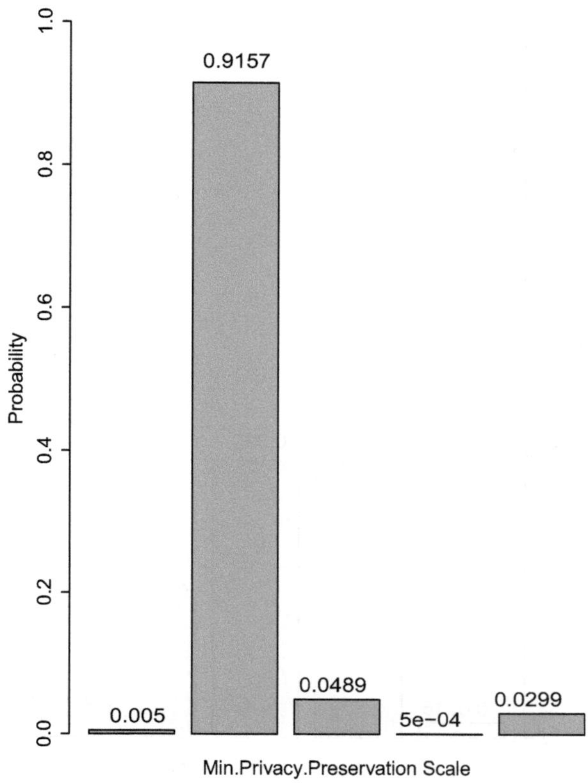

Fig. 15.12: Optimal loss assured for "Privacy Preservation"

Acknowledgements This work was supported by the European Commission's Project No. 608090, HyRiM (Hybrid Risk Management for Utility Networks) under the 7th Framework Programme (FP7-SEC-2013-1).

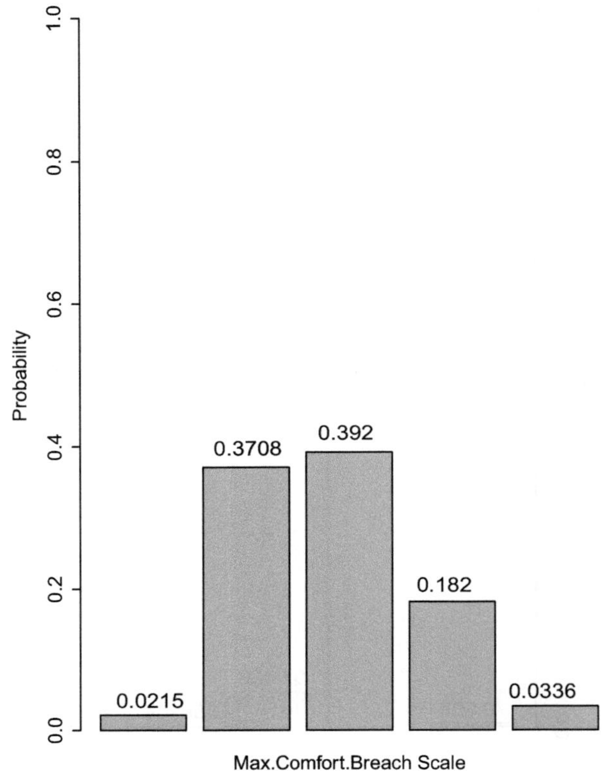

Fig. 15.13: Optimal loss assured for "Maximum Comfort Breach"

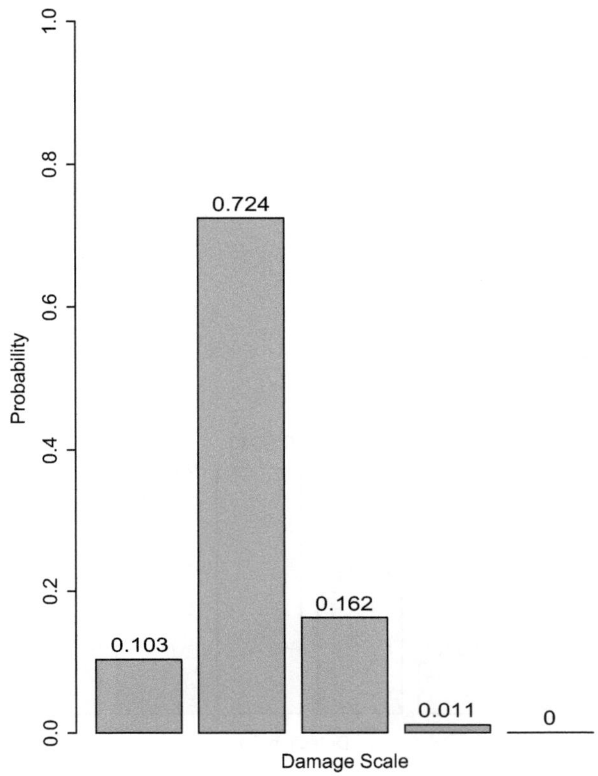

Fig. 15.14: Optimal loss assured for "Caused Damage"

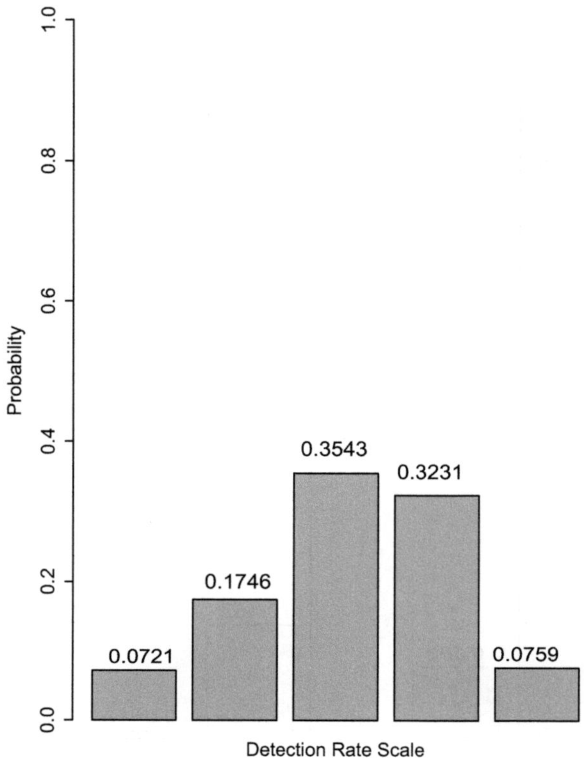

Fig. 15.15: Optimal loss assured for "Detection Rate"

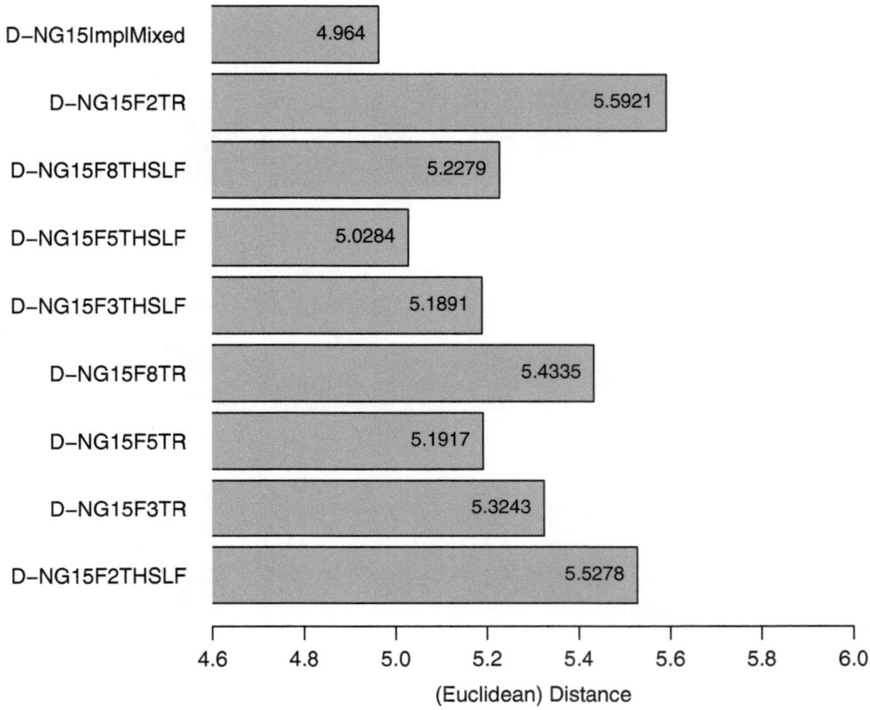

Fig. 15.16: Distance to the origin of all defense strategies

Appendix

See Figures 15.17, 15.18, 15.19, and 15.20.

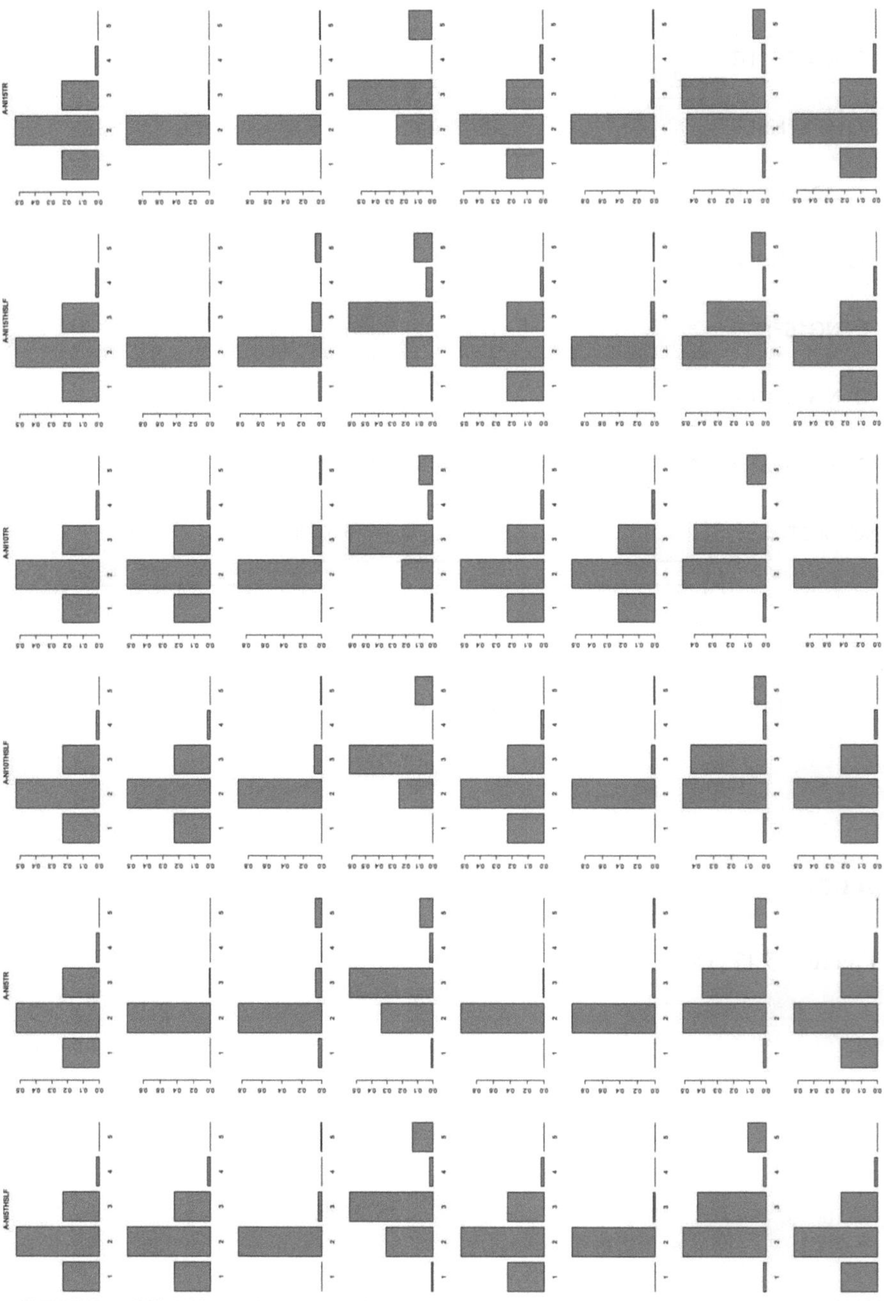

Fig. 15.17: Simulation results: payoffs of the “Max Privacy Preservation”

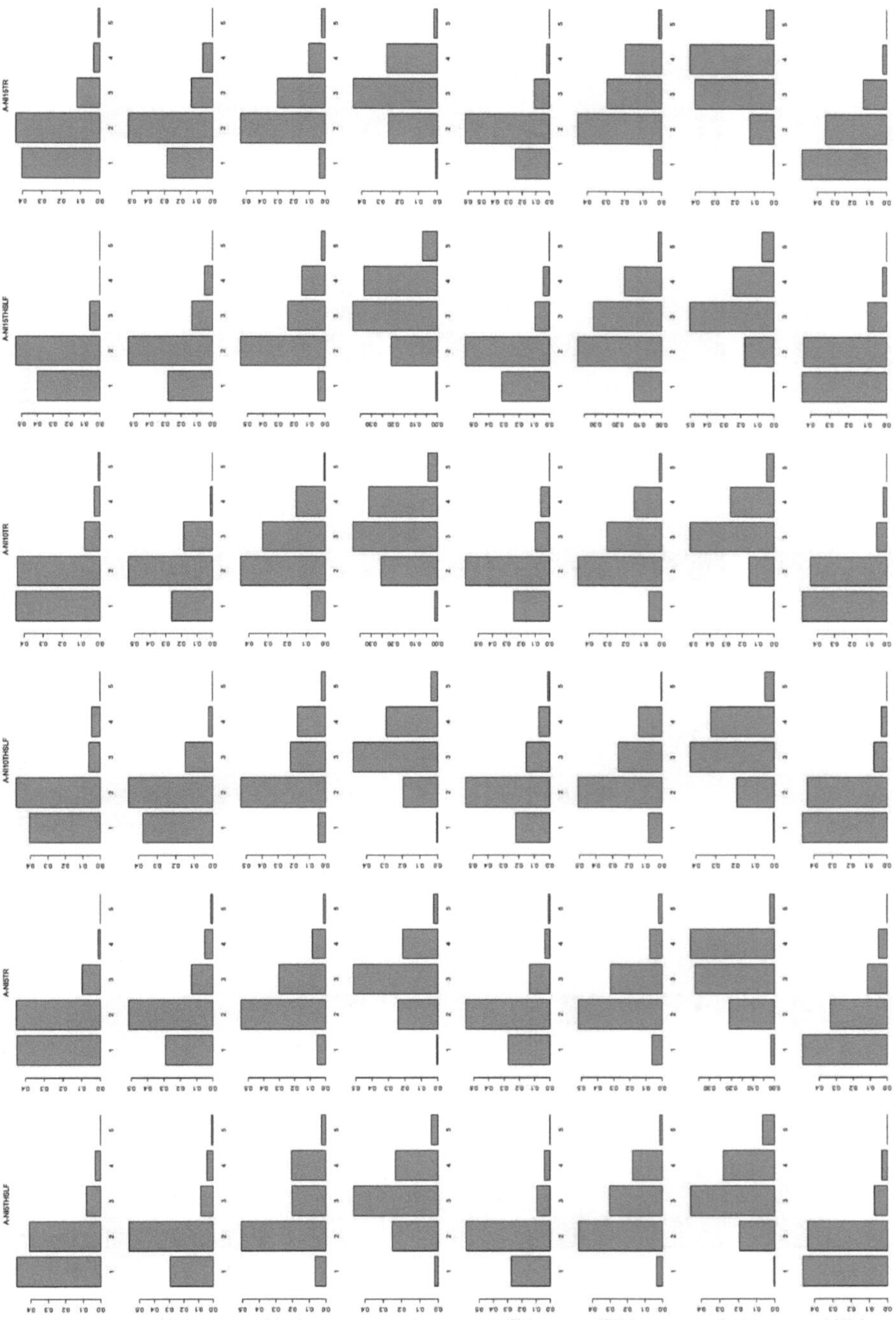

Fig. 15.18: Simulation results: payoffs of the "Min Comfort Breach"

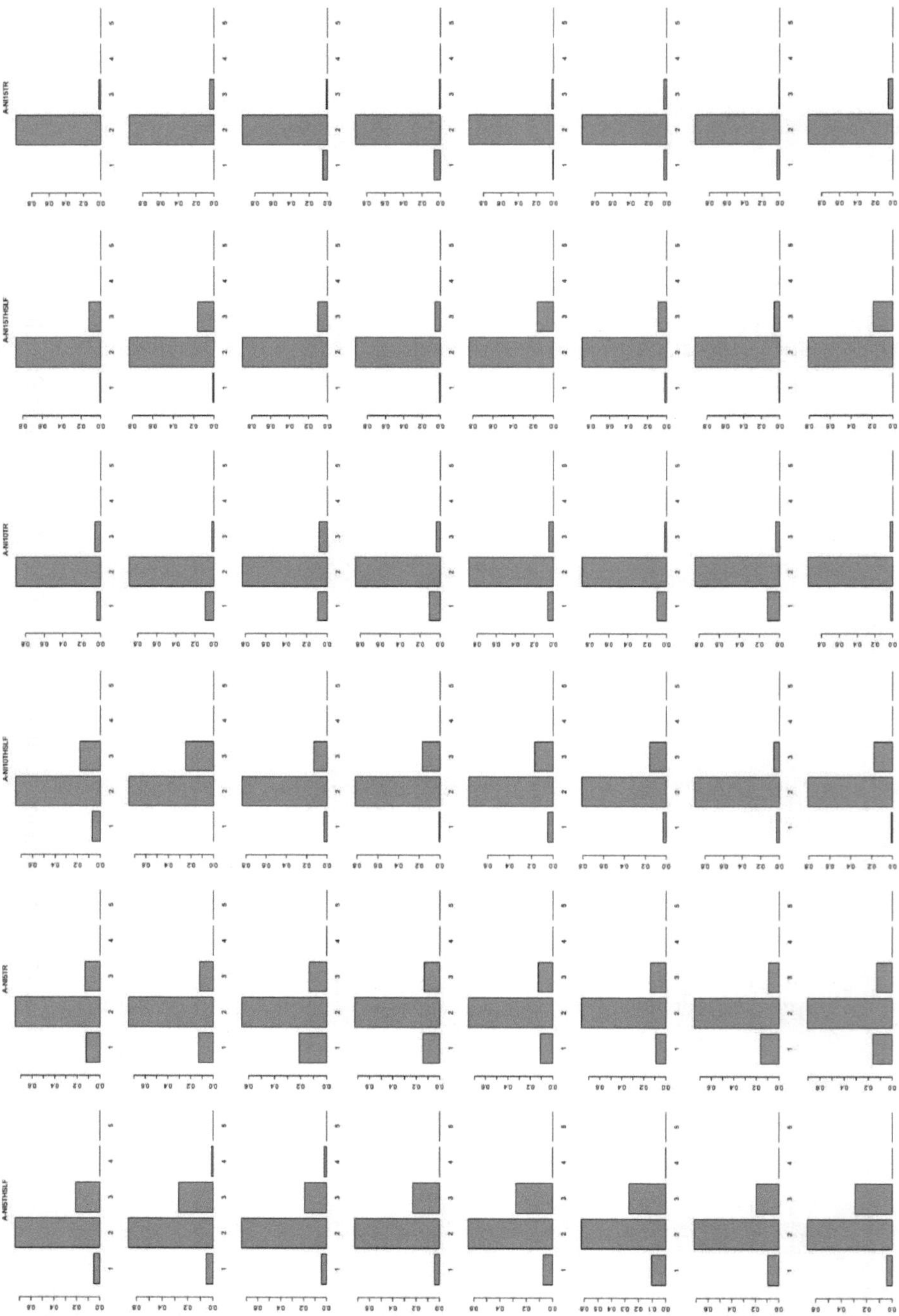

Fig. 15.19: Simulation results: payoffs of the “Caused Damage”

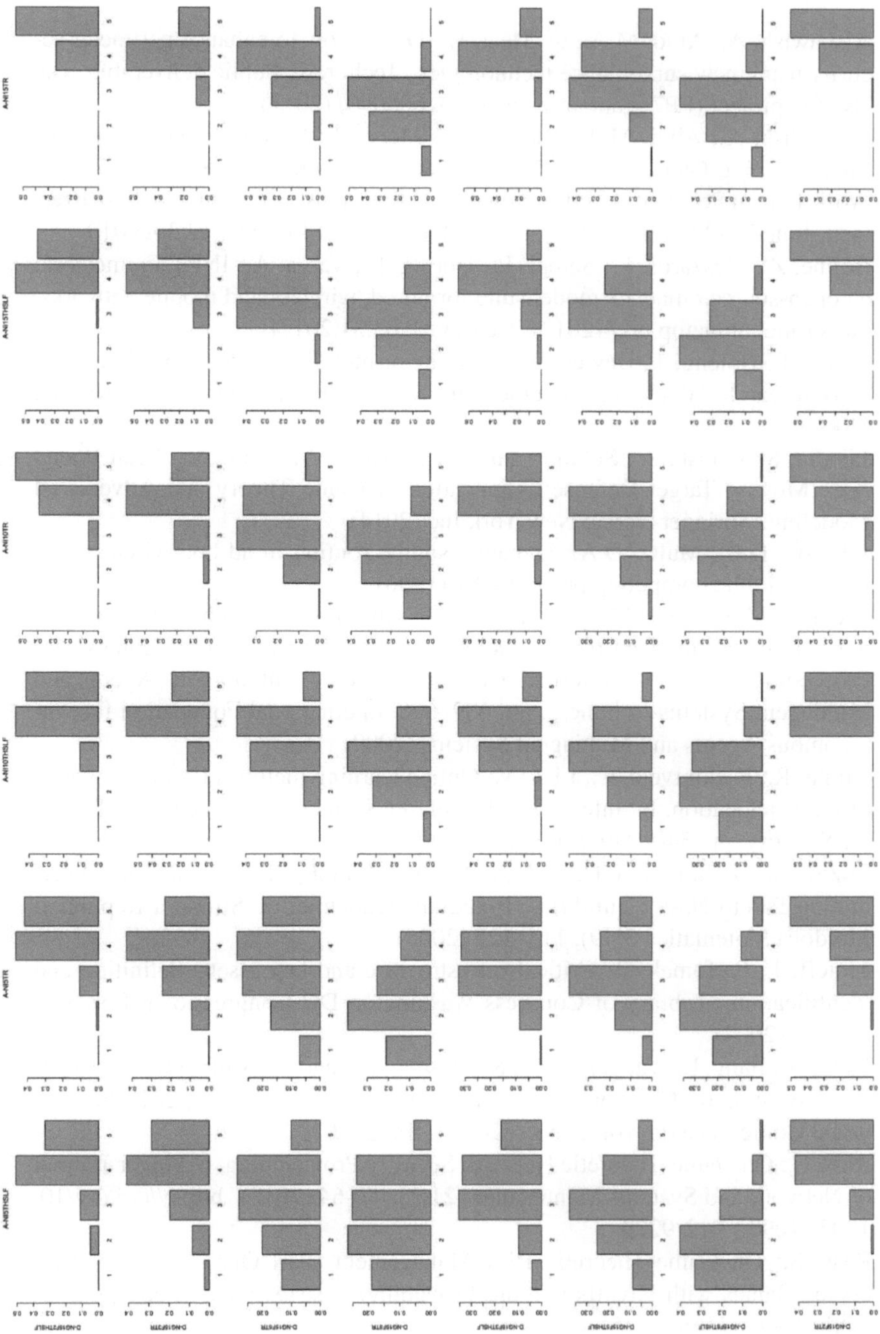

Fig. 15.20: Simulation results: payoffs of the “Detection Rate”

References

1. Alshawish, A., Abid, M.A., de Meer, H.: D4.3 - how to enhance perimeter security using new surveillance technologies. Tech. rep., Public deliverable, The HyRiM project (FP7 grant agreement no. 608090) (2017)
2. Alshawish, A., Abid, M.A., Rass, S., de Meer, H.: Playing a Multi-objective Spot-checking Game in Public Transportation Systems. In: SHCIS '17 - 4th Workshop on Security in highly connected IT systems, p. 6. Neuchâtel, Switzerland, June 21–22, 2017 (2017). https://doi.org/10.1145/3099012.3099019
3. Bojthe, Z., Meszaros, L., Seregi, B., Hornig, R., Varga, A.: INET Framework: an open-source omnet++ model suite for wired, wireless and mobile networks. https://inet.omnetpp.org/ (2017). [retrieved: 03.31.2017]
4. Camp, T., Boleng, J., Davies, V.: A survey of mobility models for ad hoc network research. Wireless communications and mobile computing **2**(5), 483–502 (2002)
5. Jajodia, S., Ghosh, A., Subrahmanian, V., Swarup, V., Wang, C., Sean Wang, X.e.: Moving Target Defense: Application of Game Theory and Adversarial Modeling. Springer-Verlag New York Inc (2014)
6. Johnson, D.B., Maltz, D.A.: Dynamic source routing in ad hoc wireless networks. Mobile computing pp. 153–181 (1996)
7. Kiekintveld, C., Jain, M., Tsai, J., Pita, J., Ordóñez, F., Tambe, M.: Computing optimal randomized resource allocations for massive security games. In: Proceedings of The 8th International Conference on Autonomous Agents and Multiagent Systems-Volume 1, pp. 689–696. International Foundation for Autonomous Agents and Multiagent Systems (2009)
8. Klíma, R., Kiekintveld, C., Lisý, V.: Online learning methods for border patrol resource allocation. In: International Conference on Decision and Game Theory for Security, pp. 340–349. Springer (2014)
9. Lozovanu, D., Solomon, D., Zelikovsky, A.: Multiobjective Games and Determining Pareto-Nash Equilibria. Buletinul Academiei de Stiinte a Republicii Moldova Matematica **3**(49), 115–122 (2005)
10. Moteff, J., Parfomak, P.: Critical infrastructure and key assets: definition and identification. Library of Congress Washington DC Congressional Research Service (2004)
11. Pang, H., Jiang, L., Yang, L., Yue, K.: Research of android smart phone surveillance system. In: Computer Design and Applications (ICCDA), 2010 International Conference on, vol. 2, pp. V2–373. IEEE (2010)
12. Rass, S.: On Game-Theoretic Network Security Provisioning. Springer Journal of Network and Systems Management **21**(1), 47–64 (2013). https://doi.org/10.1007/s10922-012-9229-1
13. Rass, S.: On Game-Theoretic Risk Management (Part One)-Toward a Theory of Games with Payoffs that are Probability-Distributions. arXiv preprint arXiv:1506.07368 (2015)

14. Rass, S., König, S.: R package 'hyrim': Multicriteria risk management using zero-sum games with vector-valued payoffs that are probability distributions (2017). URL https://hyrim.net/software/
15. Varga, A.: OMNeT++: discrete event simulator. https://omnetpp.org/ (2016). [retrieved: 06.20.2016]
16. Zhu, Q., Başar, T.: Game-theoretic approach to feedback-driven multi-stage moving target defense. In: 4th International Conference on Decision and Game Theory for Security - Volume 8252, GameSec 2013, pp. 246–263. Springer-Verlag New York, Inc, New York, NY, USA (2013). https://doi.org/10.1007//978-3-319-02786-9-15

Chapter 16
Smart SECPLAN: A Process Implementation Tool for Hybrid Risk Management

Alberto Zambrano, Santiago Caceres, and Ana Isabel Martinez

16.1 Introduction

Utilities are complex organizations composed by many technical systems like networks, devices, controllers, infrastructures, etc. spread around large areas. Despite their size, utility operators must carry out maintenance activities as well as implement security measures for their infrastructures. Most of the times, these activities are based on specific guidelines or standards the utilities need to follow or are based on expert knowledge gained during the years. Moreover, the influence of cultural and organizational factors affects the preventive activities carried out. Experience shows us that the schedule of these activities is often of an "ad hoc" nature and not optimal in terms of time or cost.

Partly, this may be attributed to risk management essentially being a matter of preventing losses, rather than creating gains. During well times where there is no immediate threat to deal with, people may not feel the necessity to become active toward security. Activities triggered by incidents are inevitably "behind" what is going on and as such may not be able to reduce damages to the minimum at the same quality or cost as would have been possible via a priori precautions and security actions scheduled in advance.

In general, there is no formal risk assessment exercise behind the scheduling of security measures or maintenance activities. The web-based tool Smart SECPLAN aims at covering this gap, analyzing in detail the infrastructure implemented and operated within a utility provider. In this way, Smart SECPLAN helps information technology (IT) and operations technology (OT) security experts to better understand their assets and take informed and objective decisions on how to spend efforts and resources for repetitive maintenance tasks given a utility infrastructure.

A. Zambrano (✉) · S. Caceres · A. I. Martinez
ETRA Investigación y Desarrollo S.A., Calle Tres Forques 147, 46014 Valencia, Spain
e-mail: azambrano.etraid@grupoetra.com; scaceres.etraid@grupoetra.com; amartinez.etraid@grupoetra.com

S. Rass, S. Schauer (eds.), *Game Theory for Security and Risk Management*, Static & Dynamic Game Theory: Foundations & Applications,
https://doi.org/10.1007/978-3-319-75268-6_16

In the following sections, a real-life scenario involving a medium-size distribution system operator (DSO) is provided, discussing each one of the steps from the HyRiM process (as described in Chapter 12) and how it is carried out in Smart SECPLAN to arrive to the results. In further detail, Section 16.2 provides a short recap of the HyRiM process, whereas Section 16.3 gives an overview on the Smart SECPLAN tool. The details of the real-life scenarios are described in Sections 16.4, and 16.5 shows the individual steps implemented by the Smart SECPLAN. The results are then closely inspected and interpreted in Section 16.6, reflecting also on the differences between a classical game and the distribution-valued approach followed in our scenario.

16.2 The HyRiM Process

The Hybrid Risk Management (HyRiM) process presented in the previous chapter, Chapter 12, is suited for organizations operating highly interconnected networks at different levels, such as utility providers or critical infrastructure operators. To achieve that, the HyRiM process is compliant with the general ISO 31000 process for risk management [7] and thus can also be integrated into existing risk management processes already running in the aforementioned organizations. It relies on and implements the similarities between game theory and risk management, as Chapter 1 outlines.

In detail, the operative risk management process of the ISO 31000 framework is adopted, and each step of the process is supported with the tools developed in the HyRiM project (cf. Figure 16.1). These tools cover different social and technical analysis techniques and simulation methodologies that facilitate the risk process. The relevant HyRiM tools have been identified and mapped onto the risk management process as shown in Figure 16.1. Since the ISO 31000 is a generic process and is often used as a template in other ISO standards (like in the ISO 27005 [8], the ISO 28001 [6], or others), the HyRiM process can also be integrated into these standards. This makes it possible to apply the HyRiM process to multiple fields of application.

As a general framework to model the interplay between different networks, game theory is applied in the HyRiM process. Game theory not only provides a well-sound mathematical foundation but can also be applied without a precise model of the adversary's intentions and goals. Therefore, a zero-sum game and a minimax approach [11] can be used, where the gain of one player is balanced with the loss of the other. This can be used to obtain a worst-case risk estimation. Further, the game-theoretic framework developed in HyRiM [13, 12] (cf. also Chapters 2 and 3 for more details) allows modeling the intrinsic randomness and uncertainty encountered in real-life scenarios. This is realized using distribution-valued payoffs for the game [14]. In the HyRiM process, these payoffs are coming from both percolation and co-simulation analyses, where the results of these stochastic processes are described as distributions.

The output of the game-theoretic framework is threefold and includes the maximum damage that can be caused by an adversary, an optimal attack strategy resulting in that damage and an optimal security strategy for the defender. The optimal defense strategy is, in general, a mixture of several defensive (i.e., mitigation) activities. These activities, if implemented correctly, provide a provable optimal defense against the adversary's worst-case attack strategy. The implementation can be simplified and guaranteed, for example, by the use of a job scheduling tool.

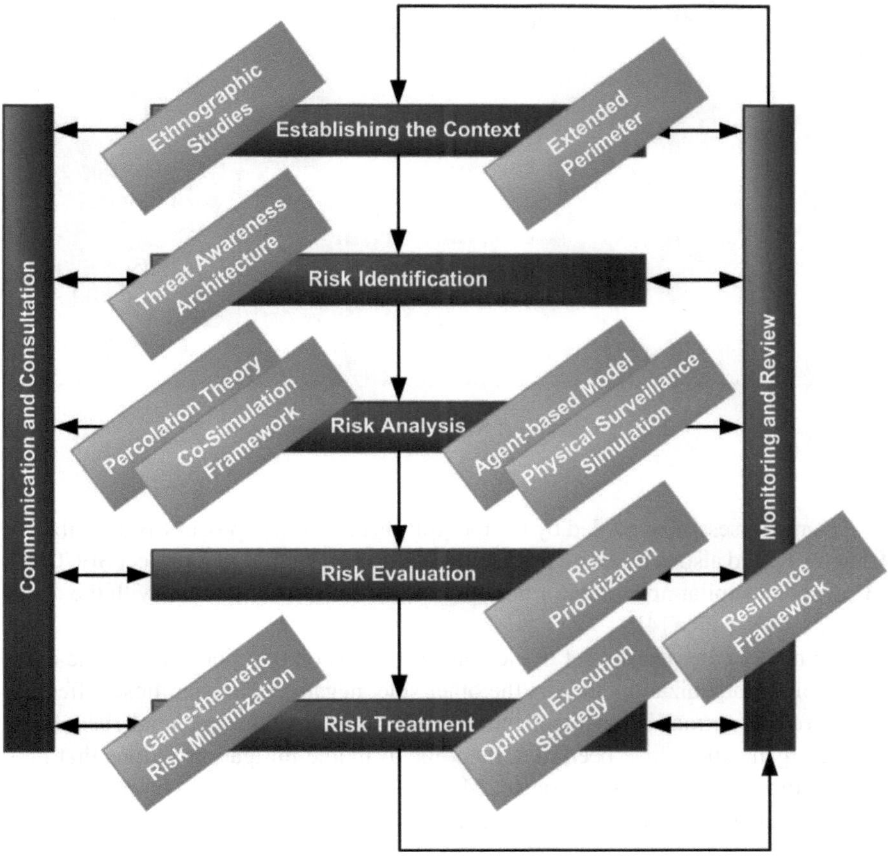

Fig. 16.1: Overview of the HyRiM process including relevant tools for each process step

16.3 The Smart SECPLAN Tool

The Smart SECPLAN tool is a web-based tool developed in the course of the HyRiM project allowing any organization to perform a risk assessment over its regular operations. Smart SECPLAN helps those involved to discover potential expositions to risks of any kind and establishes possible regular mitigation actions that help in preventing those risks. As a result, the tool generates a model of the risks to which the organization is exposed and provides the optimal strategy to perform the identified mitigation actions (Figure 16.2).

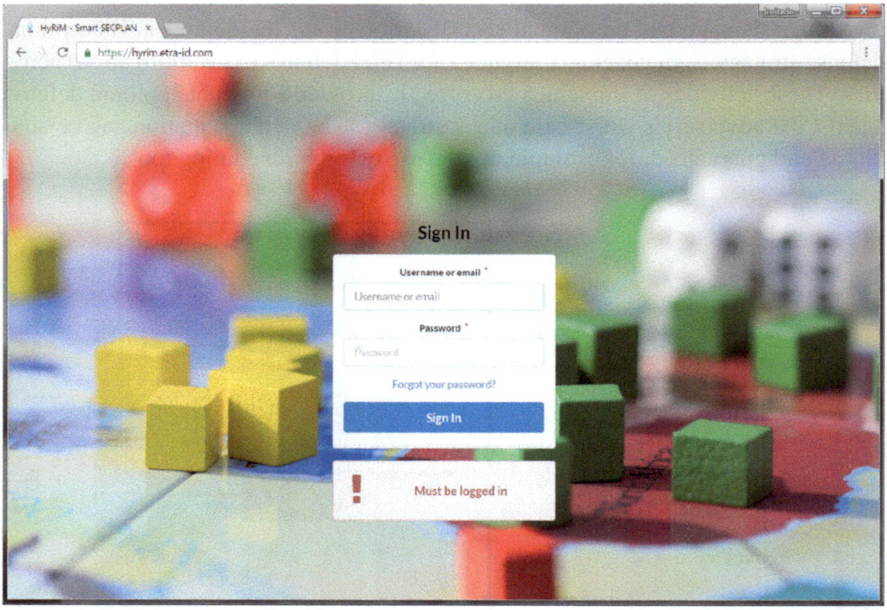

Fig. 16.2: Smart SECPLAN tool login site

The risk assessment guided by the tool fulfills the HyRiM process as described in Chapter 12 and also is based on precursor results from the EU FP7 project SECCRIT [3], where a similar approach for building the risk models was taken with the open-source tool Verinice [4].

The model consists of a set of elements that may, on one side, support the operations of the organization and, on the other side, negatively impact those, affecting therefore the normal behavior of the organization. Once the potential risks faced by the organization have been identified, the available mitigation actions that may prevent those risks are also modeled. More in detail, the model is composed of the following elements, for which the details listed below are considered:

- **Goals:** objectives to be optimized by the tool (e.g., economic costs or reputation)
- **Processes:** activities carried out by the organization in the regular performance of its business.
  - **Likelihood of Failure:** the likelihood of a failure on the process with a certain (short, mid, or long) duration
  - **Impact on Goal:** the cost the organization would confront in case of a failure of the process, in terms of the defined goals and for each temporal scope
  - **Supporting Assets (Ratio):** the relative weight of each of the supporting assets contributing to the achievement of the process

- **Assets:** the organization's supporting elements for the processes
  - **Supported Processes:** the relative importance of the contribution of the asset to each of the processes of the organization
  - **Affecting Risk Scenarios:** list of risk scenarios the asset is affected by
- **Scenarios:** possible risk scenarios which occurrence may affect the identified assets
  - **Affected Assets:** list of assets the risk scenarios affects
  - **Composing Threats:** list of particular threats that may trigger the risk scenario
- **Threats:** individual threats that compose each of the risk scenarios
  - **Likelihood:** qualitative likelihood of the threat (low-medium-high), based on previous experience or expertise
  - **Composed Scenarios:** list of risk scenarios that may be triggered by this particular threat
  - **Mitigating Actions (Ratio):** ratio of mitigation of this particular threat by a particular mitigation action (0% means no effect; 100% means the threat cannot occur if the mitigation action is performed)
- **Mitigation Actions:** regular actions the organization can perform to prevent the identified risks
  - **Mitigated Threats (Ratio):** ratio of mitigation of a particular threat by this particular mitigation action (0% means no effect; 100% means the threat cannot occur if the mitigation action is performed)
  - **Impact on Goal:** the cost of the execution of this mitigation action, in terms of the defined goals

All the elements of the model are combined in order to obtain one distribution-valued payoff matrix per goal (risk scenarios vs. mitigation actions), each cell containing the probability distribution of the impact for the organization of the occurrence of a *what-if* situation, where the risk scenarios happen while the organization is taking the corresponding mitigation action.

Specifically, if T is a threat causing some loss L and M is a mitigation action, then the game speaks about the probability distribution of the random loss L in light of the threat T becoming reality when mitigation action M is taken. Formally, the game uses the distribution function $F_{T,M}(z) = \Pr(L \leq z | T, M)$. This function is typically nontrivial to get in practice but can be estimated in various ways: by simulation (see Chapters 8, 9, 10, 14, and 15), from data due to experience or expert surveys (such as supported by the `HyRiM` package for `R` [15]), or using heuristics such as in Smart SECPLAN (Figure 16.3).

At the last step, the distribution-valued payoff matrices are fed into the HyRiM game theory framework, and from the results, we derive the main outcomes of the Smart SECPLAN for the organization, i.e.:

- **Prioritized mitigation actions:** those are given as the equilibrium in the matrix game and as such assign a probability to each mitigation action model, which gives the optimal frequency at which this action should be repeated. Under this prescription, which a task scheduling tool integrated in Smart SECPLAN converts into task assignments over time, the effect of *all* threats to the enterprise is simultaneously minimized. The fact that the mitigation tasks are scheduled according to a game-theoretic equilibrium guarantees that no alternative schedulecould perform better in terms of risk minimization. The fact that we are

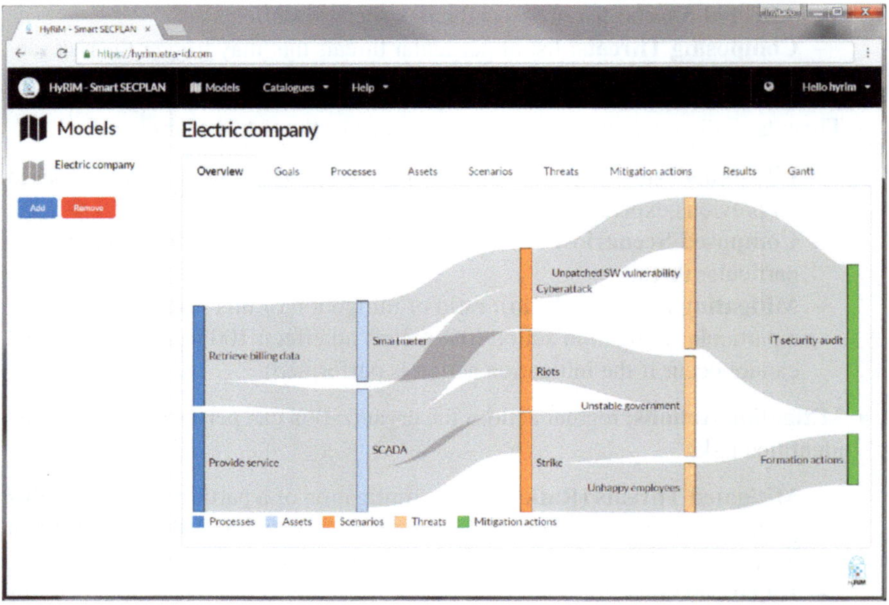

Fig. 16.3: Example of a simple complete model. Weighted relationships among elements of the model are depicted

playing a zero-sum game for that matter ensures that we do not need to assume anything about the attacker incentives, since the defense is always optimal regardless of the pattern at which the threats may occur.
Relative to each other, the frequencies (probabilities) of the mitigation actions can be interpreted as priorities.

- **Proposed Gantt chart:** taking into account the costs associated to each mitigation action, available budget, and results of the model, the tool constructs a Gantt chart with an optimum yearly maintenance plan.

In order to assist the identification of risk scenarios and threats, the Smart SECPLAN tool also integrates several threat catalogs, namely (Figure 16.4):

- ENISA's Threat Landscape 2015 [10]
- MAGERIT [1]
- NIST CVE [2]
- FP7 project SECCRIT Cloud threats catalog [3]

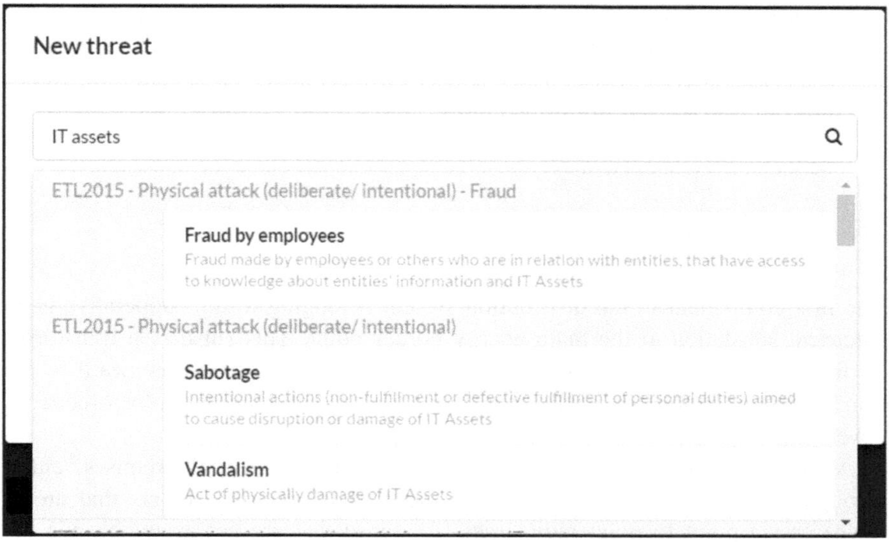

Fig. 16.4: Catalog search functionality within the Smart SECPLAN tool

16.4 Scenario Setup

The scenario discussed in this chapter is built around a medium-size electrical cooperative which manages the electricity distribution for the inhabitants of a town. In this context, a cooperative is an autonomous association of persons united voluntarily to meet their common economic, social, and cultural needs and aspirations through a jointly owned and democratically controlled enterprise. Therefore, the electrical cooperative's aim is to supply electricity to consumers that are associates at the same time, with the cheapest costs.

This cooperative can be considered as a distribution system operator (DSO). In general, a DSO is responsible for the last mile delivery of electricity to the end users. The DSO's distribution network carries electricity from the transmission system and delivers it to consumers. Besides, a DSO provides the adaption of the electricity from the high voltage used by transmission system operators (TSO). Generally speaking, most of the risks in the implementation of the smart grids relies on the competence of the DSOs; they are the responsible to manage them and to ensure the continuity of all the business processes.

Overall, the cooperative supplies about 35 million kilowatts per year, to more than 6000 end users, by means of more than 40 transformation centers, with an installed power of 18,000 kW. This company enables the capture of energy in a high voltage (132,000 kV) for further processing.

As highlights of the scenario, we enumerate the following characteristics:

1. The cooperative can collaborate with other cooperatives to exploit synergies and reduce costs, so the risk scenario is greater than in an isolated case,

2. Customers are associates at the same time, so the support to customers is one of the most important goals in the cooperative, and
3. The cooperative is always trying to reinvest its benefits into a better infrastructure and new services for its end users.

16.4.1 Component Description

The first component in the distribution system belonging to the cooperative is an electrical substation at the main energy border point. This substation transforms all the energy from high to medium voltage and is distributed downward to the transformation centers located around the town, so the medium-voltage network covers the major part of the town.

All the transformation centers are connected to the central premises, either through the private fiber optic link (for those transformation centers that are in town) or via point to point radio links (for those that are not accessible via the optical network). This private network is also used to run the supervisory control and data acquisition (SCADA) system. The SCADA control center is placed at the main premises of the company.

On the other hand, at low-voltage network, all the end users are equipped with a smart meter that connects to the concentrator, located at the nearest transformation center via a programmable logic controller (PLC). Through this connection, both the measuring of the end user's consumption and configuration and management and control of the specific smart meters, as well as the whole grid, can be done remotely from the central premises. This allows the company not only to measure the users' consumption remotely but also to manage and control the whole grid in a smart and efficient way. Besides, the electrical network is a smart grid, based on PLC technology. Furthermore, thanks to the smart metering system, all the consumption measurements are collected remotely and inserted in the enterprise resource planning (ERP) software for billing purposes.

16.5 Scenario Implementation

In the following subsections, the Hybrid Risk Management Process taken by this cooperative is presented. All information contained in the model has been extracted by directly interviewing this DSO personnel and therefore based on real experience in the field. The definition of the model and the retrieval of the results have been performed with the support of Smart SECPLAN.

16.5.1 Establishing the Context

Now, we pick up on the HyRiM process within Smart SECPLAN, starting with the first two points outlined in Section 16.2.

16.5.1.1 Goals

Failures in the organization processes have consequences which may vary quite a lot depending on where and when they appear. In addition, these consequences can be classified according to their impact, e.g., in terms of economic losses, company reputation, etc. It is hence necessary, as a first step, to define the business goals and measures thereof, together with the optimization goal being maximization (say, of revenue) or minimization (say, of damages). In the HyRiM process, as in general risk management, the focus is usually on damage prevention and hence risk *minimization*. Thus, the business goals have to be defined accordingly.

In the case of the DSO, two *goals* are defined: economic cost and reputation.

The *economic cost* is a goal to be minimized, due to the DSO's limited budget available, and also monetary investment has to be kept as low as possible, while the *reputation* is a goal to be maximized, since the DSO already counts on a good reputation from its customers and it is expected to increase over the next years.

16.5.1.2 Processes

After defining the organization goals, all regular processes carried out by the organization need to be listed in order to carry out a complete analysis. Each process will have the following properties:

- *Failure duration likelihood:* expected likelihood of the duration of the failure in such event
- *Impact on goal:* the cost the organization would face in case of a failure in terms of the defined goals
- *Supporting assets (ratio):* the relative importance of each of the supporting assets contributing to the achievement of the process

Processes relevant to the risk assessment analysis are identified by the DSO and introduced as input:

- *Billing:* retrieval of all information needed to bill the customers
- *Grid control:* sensing and control of the physical elements that compose the grid (smart meters, SCADA, high-/medium (HV/MV)-voltage substations, etc.)
- *Surveillance:* video surveillance of critical facilities
- *Customer support:* all activities related to consumers support
- *Human resources:* process related to employees of the cooperative

Their likelihood of failure and impact on objectives are shown in Tables 16.1 and 16.2. The economic impact is measured in Euro and the reputation in a scale of [-1000, 0]. As the tables show, the greatest probability of failure is in the short term, and as somehow expected, the impact on the organization is bigger as the duration of the process failure gets longer.

Table 16.1: Relationship among processes and failures in DSO implementations

	Likelihood of failure		
	Short term	Medium term	Long term
Billing	Medium	Low	Low
Grid control	Low	Low	Low
Surveillance	Medium	Low	Low
Customer support	Medium	Low	Low
Human resources	Medium	Low	Low

Table 16.2: Relationship among processes and goals in DSO implementation

	Economic cost			Reputation		
	Short term	Medium term	Long term	Short term	Medium term	Long term
Billing	1000	2000	3000	-100	-200	-500
Grid control	1000	2000	3000	-100	-200	-500
Surveillance	1000	1500	2000	-100	-200	-500
Customer support	500	600	700	-200	-500	-1000
Human resources	100	1500	1800	-100	-200	-500

16.5.2 Risk Identification and Analysis

Now, we enter the second phase of the process depicted in Figure 16.1, continuing with the third of the points outlined in Section 16.2.

16.5.2.1 Assets

At this step, all assets supporting the regular processes carried out by the organization need to be listed. Each asset will have the following properties:

- *Supported processes:* the relative importance of the contribution of the asset to each of the processes of the organization
- *Affecting risk scenarios:* list of risk scenarios the asset is affected by

Assets of the DSO are identified and listed:

- *SCADA server:* the control software in charge of monitoring and controlling the field assets of the grid
- *Smart meters:* devices that record consumption of electric energy in very short periods of time and communicate that information to the utility for monitoring and billing
- *Concentrators:* devices compiling the measurements of the smart meters of a certain area
- *ERP server:* server in charge of recovering all metering data from the smart meters and performing analysis over it as required by the DSO (mainly for billing purposes)
- *Surveillance server:* server that controls the video surveillance system
- *Office laptops*
- *Network storage server:* used at office premises for information exchange among the employees
- *HV/MV and MV/LV substations:* composed of a set of devices dedicated to the transformation of the voltage. The HM/MV substation transforms energy supplied by the transmission network at high-voltage into medium-voltage lines. The MV/LV substation transforms medium voltage into low voltage, which is supplied to the end customers

These assets are related to processes as shown in Table 16.3. A representation of these relationships, as Smart SECPLAN displays them, is shown in Figure 16.5.

Table 16.3: Percentage of assets support in each of the processes

	Billing	Grid control	Surveillance	Customer support	Human resources
SCADA server	0	15	0	0	0
Concentrators	40	10	0	0	0
Smart meters	40	18	0	19	0
ERP server	20	25	0	0	0
Surveillance server	0	0	80	0	0
Office laptops	0	0	20	61	80
Network storage server	0	0	0	20	20
HV/MV substation	0	16	0	0	0
MV/LV substation	0	16	0	0	0
Total	100	100	100	100	100

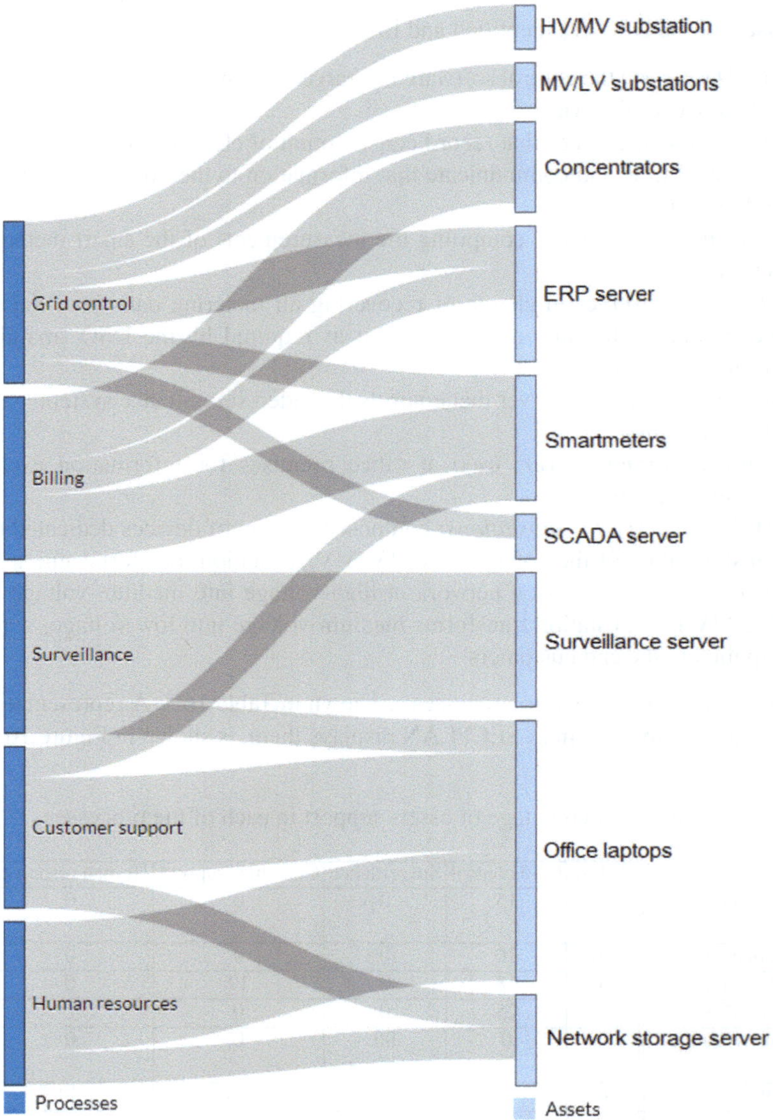

Fig. 16.5: Relationship between assets and processes

16.5.2.2 Scenarios

At this step, all possible risk scenarios to which the organization is exposed have to be listed. Those scenarios are general descriptions of events that may affect the organization. Each scenario will have the following properties:

- *Affected assets:* list of assets potentially affected by the given risk scenarios
- *Composing threats:* list of particular threats that may trigger the risk scenario

The identified scenarios are the following:

- *WannaCry:* in the first quarter of 2017, a serious worldwide cybersecurity problem arose, WannaCry [5]. This malware affected tens of thousands of devices around the world. We adopt this scenario as link between threats of malware and USB infected and assets of IT and electronic devices
- *APT:* an advanced persistent threat (APT) is a set of stealthy and continuous computer hacking processes, often orchestrated by a single person or a group of people targeting a specific entity. We consider under this scenario a targeted cyberattack onto the organization
- *Physical intrusion:* vandals or malicious people can try to enter DSO critical areas. This scenario may affect all physical assets
- *Data losses:* human errors, sabotages, or illegal manipulation of the billing infrastructure entail data losses, among other problems. This scenario links those threats with IT components

16.5.2.3 Threats

As a next step, all possible threats which may cause the defined risk scenarios need to be listed. In order to make this task easier, the Smart SECPLAN tool is preloaded with a number of threat databases. Each threat will have the following properties:

- *Likelihood:* qualitative likelihood of the threat (low-medium-high), based on previous experience or expertise
- *Composed scenarios:* list of risk scenarios that may be triggered by this particular threat
- *Mitigating actions (ratio):* ratio of mitigation of this particular threat by a particular mitigation action (0% means no effect; 100% means the threat cannot occur if the mitigation action is performed)

After some iterations of analysis, the following list comprises the most relevant threats according to experts know-how:

- *Vandalism* and *sabotage*: Vandalism involves a very high economic cost. In the same sense, a discontent employee can produce a very high economic cost or reduce the reputation of the DSO
- *Bypass the billing infrastructure:* It comprises situations like a customer deciding to pay less through the manipulation of the smart meter or an unhappy employee modifying billing database
- *Cyberattacks* and *human error:* nowadays, cyberattacks must be considered as a very important risk, especially for organizations managing critical cyber networks. In addition, workers can make mistakes during their activities, causing similar cyber effects. For the case of DSO, four threats are considered in this category:
  - *Suspicious redirects*
  - *Denial of service*

- *Malware diffusion*
- *USB infection*

Likelihoods of each threat are shown in Table 16.4.

Table 16.4: Likelihood of each threat

Threat	Likelihood
Vandalism	Low
Sabotage	Medium
Bypass the billing infrastructure	High
Suspicious redirect	Medium
Denial of service	Medium
Malware diffusion	High
USB infected	Medium

Figure 16.6 depicts a graphical relationship among scenarios, assets, and threats.

16.5.3 Risk Evaluation and Treatment

This is the stage (in Figure 16.1) where the game-theoretic analysis, decision, and algorithmic framework (Chapters 2 and 3) come into play.

16.5.3.1 Mitigation Actions

In this step, all mitigation actions needed to prevent the above identified threats are listed. A number of mitigation action databases are offered by Smart SECPLAN. Each mitigation action will have the following properties:

- *Mitigated threats (ratio):* ratio of mitigation of a particular threat by this particular mitigation action (0% means no effect; 100% means the threat cannot occur if the mitigation action is performed)
- *Impact on goal:* the cost of the execution of this mitigation action, in terms of the defined goals

Considering all threats and vulnerabilities, and removing defenses with a low-level impact on threats, the following set of *defenses* are defined:

- *Check data integrity:* this action focuses on mitigating data manipulation. For instance, identify illegal bypassing of the billing infrastructure smart meters or effects of different kinds of cyberattacks
- *Company awareness formation:* courses to employees about security

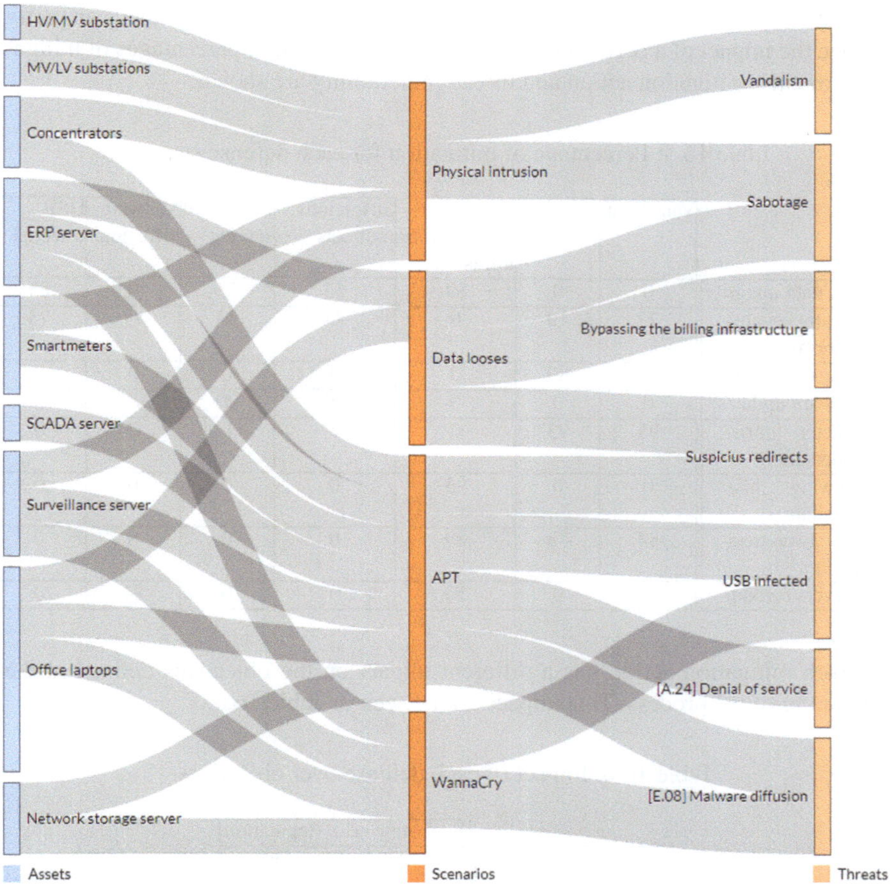

Fig. 16.6: Scenarios included in DSO model

- *Check smart meters:* periodically checking if a smart meter is working properly mitigates problems arising from physical damage
- *Software updates:* the protection of the IT infrastructure implies to have all software resources up to date. An updated software protects against cyber threats.
- *Perimeter intrusion detection:* a perimeter intrusion detector protects from vandalism or sabotages in the physical facilities
- *Reviewing historical data:* this action is oriented to detect abrupt changes in the billing process, by identifying patterns of consumption. Historical data can be checked in order to detect fraud
- *Public awareness campaign:* if the customer understands the importance of keeping facilities secure and safe, the risk of vandalism or sabotage will decrease
- *Server backups:* in cases of a software update, malicious actions, human errors, etc., data can be lost, and devices may not work properly. By performing periodic backups, servers will be protected against those kinds of threats

Each of these actions does not completely mitigate the threats, but they can partly reduce the impact of a particular set of threats. In Table 16.5, percentages of mitigation for each mitigation action and threat/vulnerability are shown.

Table 16.5: Percentage of mitigation for each defense strategy

	Vandalism	Sabotage	Bypassing the billing infras.	Suspicious redirects	Denial of service	Malware diffusion	USB infected
Check data integrity	0	0	89	53	73	0	85
Company awareness formation	22	19	0	88	53	83	75
Check smartmeters	85	80	0	0	0	0	0
Software updates	0	0	0	90	90	90	0
Perimeter intrusion detection	95	93	0	0	0	0	0
Reviewing historical data	0	0	95	0	0	0	0
Public awareness campaign	88	88	89	0	0	0	0
Servers backup	0	0	50	0	0	84	0

Each mitigation action has a different impact on the objectives defined at the beginning of the process. This impact can be seen in Table 16.6.

Table 16.6: Impact of each defense over objectives

	Economic cost	Reputation
Check data integrity	1000	50
Company awareness formation	1200	500
Check smart meters	850	500
Software updates	1200	100
Perimeter intrusion detection	4000	500
Reviewing historical	900	50
Public awareness campaign	6000	700
Servers backup	800	100

16.5.3.2 Results of the Process

In Figure 16.7, a graphical representation of the whole process including the interdependencies among assets, threats, scenarios, and mitigation actions is depicted.

Once the model is completed, a payoff matrix per goal is computed. The respective matrices for the given example can be seen in Figures 16.10 and 16.11 in the Appendix. In both cases, each combination represents the probability of facing an

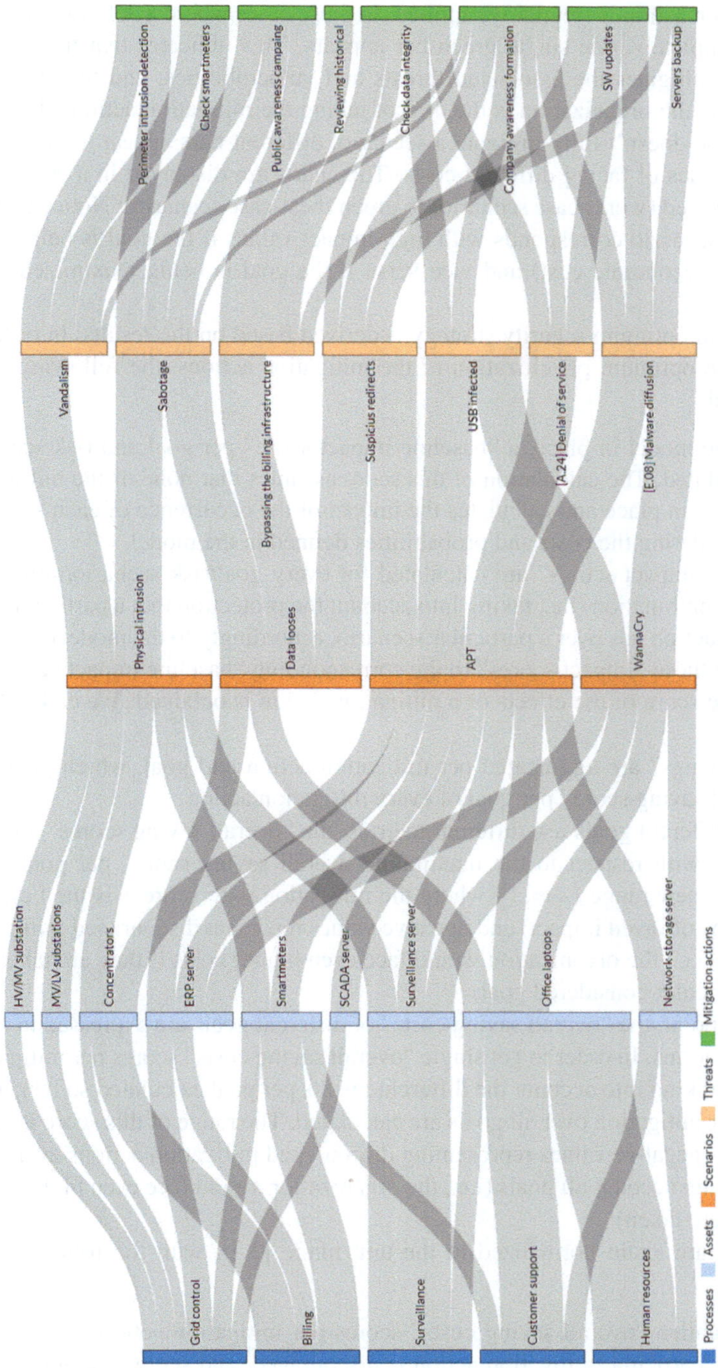

Fig. 16.7: Risk assessment model as shown in Smart SECPLAN

impact (economic cost or reputation) in case that a risk scenario is taking place when the mitigation action is (regularly) applied. In those cases where a defense mitigates a threat completely, one single green bar appears. This indicates that the defense completely mitigates the associated threats and vulnerabilities, and thus the only cost faced by the organization is the cost of taking the mitigation action. However, if the defense doesn't fully mitigate a scenario, different bars appear, representing the probabilities of facing different costs. The charts are displayed in a color code ranging from red (worst-case scenario) to green (best-case scenario). Notice that the worst-case scenario corresponds with high-impact values if the goal is being minimized (e.g., economic cost) and vice versa if the goal is being maximized (e.g., reputation).

Finally, an optimum security strategy is derived based on the results. In order to calculate the optimum prioritization of the mitigation actions, the following steps are followed:

1. With the model in place, a "baseline impact score" per goal and risk scenario is calculated. The calculation of this score assumes that none of the mitigation actions is in place and calculates the impact of the occurrence of each scenario by considering the costs and probabilities defined in the model.
2. Similar "impact scores" are calculated for every goal/risk scenario/mitigation action combination (i.e., taking into account the protection that a particular mitigation action has over a particular scenario, accordingly to the model). By subtracting these "impact scores" to the corresponding "baseline impact score," an absolute score of the effects of a mitigation action is obtained. We call it "savings."
3. All "savings" are aggregated per mitigation action and goal, which gives the "overall saving score" per goal of every mitigation action.
4. Since different goals use different metrics, the "overall saving scores" are normalized with respect to the maximum "overall saving score" per goal. This brings a new range $(-\infty, 1]$, where the mitigation with score 1 is the best one (in terms of saved impact) and negative values represent discouraged mitigation actions (i.e., the organization would face even higher costs if they are taken for the particular considered goal).
5. At this point, the "overall saving costs" of different goals are expressed in comparable terms. In order to get single "overall saving costs" scores per mitigation action (taking into account the different results per goal calculated so far), averages per mitigation over all goals are calculated. The range of this score is again $(-\infty, 1]$, negative values representing discouraged mitigation actions even after taking into account all goals (i.e., the organization would face even higher costs if they are taken).
6. Results are again normalized to the unit range $[0, 1]$, with the following approach:

    - Negative "overall saving costs" scores per mitigation action indicate that this particular mitigation action is discouraged, and scores are normalized to 0.

- Positive "overall saving costs" scores per mitigation action indicate that this particular mitigation action needs to be taken with a particular priority. Those are normalized so they all sum up to 1.

7. The normalized results directly provide the relative weights of every mitigation action, which directly represents the optimum strategy.

The results show DSO managers where to invest more effort in a usable way as given in Table 16.7 and Figure 16.8.

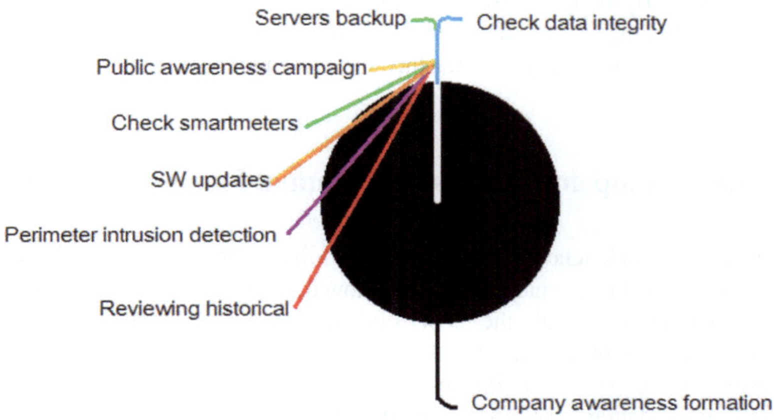

Fig. 16.8: Percentages of application for each defense (screenshot from Smart SEC-PLAN)

Table 16.7: Percentages of application for each defense

	Percentage
Company awareness formation	99,90
Public awareness campaign	0,10
Perimeter intrusion detection	0,00
Reviewing historical	0,00
Servers backup	0,00
Check smart meters	0,00
Check data integrity	0,00
Software updates	0,00

16.5.3.3 Gantt

Whenever the equilibrium is pure, the advice is obviously to install a static (ever-repeating) countermeasure. However, when the equilibrium is mixed, then defenses should be repeated at different frequencies, corresponding to the equilibrium probabilities.

The organization can use the results of Smart SECPLAN to directly design a maintenance plan. Smart SECPLAN facilitates this task by automatically taking the results of the model, and different constraints imposed by the organization can be included (as yearly budget or limits to the frequency of execution of a particular mitigation action that is not contemplated in the model). As a final result, a Gantt chart is derived with a yearly maintenance plan (Figure 16.9).

16.6 Game Setup and Analysis: A Manual Background Check

The entire process sketched up to this point is fully embodied in and supported by Smart SECPLAN. For an understanding of how the risk control according to game-theoretic methods, especially the theory outlined in Chapters 2 and 3, works, let us go back and take a closer look at the distribution-valued game model, describing the two business goals "economic impact" and "reputation" (as Table 16.2 illustrates).

Based on the data gathered in Smart SECPLAN, a multi-objective game is constructed, which focuses on economic cost and reputation. These two goals are described by the respective game matrices **A** and **B** as displayed in Figures 16.12 and 16.13. The Pareto-Nash equilibrium computed from this setup is pure and given by APT (to raise company awareness). For a visual inspection (and confirmation) to as why this is an equilibrium, we would need to form the weighted sum $0.5 \cdot \mathbf{A} + 0.5 \cdot \mathbf{B}$ and verify that the entry in row 2 and column 2 is dominant for the respective player. This is easy (yet laborious) to do on the full example model, so let us confine ourselves to only an example comparison based on the figures given.

Why is the payoff for an APT under the fifth defense strategy "perimeter intrusion detection" worse than under the second "company awareness formation"? Looking at the plots in column 2, rows 1 and 5, we see that the payoff distribution under the second defense strategy assigns far less likelihood mass to losses in the range around 80, as opposed to the intrusion detection (defense strategy d_5), under which losses around 90 are much more probable. The stochastic order introduced in Chapter 2, on which the above equilibrium is based, would thus prefer the behavior under which large losses are less likely, which is defense strategy d_2.

From a practical perspective, this appears plausible, since raising awareness will most likely entail many of the explicit defense strategies given in the game. Likewise, an APT is typically a combination of various attack techniques, to which physical intrusion and data losses and the outage of an entire operation infrastructure are examples (as seen, e.g., by this year's malware attack like WannaCry [5] or last year's hack of the Ukrainian power grid [16]). As such, an APT can be considered

Fig. 16.9: Yearly maintenance plan

to cause much more damage and is therefore the optimal choice for the attacker under the stochastic order described in Chapter 2.

Still if the outcome is implausible in light of a more detailed interpretation of the defense strategies (which may be imaginable in the context of the particular application), then the focus of the loss range which represents the basis for the decision can be changed. That is, in truncating the right tails of the loss distributions at a threshold a, we implicitly say that losses bigger than a are considered as "simply too high." In that case, the stochastic order will be computed depending on which behavior leads to losses up to a or a little below (mathematically, the left neighborhood $(a-\varepsilon, a]$ of a for some $\varepsilon > 0$ determines the order relation). Practically, the value of the threshold a could be set to the *risk acceptance level*, i.e., the value above which residual risks are simply taken or transferred to an insurance.

For the example here, rerunning the computation with the cutoff (truncation) at 80 out of the full loss range $[1, 100]$ (so that the last 20% of the loss range has its mass squeezed underneath the distribution in the range $[0, 80]$) changes the pure equilibrium into a mixed one: the optimal defense is then a mix of $\approx 52.2\%$ on company awareness formation and $\approx 47.7\%$ on software updates. Also, for the two goals, the worst-case strategies are now different, retaining the APT as most severe in terms of economic cost, but for the maximal reputation damage, the adversary would need to play a $\approx 44.55\% : 55.44\%$-mix between WannaCry and the APT.

This change provides an interesting insight, namely, the fact that the first optimal defense (with the full loss range up to 100) has its recommendation(s) based only on the last 20% of the loss range. Thus, apparently, the defense actions "raise company awareness" and "software (SW) updates" are effective in the higher loss regions, as opposed to the other defense strategies that apparently have an impact on losses with relatively lower magnitudes only. In that sense, the game's outcome points at "raise company awareness" and "SW updates" are the two defenses with the strongest impact and hence the most important. By varying the focus on the loss range where the optimal defenses are computed (by setting different cutoff points to truncate the loss distributions), we get a more differentiated picture about the effectiveness of risk mitigation actions, also with different optimal defenses. This is indeed not possible for a standard game-theoretic modeling, where there is no comparable degree of freedom in changing the "loss focus" in any similar way. The latter (second) Pareto-Nash equilibrium appears no less plausible than the former, since incidents like WannaCry may have quite an impact on the reputation, regardless of whether or not the incident was part of a (larger) APT or not. As such, the more differentiated picture obtained by changing the angle on which we look at the loss scale (by truncating the loss distributions before the analysis) can be used with other settings to get the most plausible among perhaps many possible equilibria, i.e., defense schedule.

16.7 Conclusion

For the practitioner seeking to apply a game-theoretic risk defense, our experiments deliver the lesson that the most important factor in using a distribution-valued game is careful modeling of loss distributions. This task is to a considerable extent also an art besides being a science, and finding good distribution models working as a basis for the game calls for deep domain knowledge and statistical skills. The enforcement of care in this regard can, as we believe, only be beneficial for the risk manager, since it leads to the drawing of a much more fine-grained and detailed picture about the risk situation than a classical real-valued game analysis could provide. This means that for a decent risk analysis based on game theory, much effort is required for (and shifted to) the loss modeling issues, for which a whole body of designated literature exists [9]. While such loss models are common and standard in other sectors like financial or credit risk management, their application in security is not nearly as intensively studied. The framework of distribution-valued games opens a close connection that can deliver interesting new insights, some of which have been outlined above.

The application of classical game theory in risk management is no less fruitful, and the respective models are certainly easier to set up and understand. This simplicity is traded for a loss of information in the loss models, since the full data available on a defense consequence needs to be aggregated into a single figure in the payoff matrix of the game. However, the results are in both cases to be interpreted in the same way, so that the main difference between classical and distribution-valued game theory for risk management lies in the efforts needed for the model building. Classical games are easier to set up yet contain (and deliver) less information, while distribution-valued games come with richer information and potentially more insights, yet with considerably more difficulties in setting up good models.

Acknowledgements This work was supported by the European Commission's Project No. 608090, HyRiM (Hybrid Risk Management for Utility Networks), under the Seventh Framework Programme (FP7-SEC-2013-1).

Appendix: Game Matrices

See Figures 16.10, 16.11, 16.12, and 16.13.

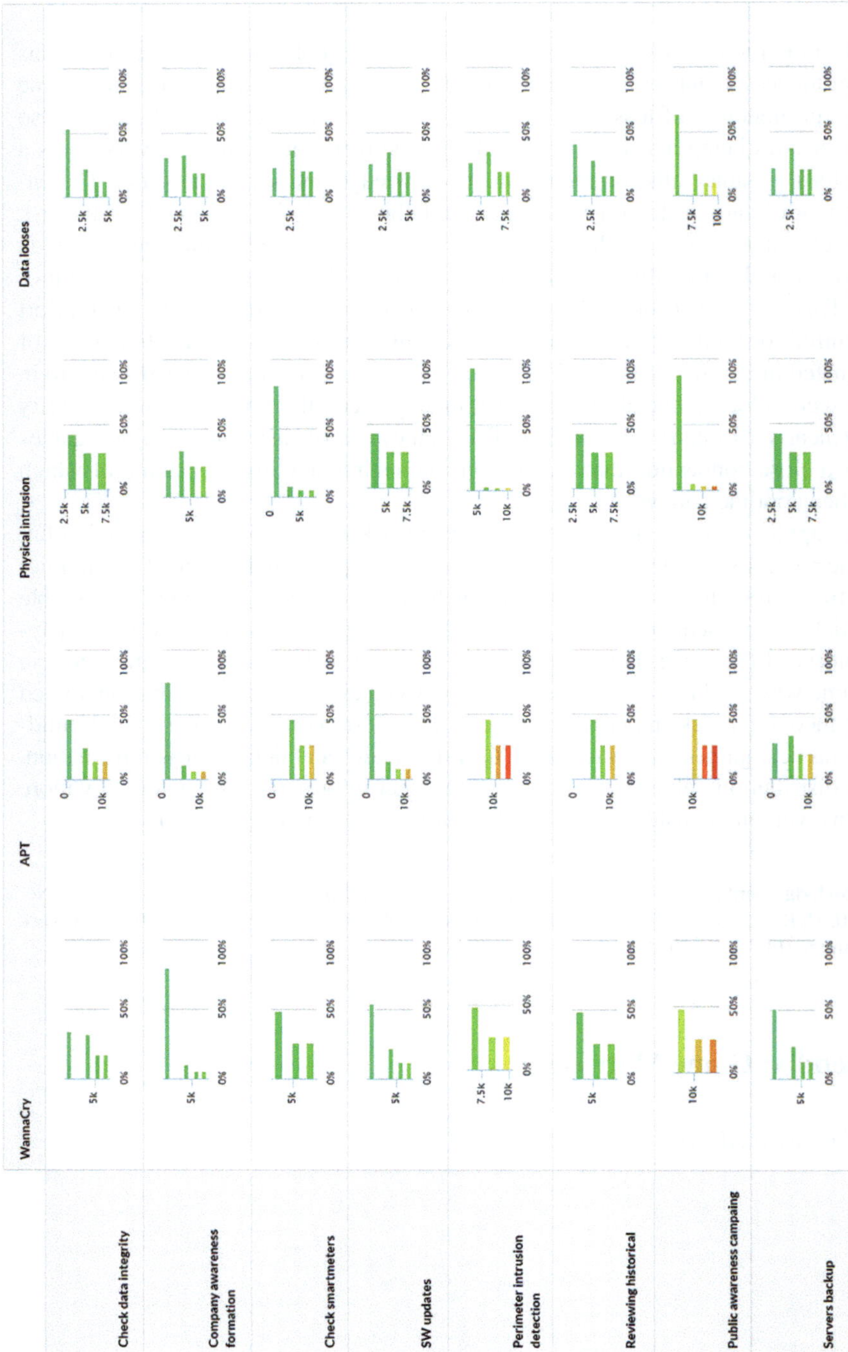

Fig. 16.10: Game matrix for economic impact (plot from Smart SECPLAN)

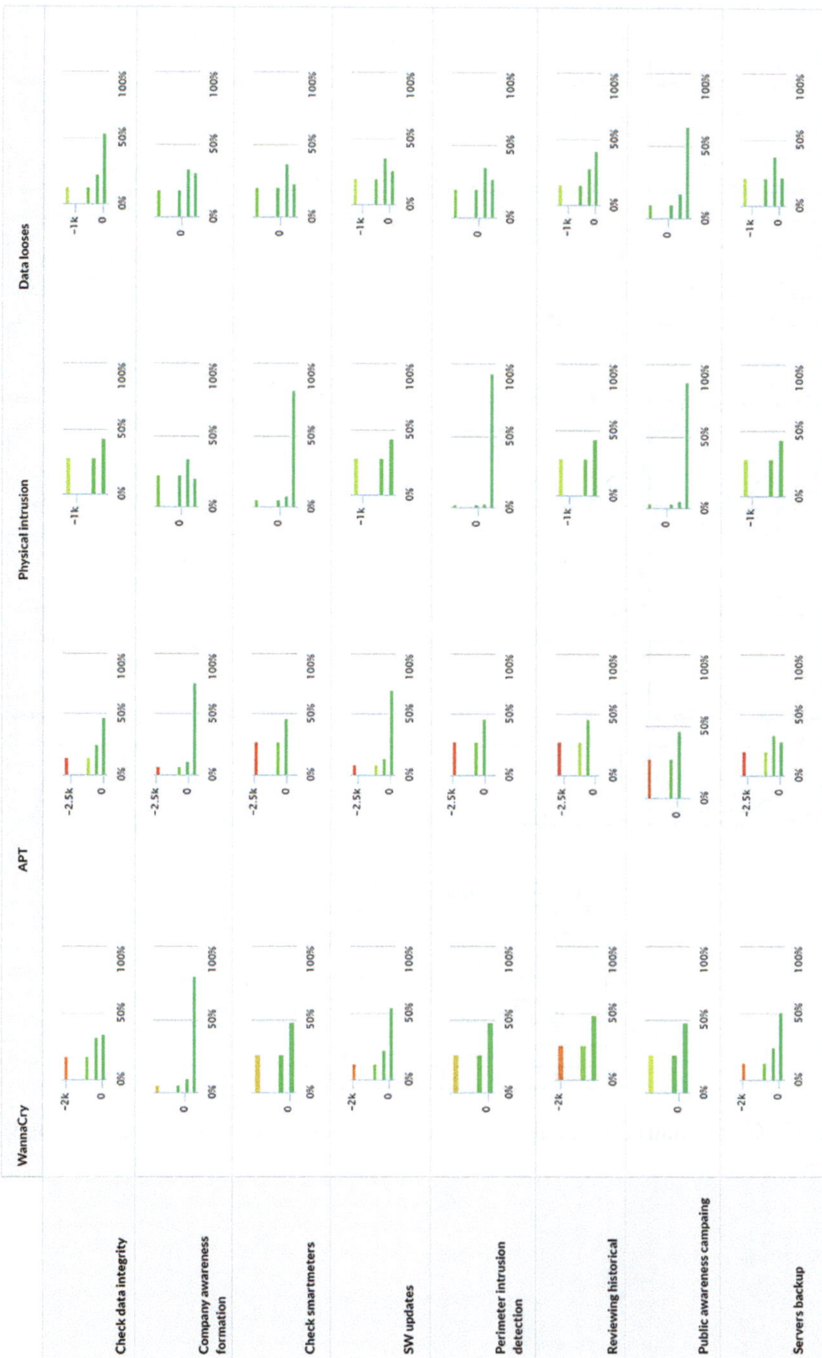

Fig. 16.11: Game matrix for reputation (plot from Smart SECPLAN)

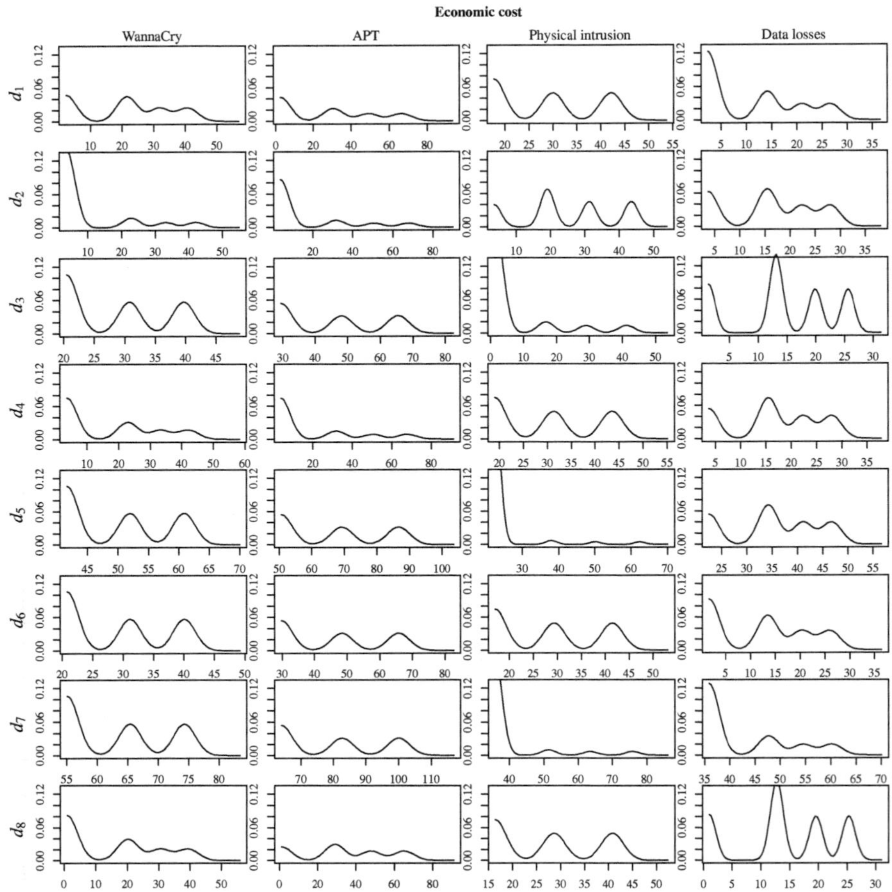

Abbreviation	Defense strategy
d_1	check data integrity
d_2	company awareness formation
d_3	check smartmeters
d_4	SW updates
d_5	perimeter intrusion detection
d_6	reviewing historical
d_7	public awareness campaign
d_8	servers backup

Fig. 16.12: Game matrix for goal "economic cost" (defense strategies abbreviated; see table)

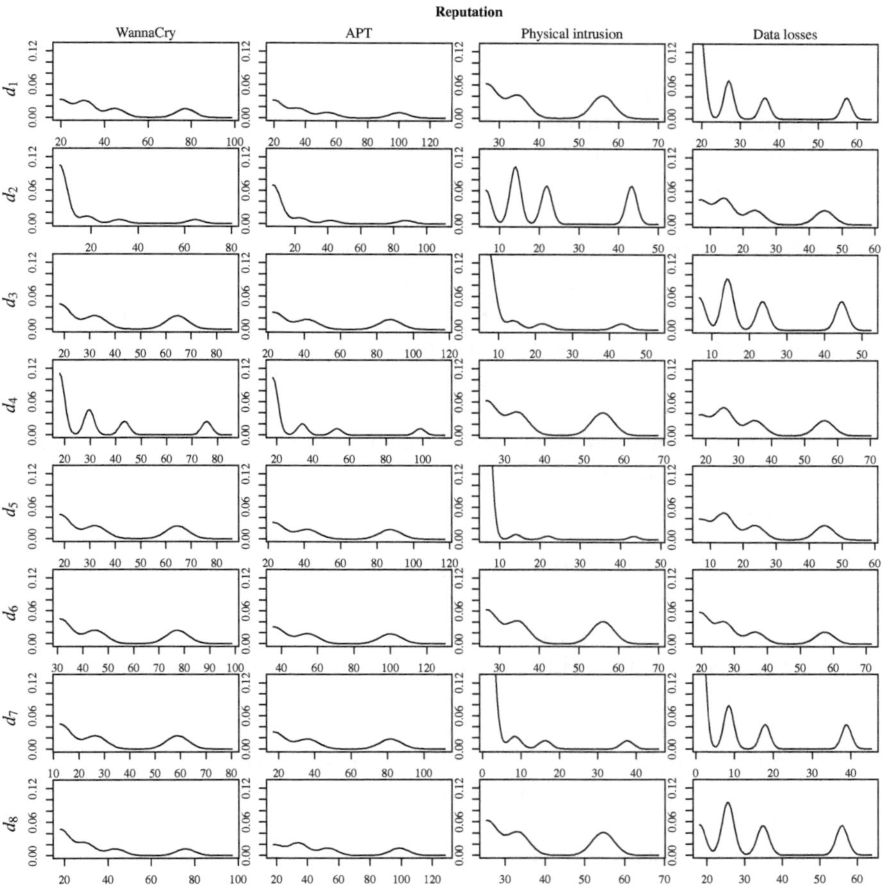

Abbreviation	Defense strategy
d_1	check data integrity
d_2	company awareness formation
d_3	check smartmeters
d_4	SW updates
d_5	perimeter intrusion detection
d_6	reviewing historical
d_7	public awareness campaign
d_8	servers backup

Fig. 16.13: Game matrix for goal "reputation cost" (defense strategies abbreviated; see table)

References

1. Magerit homepage. URL https://www.enisa.europa.eu/topics/threat-risk-management/risk-management/current-risk/risk-management-inventory/rm-ra-methods/m_magerit.html
2. National Vulnerability Database (NVD). URL https://nvd.nist.gov/
3. Seccrit homepage. URL http://www.seccrit.eu/
4. Verinice homepage. URL https://verinice.com/en/
5. Bill, B.: WannaCry: the ransomware worm that didn't arrive on a phishing hook. Tech. rep., Sophos Ltd (2017). URL https://nakedsecurity.sophos.com/2017/05/17/wannacry-the-ransomware-worm-that-didnt-arrive-on-a-phishing-hook/
6. International Standardization Organization: ISO 28001: Security management systems for the supply chain - Best practices for implementing supply chain security, assessments and plans - Requirements and guidance. Geneva, Switzerland (2007). English version
7. International Standardization Organization: ISO 31000: Risk Management - Principles and Guidelines. Geneva, Switzerland (2009). English version
8. International Standardization Organization: ISO/IEC 27005: Information technology - Security techniques - Information security risk management. Geneva, Switzerland (2011). English version
9. Klugman, S.A., Panjer, H.H., Willmot, G.E.: Loss models: From data to decisions. A Wiley-Interscience publication. Wiley, New York, NY (1998). URL http://www.loc.gov/catdir/description/wiley031/97028718.html
10. Marinos, L., Belmonte, A., Rekleitis, E.: ENISA Threat Landscape 2015. ENISA (2016). URL https://www.enisa.europa.eu/publications/etl2015/at_download/fullReport
11. Maschler, M., Solan, E., Zamir, S.: Game Theory. Cambridge University Press (2013)
12. Rass, S.: On Game-Theoretic Risk Management (Part One) - Towards a Theory of Games with Payoffs that are Probability-Distributions. ArXiv e-prints (2015)
13. Rass, S., König, S., Schauer, S.: Deliverable 1.2 - Report on Definition and Categorisation of Hybrid Risk Metrics. HyRiM Deliverable, Vienna, Austria (2015)
14. Rass, S., König, S., Schauer, S.: Uncertainty in Games: Using Probability-Distributions as Payoffs. In: Decision and Game Theory for Security, no. 9406 in Lecture Notes inComputer Science, pp. 346–357. Springer, London, UK (2015)
15. Rass, S., König, S.: R package 'hyrim': Multicriteria risk management using zero-sum games with vector-valued payoffs that are probability distributions (2017). URL https://hyrim.net/software/
16. Zetter, K.: Everything We Know About Ukraine's Power Plant Hack | WIRED (2016). URL https://www.wired.com/2016/01/everything-we-know-about-ukraines-power-plant-hack/

MIX
Papier aus verantwortungsvollen Quellen
Paper from responsible sources
FSC® C105338

If you have any concerns about our products,
you can contact us on
ProductSafety@springernature.com

In case Publisher is established outside the EU,
the EU authorized representative is:
Springer Nature Customer Service Center GmbH
Europaplatz 3, 69115 Heidelberg, Germany

Printed by Libri Plureos GmbH
in Hamburg, Germany